Darwin's Finches

Darwin's Finches

READINGS IN THE EVOLUTION
OF A SCIENTIFIC PARADIGM

Edited, with commentary, by Kathleen Donohue

The University of Chicago Press   Chicago and London

Kathleen Donohue is associate professor of biology at Duke University.

The University of Chicago Press, Chicago 60637
The University of Chicago Press, Ltd., London
© 2011 by The University of Chicago
All rights reserved. Published 2011
Printed and bound by CPI Group (UK) Ltd,
Croydon, CR0 4YY

20 19 18 17 16 15 14 13 12 11   1 2 3 4 5

ISBN-13: 978-0-226-15770-2 (cloth)
ISBN-13: 978-0-226-15771-9 (paper)

Library of Congress Cataloging-in-Publication Data

Darwin's finches : readings in the evolution of a scientific paradigm / edited, with commentary,
by Kathleen Donohue.
      p. cm.
    Includes bibliographical references.
    ISBN-13: 978-0-226-15770-2 (cloth : alk. paper)
    ISBN-10: 0-226-15770-9 (cloth : alk. paper)
    ISBN-13: 978-0-226-15771-9 (pbk. : alk. paper)
    ISBN-10: 0-226-15771-7 (pbk. : alk. paper)   1. Finches—Evolution—Galapagos Islands.
2. Evolution (Biology)   I. Donohue, Kathleen, 1963–   II. Title.

    QL696.P246D37   2011
    576.8—dc22

                                                                                2010036542

Contents

Acknowledgments

This collection began as a part of an undergraduate seminar on Darwin's finches, as an introduction and enticement to the field of evolutionary biology. As such, my very first thanks must go to the students of that course, who helped me hone the collection and its presentation. Of course, I am also grateful to all the authors who wrote the engaging articles that make up this volume, and in particular Rosemary and Peter Grant for reviewing the collection and suggesting some excellent recent papers for inclusion. I especially thank Jonathan Losos for his careful and insightful comments on several of the chapters, and Janet Browne for her comments on some of the historical chapters. Two anonymous reviewers were also extremely helpful in making suggestions that improved the volume. My gratitude goes to Charles T. Donohue for his careful and perceptive editorial eye. I thank Mark Adams, curator of birds in the Natural History Museum at Tring, UK, for clarifying the modern placement of some previous species designations of Darwin's finches.

For assistance with the production of this volume, I thank Alicia Frank, Katie Parodi, Katie Kovach, Jonathan Schwartz, Abby Collier, Jennifer Howard, Joann Hoy, and my editor, Christie Henry. The Baker Foundation contributed funds for preparation of this volume. Duke University provided some funds to defray publication costs and also provided the most valuable commodity of all—time.

General Introduction

THE BIRDS AND THE ISLANDS

This collection of papers focuses on one of the most charismatic research systems developed in the field of evolutionary biology: Darwin's finches. The diversification of these finches on the Galápagos Islands has become one the most convincing examples of Darwinian evolution in progress. This collection illustrates why the finches were so important to the development of evolutionary theory and how they are still being used to test fundamental hypotheses of evolutionary biology today. Because of their long history of importance in the field, the finches have been studied using eclectic methodologies to address a broad range of evolutionary questions within a single, fascinating natural-history framework.

If you think the finches are colorful jewels of diverse greens and reds, they will disappoint. They are not. Their dull plumage blends perfectly with the blasted volcanic landscape the birds inhabit. If you think of them as mellifluous as a bobolink or skylark, sorry again. "Tür–tür," or "chöök" is their call, or even, embarrassingly, "cheep."[1] Theirs is an entirely different aesthetic. It is an aesthetic of subtlety and smallness on a backdrop of salty, wet enormity.

Darwin's finches are muddy-looking little birds that comprise the subfamily Geospizinae, a group of uncertain taxonomic placement. They include fourteen species that differ from each other primarily in body size and the shape and size of their bills. In these, they differ on the order of milligrams and millimeters. It is difficult to believe that a millimeter here or there can make much difference to evolutionary outcomes, much less major intellectual trajectories, but it does. After all, these birds are named after a man whose final book was about the turds of worms. Not exactly a world-rocking topic, you may think. But they are worms, and move the world they do—turning its surface over to a depth of 1.5 inches every ten years; burying ancient buildings and old pipe stems by a full foot every eighty years or so. Weigh a single worm casting, and it is not much. But it is not irrelevant. Scientists devote an enormous amount of time and effort to studying tiny things, not only because

1. Birdsong renditions are according to Snodgrass and Heller's 1904 summary of the Hopkins-Stanford expedition to the Galapagos.

these details reflect larger processes at work, but because in evolution, small differences are important. Small differences, added up, have big outcomes.

Darwin's finches differ not only in size and bill shape, but they differ in feeding behavior too, some eating small seeds, some eating large seeds, some eating cactus nectar and pollen, some stripping the bark of twigs to get at the softer, sweeter phloem. Some species practice remarkably innovative foraging habits: they break large eggs open by kicking them, they use twigs to chase larvae out of little cracks, they pick insects off iguanas, they sip blood from the holes they have poked into the tails of boobies.

The finches fascinate evolutionary biologists not only because of their differences but also because of their similarities. In truth, it is exceedingly hard to tell some of them apart. Yet they are, upon close study and taking geography into account, identifiably different from each other. Within this single endemic group are species that barely differ and those that differ as much as warblers do from grosbeaks. The whole range. Collectively, they provide snapshots of the entire process of diversification, starting with variation within a single species, to subtle differences between closely related species, to impressive extremes that resemble differences found between families in other bird taxa.

The papers here are about the birds and the islands—the space they inhabit. The finches reside in the Galápagos Archipelago, 525 nautical miles from Ecuador. The islands are of volcanic origin, the present ones 3 million to less than 1 million years old. They differ in age, size, and degree of isolation from other islands. Early in their history, the islands were warmer and wetter than they are now, resembling the tropical Cocos Island, 432 miles to the north. Around 1.7 million years ago, sea temperatures began to cool in the eastern Pacific, leading to a cooler, drier, and more variable climate in the Galápagos, with cyclic El Niño fluctuations between warm, wet years and cool, dry years. Altitudinal variation on the larger islands created distinct vegetative and climatic habitats as the sea cooled. The islands today are far different from what they were when the first birds arrived.

In the Galápagos, not only have the finches diversified, but so too have some other organisms. This is not uncommon in isolated archipelagos, which appear especially well suited for speciation. In the words of David Lack (1947, 158), one of the prominent researchers on Darwin's finches, and, indeed, the man who gave them their name: "That on oceanic islands radiations have occurred not only among the land birds, but also among insects, land mollusks and land plants, suggests that the evolutionary factors involved in these other groups are fundamentally the same as they are in Darwin's finches. This is one of the chief reasons why I hope this book [*Darwin's Finches*] may have been worth writing."

And this is one of the chief reasons the finches continue to be studied.

Because the finches have so much company in their evolutionary enthusiasm, we know that it is not something special about the finches that enabled them to diversify; perhaps it is something special about the place. If we can understand what it is that is special, we may recognize it elsewhere.

Place determines who encounters whom and how frequently. It determines how different habitats are, how isolated populations are, and how large populations are. Island archipelagos, moreover, give a manageable microcosm in which we can cover most of the ground, learn many of the organisms, and in the case of the finches, sometimes even know all the individuals. And it's thought that the processes of evolution there are essentially the same as they are elsewhere, except perhaps simplified by the simplicity of the biological community; that the globe in essence is like an archipelago, with its inhabitants both near and far from one another, both mixing and remaining distinct.

And yet, not all organisms on archipelagos diversify, and not all on the Galápagos kept evolutionary pace with the finches. It is not only something special about the place that enables diversification, but something special about how some organisms inhabit that space—how biology inhabits geography. So this book is about Darwin's finches and the Galápagos islands. It is about the birds and about the place, and about how the birds interact with the place and evolve.

A HISTORICAL PERSPECTIVE ON SCIENTIFIC INQUIRY

Practicing science is like entering a dialog that has been in progress for decades. While it is possible to jump in and respond to the most recent statement, historical context enriches the dialog and deepens the interpretations. One learns why the present issues are issues, why some lines of inquiry are today more compelling than others, and which ones may have been orphaned and could be adopted anew to good result.

The readings in this book include historical and contemporary texts. The introductions to each chapter give historical background to the particular topic and provide background information to make the readings more accessible. Following this is a brief description of each of the readings included in the chapter. The readings themselves follow.

One goal of this selection of readings is to introduce some of the most engaging issues currently being pursued within the field of evolutionary biology and to illustrate how they have been and are being addressed empirically. The readings include classic and recent examples of how hypotheses are articulated and tested. The readings are typically journal articles or book chapters—primary literature written for scientists. They present new observations, analyses, and interpretations as they were first published.

The readings also illustrate how hypotheses are formed and revised. By following lines of inquiry from first inception to present day, one sees how the questions themselves change over time, with new data and with new dialog and interpretations. Present in the readings are examples of great thinkers changing their minds completely.

The collection should also therefore be of interest to readers who want to examine the scientific process itself. Historical texts are frequently more accessible than contemporary texts. This, too, is interesting. Technology plays some role. In the time of Darwin, the subjects of study were apparent using the technology of the day: microscopes and sailing vessels. When evolution was a new concept, questions about it tended to be broader, more general ones, and questions and answers were expressed in language suitable to general discourse. Today's technology, both computational and instrumental, has resulted in such specialization that some phenomena are made visible to others only through technical manipulation and the labored communication required to accurately register the results of those manipulations. Influenced by decades of shared dialog and common references, we now discuss evolution in a highly technical and specific language, and our questions are frequently more precise and consequently narrower than they were in the past. Modes of inquiry have also changed. Many of us today practice a science of representation, where specific phenomena represent larger processes; a science of illustration, where a single well-worked example illustrates a generality. It can often be difficult to tell the particulars from the generalities, so much of the art of scientific inquiry depends on developing an eye to distinguish the two. These readings will give some practice with that, for certain.

PROBABILITY AND PLAUSIBILITY

One of the most interesting historical patterns apparent in this collection is the changing relationship between probability and plausibility. How plausible a cause is an unlikely event?

Darwin thought much about how much time the process of transmutation—evolution—would take. Sometimes it seemed there was adequate time, sometimes not. "Every species is due to adaptation and hereditary structure . . . It leads you to believe the world older than GEOLOGISTS think," he wrote in his notebook (Darwin 1837). Around Darwin's time, Lord Kelvin calculated the age of the earth to be around 20 million years. This was not long enough for the slow process of transmutation that Darwin envisioned. Other estimates included 75,000 years (comte de Buffon), 18 million years (Simon Newcomb), 22 million years (Hermann von Helmholtz), and 96 million years (John Phillips). Darwin was disturbed by the lack of reconciliation between his concept of slow evolution, which required the accumulation of many rare events, and geological estimates of the age of the earth, and remained open-minded to revisions of those estimates. The current estimated age is over 4.5 billion years. That is a long time.

Not only is the earth old, the earth is also a big place. It took Darwin five years to circle it. It supports a lot of vegetation and a whopping lot of insects and microorganisms. Things live everywhere and in the most unlikely places: in ice in the Antarctic, in boiling sulfurous springs, in hot vents at the bottom of the ocean.

Not only that, within each organism, there are thousands of genes. Each of an organism's traits is determined by several genes, mixing to create a continuous diversity of form. In the late 1960s the new technology of protein electrophoresis revealed far more genetic variation in natural populations than anyone expected or could easily explain. It found that each gene has several different versions, or alleles. And different alleles of different genes combine to further increase the possibilities of variation in form. Now we know that some genes also have several ways for their transcripts[2] to be stitched or spliced together to create different final products from a single DNA sequence.

And each gene also can be expressed, turned on or off, in different places and at different times. This creates additional diversity of form and function. Small differences in the location of expression of the very same gene can lead to impressive variation in form, sometimes converting one organ into another. Differences in the timing and duration of gene expression can lead to exaggerations or otherwise modified proportions of form.

The point here is that there is abundant raw material for evolution to work with: many individuals, many genes within individuals, many versions of each gene, and many ways to put genes together and turn them off and on. And there is plenty of time for things to come and go, for opportunities to happen. If something doesn't work the first time, try, try again. Something else may work eventually.

The improbable, given enough chances, becomes eminently plausible. A little brown bird, blown away in a gale, may perish in the ocean. Another may, against daunting odds, find the speck of an island to rest on, and then perish alone. But another may find that speck in the ocean, then find a seed and a sip of water, and make it. Not likely, not often, but it doesn't need to be likely or happen often. Indeed, it is the sheer unlikelihood of events, their improbability, that makes things different from each other. The improbabil-

---

2. Transcripts are messenger RNA strands that act as the template for translation of DNA sequences into sequences of amino acids. These amino acid sequences are proteins—the "gene products"—that perform various functions in organisms, including giving physical structure, catalyzing biochemical reactions, relaying environmental signals, and regulating the expression of other genes.

ity of it all, then, shouldn't bother us so much. We have the time. We have the opportunities, the chances.

What emerges from these readings is a struggle with the improbable. At first, rare events are interpreted as too unlikely to explain anything at all. The outrage during Darwin's time (and among the proponents of intelligent design in our own) that such an intricacy as life could emerge from an undirected process testifies to this struggle. This struggle is reflected in early thoughts about the biogeography of the Galápagos, and later in debates about chance genetic drift versus natural selection. Truth is, current evolutionary theory is based on totally improbable sequences of events. A random mutation arises—not frequently. It confers a little edge in something—small chance of that. The carrier manages not to get squashed by a falling rock, washed away in a flash flood, or burned up in a wildfire. It manages to find a mate if it needs one. Its offspring, too, luck out. None of these events is very likely, and the chance of them all happening in a sequence is vanishingly small. But multiply that miniscule chance by the number of mutations, and the number of individuals, and the number of millennia, and you find there is a chance of the sequence happening at least once, and a pretty good one at that. Evolutionists have become completely comfortable with invoking the highly improbable as the most plausible scenario. And we stand by that.

But the globe it took Darwin five years to circumnavigate can now be circled in days. It's a small world after all. Species are getting around as never before. Previously drifting along on slow continents or carried away by a rare cyclone, they are now hitching rides in ballast, agricultural shipments, construction soil, and illicit pockets. Goats, sheep, and rats inhabit most habitable latitudes. Newcomers, like kudzu in the southern United States, take over, wiping out the diversity of vegetation beneath them. East meets West and North meets South. No longer are things so isolated. No longer are they so protected from each other.

The land itself now has corridors and zones of homogeneity. More fields, less forest; more asphalt, less wetland. And many populations themselves are becoming smaller. And smaller. Not only are there smaller populations, fewer individuals of many species, but also less variety within species. Is the raw material of evolution diminishing? Could our chances be getting fewer? And time? How much time do we have? Since the environmental movements of the 1960s and 1970s, there is a sentiment that time is not on our side.

This is a new perspective for evolutionary biologists. With a fairly new interest in evolutionary processes operating in ecological, as opposed to geological, time frames, and a willingness to make ourselves useful in solving ecological problems, it is also an interesting challenge. Previously reveling in an overabundance of time and chances, now it appears there is a shortage of both. What will be the new scientific philosophy of rarity, I wonder?

PART I

The Finches in Place

The theory of evolution emerged from the study of comparative anatomy and biogeography. By comparing specimens, natural historians characterized and interpreted the affinities among organisms. By characterizing the geographic distributions of related groups, they discovered the geography that underlay species affinities. This geography suggested that biological processes of dispersal, rather than events of independent creation, linked related species.

The first chapter recounts the popular legend that Darwin's finches were the source of Darwin's inspiration and sudden insights on the process of evolution. It then presents Darwin's own writings on the finches and the Galápagos, which demonstrate, far more accurately than the legend, how the finches figured in his developing thoughts on evolution. Frank Sulloway's article then provides a coherent and realistic account of how the finches shaped Darwin's ideas.

In chapter 2, the reader sees the Galápagos Archipelago through the eyes of very different travelers over the decades. Beginning with the first surviving description of the islands, it includes accounts from the bishop of Panama, a buccaneer, a whaler, and a journalist. Each writer encounters the same land but perceives very different things there.

Chapter 3 again focuses on the Galápagos Archipelago, but the readings included here center on the question of its origin. How the land was formed relates to how the species were formed, and debates on their origin reflect debates over the mechanisms of isolation and the rate of evolutionary change.

Chapter 4 presents the different taxonomic classifications of the finches over time. The reshuffling of genera reflects theories of the number of founding species and the extent of their divergence. The instability of the numbers of species, in turn, reflects the prevailing uncertainty about concepts of species themselves. This difficulty in species designations comes directly from the close correspondence of the finches to their islands—that is, the fact that a single species frequently differs from island to island.

Combined, these readings show that geological processes and biological processes correspond to each other, that evolutionary history is reflected in geological history, and that geology itself influences evolutionary processes.

* 1 *
Grounding a Legend

We've probably all heard the legend of Darwin and his finches: Darwin sails into the Galápagos, notices a little bird, then another that looks like it but doesn't look like it because it has a huge bill and is cracking huge seeds with it. Then he sails off to another island, sees yet another bird, looking like the first two but not looking like the first two because this one is probing cactus flowers with its straight, pointy bill. Then he sails off to another island, sees another bird looking like the others, but not looking like the others, because this one, astonishingly, has picked up a small twig and is probing the holes and crevices in the trees looking to evict for itself a larval snack. There on the spot, or perhaps after some quiet time examining his neatly organized and meticulously labeled specimens, or perhaps in conversation with the *Beagle*'s Captain Fitzroy, Darwin realizes he is looking at the perfect evidence for his revolutionary theory of Evolution by Natural Selection. In all versions of the myth, one assumption is salient: the evidence the birds present, particularly their beaks, was the most important source of inspiration for Darwin's ideas, and none was more important to the development of his theory of natural selection. That's the legend as presented in several elementary biology text-books from earlier decades. We do love a good story.

In myth and legend birds talk to people. They may even say, "I'm very closely related to that other bird but morphologically and behaviorally differentiated in a most fascinating manner." But in real life one needs time and patience to watch the birds eat, sing, court, raise young, and die. If all the birds behave alike at first glance, we must stay for another six months until vegetation is scarcer, or more abundant, and keep watching. If we are in the right spot at the right time in the right weather, we may glimpse a wood-pecker finch, and if we are lucky, we may see it pick up a twig. If we are very lucky, we'll see it poke the twig into a hole and a grub come squirming out to be picked off for lunch. Then we'll be happily excited and think: "That bird just poked with a stick, and a grub came running out, and the bird ate it!" And after the first delight at witnessing any odd action at all, we may realize that the bird just used a tool. And this excitement may get us thinking.

The legend edits out these intervening steps—the learning—in favor of creating another mythical creature—The Scientist. By doing so, it makes Darwin's conclusions and insights that much more inaccessible. While making a dramatic read, such legends can be a rather demoralizing form of pedagogy. In truth, Darwin's most important ideas were less the product of sudden inspiration than of slow gestation, and his crucial insights came long after he left the islands, not to be published until two decades after his voyage. The readings included in this chapter disclose in Darwin's own words the chronology of his reflections on the finches and how these birds informed his major theories. The readings depict his first encounters with the finches, then follow his references to them in his notes and letters and into his published works, early and late. Dealing a final blow to the legend of epiphany in the voices of finches is Frank Sulloway's engaging detective work. His article, in which he hunts along the threads of a richer narrative, chasing bird skin after bird skin, synthesizes Darwin's notes, labels, and correspondence into a coherent narrative of his changing interpretations of the Galápagos finches and how they figured in his major theories. Together, the excerpts in Darwin's words and Sulloway's article give a detailed picture of how Darwin initially collected his finches—not in any very organized manner; what he initially thought of them—not very much; and two developments that changed his mind about them—John Gould's classification of the birds into distinct species as opposed to varieties, and new discoveries of interisland variation in taxa other than the finches.

READINGS

To begin are Darwin's earliest impressions of the finches, formed while he collected them and shortly thereafter. They include brief passages from his "*Beagle* Diary," field notebooks, and his "Ornithological Notes," which, according to Nora Barlow, were probably written aboard the *Beagle* between early 1834 and sometime prior to March 1837. Excerpts from the "Ornithological Notes" presented here include a list of Darwin's collected specimens (including the misidentified "wren," which John Gould later classified as *Certhidea olivacea*, the warbler finch), and some observations on the finches' appearance and habits. Following these is a brief communication to the Zoological Society of London. All references to the finches here are conspicuously brief. As a hint of future interest, however, Darwin casually mentions in a letter written in Sydney in January 1836, to his mentor, J. S. Henslow, that while in the Galápagos he "paid also much attention to the Birds, which I suspect are very curious."

It becomes clear somewhat later that the mockingbirds, not the finches, alerted Darwin to the possibility that species vary from island to island. At the time Darwin was collecting his specimens, he did not recognize these interisland differences, and in his *Journal of Researches* he publicly bemoans his

failure to heed the fine-scale variations among the island taxa. His new preoccupation with interisland variation is reflected also in his repeated pestering
of Henslow, one such letter being included here, and his inquiries to J. D.
Hooker about whether the plants also indicate that closely related species are
found on different islands. Indeed, he had guessed as much and was rewarded
by Hooker's affirmation and insightful interpretation of the finding: "I was
quite prepared to see the extraordinary difference between the plants of the
separate Islands from your journal, a most strange fact, & one which quite
overturns all our preconceived notions of species radiating from a centre &
migrating to any extent from one focus of greater development" (Hooker
1844). Interisland differences among taxa were a preoccupation of Darwin
and his colleagues because they challenged the notion of a restricted number
of locations for creation (to be discussed in chapter 3).

Darwin's cogitations on how differences among taxa arise and are maintained are inextricably tied to his analysis of the geographic distribution of
taxa. Next are excerpts from Darwin's notebooks on the transmutation of
species, thought to be composed upon returning from his voyage in 1836.
These notes reveal some of Darwin's earlier thoughts on how geographic
isolation maintains differences that would otherwise disappear through interbreeding. Rather than focusing on interisland variation, the first excerpt
shows the beginnings of his analysis of the relationship between continental
and island affinities and differences, which was to figure so prominently in his
major works. The remaining excerpts from these notebooks elaborate upon
the causes of the changes that occur in isolation, as well as causes of the isolation itself. In particular, he focuses on dispersal and isolation, in contrast to
divine creation, as causes of the geographic patterns of affinity and difference among taxa—the "horizontal history" that he was able to read so fluently
from the spatial distribution of species. This notion of "horizontal history"
marks a major insight, as it reveals Darwin's historical (eventually genealogical) interpretation of spatial patterns of distribution. While the Galápagos
biota are mentioned several times, the finches do not figure in these notes.

The year before the second edition of his *Journal of Researches*, Darwin
composed a 250-page essay (his *1844 Essay*) in which he described the process
of "a natural means of selection." An excerpt from this essay appears next.
Here Darwin gives considerable attention to the geographic distribution of
variation among taxa and postulates geographic isolation as a central factor
in the accumulation of species differences. He cites affinities and differences
between the Galápagos and the American mainland as evidence, but now he
also cites smaller differences between closely related species from different
islands, appearing to recognize these two patterns as reflections of the same
process on different scales. Linking these spatial scales was major progress
toward considering evolution to be a continuous process. In this narrative

of the evolution of species differences, Darwin masterfully describes the role of isolation first as it occurs on oceanic islands, and then as it occurs on continents. In this manner he spells out explicitly how evolution as observed on islands pertains to the whole of the globe.

This same year, Darwin sent his "confession" to Hooker in his beliefs in transmutation, included as the next selection, crediting the Galápagos for his insight into the transmutation of species. Apparently, he had become comfortable enough with his theory by this time to discuss it with other correspondents, including Leonard Jenyns, a letter to whom is included here. Significantly, though, in this letter he tempers his theory of species transmutation via the differential death of individuals by describing it as one of the "intermediate laws" of creation.

Differences between the 1837 and 1845 editions of his *Journal of Researches* are revealing, so the Galápagos chapters from both editions are presented here in tandem, the later in excerpted form. By 1837 Darwin had been enlightened in much of his thinking about the finches by John Gould, who by that time had published his description of the finches and other birds of the Galápagos. Darwin knew that, according to the most eminent ornithologist of his time, the finches represented good species, rather than being, as he had earlier thought, "only varieties." He knew also that they were indeed very closely related, though vastly different in morphological characters that typically define genera—that is, their bill morphology. Between the two editions not only had Hooker completed an initial analysis of the plants of the Galápagos, but, as his notebooks reveal, Darwin had been struggling also with the interpretation of these patterns. What he chose to add and change in the 1845 edition shows the progression of his thoughts and his timid sharing of this progress with a public audience. By then, the Galápagos finches had impressed Darwin enough that he included another plate—at his own expense—depicting the heads of four of them.

When Darwin discusses the Galápagos in his major published works, however, his emphasis is on the overall relation of the biota of the Galápagos with that of South America—not the interisland variation. Although the Galápagos make it into his stunning *Origin of the Species*, the finches do not, nor do they make it into any of his other major works.

Darwin's later publications, correspondence, and memoirs highlight the Galápagos flora and fauna as influential to his thinking about the transmutation of species. When writing to his various correspondents, he repeatedly invoked the Galápagos as a source of the crystallization of his ideas on the transmutation of species. For example, in reference to "the law of succession"—the observation that fossil organisms resemble, but are not identical to, extant ones—he writes: "In fact this law with the Galapagos Distribution first turned my mind on origin of species" (C. Darwin 1859) and "It would have been a strange fact if I had overlooked the importance of isolation, seeing that it was such cases as that of the Galapagos Archipelago, which chiefly led me to study the origin of species" (Charles R. Darwin to Moritz Wagner, Down, October 13, 1876).

This tendency to refer to the Galápagos as the source of his insights is reflected in his "little history of the Origin" that he sent to Ernst Haeckel and in his autobiography (cited in F. Darwin 1909), excerpts of which are included. These publications and communications contribute in their own way to the legend that the Galápagos species held a special and rather immediate place in the development of his theory of evolution by natural selection, but, again, they contain no specific mention of the finches.

In fact, Darwin himself never used the Galápagos finches as a case example for any of his major principles of the transmutation of species by natural selection. Instead, he frequently cited the more general pattern of the Galápagos flora and fauna resembling the South American biota as central to his theories. This larger pattern illustrated affinity through descent and divergence from an ancestral stock—his major message. The finer pattern of interisland variation was suggestive of the ubiquity of the transmutation process itself, as he implies in his *1844 Essay*: indispensable for insight, but far more difficult to show definitively. Sulloway speculates that the finches never figured in Darwin's discussion because of the incomplete information he had on their geographic distribution and the guesswork involved in assigning the specimens to different islands. No doubt this is true, but the fine variation among the taxa, and the difficulty of their classification into species versus varieties, was problematic in itself. Darwin simply didn't have the answers to explain how interspecific differences among the finches arose and were maintained.

Why then have the finches become such an enduringly important example of the evolutionary process? In part because they surprised even Darwin and challenged him directly. A letter from Darwin to Osbert Salvin in 1875, included last, expresses Darwin's surprise that the finches from different islands do not vary in the manner he predicted, with a single species isolated on each island. Yet, he says, studying the birds "on the spot" would "throw a flood of light upon variation." Still very much unanswered for Darwin is how species differences in these finches are maintained at such close quarters. Enough said. If these finches can illuminate the problem that stumped the master, then they are worth the trouble we take to study them.

NOTE ON TRANSCRIPTIONS

I have omitted strike-throughs and line breaks from the original transcriptions to facilitate reading. I have also inserted periods in place of transcriptional dashes where it seemed there was an end to the sentence or thought. Transcriptional corrections were also incorporated without brackets. The original transcriptions, without these molestations, can be obtained from the sources cited.

Many of these readings are available online on the Web site of the University of Cambridge's "Complete Works of Charles Darwin Online" (http://darwin-online.org.uk/) and are reproduced with their permission, as indicated. All correspondence is also available on the Web site of the University of Cambridge's "Darwin's Correspondence Project" (http://www.darwin project.ac.uk/).

From *Darwin's* Beagle *Diary* (1831–1836)

❋ CHARLES R. DARWIN

* 1835 Sept: 17th [page 606]
* Galapagos Isds

The birds are Strangers to Man & think him as innocent as their countrymen the huge Tortoises. Little birds within 3 & four feet, quietly hopped about the Bushes & were not frightened by stones being thrown at them. Mr King killed one with his hat & I pushed off a branch with the end of my gun a large Hawk.

* Oct 1st [page 614]

Since leaving the last Island, owing to the small quantity of water on board, only half allowance of water has been served out (ie 1/2 a Gallon for cooking & all purposes). This under the line with a Vertical sun is a sad drawback to the few comforts which a Ship possesses. From different accounts, we had hoped to have found water here. To our disappointment the little pits in the Sandstone contained scarcely a Gallon & that not good, it was however sufficient to draw together all the little birds in the country. Doves & Finches swarmed round its margin. I was reminded of the manner in which I saw at Charles Isd a boy procuring dinner for his family. Sitting by the side of the Well with a long stick in his hand, as the doves came to drink he killed as many as he wanted & in half an hour collected them together & carried them to the house.

Transcribed by Kees Rookmaaker. Reproduced from "The Complete Work of Charles Darwin Online," edited by John van Wyhe, (http://darwin-online.org.uk/).

From *The Galapagos Notebook* (1835)

❊ CHARLES R. DARWIN

* 14th October 1835 [page 43b]

Wandered about Bird collecting. Iguana shakes head vertically; sea—one . . . dozes hind legs stretched out walks very slowly—sleeps—closes eyes—Eats much Cactus:

    Mr Bynoe* saw one walking from two other carrying it in mouth. Eats very deliberately, without chewing. Small Finch picking from same piece after alights on back.

Edited by Gordon Chancellor and John van Wyhe. Reproduced from "The Complete Work of Charles Darwin Online," edited by John van Wyhe (http://darwin-online.org.uk/).

From *Darwin's Ornithological Notes* (1835)

❊ CHARLES R. DARWIN

* MS 72

Annually, heavy torrents of rain at one particular season fall; grasses and other plants rapidly shoot up, flower, & as rapidly disappear. The seeds however lie dormant, till the next year, buried in the cindery soil. Hence these Finches are in number of species & individuals far preponderant over any other family of birds. Amongst the species of this family there reigns (to me) an inexplicable confusion. Of each kind, some are jet black, & from this, by intermediate shades, to brown; the proportional number, in all the black kinds is *exceedingly* small; yet my series of specimens would go to show, that, that color is proper to the old cock birds alone. On the other hand Mr. Bynoe & Fuller assert, they have each a small jet black bird of the female sex. Moreover a gradation in form of the bill, appears to me to exist. There is no possibility of distinguishing the species by their habits, as they are all similar, & they feed together (also with doves) in large irregular flocks. I should observe, that with respect to the probable age of the smaller birds, that in no case were any of the feathers imperfect, or bill soft, so as to indicate immaturity, & on the other hand in no case were the eggs in the ovarium of the hen birds much developed. I should suppose the season of incubation would be two or *three months* later . . .

---

* Benjamin Bynoe (1803–1865), assistant Surgeon on the Beagle.

* MS 73–74

[In reference to a Chatham Island Thenca (mockingbird) specimen. Numbers refer to specimen numbers.]

These birds are closely allied in appearance to the Thenca of Chile (2169) or Callandra of la Plata (1216). In their habits I cannot point out a single difference; They are lively inquisitive, active *run fast*, frequent houses to pick the meat of the Tortoise, which is hung up, sing tolerably well; are said to build a simple open nest. Are *very* tame, a character in common with the other birds: I *imagined* however its note or cry was rather different from the Thenca of Chile? Are very abundant, over the whole Island; are chiefly tempted up into the high & damp parts, by the houses & cleared ground.

I have specimens from four of the larger Islands; the two above enumerated, and (3349: female. Albermarle Isd.) & (3350: male: James Isd). The specimens from Chatham & Albermarle Isd appear to be the same; but the other two are different. In each Isld. each kind is *exclusively* found: habits of all are indistinguishable. When I recollect, the fact that the form of the body, shape of scales & general size, the Spaniards can at once pronounce, from which Island any Tortoise may have been brought. When I see these Islands in sight of each other, & possessed of but a scanty stock of animals, tenanted by these birds, but slightly differing in structure & filling the same place in Nature, I must suspect they are only varieties. The only fact of a similar kind of which I am aware, is the constant asserted difference between the wolf-like Fox of East & West Falkland Islds. If there is the slightest foundation for these remarks the zoology of Archipelagoes will be well worth examining; for such facts would undermine the stability of Species.

* MS 74–75 [page 263]

[This table lists Darwin's collection of the finches.]

Editor's note: "do" indicates "ditto." Numbers refer to specimen numbers.

3310	Wren		Male
3312	Fringilla		Male
3313	do.		(Sex unknown)
3314	do.		Female
3315	do.	—	do
3316	do.	—	Male
3317	do.		Male
3318	do.	—	Male
3319		—	Male

(3314–3319 bracketed) V. suprà

3320   (Icterus 3320: Male. jet black). (3321: 3322 Males) (3323 Female).
3321*   *This is the only bird, out of the number which compose the large irregular
3322   flocks, which can be distinguished from its habits. Its most frequent resort is
3323   hopping & *climbing* about the great Cacti, to feed with its sharp beak, on the

fruit & flowers. Commonly however it alights on the ground & with the Fringilla in the same manner, seeks for seeds. The rarity of the jet black specimens is well exemplified in this case; out of the many brown ones which I daily saw, I never could observe a single black one, besides the one preserved. Mr. Bynoe however has another Specimen; Fuller in vain tried to procure one. I should add that Specimen (3320) was shot when picking together with a brown one, the fruit of a Cactus.

3324	Fringilla.	Male. (Young?)
3325	do	Female.
3326	Fringilla: Female: there were very many individuals of exactly the same plumage.	
3327	Fringilla	Male
3328	do	Female
3329	do	do

3330 . . . Male:) (3331 Female) (3332 Male). This species is well characterized by its
3331 curious beak. Is a true Fringilla in its habits. I only saw this bird in one Island.
3332 James Isld & in one part alone of it. Was feeding in considerable numbers with the other species. Mr. Bynoe has a much blacker variety. [Capt. FitzRoy's specimen comes from same isld — *written in margin*.]

3333	Fringilla.	Male
3334	do	do
3335	do	do
3336	do	do

N B. The Gross-beaks are very injurious to the cultivated land; they stock up seeds & plants, buried six inches beneath the surface.

3337	do	Female. Upper Mandible is in Pill Box. (3361)
3338	do	do
3339	do	do
3340	Male	

3341. Fringilla. Male. I saw specimens with *precisely* similar plumage, which were females

Edited by Nora Barlow, *Bulletin of the British Museum (Natural History)*, Historical Series, 2 (1963), no. 7:201–278; with introduction, notes, and appendix by the editor. Reproduced from "The Complete Work of Charles Darwin Online," edited by John van Wyhe (http://darwin-online.org.uk/).

Remarks upon the Habits of the Genera Geospiza, Camarhynchus, Cactornis *and* Certhidea *of Gould* (1837)

※ CHARLES R. DARWIN

* May 10th, 1837

The group of groundfinches, characterised, at a previous meeting, by Mr. Gould, under the generic appellations of *Geospiza, Camarhynchus, Certhidea*, and *Cactornis*, were upon the table; and Mr. Darwin being present, remarked that these birds were exclusively confined to the Gallapagos Islands; but their general resemblance in character, and the circumstance of their indiscriminately associating in large flocks, rendered it almost impossible to study the habits of particular species. In common with nearly all the birds of these islands, they were so tame that the use of the fowling-piece in procuring specimens was quite unnecessary. They appeared to subsist on seeds, deposited on the ground in great abundance by a rich annual crop of herbage.

Read 10 May. *Proceedings of the Zoological Society of London* 5, no. 53 (1837): 49.

Letter to John S. Henslow (1838)

❋ CHARLES R. DARWIN

∗ Saturday, 3 Novr 1838

My dear Henslow,

I am preparing an appendix to my Journal, in which I mean to add a few remarks on some of the subjects, which I have there discussed. You may recollect how often I have talked over the marvellous fact of the species of birds being different, in those different islands of the Galapagos. Lately I have gained some curious facts, bearing on the same points, regarding the lizards & tortoises of those same islands; & now I want to know whether you can tell me anything about the plants. Pray understand, I do not want you to take any trouble in giving me names &c &c. all I want is to know whether in casting your eye over my plants, how many cases (for you told me of some one or two) there are of *near* species, of the same genus; one species coming from one island, & the other from a second island.

If there are any number of these cases you can mention to what natural families they belong. (or the genera if you happen to know them). Also how near the species are; whether they require comparison to be distinguished; or whether, merely in any natural classification, they would follow one after another. I have already given you trouble enough about this same Galap: Arch: but all I now want is for you to answer me, as far as your present knowledge goes; & in doing this, I apprehend it will not take up more than a hour, & it will (if such be the fact with the plants) support my case of the birds & Tortoises in a *glorious* manner.

Many thanks for your letter, which I received some time. Leonard Jenyns will see about the Savings Bank.

Will you have the kindness to write pretty soon, as I believe (but am not sure) I shall go to press with the appendix immediately,

Ever yours most truly
Chas. Darwin

I have written to you on a torn piece of Paper without perceiving it.

Letter 429, 3 November 1838. From Charles R. Darwin, *The Correspondence of Charles Darwin*, volume 2, *1837–1843*, edited by Frederick Burkhardt and Sydney Smith (Cambridge: Cambridge University Press, 1986). Reprinted with permission from Cambridge University Press.

From *Notebook [B] on "Transmutation of Species"* (1837–1838)

✳ CHARLES R. DARWIN

* Page 5

With this tendency to vary by generation, why are species are constant over whole country; beautiful law of intermarriages (separating) partaking of characters of both parents, and then infinite in number. In man it has been said, there is instinct for opposites to like each other. Aegyptian cats and dogs, ibis, same as formerly, but separate a pair and place them on a fresh [i.e., geologically new] island. It is very doubtful whether they would remain constant. Is it not said that marrying in *deteriorates* a race; that is, alters it from some end which is good for man? Let a pair be introduced and increase slowly, [away] from many enemies, so as often to intermarry who will dare say what result? According to this view animals on separate islands ought to become different if kept long enough apart with slightly differing circumstances. Now Galapagos Tortoises, Mocking birds, Falkland Fox, Chiloe fox, English and Irish Hare.

As we thus believe species vary, in changing climate, we ought to find representative species; this we do in South America (closely approaching), but as they inosculate, we must suppose the change is effected at once, something like a variety produced (every grade in that case surely is not produced?). Species according to Lamarck disappear as collection made perfect; truer even than in Lamarck's time. Gray's remark, best known species (as some common land shells) most difficult to separate.

Every character continues to vanish: bones, instinct, etc., etc., etc.

Nonfertility of hybridity, etc., etc.

If species (1) may be derived from form (2), etc., then (remembering Lyell's arguments of transportal) island near continent might have some species same as nearest land, which were late arrivals; others old ones (of which none of same kind had in interval arrived) might have grown altered. Hence the type would be of the continent, though species all different. In cases as Galapagos and Juan Fernandez, when continent of Pacific existed, might have been monsoons . . . when they ceased, importation ceased, and changes commenced; or intermediate land existed, or they may represent some large country long separated. On this idea of propagation of species we can see why a form peculiar to continents, all bred in from one parent why.

* Page 98

The question if creative power acted at Galapagos it so acted that birds with plumage and tone of voice partly American, North and South. (And geographical divisions are arbitrary and not permanent. This might be made very strong. If we believe the Creator creates by any laws. Which, I think is shown by the very facts of the Zoological character of these islands) so permanent a breath cannot reside in space before island existed. Such an influence must exist in such spots. We know birds do arrive, and seeds.

The same remarks applicable to fossil animals same type; armadillo-like covering created; passage for vertebrae in neck same cause, such beautiful adaptations, yet other animals live so well. This view of propagation gives a hiding place for many unintelligible structures; it might have been of use in progenitor, or it may be of use, like mammae on man's breast.

How does it come wandering birds such, [as] sandpipers, not new at Galapagos? Did the creative force know that these species could arrive? Did it only create those kinds not so likely to wander? Did it create two species closely allied to Mus. coronata, but not coronata? We know that domestic animals vary in countries without any assignable reason.

Astronomers might formerly have said that God ordered each planet to move in its particular destiny. In same manner God orders each animal created with certain form in certain country, but how much more simple and sublime power let attraction act according to certain laws are inevitable consequences. Let animals be created, then by the fixed laws of generation, such will be their successors.

Let the powers of transportal be such, and so will be the forms of one country to another. Let geological changes go at such a rate, so will be the number and distribution of the species!!

It may be argued representative species chiefly found where barriers (and what are barriers but), interruption of communication, or when country changes. Will it be said that volcanic soil of Galapagos under equator, that external conditions would produce species so close as Patagonian Chat and Galapagos Orpheus? Put this strong, so many thousand miles distant.

Absolute knowledge that species die and others replace them. Two hypotheses, fresh creations is mere assumption, it explains nothing further; points gained if any facts are connected.

No doubt in birds, mundine genera (Bats, Foxes, Mus) are birds that are apt to wander and of easy transportal. Waders and waterfowl. Scrutinize genera, and draw up tables. Instinct may confine certain birds which have wide powers of flight; but are there any genera mundine which cannot transport easily (it would have been wonderful if the two Rheas had existed in different continents)? In plants I believe not.

* Page 224

If my theory true, we get: (I) a *horizontal* history of earth within recent times and many curious points of speculation; for having ascertained means of transport, we should then know whether former lands intervened; (2nd.) by character of any two ancient fauna, we may form some idea of origin under connection of those two countries (hence India, Mexico, and Europe, one great sea; coral reefs; therefore shallow water at Melville island); (3rd.) We know that structure of every organ in A.B.C., three species of one genus, can pass into each other by steps we see; but this cannot be predicated of genus structures in two genera. We then cease to know the steps. Although D.E.F. follow close to A.B.C., we cannot be sure that structure (C) could pass into (D). We may foretell species, limits of good species being known. It explains the blending of two genera. It explains typical structure. Every species is due to adaptation and hereditary structure (latter far chief element, therefore little service habits in classification, or rather the fact that they are not far the most serviceable). We may speculate of durability of succession from what we have seen in old world and on amount [of] changes which may happen. *It leads you to believe the world older than* GEOLOGISTS *think. It agrees with excessive inequality of numbers of species in divisions. Look at articulata!!!*

It leads to nature of physical change between one group of animals and a successive one. It leads to knowledge what kinds of structure may pass into each other; now on this view no one need look for intermediate structures, between say in brain, between lowest Mammal and Reptile, or between extremities of any great divisions. Thus a knowledge of possible changes is discovered, for speculating on future.

Fish never become a man. Does not require fresh creation! If continent had sprung up round Galapagos on Pacific side, the Oolite order of things might have easily been formed. With belief of change, transmutation and geographical grouping we are led to endeavour to discover *causes* of change, the manner of adaptation (wish of parents??); instinct and structure become full of speculation and line of observation.

View of generation being condensation, test of highest organization intelligible; may look to first germ, led to comprehend two affinities. My theory would give zest to recent and fossil Comparative Anatomy, it would lead to study of instincts, heredity and mind heredity, whole metaphysics. It would lead to closest examination of hybridity (to what circumstances favour crossing and what prevents it) and generation, causes of change in order to know what we have come from and to what we tend; this, and direct examination of direct passages of structures in species, might lead to laws of change, which would then be main object of study, to guide our past speculations with respect to past and future. (The Grand Question which every naturalist ought to have before him when dissecting a whale or classifying a mite, a fungus, or an infusorian is, "What are the Laws of Life?")

Transcribed by David Kohn. In *Charles Darwin's Notebooks, 1836–1844*, edited by Paul H. Barrett, Peter J. Gautrey, Sandra Herbert, David Kohn, and Sydney Smith (Cambridge: Cambridge University Press, 2008). © The Committee for the Publication of the Charles Darwin's Notebooks 2009. Reprinted with the permission of Cambridge University Press.

From *Notebook [C] on "Transmutation of Species"* (1838)

❋ CHARLES R. DARWIN

* Page 145

My theory agrees with unequal distances between species some fine & some wide which is strange if creator had so created them.

People will argue & fortify their minds with such sentences as "oh turn a Buccinum into a Tiger" but perhaps I feel the impossibility of this more than any one—no turn the Zebra into the Quagga—let them be wild in same country with their own instinct & (even though fertile when compelled to breed) & then all that I want is granted.

For at Galapagos make ten species of Orpheus, one of which has very short legs & long tail, short much curved beak—other very long beak with short [tail]; let these only have progeny with species & there will be two genera— let short billed one be exaggerated & all rest destroyed, far remote genera will be produced.

Transcribed by David Kohn. In *Charles Darwin's Notebooks, 1836–1844*, edited by Paul H. Barrett, Peter J. Gautrey, Sandra Herbert, David Kohn, and Sydney Smith (Cambridge: Cambridge University Press, 2008). © The Committee for the Publication of the Charles Darwin's Notebooks 2009. Reprinted with the permission of Cambridge University Press.

From *The Foundations of "The Origin of Species," Two Essays Written in 1842 and 1844* (1909)

❋ FRANCIS DARWIN

Editor's note: Footnotes are originals from F. Darwin.

* Pages xii–xiii

In his Autobiography he [Charles R. Darwin] wrote:—"During the voyage of the *Beagle* I had been deeply impressed by discovering in the Pampean formation great fossil animals covered with armour like that on the existing armadillos; secondly, by the manner in which closely allied animals replace one another in proceeding southward over the Continent; and thirdly, by the South American character of most of the productions of the Galapagos archipelago, and more especially by the manner in which they differ slightly on each island of the group; none of the islands appearing to be very ancient in a geological sense. It was evident that such facts as these, as well as many others, could only be explained on the supposition that species gradually become modified; and the subject haunted me."

Again we have to ask: how soon did any of these influences produce an effect on Darwin's mind? Different answers have been attempted. Huxley held that these facts could not have produced their essential effect until the voyage had come to an end, and the "relations of the existing with the extinct species and of the species of the different geographical areas with one another were determined with some exactness." He does not therefore allow that any appreciable advance toward evolution was made during the actual voyage of the *Beagle*.

Professor Judd takes a very different view. He holds that November 1832 may be given with some confidence as the "date at which Darwin commenced that long series of observations and reasonings which eventually culminated in the preparation of the *Origin of Species*." Though I think these words suggest a more direct and continuous march than really existed between fossil-collecting in 1832 and writing the *Origin of Species* in 1859, yet I hold that it was during the voyage that Darwin's mind began to be turned in the direction of Evolution, and I am therefore in essential agreement with Prof. Judd, although I lay more stress than he does on the latter part of the voyage.

THE ESSAY OF 1844 [PP. 183–194, WRITTEN BY CHARLES R. DARWIN]

Section Third.

An attempt to explain the foregoing laws of geographical distribution, on the theory of allied species having a common descent.

First let us recall the circumstances most favourable for variation under domestication, as given in the first chapter—viz. 1st, a change, or repeated changes, in the conditions to which the organism has been exposed, continued through several seminal (*i.e.* not by buds or divisions) generations: 2nd, steady selection of the slight varieties thus generated with a fixed end in view: 3rd, isolation as perfect as possible of such selected varieties; that is, the preventing their crossing with other forms; this latter condition applies to all terrestrial animals, to most if not all plants and perhaps even to most (or all) aquatic organisms. It will be convenient here to show the advantage of isolation in the formation of a new breed, by comparing the progress of two persons (to neither of whom let time be of any consequence) endeavouring to select and form some very peculiar new breed. Let one of these persons work on the vast herds of cattle in the plains of La Plata,[1] and the other on a small stock of 20 or 30 animals in an island. The latter might have to wait centuries (by the hypothesis of no importance)[2] before he obtained a "sport" approaching to what he wanted; but when he did and saved the greater number of its offspring and their offspring again, he might hope that his whole little stock would be in some degree affected, so that by continued selection he might gain his end. But on the Pampas, though the man might get his first approach to his desired form sooner, how hopeless would it be to attempt, by saving its offspring amongst so many of the common kind, to affect the whole herd: the effect of this one peculiar "sport"[3] would be quite lost before he could obtain a second original sport of the same kind. If, however, he could separate a small number of cattle, including the offspring of the desirable "sport," he might hope, like the man on the island, to effect his end. If there be organic beings of which two individuals *never* unite, then simple selection whether on a continent or island would be equally serviceable to make a new and desirable breed; and this new breed might be made in surprisingly few years from the great and geometrical powers of propagation to beat out the old breed; as has happened (notwithstanding crossing) where good breeds of dogs and pigs have been introduced into a limited country,—for instance, into the islands of the Pacific.

Let us now take the simplest natural case of an islet upheaved by the volcanic or subterranean forces in a deep sea, at such a distance from other land that only a few organic beings at rare intervals were transported to it, whether borne by the sea[4] (like the seeds of plants to coral-reefs), or by hurricanes, or by floods, or on rafts, or in roots of large trees, or the germs of one plant or animal attached to or in the stomach of some other animal, or by the intervention (in most cases the most probable means) of other islands since sunk or destroyed. It may be remarked that when one part of the earth's crust is raised it is probably the general rule that another part sinks. Let this island go on slowly, century after century, rising foot by foot; and in the course of time we shall have instead of a small mass of rock,[5] lowland and highland, moist woods and dry sandy spots, various soils, marshes, streams and pools: under water on

1. This instance occurs in the Essay of 1842, p. 32, but not in the *Origin*; though the importance of isolation is discussed (*Origin*, Ed. i. p. 104, vi. p. 127).
2. The meaning of the words within parenthesis is obscure.
3. It is unusual to find the author speaking of the selection of *sports* rather than small variations.
4. This brief discussion is represented in the *Origin*, Ed. i. by a much fuller one (pp. 356, 383, vi. pp. 504, 535). See, however, the section in the present Essay, p. 168.
5. On the formation of new stations, see *Origin*, Ed. i. p. 292, vi. p. 429.

the sea shore, instead of a rocky steeply shelving coast we shall have in some parts bays with mud, sandy beaches and rocky shoals. The formation of the island by itself must often slightly affect the surrounding climate. It is impossible that the first few transported organisms could be perfectly adapted to all these stations; and it will be a chance if those successively transported will be so adapted. The greater number would probably come from the lowlands of the nearest country; and not even all these would be perfectly adapted to the new islet whilst it continued low and exposed to coast influences. Moreover, as it is certain that all organisms are nearly as much adapted in their structure to the other inhabitants of their country as they are to its physical conditions, so the mere fact that a *few* beings (and these taken in great degree by chance) were in the first case transported to the islet, would in itself greatly modify their conditions.[6] As the island continued rising we might also expect an occasional new visitant; and I repeat that even one new being must often affect beyond our calculation by occupying the room and taking part of the subsistence of another (and this again from another and so on), several or many other organisms. Now as the first transported and any occasional successive visitants spread or tended to spread over the growing island, they would undoubtedly be exposed through several generations to new and varying conditions: it might also easily happen that some of the species *on an average* might obtain an increase of food, or food of a more nourishing quality.[7] According then to every analogy with what we have seen takes place in every country, with nearly every organic being under domestication, we might expect that some of the inhabitants of the island would "sport," or have their organization rendered in some degree plastic. As the number of the inhabitants are supposed to be few and as all these cannot be so well adapted to their new and varying conditions as they were in their native country and habitat, we cannot believe that every place or office in the economy of the island would be as well filled as on a continent where the number of aboriginal species is far greater and where they consequently hold a more strictly limited place. We might therefore expect on our island that although very many slight variations were of no use to the plastic individuals, yet that occasionally in the course of a century an individual might be born[8] of which the structure or constitution in some slight degree would allow it better to fill up some office in the insular economy and to struggle against other species. If such were the case the individual and its offspring would have a better *chance* of surviving

---

6. *Origin*, Ed. i. pp. 390, 400, vi. pp. 543, 554.

7. In the MS. *some of the species . . . nourishing quality* is doubtfully erased. It seems clear that he doubted whether such a problematical supply of food would be likely to cause variation.

8. At this time the author clearly put more faith in the importance of sport-like variation than in later years.

and of beating out its parent form; and if (as is probable) it and its offspring crossed with the unvaried parent form, yet the number of the individuals being not very great, there would be a chance of the new and more serviceable form being nevertheless in some slight degree preserved. The struggle for existence would go on annually selecting such individuals until a new race or species was formed. Either few or all the first visitants to the island might become modified, according as the physical conditions of the island and those resulting from the kind and number of other transported species were different from those of the parent country—according to the difficulties offered to fresh immigration—and according to the length of time since the first inhabitants were introduced. It is obvious that whatever was the country, generally the nearest from which the first tenants were transported, they would show an affinity, even if all had become modified, to the natives of that country and even if the inhabitants of the same source (?) had been modified. On this view we can at once understand the cause and meaning of the affinity of the fauna and flora of the Galapagos Islands with that of the coast of S. America; and consequently why the inhabitants of these islands show not the smallest affinity with those inhabiting other volcanic islands, with a very similar climate and soil, near the coast of Africa.[9]

To return once again to our island, if by the continued action of the subterranean forces other neighbouring islands were formed, these would generally be stocked by the inhabitants of the first island, or by a few immigrants from the neighbouring mainland; but if considerable obstacles were interposed to any communication between the terrestrial productions of these islands, and their conditions were different (perhaps only by the number of different species on each island), a form transported from one island to another might become altered in the same manner as one from the continent; and we should have several of the islands tenanted by representative races or species, as is so wonderfully the case with the different islands of the Galapagos Archipelago. As the islands become mountainous, if mountain-species were not introduced, as could rarely happen, a greater amount of variation and selection would be requisite to adapt the species, which originally came from the lowlands of the nearest continent, to the mountain-summits than to the lower districts of our islands. For the lowland species from the continent would have first to struggle against other species and other conditions on the coast-land of the island, and so probably become modified by the selection of its best fitted varieties, then to undergo the same process when the land had attained a moderate elevation; and then lastly when it had become Alpine. Hence we can understand why the faunas of insular mountain-summits are, as in the case of Teneriffe, eminently peculiar. Putting on one side the case

9. *Origin*, Ed. i. p. 398, vi. p. 553.

of a widely extended flora being driven up the mountain-summits, during a change of climate from cold to temperate, we can see why in other cases the floras of mountain-summits (or as I have called them islands in a sea of land) should be tenanted by peculiar species, but related to those of the surrounding lowlands, as are the inhabitants of a real island in the sea to those of the nearest continent.[10]

Let us now consider the effect of a change of climate or of other conditions on the inhabitants of a continent and of an isolated island without any great change of level. On a continent the chief effects would be changes in the numerical proportion of the individuals of the different species; for whether the climate became warmer or colder, drier or damper, more uniform or extreme, some species are at present adapted to its diversified districts; if for instance it became cooler, species would migrate from its more temperate parts and from its higher land; if damper, from its damper regions, &c. On a small and isolated island, however, with few species, and these not adapted to much diversified conditions, such changes instead of merely increasing the number of certain species already adapted to such conditions, and decreasing the number of other species, would be apt to affect the constitutions of some of the insular species: thus if the island became damper it might well happen that there were no species living in any part of it adapted to the consequences resulting from more moisture. In this case therefore, and still more (as we have seen) during the production of new stations from the elevation of the land, an island would be a far more fertile source, as far as we can judge, of new specific forms than a continent. The new forms thus generated on an island, we might expect, would occasionally be transported by accident, or through long-continued geographical changes be enabled to emigrate and thus become slowly diffused.

But if we look to the origin of a continent; almost every geologist will admit that in most cases it will have first existed as separate islands which gradually increased in size;[11] and therefore all that which has been said concerning the probable changes of the forms tenanting a small archipelago is applicable to a continent in its early state. Furthermore, a geologist who reflects on the geological history of Europe (the only region well known) will admit that it has been many times depressed, raised and left stationary. During the sinking of a continent and the probable generally accompanying changes of climate

10. See *Origin*, Ed. i. p. 403, vi. p. 558, where the author speaks of Alpine humming birds, rodents, plants, &c. in S. America, all of strictly American forms. In the MS. the author has added between the lines "As world has been getting hotter, there has been radiation from high-lands,—old view?—curious; I presume Diluvian in origin."
11. See the comparison between the Malay Archipelago and the probable former state of Europe, *Origin*, Ed. i. p. 299, vi. p. 438, also *Origin*, Ed. i. p. 292, vi. p. 429.

the effect would be little, *except* on the numerical proportions and in the extinction (from the lessening of rivers, the drying of marshes and the conversion of high-lands into low &c.) of some or of many of the species. As soon however as the continent became divided into many isolated portions or islands, preventing free immigration from one part to another, the effect of climatic and other changes on the species would be greater. But let the now broken continent, forming isolated islands, begin to rise and new stations thus to be formed, exactly as in the first case of the upheaved volcanic islet, and we shall have equally favourable conditions for the modification of old forms, that is the formation of new races or species. Let the islands become reunited into a continent; and then the new and old forms would all spread, as far as barriers, the means of transportal, and the pre-occupation of the land by other species, would permit. Some of the new species or races would probably become extinct, and some perhaps would cross and blend together. We should thus have a multitude of forms, adapted to all kinds of slightly different stations, and to diverse groups of either antagonist or food-serving species. The oftener these oscillations of level had taken place (and therefore generally the older the land) the greater the number of species [which] would tend to be formed. The inhabitants of a continent being thus derived in the first stage from the same original parents, and subsequently from the inhabitants of one wide area, since often broken up and reunited, all would be obviously related together and the inhabitants of the most *dissimilar* stations on the same continent would be more closely allied than the inhabitants of two very *similar* stations on two of the main divisions of the world.[12]

I need hardly point out that we now can obviously see why the number of species in two districts, independently of the number of stations in such districts, should be in some cases as widely different as in New Zealand and the Cape of Good Hope.[13] We can see, knowing the difficulty in the transport of terrestrial mammals, why islands far from mainlands do not possess them;[14] we see the general reason, namely accidental transport (though not the precise reason), why certain islands should, and others should not, possess members of the class of reptiles. We can see why an ancient channel of communication between two distant points, as the Cordillera probably was between southern Chile and the United States during the former cold periods; and icebergs between the Falkland Islands and Tierra del Fuego; and gales, at a former or present time, between the Asiatic shores of the Pacific and eastern

12. *Origin*, Ed. i. p. 349, vi. p. 496. The arrangement of the argument in the present Essay leads to repetition of statements made in the earlier part of the book: in the *Origin* this is avoided.

13. *Origin*, Ed. i. p. 389, vi. p. 542.

14. *Origin*, Ed. i. p. 393, vi. p. 547.

islands in this ocean; is connected with (or we may now say causes) an affinity between the species, though distinct, in two such districts. We can see how the better chance of diffusion, from several of the species of any genus having wide ranges in their own countries, explains the presence of other species of the same genus in other countries;[15] and on the other hand, of species of restricted powers of ranging, forming genera with restricted ranges.

As every one would be surprised if two exactly similar but peculiar varieties[16] of any species were raised by man by long continued selection, in two different countries, or at two very different periods, so we ought not to expect that an exactly similar form would be produced from the modification of an old one in two distinct countries or at two distinct periods. For in such places and times they would probably be exposed to somewhat different climates and almost certainly to different associates. Hence we can see why each species appears to have been produced singly, in space and in time. I need hardly remark that, according to this theory of descent, there is no necessity of modification in a species, when it reaches a new and isolated country. If it be able to survive and if slight variations better adapted to the new conditions are not selected, it might retain (as far as we can see) its old form for an indefinite time. As we see that some sub-varieties produced under domestication are more variable than others, so in nature, perhaps, some species and genera are more variable than others. The same precise form, however, would probably be seldom preserved through successive geological periods, or in widely and differently conditioned countries.[17]

Finally, during the long periods of time and probably of oscillations of level, necessary for the formation of a continent, we may conclude (as above explained) that many forms would become extinct. These extinct forms, and those surviving (whether or not modified and changed in structure), will all be related in each continent in the same manner and degree, as are the inhabitants of any two different sub-regions in that same continent. I do not mean to say that, for instance, the present Marsupials of Australia or Edentata and rodents of S. America have descended from any one of the few fossils of the same orders which have been discovered in these countries. It is possible that, in a very few instances, this may be the case; but generally they must be considered as merely codescendants of common stocks.[18] I believe in this, from the improbability, considering the vast number of species, which (as explained in the last chapter) must by our theory have existed, that the *comparatively* few fossils which have been found should chance to be the immedi-

15. *Origin*, Ed. i. pp. 350, 404, vi. pp. 498, 559.
16. *Origin*, Ed. i. p. 352, vi. p. 500.
17. *Origin*, Ed. i. p. 313, vi. p. 454.
18. *Origin*, Ed. i. p. 341, vi. p. 487.

ate and linear progenitors of those now existing. Recent as the yet discovered fossil mammifers of S. America are, who will pretend to say that very many intermediate forms may not have existed? Moreover, we shall see in the ensuing chapter that the very existence of genera and species can be explained only by a few species of each epoch leaving modified successors or new species to a future period; and the more distant that future period, the fewer will be the *linear* heirs of the former epoch. As by our theory, all mammifers must have descended from the same parent stock, so is it necessary that each land now possessing terrestrial mammifers shall at some time have been so far united to other land as to permit the passage of mammifers;[19] and it accords with this necessity, that in looking far back into the earth's history we find, first changes in the geographical distribution, and secondly a period when the mammiferous forms most distinctive of two of the present main divisions of the world were living together.[20]

I think then I am justified in asserting that most of the above enumerated and often trivial points in the geographical distribution of past and present organisms (which points must be viewed by the creationists as so many ultimate facts) follow as a simple consequence of specific forms being mutable and of their being adapted by natural selection to diverse ends, conjoined with their powers of dispersal, and the geologico-geographical changes now in slow progress and which undoubtedly have taken place. This large class of facts being thus explained, far more than counterbalances many separate difficulties and apparent objections in convincing my mind of the truth of this theory of common descent.

Cambridge: Cambridge University Press, 1909. Reprinted with the permission of Cambridge University Press.

Letter to Joseph D. Hooker (1844)

✻ CHARLES R. DARWIN

* Down. Bromley Kent
* Thursday

My dear Sir

I must write to thank you for your last letter; I to tell you how much all your views & facts interest me. I must be allowed to put my own interpretation on what you say of "not being a good arranger of extended views" which is, that you do not indulge in the loose speculations so easily started by every

19. *Origin*, Ed. i. p. 396, vi. p. 549.
20. *Origin*, Ed. i. p. 340, vi. p. 486.

smatterer & wandering collector. I look at a strong tendency to generalize as an entire evil.

What limit shall you take on the Patagonian side has d'Orbigny published, I believe he made a large collection at the R. Negro, where Patagonia retains its usual forlorn appearance; at Bahia Blanca & northward the features of Patagonia insensibly blend into the savannahs of La Plata. The Botany of S. Patagonia (& I collected *every* plant in flower at the season when there) would be worth comparison with the N. Patagonian collection by d'Orbigny. I do not know anything about King's plants, but his birds were so inaccurately habitated, that I have seen specimen from Brazil, Tierra del & *the Cape de Verde Isd* all said to come from the St. Magellan. What you say of Mr Brown is humiliating; I had suspected it, but cd not allow myself to believe in such heresy. FitzRoy gave him a rap in his Preface, & made me very indignant, but it seems a much harder one wd not have been wasted. My crptogamic collection was sent to Berkeley; it was not large; I do not believe he has yet published an account, but he wrote to me some year ago that he had described & mislaid all his descriptions. Wd it not be well for you to put yourself in communication with him; as otherwise some things will perhaps be twice laboured over. My best (though poor) collection of the Crptogam. was from the Chonos Islands.

Would you kindly observe one little fact for me, whether any species of plant, *peculiar* to any isld, as Galapagos, St. Helena or New Zealand, where there are no large quadrupeds, have hooked seeds, such hooks as if observed here would be thought with justness to be adapted to catch into wool of animals.

Would you further oblige me some time by informing me (though I forget this will certainly appear in your Antarctic Flora) whether in isld like St. Helena, Galapagos, & New Zealand, the number of families & genera are large compared with the number of species, as happens in coral-isld, & as I *believe* ? in the extreme Arctic land. Certainly this is case with Marine shells in extreme Arctic seas. Do you suppose the fewness of species in proportion to number of large groups in *Coral-islets.*, is owing to the chance of seeds from all orders, getting drifted to such new spots? as I have supposed.

Did you collect sea-shells in Kerguelen land, I shd like to know their character?

Your interesting letters tempt me to be very unreasonable in asking you questions; but you must not give yourself any trouble about them, for I know how fully & worthily you are employed.

Besides a general interest about the Southern lands, I have been now ever since my return engaged in a very presumptuous work & which I know no one individual who wd not say a very foolish one. I was so struck with distribution of Galapagos organisms &c &c & with the character of the American fossil mammifers, &c &c that I determined to collect blindly every sort of fact, which cd bear any way on what are species. I have read heaps of agricultural & horticultural books, & have never ceased collecting facts. At last gleams of light have come, & I am almost convinced (quite contrary to opinion I started with) that species are not (it is like confessing a murder) immutable. Heaven forfend me from Lamarck nonsense of a "tendency to progression" "adaptations from the slow willing of animals" &c, but the conclusions I am led to are not widely different from his—though the means of change are wholly so—I think I have found out (here's presumption!) the simple way by which species become exquisitely adapted to various ends. You will now groan, & think to yourself "on what a man have I been wasting my time in writing to." I shd, five years ago, have thought so. I fear you will also groan at the length of this letter—excuse me, I did not begin with malice prepense.

Believe me my dear
Sir Very truly your's
C. Darwin

Letter 729, [11 January 1844]. From Charles R. Darwin, *The Correspondence of Charles Darwin*, volume 3, *1844–1846*, edited by Frederick Burkhardt and Sydney Smith (Cambridge: Cambridge University Press, 1987). Reprinted with the permission of Cambridge University Press.

Letter to Leonard Jenyns (1844)

✳ CHARLES R. DARWIN

My dear Jenyns,

I am very much obliged to you for the trouble you have taken in having written me so long a note. The question of where, when, & how, the check to the increase of a given species falls appears to me particularly interesting; & our difficulty in answering it, shows how really ignorant we are of the lives & habits of our most familiar species. I was aware of the bare fact of old Birds driving away their young, but had never thought of the effect, you so clearly point out, of local gaps in number being thus immediately filled up. But the original difficulty remains, for if your farmers had not killed your sparrows & rooks, what would have become of those, which now immigrate into your Parish: in the middle of England one is too far distant from the natural limits of the Rook & sparrow, to suppose that the young are thus far expelled from Cambridgeshire. The check must fall heavily at some time of each species's life, for if one calculates that only half the progeny are reared & breed, how enormous is the increase! One has, however, no business to feel so much surprise at one's ignorance, when one knows how impossible, it is, without statistics, to conjecture the duration of life & percentage of deaths to births in mankind.

If it could be shown that apparently the birds of passage, *which breed here* & increase return in the succeeding years in about the same number, whereas those that come here for their winter & non-breeding season, annually come here with the same numbers, but return with greatly decreased numbers, one would know (as indeed seems probable) that the check fell chiefly on full-grown birds in the winter season, & not on the eggs & *very young birds*, which has appeared to me often the most probable period. If at any time any remarks on this subject should occur to you, I shd be most grateful for the benefit of them.

With respect to my far-distant work on species, I must have expressed myself with singular inaccuracy, if I led you to suppose that I meant to say that my conclusions were inevitable. They have become so, after years of weighing puzzles, to myself *alone*; but in my wildest day-dream, I never expect more than to be able to show that there are two sides to the question of the immutability of species, ie whether species are *directly* created, or by intermediate laws, (as with the life & death of individuals). I did not approach the subject on the side of the difficulty in determining what are species & what are varieties, but (though, why I shd give you such a *history* of my doings, it wd be hard to say) from such facts, as the relationship between the living & extinct mammifers in S. America, & between those living on the continent & on adjoining islands, such as the Galapagos. It occurred to me, that a collection of all such analogous facts would throw light either for or against the view of related species, being co-descendants from a common stock. A long searching amongst agricultural & horticultural books & people, makes me believe (I well know how absurdly presumptuous this must appear) that I see the way in which new varieties become exquisitely adapted to the external conditions of life, & to other surrounding beings. I am a bold man to lay myself open to being thought a complete fool, & a most deliberate one. From the nature of the grounds, which make me believe that species are mutable in form, these grounds cannot be restricted to the closest-allied species; but how far they extend, I cannot tell, as my reasons fall away by degrees, when applied to species more & more remote from each other.

Pray do not think, that I am so blind as not to see that there are numerous immense difficulties on my notions, but they appear to me less than on the common view. I have drawn up a sketch & had it copied (in 200 pages) of my conclusions; & if I thought at some future time, that you would think it worth reading, I shd. of course be most thankful to have the criticism of so competent a critic.

Excuse this very long & egotistical & ill written letter, which by your remarks you have led me into, & believe me,

Yours very truly
C. Darwin

Letter 793, [25 November 1844]. From Charles R. Darwin, *The Correspondence of Charles Darwin*, volume 3, *1844–1846*, edited by Frederick Burkhardt and Sydney Smith (Cambridge: Cambridge University Press, 1987). Reprinted with the permission of Cambridge University Press.

From *Journal of Researches into the Geology and Natural History of the Various Countries Visited by H.M.S.* Beagle (1839)

❋ CHARLES R. DARWIN

* Chapter XIX

GALAPAGOS ARCHIPELAGO

SEPTEMBER 15TH.—The Beagle arrived at the southernmost of the Galapagos islands. This archipelago consists of ten principal islands, of which five much exceed the others in size. They are situated under the equatorial line, and between five and six hundred miles to the westward of the coast of America. The constitution of the whole is volcanic. With the exception of some ejected fragments of granite, which have been most curiously glazed and altered by the heat, every part consists of lava, or of sandstone resulting from the attrition of such materials. The higher islands, (which attain an elevation of three, and even four thousand feet) generally have one or more principal craters towards their centre, and on their flanks smaller orifices. I have no exact data from which to calculate, but I do not hesitate to affirm, that there must be, in all the islands of the archipelago, at least two thousand craters. These are of two kinds; one, as in ordinary cases, consisting of scoriae and lava, the other of finely-stratified volcanic sandstone. The latter in most instances have a form beautifully symmetrical: their origin is due to the ejection of mud,—that is, fine volcanic ashes and water,—without any lava.

Considering that these islands are placed directly under the equator, the climate is far from being excessively hot; a circumstance which, perhaps, is chiefly owing to the singularly low temperature of the surrounding sea. Excepting during one short season, very little rain falls, and even then it is not regular: but the clouds generally hang low. From these circumstances the lower parts of the islands are extremely arid, whilst the summits, at an elevation of a thousand feet or more, possess a tolerably luxuriant vegetation. This is especially the case on the windward side, which first receives and condenses the moisture from the atmosphere.

In the morning (17th,) we landed on Chatham Island, which, like the others, rises with a tame and rounded outline, interrupted only here and there by scattered hillocks—the remains of former craters. Nothing could be less inviting than the first appearance. A broken field of black basaltic lava is every where covered by astunted brushwood, which shows little signs of life. The dry and parched surface, having been heated by the noonday sun, gave the air a close and sultry feeling, like that from a stove: we fancied even the bushes smelt unpleasantly. Although I diligently tried to collect as many plants as possible, I succeeded in getting only ten kinds; and such wretched-looking little weeds would have better become an arctic, than an equatorial Flora.

The thin woods, which cover the lower parts of all the islands, excepting where the lava has recently flowed, appear from a short distance quite leafless, like the deciduous trees of the northern hemisphere in winter. It was some time before I discovered, that not only almost every plant was in full leaf, but that the greater number were now in flower. After the period of heavy rains, the islands are said to appear for a short time partially green. The only other country, in which I have seen a vegetation with a character at all approaching to this, is at the volcanic island of Fernando Noronha, placed in many respects under similar conditions.

The natural history of this archipelago is very remarkable: it seems to be a little world within itself; the greater number of its inhabitants, both vegetable and animal, being found nowhere else. As I shall refer to this subject again, I will only here remark, as forming a striking character on first landing, that the birds are strangers to man. So tame and unsuspecting were they, that they did not even understand what was meant by stones being thrown at them; and quite regardless of us, they approached so close that any number might have been killed with a stick.

The Beagle sailed round Chatham Island, and anchored in several bays. One night I slept on shore, on a part of the island where some black cones— the former chimneys of the subterranean heated fluids—were extraordinarily numerous. From one small eminence, I counted sixty of these truncated hillocks, which were all surmounted by a more or less perfect crater. The greater number consisted merely of a ring of red scoriae, or slags, cemented together: and their height above the plain of lava, was not more than from fifty to a hundred feet. From their regular form, they gave the country a *workshop* appearance, which strongly reminded me of those parts of Staffordshire where the great iron-foundries are most numerous.

The age of the various beds of lava was distinctly marked by the comparative growth, or entire absence, of vegetation. Nothing can be imagined more rough and horrid than the surface of the more modern streams. These have been aptly compared to the sea petrified in its most boisterous moments: no sea, however, would present such irregular undulations, or would be traversed by such deep chasms. All the craters are in an extinct condition; and although the age of the different streams of lava could be so clearly distinguished, it is probable they have remained so for many centuries. There is no account in any of the old voyagers of any volcano on this island having been seen in activity; yet since the time of Dampier (1684), there must have been some increase in the quantity of vegetation, otherwise so accurate a person would not have expressed himself thus: "Four or five of the easternmost islands are rocky, barren, and hilly, producing neither tree, herb, nor grass, but a few dildoe (cactus) trees, except by the sea-side."[1] This description is at present applicable only to the western islands, where the volcanic forces are in frequent activity.

The day, on which I visited the little craters, was glowing hot, and the scrambling over the rough surface, and through the intricate thickets, was very fatiguing; but I was well repaid by the Cyclopian scene. In my walk I met two large tortoises, each of which must have weighed at least two hundred pounds. One was eating a piece of cactus, and when I approached, it looked at me, and then quietly walked away: the other gave a deep hiss and drew in its head. These huge reptiles, surrounded by the black lava, the leafless shrubs, and large cacti, appeared to my fancy like some antediluvian animals.

SEPTEMBER 23D. — The Beagle proceeded to Charles Island. This archipelago has long been frequented, first by the Bucaniers, and latterly by whalers, but it is only within the last six years, that a small colony has been established on it. The inhabitants are between two and three hundred in number: they nearly all consist of people of colour, who have been banished for political crimes from the Republic of the Equator (Quito is the capital of this state) to which these islands belong. The settlement is placed about four and a half miles inland, and at an elevation probably of a thousand feet. In the first part of the road we passed through leafless thickets, as in Chatham Island. Higher up, the wood gradually became greener; and immediately we had crossed the ridge of the island, our bodies were cooled by the fine southerly trade-wind, and our senses refreshed by the sight of a green and thriving vegetation. The houses are irregularly scattered over a flat space of ground, which is cultivated with sweet potatoes and bananas. It will not easily be imagined how pleasant the sight of black mud was to us, after having been so long accustomed to the parched soil of Peru and Chile.

The inhabitants, although complaining of poverty, gain, without much trouble, the means of subsistence from the fertile soil. In the woods there are many wild pigs and goats, but the main article of animal food is derived from the tortoise. Their numbers in this island have of course been greatly reduced, but the people yet reckon on two days' hunting supplying food for the rest of the week. It is said that formerly single vessels have taken away as many as seven hundred of these animals, and that the ship's company of a frigate some years since brought down two hundred to the beach in one day.

We staid at this island four days, during which time I collected many plants and birds. One morning I ascended the highest hill, which has an altitude of nearly 1800 feet. The summit consists of a broken-down crater, thickly clothed with coarse grass and brushwood. Even in this one island, I counted

1. Dampier's Voyage, vol. i., p. 101.

thirty-nine hills, each of which was terminated by a more or less perfect circular depression.

SEPTEMBER 29TH. — We doubled the south-west extremity of Albermarle Island, and the next day were nearly becalmed between it and Narborough Island. Both are covered with immense streams of black naked lava; which, having either flowed over the rims of the great caldrons, or having burst forth from the smaller orifices on the flanks, have in their descent spread over miles of the sea-coast. On both of these islands eruptions are known occasionally to take place; and in Albermarle we saw a small jet of smoke curling from the summit of one of the more lofty craters. In the evening we anchored in Bank's Cove, in Albermarle Island.

When morning came, we found that the harbour in which we were at anchor was formed by a broken-down crater, composed of volcanic sandstone. After breakfast I went out walking. To the southward of this first crater, there was another of similar composition, and beautifully symmetrical. It was elliptic in form; the longer axis being less than a mile, and its depth about 500 feet. The bottom was occupied by a shallow lake, and in its centre a tiny crater formed an islet. The day was overpoweringly hot, and the lake looked clear and blue. I hurried down the cindery slope, and choked with dust eagerly tasted the water—but to my sorrow I found it salt as brine.

The rocks on the coast abounded with great black lizards, between three and four feet long; and on the hills, another species was equally common. We saw several of the latter, some clumsily running out of our way, and others shuffling into their burrows. I shall presently describe in more detail the habits of both these reptiles.

OCTOBER 3D. — We sailed round the northern end of Albermarle Island. Nearly the whole of this side is covered with recent streams of dark-coloured lavas, and is studded with craters. I should think it would be difficult to find in any other part of the world, an island situated within the tropics, and of such considerable size (namely 75 miles long), so sterile and incapable of supporting life.

On the 8th we reached James Island. Captain FitzRoy put Mr. Bynoe, myself, and three others on shore, leaving with us a tent and provisions, to wait there till the vessel returned from watering. This was an admirable plan for the collections, as we had an entire week of hard work. We found here a party of Spaniards, who had been sent from Charles Island to dry fish, and to salt tortoise-meat.

At the distance of about six miles, and at the height of nearly 2000 feet, the Spaniards had erected a hovel in which two men lived, who were employed in catching tortoises, whilst the others were fishing on the coast. I paid this party two visits, and slept there one night. In the same manner as in the other islands, the lower region is covered by nearly leafless bushes: but here many of them grow to the size of trees. I measured several which were two feet in diameter, and some even two feet nine inches. The upper region being kept damp, from the moisture of the condensed clouds, supports a green and flourishing vegetation. So damp was the ground, that there were large beds of a coarse carex, in which great numbers of a very small water-rail lived and bred. While staying in this upper region, we lived entirely upon tortoise-meat. The breastplate roasted (as the Gauchos do *carne con cuero*), with the flesh attached to it, is very good; and the young tortoises make excellent soup; but otherwise the meat to my taste is very indifferent.

During another day we accompanied a party of the Spaniards in their whale-boat to a salina, or lake from which salt is procured. After landing, we had a very rough walk over a rugged field of recent lava, which has almost surrounded a sandstone crater, at the bottom of which the salt-lake is situated. The water was only three or four inches deep, and rested on a layer of beautifully crystallized white salt. The lake was quite circular, and fringed with a border of brightly green succulent plants: the precipitous walls of the crater were also clothed with wood, so that the scene was both picturesque and curious. A few years since, the sailors belonging to a sealing-vessel murdered their captain in this quiet spot; and we saw his skull lying among the bushes.

During the greater part of our week on shore, the sky was cloudless, and if the trade-wind failed for an hour, the heat became very oppressive. On two days, the thermometer within the tent stood for some hours at 93°; but in the open air, in the wind and sun, at only 85°. The sand was extremely hot; the thermometer placed in some of a brown colour immediately rose to 137°, and how much higher it would have risen, I do not know, for it was not graduated above that number. The *black* sand felt much hotter, so that even in thick boots it was disagreeable, on this account, to walk over it.

I will now offer a few general observations on the natural history of these islands. I endeavoured to make as nearly a perfect collection in every branch as time permitted. The plants have not yet been examined, but Professor Henslow, who has kindly undertaken the description of them, informs me that there are probably many new species, and perhaps even some new genera. They all have an extremely weedy character, and it would scarcely have been supposed, that they had grown at an inconsiderable elevation directly under the equator. In the lower and sterile parts, the bush, which from its minute brown leaves chiefly gives the leafless appearance to the brushwood, is one of the Euphorbiaceae. In the same region an acacia and a cactus (*Opuntia Galapageia*[2]), with large oval compressed articulations, springing from a cylindrical stem, are in some parts common. These are the only trees which

2. Magazine of Zoology and Botany, vol. i., p. 466.

in that part afford any shade. Near the summits of the different islands, the vegetation has a very different character; ferns and coarse grasses are abundant; and the commonest tree is one of the Compositae. Tree-ferns are not present. One of the most singular characters of the Flora, considering the position of this archipelago, is the absence of every member of the palm family. Cocos Island, on the other hand, which is the nearest point of land, takes its name from the great number of cocoa-nut trees on it. From the presence of the Opuntias and some other plants, the vegetation partakes more of the character of that of America than of any other country.

Of mammalia a large kind of mouse forms a well-marked species. From its large thin ears, and other characters, it approaches in form a section of the genus, which is confined to the sterile regions of South America. There is also a rat which Mr. Waterhouse believes is probably distinct from the English kind; but I cannot help suspecting that it is only the same altered by the peculiar conditions of its new country.

In my collections from these islands, Mr. Gould considers that there are twenty-six different species of land birds. With the exception of one, all probably are undescribed kinds, which inhabit this archipelago, and no other part of the world. Among the waders and waterfowl it is more difficult, without detailed comparison, to say what are new. But a water-sail which lives near the summits of the mountains, is undescribed, as perhaps is a Totanus and a heron. The only kind of gull which is found among these islands, is also new; when the wandering habits of this genus are considered, this is a very remarkable circumstance. The species most closely allied to it, comes from the Strait of Magellan. Of the other aquatic birds, the species appear the same with well-known American birds.

The general character of the plumage of these birds is extremely plain, and like the Flora possesses little beauty. Although the species are thus peculiar to the archipelago, yet nearly all in their general structure, habits, colour of feathers, and even tone of voice, are strictly American. The following brief list will give an idea of their kinds. 1st. A buzzard, having many of the characters of Polyborus or Caracara; and in its habits not to be distinguished from that peculiar South American genus; 2d. Two owls; 3d. Three species of tyrant-flycatchers—a form strictly American. One of these appears identical with a common kind (*Muscicapa coronata?* Lath.), which has a very wide range, from La Plata throughout Brazil to Mexico; 4th. A sylvicola, an American form, and especially common in the northern division of the continent; 5th. Three species of mocking-birds, a genus common to both Americas; 6th. A finch, with a stiff tail and a long claw to its hinder toe, closely allied to a North American genus; 7th. A swallow belonging to the American division of that genus; 8th. A dove, like, but distinct from, the Chilian species; 9th. A group of finches, of which Mr. Gould considers there are thirteen species; and these he has distributed into four new sub-genera. These birds are the most singular of any in the archipelago. They all agree in many points; namely, in a peculiar structure of their bill, short tails, general form, and in their plumage. The females are gray or brown, but the old cocks jet-black. All the species, excepting two, feed in flocks on the ground, and have very similar habits. It is very remarkable that a nearly perfect gradation of structure in this one group can be traced in the form of the beak, from one exceeding in dimensions that of the largest gros-beak, to another differing but little from that of a warbler. Of the aquatic birds I have already remarked that some are peculiar to these islands, and some common to North and South America.

We will now turn to the order of reptiles, which forms, perhaps, the most striking feature in the zoology of these islands. The species are not numerous, but the number of individuals of each kind, is extraordinarily great. There is one kind both of the turtle and tortoise; of lizards four; and of snakes about the same number.

I will first describe the habits of the tortoise (*Testudo Indicus*) which has been so frequently alluded to. These animals are found, I believe, in all the islands of the Archipelago; certainly in the greater number. They frequent in preference the high damp parts, but likewise inhabit the lower and arid districts. I have already mentioned[3] proofs, from the numbers which have been taken in a single day, how very numerous they must be. Some individuals grow to an immense size: Mr. Lawson, an Englishman, who had at the time of our visit charge of the colony, told us that he had seen several so large, that it required six or eight men to lift them from the ground; and that some had afforded as much as two hundred pounds of meat. The old males are the largest, the females rarely growing to so great a size. The male can readily be distinguished from the female by the greater length of its tail. The tortoises which live on those islands where there is no water, or in the lower and arid parts of the others, chiefly feed on the succulent cactus. Those which frequent the higher and damp regions, eat the leaves of various trees, a kind of berry (called guayavita) which is acid and austere, and likewise a pale green filamentous lichen, that hangs in tresses from the boughs of the trees.

The tortoise is very fond of water, drinking large quantities, and wallowing in the mud. The larger islands alone possess springs, and these are always situated towards the central parts, and at a considerable elevation. The tortoises, therefore, which frequent the lower districts, when thirsty, are obliged to travel from a long distance. Hence broad and well-beaten paths radiate off in every direction from the wells even down to the sea-coast; and the Spaniards by following them up, first discovered the watering-places. When I landed at

3. Dampier says, "The land-turtles are here so numerous, that five or six hundred men might subsist on them for several months without any other sort of provisions. They are so extraordinarily large and fat, and so sweet, that no pullet eats more pleasantly."—Vol. i., p. 110.

Chatham Island, I could not imagine what animal travelled so methodically along the well-chosen tracks. Near the springs it was a curious spectacle to behold many of these great monsters; one set eagerly travelling onwards with outstretched necks, and another set returning, after having drunk their fill. When the tortoise arrives at the spring, quite regardless of any spectator, it buries its head in the water above its eyes, and greedily swallows great mouthfuls, at the rate of about ten in a minute. The inhabitants say each animal stays three or four days in the neighbourhood of the water, and then returns to the lower country; but they differed in their accounts respecting the frequency of these visits. The animal probably regulates them according to the nature of the food which it has consumed. It is, however, certain, that tortoises can subsist even on those islands where there is no other water, than what falls during a few rainy days in the year.

I believe it is well ascertained, that the bladder of the frog acts as a reservoir for the moisture necessary to its existence: such seems to be the case with the tortoise. For some time after a visit to the springs, the urinary bladder of these animals is distended with fluid, which is said gradually to decrease in volume, and to become less pure. The inhabitants, when walking in the lower district, and overcome with thirst, often take advantage of this circumstance, by killing a tortoise, and if the bladder is full, drinking its contents. In one I saw killed, the fluid was quite limpid, and had only a *very slightly* bitter taste. The inhabitants, however, always drink first the water in the pericardium, which is described as being best.

The tortoises, when moving towards any definite point, travel by night and day, and arrive at their journey's end much sooner than would be expected. The inhabitants, from observations on marked individuals, consider that they can move a distance of about eight miles in two or three days. One large tortoise, which I watched, I found walked at the rate of sixty yards in ten minutes, that is 360 in the hour, or four miles a day, allowing also a little time for it to eat on the road.

During the breeding season, when the male and female are together, the male utters a hoarse roar or bellowing, which it is said, can be heard at the distance of more than a hundred yards. The female never uses her voice, and the male only at such times; so that when the people hear this noise, they know the two are together. They were at this time (October) laying their eggs. The female, where the soil is sandy, deposits them together, and covers them up with sand; but where the ground is rocky she drops them indiscriminately in any hollow. Mr. Bynoe found seven placed in a line in a fissure. The egg is white and spherical; one which I measured was seven inches and three-eighths in circumference. The young animals, as soon as they are hatched, fall a prey in great numbers to the buzzard, with the habits of the Caracara. The old ones seem generally to die from accidents, as from falling down precipices. At least several of the inhabitants told me, they had never found one dead without some such apparent cause.

The inhabitants believe that these animals are absolutely deaf; certainly they do not overhear a person walking close behind them. I was always amused, when overtaking one of these great monsters as it was quietly pacing along, to see how suddenly, the instant I passed, it would draw in its head and legs, and uttering a deep hiss fall to the ground with a heavy sound, as if struck dead. I frequently got on their backs, and then, upon giving a few raps on the hinder part of the shell, they would rise up and walk away; but I found it very difficult to keep my balance.

The flesh of this animal is largely employed, both fresh and salted; and a beautifully clear oil is prepared from the fat. When a tortoise is caught, the man makes a slit in the skin near its tail, so as to see inside its body, whether the fat under the dorsal plate is thick. If it is not, the animal is liberated; and it is said to recover soon from this strange operation. In order to secure the tortoises, it is not sufficient to turn them like turtle, for they are often able to regain their upright position.

It was confidently asserted, that the tortoises coming from different islands in the archipelago were slightly different in form; and that in certain islands they attained a larger average size than in others. Mr. Lawson maintained that he could at once tell from which island any one was brought. Unfortunately, the specimens which came home in the Beagle were too small to institute any certain comparison. This tortoise, which goes by the name of *Testudo Indicus*, is at present found in many parts of the world. It is the opinion of Mr. Bell, and some others who have studied reptiles, that it is not improbable that they all originally came from this archipelago. When it is known how long these islands have been frequented by the bucaniers, and that they constantly took away numbers of these animals alive, it seems very probable that they should have distributed them in different parts of the world. If this tortoise does not originally come from these islands, it is a remarkable anomaly; inasmuch as nearly all the other land inhabitants seem to have had their birthplace here.

Of lizards there are four or five species; two probably belong to the South American genus Leiocephalus, and two to Amblyrhyncus. This remarkable genus was characterized by Mr. Bell,[4] from a stuffed specimen sent from Mexico, but which I conceive there can be little doubt originally came through some whaling ship from these islands. The two species agree pretty closely in general appearance; but one is aquatic and the other terrestrial in its habits. Mr. Bell thus concludes his description of *Amb. cristatus*: "On a comparison of this animal with the true Iguanas, the most striking and important discrepancy is in the form of the head. Instead of the long, pointed, narrow muzzle

4. Zoological Journal, July, 1835.

of those species, we have here a short, obtusely truncated head, not so long as it is broad, the mouth consequently only capable of being opened to a very short space. These circumstances, with the shortness and equality of the toes, and the strength and curvature of the claws, evidently indicate some striking peculiarity in its food and general habits, on which, however, in the absence of all certain information, I shall abstain from offering any conjecture." The following account of these two lizards, will, I think, show with what judgment Mr. Bell foresaw a variation in habit, accompanying change in structure.

First for the aquatic kind (*Amb. cristatus*). This lizard is extremely common on all the islands throughout the Archipelago. It lives exclusively on the rocky sea-beaches, and is never found, at least I never saw one, even ten yards inshore. It is a hideous-looking creature, of a dirty black colour, stupid and sluggish in its movements. The usual length of a full-grown one is about a yard, but there are some even four feet long: I have seen a large one which weighed twenty pounds. On the island of Albemarle they seem to grow to a greater size than on any other. These lizards were occasionally seen some hundred yards from the shore swimming about; and Captain Collnett, in his Voyage, says, "they go out to sea in shoals to fish." With respect to the object, I believe he is mistaken; but the fact stated on such good authority cannot be doubted. When in the water the animal swims with perfect ease and quickness, by a serpentine movement of its body and flattened tail,—the legs, during this time, being motionless and closely collapsed on its sides. A seaman on board sank one, with a heavy weight attached to it, thinking thus to kill it directly; but when an hour afterwards he drew up the line, the lizard was quite active. Their limbs and strong claws are admirably adapted for crawling over the rugged and fissured masses of lava, which every where form the coast. In such situations, a group of six or seven of these hideous reptiles may oftentimes be seen on the black rocks, a few feet above the surf, basking in the sun with outstretched legs.

I opened the stomach of several, and in each case found it largely distended with minced sea-weed, of that kind which grows in thin foliaceous expansions of a bright green or dull red colour. I do not recollect having observed this sea-weed in any quantity on the tidal rocks; and I have reason to believe it grows at the bottom of the sea, at some little distance from the coast. If such is the case, the object of these animals occasionally going out to sea is explained. The stomach contained nothing but the seaweed. Mr. Bynoe, however, found a piece of a crab in one; but this might have got in accidentally, in the same manner as I have seen a caterpillar, in the midst of some lichen, in the paunch of a tortoise. The intestines were large, as in other herbivorous animals.

The nature of this lizard's food, as well as the structure of its tail, and the certain fact of its having been seen voluntarily swimming out at sea, absolutely prove its aquatic habits; yet there is in this respect one strange anomaly;

namely, that when frightened it will not enter the water. From this cause, it is easy to drive these lizards down to any little point overhanging the sea, where they will sooner allow a person to catch hold of their tail than jump into the water. They do not seem to have any notion of biting; but when much frightened they squirt a drop of fluid from each nostril. One day I carried one to a deep pool left by the retiring tide, and threw it in several times as far as I was able. It invariably returned in a direct line to the spot where I stood. It swam near the bottom, with a very graceful and rapid movement, and occasionally aided itself over the uneven ground with its feet. As soon as it arrived near the margin, but still being under water, it either tried to conceal itself in the tufts of sea-weed, or it entered some crevice. As soon as it thought the danger was past, it crawled out on the dry rocks, and shuffled away as quickly as it could. I several times caught this same lizard, by driving it down to a point, and though possessed of such perfect powers of diving and swimming, nothing would induce it to enter the water; and as often as I threw it in, it returned in the manner above described. Perhaps this singular piece of apparent stupidity may be accounted for by the circumstance, that this reptile has no enemy whatever on shore, whereas at sea it must often fall a prey to the numerous sharks. Hence, probably urged by a fixed and hereditary instinct that the shore is its place of safety, whatever the emergency may be, it there takes refuge.

During our visit (in October) I saw extremely few small individuals of this species, and none I should think under a year old. From this circumstance it seems probable that the breeding season had not commenced. I asked several of the inhabitants if they knew where it laid its eggs: they said, that although well acquainted with the eggs of the other kind, they had not the least knowledge of the manner in which this species is propagated;—a fact, considering how common an animal this lizard is, not a little extraordinary.

We will now turn to the terrestrial species (*Amb. subcristatus* of Gray).[5] This species, differently from the last, is confined to the central islands of the Archipelago, namely to Albemarle, James, Barrington, and Indefatigable. To the southward, in Charles, Hood, and Chatham islands, and to the northward, in Towers, Bindloes, and Abington, I neither saw nor heard of any. It would appear as if this species had been created in the centre of the Archipelago, and thence had been dispersed only to a certain distance.

5. Briefly characterized by Mr. Gray in the Zoological Miscellany, from a specimen badly stuffed; from which cause one of its most important characters (the rounded tail, compared to the flattened one of the aquatic kind) was overlooked. Captain FitzRoy has presented some fine specimens of both species to the British Museum. I cannot omit here returning my thanks to Mr. Gray, for the kind manner in which he has afforded me every facility as often as I have visited the British Museum.

In the central islands they inhabit both the higher and damp, as well as the lower and sterile parts; but in the latter they are much the most numerous. I cannot give a more forcible proof of their numbers, than by stating, that when we were left at James Island, we could not for some time find a spot free from their burrows, on which to pitch our tent. These lizards, like their brothers the sea-kind, are ugly animals; and from their low facial angle have a singularly stupid appearance. In size perhaps they are a little inferior to the latter, but several of them weighed between ten and fifteen pounds each. The colour of their belly, front legs, and head (excepting the crown which is nearly white), is a dirty yellowish-orange: the back is a brownish-red, which in the younger specimens is darker. In their movements they are lazy and half torpid. When not frightened, they slowly crawl along with their tails and bellies dragging on the ground. They often stop, and doze for a minute with closed eyes, and hind legs spread out on the parched soil.

They inhabit burrows; which they sometimes excavate between fragments of lava, but more generally on level patches of the soft volcanic sandstone. The holes do not appear to be very deep, and they enter the ground at a small angle; so that when walking over these lizard *warrens*, the soil is constantly giving way, much to the annoyance of the tired walker. This animal when excavating its burrow, alternately works the opposite sides of its body. One front leg for a short time scratches up the soil, and throws it towards the hind foot, which is well placed so as to heave it beyond the mouth of the hole. This side of the body being tired, the other takes up the task, and so on alternately. I watched one for a long time, till half its body was buried; I then walked up and pulled it by the tail; at this it was greatly astonished, and soon shuffled up to see what was the matter; and then stared me in the face, as much as to say, "What made you pull my tail?"

They feed by day, and do not wander far from their burrows; and if frightened they rush to them with a most awkward gait. Except when running down hill, they cannot move very fast; which appears chiefly owing to the lateral position of their legs.

They are not at all timorous: when attentively watching any one, they curl their tails, and raising themselves on their front legs, nod their heads vertically, with a quick movement, and try to look very fierce: but in reality they are not at all so; if one just stamps the ground, down go their tails, and off they shuffle as quickly as they can. I have frequently observed small muscivorous lizards, when watching any thing, nod their heads in precisely the same manner; but I do not at all know for what purpose. If this Amblyrhyncus is held, and plagued with a stick, it will bite it very severely; but I caught many by the tail, and they never tried to bite me. If two are placed on the ground and held together, they will fight and bite each other till blood is drawn.

The individuals (and they are the greater number) which inhabit the lower country, can scarcely taste a drop of water throughout the year; but they consume much of the succulent cactus, the branches of which are occasionally broken off by the wind. I have sometimes thrown a piece to two or three when together; and it was amusing enough to see each trying to seize and carry it away in its mouth, like so many hungry dogs with a bone. They eat very deliberately, but do not chew their food. The little birds are aware how harmless these creatures are: I have seen one of the thick-billed finches picking at one end of a piece of cactus (which is in request among all the animals of the lower region), whilst a lizard was eating at the other; and afterwards the little bird with the utmost indifference hopped on the back of the reptile.

I opened the stomachs of several, and found them full of vegetable fibres, and leaves of different trees, especially of a species of acacia. In the upper region they live chiefly on the acid and astringent berries of the guayavita, under which trees I have seen these lizards and the huge tortoises feeding together. To obtain the acacia-leaves, they crawl up the low stunted trees; and it is not uncommon to see one or a pair quietly browsing, whilst seated on a branch several feet above the ground.

The meat of these animals when cooked is white, and by those whose stomachs rise above all prejudices, it is relished as very good food. Humboldt has remarked that in intertropical South America, all lizards which inhabit *dry* regions are esteemed delicacies for the table. The inhabitants say, that those inhabiting the damp region drink water, but that the others do not travel up for it from the sterile country like the tortoises. At the time of our visit, the females had within their bodies numerous large elongated eggs. These they lay in their burrows, and the inhabitants seek them for food.

These two species of Amblyrhyncus agree, as I have already stated, in general structure, and in many of their habits. Neither have that rapid movement, so characteristic of true Lacerta and Iguana. They are both herbivorous, although the kind of vegetation consumed in each case is so very different. Mr. Bell has given the name to the genus from the shortness of the snout: indeed, the form of the mouth may almost be compared to that of the tortoise. One is tempted to suppose this is an adaptation to their herbivorous appetites. It is very interesting thus to find a well-characterized genus, having its aquatic and terrestrial species, belonging to so confined a portion of the world. The former species is by far the most remarkable, because it is the only existing Saurian, which can properly be said to be a maritime animal. I should perhaps have mentioned earlier, that in the whole archipelago, there is only one rill of fresh water that reaches the coast; yet these reptiles frequent the sea-beaches, and no other parts in all the islands. Moreover, there is no existing lizard, as far as I am aware, excepting this Amblyrhyncus, that feeds exclusively on aquatic productions. If, however, we refer to epochs long past, we shall find such habits common to several gigantic animals of the Saurian race.

To conclude with the order of reptiles. Of snakes there are several species, but all harmless. Of toads and frogs there are none. I was surprised at this, considering how well the temperate and damp woods in the elevated parts appeared adapted for their habits. It recalled to my mind the singular statement made by Bory St. Vincent,[6] namely, that none of this family are to be found on the volcanic islands in the great oceans. There certainly appears to be some foundation for this observation; which is the more remarkable, when compared with the case of lizards, which are generally among the earliest colonists of the smallest islet. It may be asked, whether this is not owing to the different facilities of transport through salt-water, of the eggs of the latter protected by a calcareous coat, and of the slimy spawn of the former?

As I at first observed, these islands are not so remarkable for the number of species of reptiles, as for that of individuals; when we remember the well-beaten paths made by the many hundred great tortoises—the warrens of the terrestrial Amblyrhyncus—and the groups of the aquatic species basking on the coast-rocks—we must admit that there is no other quarter of the world, where this order replaces the herbivorous mammalia in so extraordinary a manner. It is worthy of observation by the geologist (who will probably refer back in his mind to the secondary periods, when the Saurians were developed with dimensions, which at the present day can be compared only to the cetaceous mammalia), that this archipelago, instead of possessing a humid climate and rank vegetation, cannot be considered otherwise than extremely arid, and for an equatorial region, remarkably temperate.

To finish with the zoology: I took great pains in collecting the insects, but I was surprised to find, even in the high and damp region, how exceedingly few they were in number. The forests of Tierra del Fuego are certainly much more barren; but with that exception I never collected in so poor a country. In the lower and sterile land I took seven species of Heteromera, and a few other insects; but in the fine thriving woods towards the centre of the islands, although I perseveringly swept under the bushes during all kinds of weather, I obtained only a few minute Diptera and Hymenoptera. Owing to this scarcity of insects, nearly all the birds live in the lower country; and the part which any one would have thought much the most favourable for them, is frequented only by a few of the small tyrant-flycatchers. I do not believe a single bird, excepting the water-rail, is confined to the damp region. Mr. Waterhouse informs me that nearly all the insects belong to European forms, and that they do not by any means possess an equatorial character. I did not take a single one of large size, or of bright colours. This last observation applies equally to the birds and flowers. It is worthy of remark, that the only land-bird with bright colours, is that species of tyrant-flycatcher, which seems to be a wan-

6. Voyage aux quatre Iles d'Afrique.

derer from the continent. Of shells, there are a considerable number of land kinds, all of which, I believe are confined to this archipelago. Even of marine species, a large proportion were not known, before the collection made by Mr. Cuming on these islands was brought to England.

I will not here attempt to come to any definite conclusions, as the species have not been accurately examined; but we may infer, that, with the exception of a few wanderers, the organic beings found on this archipelago are peculiar to it; and yet that their general form strongly partakes of an American character. It would be impossible for any one accustomed to the birds of Chile and La Plata to be placed on these islands, and not to feel convinced that he was, as far as the organic world was concerned, on American ground. This similarity in type, between distant islands and continents, while the species are distinct, has scarcely been sufficiently noticed. The circumstance would be explained, according to the views of some authors, by saying that the creative power had acted according to the same law over a wide area.

It has been mentioned, that the inhabitants can distinguish the tortoises, according to the islands whence they are brought. I was also informed that many of the islands possess trees and plants which do not occur on the others. For instance the berry-bearing tree, called Guyavita, which is common on James Island, certainly is not found on Charles Island, though appearing equally well fitted for it. Unfortunately, I was not aware of these facts till my collection was nearly completed: it never occurred to me, that the productions of islands only a few miles apart, and placed under the same physical conditions, would be dissimilar. I therefore did not attempt to make a series of specimens from the separate islands. It is the fate of every voyager, when he has just discovered what object in any place is more particularly worthy of his attention, to be hurried from it. In the case of the mocking-bird, I ascertained (and have brought home the specimens) that one species (*Orpheus trifasciatus*, Gould) is exclusively found in Charles Island; a second (*O. parvulus*) on Albemarle Island; and a third (*O. melanotus*) common to James and Chatham Islands. The two last species are closely allied, but the first would be considered by every naturalist as quite distinct. I examined many specimens in the different islands, and in each the respective kind was *alone* present. These birds agree in general plumage, structure, and habits; so that the different species replace each other in the economy of the different islands. These species are not characterized by the markings on the plumage alone, but likewise by the size and form of the bill, and other differences. I have stated, that in the thirteen species of ground-finches, a nearly perfect gradation may be traced, from a beak extraordinarily thick, to one so fine, that it may be compared to that of a warbler. I very much suspect, that certain members of the series are confined to different islands; therefore, if the collection had been made on any *one* island, it would not have presented so perfect a gradation. It is clear, that if several islands have each their peculiar species of the same genera, when these are placed together, they will have a wide range of character. But there is not space in this work, to enter on this curious subject.

Before concluding my account of the zoology of these islands, I must describe more in detail the tameness of the birds. This disposition is common to all the terrestrial species; namely, to the mocking-birds, the finches, sylvicolae, tyrant-flycatchers, doves, and hawks. There is not one which will not approach sufficiently near to be killed with a switch, and sometimes, as I have myself tried, with a cap or hat. A gun is here almost superfluous; for with the muzzle of one I pushed a hawk off the branch of a tree. One day a mocking-bird alighted on the edge of a pitcher (made of the shell of a tortoise), which I held in my hand whilst lying down. It began very quietly to sip the water, and allowed me to lift it with the vessel from the ground.

I often tried, and very nearly succeeded, in catching these birds by their legs. Formerly the birds appear to have been even tamer than at present. Cowley[7] (in the year 1684) says that the "Turtle-doves were so tame that they would often alight upon our hats and arms, so as that we could take them alive: they not fearing man, until such time as some of our company did fire at them, whereby they were rendered more shy." Dampier[8] (in the same year) also says that a man in a morning's walk might kill six or seven dozen of these birds. At present, although certainly very tame, they do not alight on people's arms; nor do they suffer themselves to be killed in such numbers. It is surprising that the change has not been greater; for these islands during the last hundred and fifty years, have been frequently visited by bucaniers and whalers; and the sailors, wandering through the woods in search of tortoises, always take delight in knocking down the little birds.

These birds, although much persecuted, do not become wild in a short time: in Charles Island, which had then been colonized about six years, I saw a boy sitting by a well with a switch in his hand, with which he killed the doves and finches as they came to drink. He had already procured a little heap of them for his dinner; and he said he had constantly been in the habit of waiting there for the same purpose. We must conclude that the birds, not having as yet learnt that man is a more dangerous animal than the tortoise, or the amblyrhyncus, disregard us, in the same manner as magpies in England do the cows and horses grazing in the fields.

The Falkland Islands offer a second instance of this disposition among its birds. The extraordinary tameness of the dark-coloured Furnarius has been remarked by Pernety, Lesson, and other voyagers. It is not, however, peculiar to that bird: the Caracara, snipe, upland and lowland goose, thrush,

7. Cowley's Voyage, p. 10, in Dampier's Collection of Voyages.
8. Dampier's Voyage, vol. i., p. 103.

Emberiza, and even some true hawks, are all more or less tame. Both hawks and foxes are present; and as the birds are so tame, we may infer that the absence of all rapacious animals at the Galapagos, is not the cause of their tameness there. The geese at the Falklands, by the precaution they take in building on the islets, show that they are aware of their danger from the foxes; but they are not by this rendered wild towards man. This tameness of the birds, especially the waterfowl, is strongly contrasted with the habits of the same species in Tierra del Fuego, where for ages past they have been persecuted by the wild inhabitants. In the Falklands, the sportsman may sometimes kill more of the upland geese in one day, than he is able to carry home; whereas in Tierra del Fuego, it is nearly as difficult to kill one, as it is in England of the common wild species.

In the time of Pernety[9] (1763), all the birds appear to have been much tamer than at present. Pernety states that the Furnarius would almost perch on his finger; and that with a wand he killed ten in half an hour. At that period, the birds must have been about as tame as they now are at the Galapagos. They appear to have learnt caution more quickly at the Falklands than at the latter place, and they have had proportionate means of experience; for besides frequent visits from vessels, the islands have been at intervals colonized during the whole period.

Even formerly, when all the birds were so tame, by Pernety's account it was impossible to kill the black-necked swan. It is rather an interesting fact, that this is a bird of passage, and therefore brings with it the wisdom learnt in foreign countries.

I have not met with any account of the *land* birds being so tame, in any other quarter of the world, as at the Galapagos and Falkland Islands. And it may be observed that of the few archipelagoes of any size, which when discovered were uninhabited by man, these two are among the most important. From the foregoing statements we may, I think, conclude; first, that the wildness of birds with regard to man, is a particular instinct directed against *him*, and not dependent on any general degree of caution arising from other sources of danger; secondly, that it is not acquired by them in a short time, even when much persecuted; but that in the course of successive generations it becomes hereditary. With domesticated animals we are accustomed to see instincts becoming hereditary; but with those in a state of nature, it is more rare to discover instances of such acquired knowledge. In regard to the wildness of birds towards men, there is no other way of accounting for it. Few young birds in England have been injured by man, yet all are afraid of him: many individuals, on the other hand, both at the Galapagos and at the Falklands, have been injured, but yet have not learned that salutary dread. We

9. Pernety, Voyage aux Iles Malouines, vol. ii., p. 20.

may infer from these facts, what havoc the introduction of any new beast of prey must cause in a country, before the instincts of the aborigines become adapted to the stranger's craft or power.

London: Henry Colburn, 1839.

From *The Voyage of the* Beagle (1845)

✳ CHARLES R. DARWIN

* Chapter XVII

GALAPAGOS ARCHIPELAGO

* Pages 377–381

The natural history of these islands is eminently curious, and well deserves attention. Most of the organic productions are aboriginal creations, found nowhere else; there is even a difference between the inhabitants of the different islands; yet all show a marked relationship with those of America, though separated from that continent by an open space of ocean, between 500 and 600 miles in width. The archipelago is a little world within itself, or rather a satellite attached to America, whence it has derived a few stray colonists, and has received the general character of its indigenous productions. Considering the small size of these islands, we feel the more astonished at the number of their aboriginal beings, and at their confined range. Seeing every height crowned with its crater, and the boundaries of most of the lava-streams still distinct, we are led to believe that within a period, geologically recent, the unbroken ocean was here spread out. Hence, both in space and time, we seem to be brought somewhat near to that great fact—that mystery of mysteries—the first appearance of new beings on this earth.

Of terrestrial mammals, there is only one which must be considered as indigenous, namely, a mouse (Mus Galapagoensis), and this is confined, as far as I could ascertain, to Chatham island, the most easterly island of the group. It belongs, as I am informed by Mr. Waterhouse, to a division of the family of mice characteristic of America. At James island, there is a rat sufficiently distinct from the common kind to have been named and described by Mr. Waterhouse; but as it belongs to the old-world division of the family, and as this island has been frequented by ships for the last hundred and fifty years, I can hardly doubt that this rat is merely a variety, produced by the new and peculiar climate, food, and soil, to which it has been subjected. Although no one has a right to speculate without distinct facts, yet even with respect to the Chatham island mouse, it should be borne in mind, that it may possibly be an

American species imported here; for I have seen, in a most unfrequented part of the Pampas, a native mouse living in the roof of a newly-built hovel, and therefore its transportation in a vessel is not improbable: analogous facts have been observed by Dr. Richardson in North America.

Of land-birds I obtained twenty-six kinds, all peculiar to the group and found nowhere else, with the exception of one lark-like finch from North America (Dolichonyx oryzivorus), which ranges on that continent as far north as 54°, and generally frequents marshes. The other twenty-five birds consist, firstly, of a hawk, curiously intermediate in structure between a Buzzard and the American group of carrion-feeding Polybori; and with these latter birds it agrees most closely in every habit and even tone of voice. Secondly, there are two owls, representing the short-eared and white barn-owls of Europe. Thirdly, a wren, three tyrant fly-catchers (two of them species of Pyrocephalus, one or both of which would be ranked by some ornithologists as only varieties), and a dove—all analogous to, but distinct from, American species. Fourthly, a swallow, which though differing from the Progne purpurea of both Americas, only in being rather duller coloured, smaller, and slenderer, is considered by Mr. Gould as specifically distinct. Fifthly, there are three species of mocking-thrush—a form highly characteristic of America. The remaining land-birds form a most singular group of finches, related to each other in the structure of their beaks, short tails, form of body, and plumage: there are thirteen species, which Mr. Gould has divided into four sub-groups. All these species are peculiar to this archipelago; and so is the whole group, with the exception of one species of the subgroup Cactornis, lately brought from Bow island, in the Low Archipelago. Of Cactornis, the two species may be often seen climbing about the flowers of the great cactus-trees; but all the other species of this group of finches, mingled together in flocks, feed on the dry and sterile ground of the lower districts. The males of all, or certainly of the greater number, are jet black; and the females (with perhaps one or two exceptions) are brown. The most curious fact is the perfect gradation in the size of the beaks in the different species of Geospiza, from one as large as that of a hawfinch to that of a chaffinch, and (if Mr. Gould is right in including his sub-group, Certhidea, in the main group), even to that of a warbler. The largest beak in the genus Geospiza is shown in Fig. 1, and the smallest in Fig. 3; but instead of there being only one intermediate species, with a beak of the size shown in Fig. 2, there are no less than six species with insensibly graduated beaks. The beak of the sub-group Certhidea, is shown in Fig. 4. The beak of Cactornis is somewhat like that of a starling; and that of the fourth sub-group, Camarhynchus, is slightly parrot-shaped. Seeing this gradation and diversity of structure in one small, intimately related group of birds, one might really fancy that from an original paucity of birds in this archipelago, one species had been taken and modified for different ends. In

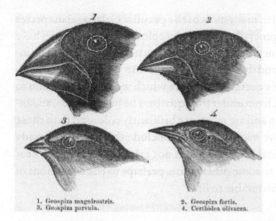

1. Geospiza magnirostris.
3. Geospiza parvula.

2. Geospiza fortis.
4. Certhidea olivacea.

a like manner it might be fancied that a bird originally a buzzard, had been induced here to undertake the office of the carrion-feeding Polybori of the American continent.

Of waders and water-birds I was able to get only eleven kinds, and of these only three (including a rail confined to the damp summits of the islands) are new species. Considering the wandering habits of the gulls, I was surprised to find that the species inhabiting these islands is peculiar, but allied to one from the southern parts of South America. The far greater peculiarity of the land-birds, namely, twenty-five out of twenty-six being new species or at least new races, compared with the waders and web-footed birds, is in accordance with the greater range which these latter orders have in all parts of the world. We shall hereafter see this law of aquatic forms, whether marine or fresh-water, being less peculiar at any given point of the earth's surface than the terrestrial forms of the same classes, strikingly illustrated in the shells, and in a lesser degree in the insects of this archipelago.

Two of the waders are rather smaller than the same species brought from other places: the swallow is also smaller, though it is doubtful whether or not it is distinct from its analogue. The two owls, the two tyrant fly-catchers (Pyrocephalus) and the dove, are also smaller than the analogous but distinct species, to which they are most nearly related; on the other hand, the gull is rather larger. The two owls, the swallow, all three species of mocking-thrush, the dove in its separate colours though not in its whole plumage, the Totanus, and the gull, are likewise duskier coloured than their analogous species; and in the case of the mocking-thrush and Totanus, than any other species of the two genera. With the exception of a wren with a fine yellow breast, and of a tyrant fly-catcher with a scarlet tuft and breast, none of the birds are brilliantly coloured, as might have been expected in an equatorial district. Hence it would appear probable, that the same causes which here make the immigrants of some species smaller, make most of the peculiar Galapageian species also smaller, as well as very generally more dusky coloured. All the plants have a wretched, weedy appearance, and I did not see one beautiful flower. The insects, again, are small sized and dull coloured, and, as Mr. Waterhouse informs me, there is nothing in their general appearance which would have led him to imagine that they had come from under the equator. The birds, plants, and insects have a desert character, and are not more brilliantly coloured than those from southern Patagonia; we may, therefore, conclude that the usual gaudy colouring of the intertropical productions, is not related either to the heat or light of those zones, but to some other cause, perhaps to the conditions of existence being generally favourable to life.

* Pages 384–385

There can be little doubt that this tortoise is an aboriginal inhabitant of the Galapagos; for it is found on all, or nearly all, the islands, even on some of the smaller ones where there is no water; had it been an imported species, this would hardly have been the case in a group which has been so little frequented. Moreover, the old Bucaniers found this tortoise in greater numbers even than at present: Wood and Rogers also, in 1708, say that it is the opinion of the Spaniards, that it is found nowhere else in this quarter of the world. It is now widely distributed; but it may be questioned whether it is in any other place an aboriginal. The bones of a tortoise at Mauritius, associated with those of the extinct Dodo, have generally been considered as belonging to this tortoise: if this had been so, undoubtedly it must have been there indigenous; but M. Bibron informs me that he believes that it was distinct, as the species now living there certainly is.

* Pages 392–398

The botany of this group is fully as interesting as the zoology. Dr. J. Hooker will soon publish in the "Linnean Transactions" a full account of the Flora, and I am much indebted to him for the following details. Of flowering plants there are, as far as at present is known, 185 species, and 40 cryptogamic species, making together 225; of this number I was fortunate enough to bring home 193. Of the flowering plants, 100 are new species, and are probably confined to this archipelago. Dr. Hooker conceives that, of the plants not so confined, at least 10 species found near the cultivated ground at Charles Island, have been imported. It is, I think, surprising that more American species have not been introduced naturally, considering that the distance is only between 500 and 600 miles from the continent; and that (according to Collnett, p. 58) drift-wood, bamboos, canes, and the nuts of a palm, are often washed on the south-eastern shores. The proportion of 100 flowering plants out of 185 (or 175 excluding the imported weeds) being new, is sufficient, I conceive, to make the Galapagos Archipelago a distinct botanical province; but this Flora is not nearly so peculiar as that of St. Helena, nor, as I am informed by Dr. Hooker, of Juan Fernandez. The peculiarity of the Galapageian Flora is best shown in certain families;—thus there are 21 species of Compositae, of which 20 are peculiar to this archipelago; these belong to twelve genera, and of these genera no less than ten are confined to the archipelago! Dr. Hooker informs me that the Flora has an undoubted Western American character; nor can he detect in it any affinity with that of the Pacific. If, therefore, we except the eighteen marine, the one fresh-water, and one land-shell, which have apparently come here as colonists from the central islands of the Pacific, and likewise the one distinct Pacific species of the Galapageian group of finches, we see that this archipelago, though standing in the Pacific Ocean, is zoologically part of America.

If this character were owing merely to immigrants from America, there would be little remarkable in it; but we see that a vast majority of all the land animals, and that more than half of the flowering plants, are aboriginal productions. It was most striking to be surrounded by new birds, new reptiles, new shells, new insects, new plants, and yet by innumerable trifling details of structure, and even by the tones of voice and plumage of the birds, to have the temperate plains of Patagonia, or the hot dry deserts of Northern Chile, vividly brought before my eyes. Why, on these small points of land, which within a late geological period must have been covered by the ocean, which are formed of basaltic lava, and therefore differ in geological character from the American continent, and which are placed under a peculiar climate,—why were their aboriginal inhabitants, associated, I may add, in different proportions both in kind and number from those on the continent, and therefore acting on each other in a different manner—why were they created on American types of organization? It is probable that the islands of the Cape de Verd group resemble, in all their physical conditions, far more closely the Galapagos Islands than these latter physically resemble the coast of America; yet the aboriginal inhabitants of the two groups are totally unlike; those of the Cape de Verd Islands bearing the impress of Africa, as the inhabitants of the Galapagos Archipelago are stamped with that of America.

I have not as yet noticed by far the most remarkable feature in the natural history of this archipelago; it is, that the different islands to a considerable extent are inhabited by a different set of beings. My attention was first called to this fact by the Vice-Governor, Mr. Lawson, declaring that the tortoises differed from the different islands, and that he could with certainty tell from which island any one was brought. I did not for some time pay sufficient attention to this statement, and I had already partially mingled together the collections from two of the islands. I never dreamed that islands, about fifty or sixty miles apart, and most of them in sight of each other, formed of precisely the same rocks, placed under a quite similar climate, rising to a nearly equal height, would have been differently tenanted; but we shall soon see that this is the case. It is the fate of most voyagers, no sooner to discover what is most interesting in any locality, than they are hurried from it; but I ought, perhaps, to be thankful that I obtained sufficient materials to establish this most remarkable fact in the distribution of organic beings.

The inhabitants, as I have said, state that they can distinguish the tortoises from the different islands; and that they differ not only in size, but in other characters. Captain Porter has described[1] those from Charles and from the nearest island to it, namely, Hood Island, as having their shells in front thick and turned up like a Spanish saddle, whilst the tortoises from James Island are rounder, blacker, and have a better taste when cooked. M. Bibron, moreover, informs me that he has seen what he considers two distinct species of tortoise from the Galapagos, but he does not know from which islands. The specimens that I brought from three islands were young ones; and probably owing to this cause, neither Mr. Gray nor myself could find in them any specific differences. I have remarked that the marine Amblyrhynchus was larger at Albemarle Island than elsewhere; and M. Bibron informs me that he has seen two distinct aquatic species of this genus; so that the different islands probably have their representative species or races of the Amblyrhynchus, as well as of the tortoise. My attention was first thoroughly aroused, by comparing together the numerous specimens, shot by myself and several other parties on board, of the mocking-thrushes, when, to my astonishment, I discovered that all those from Charles Island belonged to one species (Mimus trifasciatus); all from Albemarle Island to M. parvulus; and all from James and Chatham Islands (between which two other islands are situated, as connecting links) belonged to M. melanotis. These two latter species are closely allied, and would by some ornithologists be considered as only well-marked races or varieties; but the Mimus trifasciatus is very distinct. Unfortunately most of the specimens of the finch tribe were mingled together; but I have strong reasons to suspect that some of the species of the sub-group Geospiza are confined to separate islands. If the different islands have their representatives of Geospiza, it may help to explain the singularly large number of the species of this sub-group in this one small archipelago, and as a probable consequence of their numbers, the perfectly graduated series in the size of their beaks. Two species of the sub-group Cactornis, and two of Camarhynchus, were procured in the archipelago; and of the numerous specimens of these two sub-groups shot by four collectors at James Island, all were found to belong to one species of each; whereas the numerous specimens shot either on Chatham

1. Voyage in the U. S. ship Essex, vol. i. p. 215.

or Charles Island (for the two sets were mingled together) all belonged to the two other species: hence we may feel almost sure that these islands possess their representative species of these two sub-groups. In land-shells this law of distribution does not appear to hold good. In my very small collection of insects, Mr. Waterhouse remarks, that of those which were ticketed with their locality, not one was common to any two of the islands.

If we now turn to the Flora, we shall find the aboriginal plants of the different islands wonderfully different. I give all the following results on the high authority of my friend Dr. J. Hooker. I may premise that I indiscriminately collected everything in flower on the different islands, and fortunately kept my collections separate. Too much confidence, however, must not be placed in the proportional results, as the small collections brought home by some other naturalists, though in some respects confirming the results, plainly show that much remains to be done in the botany of this group: the Leguminosae, moreover, have as yet been only approximately worked out:

Name of Island	Total No. of Species	No. of Species found in other parts of the world	No. of Species confined to the Galapagos Archipelago	No. confined to the one Island	No. of Species confined to the Galapagos Archipelago, but found on more than the one Island
James Island.	71	33	38	30	8
Albemarle Island.	46	18	26	22	4
Chatham Island.	32	16	16	12	4
Charles Island.	68	39 (or 29, if the probably imported plants be subtracted)	29	21	8

Hence we have the truly wonderful fact, that in James Island, of the thirty-eight Galapageian plants, or those found in no other part of the world, thirty are exclusively confined to this one island; and in Albemarle Island, of the twenty-six aboriginal Galapageian plants, twenty-two are confined to this one island, that is, only four are at present known to grow in the other islands of the archipelago; and so on, as shown in the above table, with the plants from Chatham and Charles Islands. This fact will, perhaps, be rendered even more striking, by giving a few illustrations:—thus, Scalesia, a remarkable arborescent genus of the Compositae, is confined to the archipelago: it has six species; one from Chatham, one from Albemarle, one from Charles Island, two from James Island, and the sixth from one of the three latter islands, but it is not known from which: not one of these six species grows on any two islands. Again, Euphorbia, a mundane or widely distributed genus, has here eight species, of which seven are confined to the archipelago, and not one found on any two islands: Acalypha and Borreria, both mundane genera, have respectively six and seven species, none of which have the same species on two islands, with the exception of one Borreria, which does occur on two islands. The species of the Compositae are particularly local; and Dr. Hooker has furnished me with several other most striking illustrations of the difference of the species on the different islands. He remarks that this law of distribution holds good both with those genera confined to the archipelago, and those distributed in other quarters of the world: in like manner we have seen that the different islands have their proper species of the mundane genus of tortoise, and of the widely distributed American genus of the mocking-thrush, as well as of two of the Galapageian sub-groups of finches, and almost certainly of the Galapageian genus Amblyrhynchus.

The distribution of the tenants of this archipelago would not be nearly so wonderful, if, for instance, one island had a mocking-thrush, and a second island some other quite distinct genus;—if one island had its genus of lizard, and a second island another distinct genus, or none whatever;—or if the different islands were inhabited, not by representative species of the same genera of plants, but by totally different genera, as does to a certain extent hold good; for, to give one instance, a large berry-bearing tree at James Island has no representative species in Charles Island. But it is the circumstance, that several of the islands possess their own species of the tortoise, mocking-thrush, finches, and numerous plants, these species having the same general habits, occupying analogous situations, and obviously filling the same place in the natural economy of this archipelago, that strikes me with wonder. It may be suspected that some of these representative species, at least in the case of the tortoise and of some of the birds, may hereafter prove to be only well-marked races; but this would be of equally great interest to the philosophical naturalist. I have said that most of the islands are in sight of each other: I may specify that Charles Island is fifty miles from the nearest part of Chatham Island, and thirty-three miles from the nearest part of Albemarle Island. Chatham Island is sixty miles from the nearest part of James Island, but there are two intermediate islands between them which were not visited by me. James Island is only ten miles from the nearest part of Albemarle Island, but the two points where the collections were made are thirty-two miles apart. I must repeat, that neither the nature of the soil, nor height of the land, nor the climate, nor the general character of the associated beings, and therefore their action one on another, can differ much in the different islands. If there be any sensible difference in their climates, it must be between the windward group (namely Charles and Chatham Islands), and that to leeward; but there seems to be no corresponding difference in the productions of these two halves of the archipelago.

The only light which I can throw on this remarkable difference in the inhabitants of the different islands, is, that very strong currents of the sea running in a westerly and W.N.W. direction must separate, as far as transportal by the sea is concerned, the southern islands from the northern ones; and between these northern islands a strong N.W. current was observed, which must effectually separate James and Albemarle Islands. As the archipelago is free to a most remarkable degree from gales of wind, neither the birds, insects, nor lighter seeds, would be blown from island to island. And lastly, the profound depth of the ocean between the islands, and their apparently recent (in a geological sense) volcanic origin, render it highly unlikely that they were ever united; and this, probably, is a far more important consideration than any other, with respect to the geographical distribution of their inhabitants. Reviewing the facts here given, one is astonished at the amount of creative force, if such an expression may be used, displayed on these small, barren, and rocky islands; and still more so, at its diverse yet analogous action on points so near each other. I have said that the Galapagos Archipelago might be called a satellite attached to America, but it should rather be called a group of satellites, physically similar, organically distinct, yet intimately related to each other, and all related in a marked, though much lesser degree, to the great American continent.

I will conclude my description of the natural history of these islands, by giving an account of the extreme tameness of the birds.

Second edition. New York: Modern Library, 1845.

Letter to Ernst P. A. Haeckel (1864)

❈ CHARLES R. DARWIN

Dear Sir

I thank you sincerely for your letter & the confidence you repose in me. I have been deeply interested in what you say about your poor wife. Her expression in the photograph is charming. I can to a certain extent understand what your feelings are, for I am fortunate enough to know what a treasure a wife can be & no one thought is so painful to me as the possibility of surviving her.

As you seem interested about the origin of the "Origin" & I believe do not say so out of mere compliment, I will mention a few points. When I joined the "Beagle" as Naturalist I knew extremely little about Natural History, but I worked hard. In South America three classes of facts were brought strongly before my mind: 1stly the manner in which closely allied species replace species in going Southward.

2ndly the close affinity of the species inhabiting the Islands near to S. America to those proper to the Continent. This struck me profoundly, especially the difference of the species in the adjoining islets in the Galapagos Archipelago. 3rdly the relation of the living Edentata & Rodentia to the extinct species. I shall never forget my astonishment when I dug out a gigantic piece of armour like that of the living Armadillo.

Reflecting on these facts & collecting analogous ones, it seemed to me probable that allied species were descended from a common parent. But for some years I could not conceive how each form became so excellently adapted to its habits of life. I then began systematically to study domestic productions, & after a time saw clearly that man's selective power was the most important agent. I was prepared from having studied the habits of animals to appreciate the struggle for existence, & my work in Geology gave me some idea of the lapse of past time. Therefore when I happened to read "Malthus on population" the idea of Natural selection flashed on me. Of all the minor points, the last which I appreciated was the importance & cause of the principle of Divergence. I hope I have not wearied you with this little history of the "Origin."

I quite agree with what you say about Kölliker; there is a capital review of him by Huxley in the Number just published of the "Natural History Review." This letter was begun several weeks ago, but I have delayed finishing it from having little strength & other things to do. Will you have the kindness to tell this to Prof. Gegenbaur as an apology for not having thanked him for the honour he has done me in sending me his work. By a strange chance I dissected several months ago the hind foot of a toad & was particularly curious to understand what the additional bones were, & this point I see will now be explained to me. As I know from one of the papers which you have sent me that you have attended to Entomostraca it has occurred to me that you might like to have a copy of my Vol. on the Balanidae, of which I have a spare copy & would with pleasure send it if you wish for it, & will tell me how to forward it.

With sincere respect
Believe me my dear Sir
yours very faithfully
Charles Darwin

Letter 4631, [after 10 August–8 October 1864]. From Charles R. Darwin, *The Correspondence of Charles Darwin*, volume 2, *1864*, edited by Frederick Burkhardt, Duncan M. Porter, and Sheila Ann Dean (Cambridge: Cambridge University Press, 2001). Reprinted with the permission of Cambridge University Press.

Letter to Osbert Salvin (1875)

❉ CHARLES R. DARWIN

* Down
* Beckenham, Kent.
* Railway Station
* Orpington. S.E.R.
* Aug. 22nd

My dear Mr. Salvin

I am very much obliged for your memoir which I have read with very great interest. I am surprised that the birds from the different islands prove so similar. How much has been done since my days. It would, indeed, as you say be a fine work for anyone to investigate the perplexing forms on the spot.

Yours very sincerely,
Ch. Darwin

P.S You will see in my Journal, that several specimens ought, if possible, to be collected, of each of the very commonest species from each island, & their habits, nests, eggs, young &c, compared.—What a flood of light would be thus thrown on Variation!

[22 August 1875]. For this letter, grateful acknowledgment is made to the editors of the Darwin Correspondence Project for access to unpublished material. This letter is at a prepublication stage, and the Project cannot be held responsible for any errors of transcription remaining.

Darwin and His Finches: The Evolution of a Legend (1982)

❉ FRANK J. SULLOWAY

Editor's note: Some figures have been omitted, as they are reproduced elsewhere in this volume in the original sources. Some photographs have been omitted for copyright reasons.

First collected by Charles Darwin in the Galapagos Archipelago, the Geospizinae, or "Darwin's finches," have rightly been celebrated as a classic instance of the workings of evolution through natural selection. Among birds, Darwin's finches are rivaled only by the Hawaiian honeycreepers (Drepanididae) as a microcosmic exemplification of the principle of adaptive evolutionary radiation. Although the Drepanididae have undergone more evolution and adaptive radiation than the Geospizinae, the latter are in some ways more valuable to ornithologists. "Their special interest today," writes David Lack, "is in providing the best example, in birds, of an adaptive radiation into different ecological niches that is sufficiently recent, geologically speaking, for intermediate and transitional forms to have survived."

The Galapagos Archipelago, where Darwin spent five weeks collecting these finches during the voyage of the H.M.S. *Beagle* (1831–1836), comprises sixteen principal islands located on the equator some six hundred miles west of Ecuador (Fig. 1 [see figure 1 in the Lack selection in chapter 5, this volume]). The islands, most of which are several million years old, are wholly volcanic in origin and have never been connected to the mainland. Darwin's finches were evidently one of the earliest colonists of the Galapagos, since their degree of evolutionary complexity—thirteen species distributed among four genera—is unmatched by any other avian group in this archipelago. A fourteenth species, belonging to yet another genus, inhabits Cocos Island, four hundred miles to the northeast. Unlike other endemic species of Galapagos birds, the Geospizinae no longer have any close relatives on the American mainland. They are therefore classed in their own separate tribe or subfamily, which is placed with the Emberizidae.[1]

Being one of the earliest colonists of the Galapagos Islands, the ancestral form of Darwin's finches found an environment in which the types of niches occupied by other, diverse birds on the continent were largely vacant. After becoming isolated from one another on the different islands of the archipelago, various finch populations gradually evolved reproductive isolation and hence status as separate species. Certain of these species then successfully recolonized neighboring islands, and the ensuing competition between closely related forms encouraged divergence and increasing specialization in the

---

1. Darwin's finches have been the subject of numerous systematic treatments, of which the most important are by Gould (1837a, 1841, 1843), Salvin (1876), Ridgway (1890, 1897), Rothschild and Hartert (1899,1902), Snodgrass and Heller (1904), Swarth (1931), Hellmayr (1938:130–146), Lack (1945, 1947, 1969), Bowman (1961, 1963), Harris (1974), and Steadman (in press). Monographic works, such as those by Swarth (1931) and Lack (1945, 1947), have usually given Darwin's finches family or subfamily status—the latter being the general consensus. Nevertheless, some authors have recommended that they be accorded only tribal status within the Emberizinae subfamily (Paynter and Storer 1970:160–168). Differentiation between subfamilies and tribes is a subjective matter, and I have preferred to follow the monographic tradition on this point. Species and genus names of certain forms of the Geospizinae have changed over the years, making for some minor inconsistencies in terminology in discussions of the literature. For example, *Cactornis scandens* (Gould 1837a) is no longer given separate generic status, but is classified instead with the other species of *Geospiza*. I have followed the policy of using the original names proposed by Gould (1837a, 1841) when discussing individual *Beagle* specimens or Darwin's views about them. Otherwise, the current nomenclature has been followed, with the exception that I recognize *Geospiza magnirostris magnirostris* and *G. magnirostris strenua* as valid trinomials and also recognize the name *G. nebulosa* as having priority over *G. difficilis*. See note 34 and Sulloway (1982b).

many unoccupied niches presented by the Galapagos environment. Through this four-part process of geographic isolation, speciation, recolonization, and ensuing adaptive radiation, the Geospizinae have evolved a remarkable disparity in the form of their beaks, from one as massive as that of a grosbeak to one as small as that of a warbler. There are three species of seed-eating ground finches with large, medium, and small beaks; another ground finch with a sharp, pointed beak; two species of ground finches that feed on cactus; a vegetarian tree finch; three species of insectivorous tree finches; a mangrove finch; a finch that closely resembles a warbler in both habits and morphology; and finally a "tool-using" "woodpecker" finch, which employs twigs and cactus spines to extract its prey from crevices in tree trunks (Fig. 2 [see figure 3 in the Lack selection in chapter 4, this volume). As Darwin remarked in the second edition of his *Journal of Researches*, "Seeing this gradation and diversity of structure in one small, intimately related group of birds, one might really fancy that from an original paucity of birds in this archipelago, one species had been taken and modified for different ends" (1845:380).

Given the remarkable nature of these birds, it is of considerable historical interest to reconstruct the role they played in Darwin's intellectual development. This problem really involves three separate questions. First, how did Darwin initially interpret the morphology and behavior of the various species of this unusual avian group while he was in the Galapagos Archipelago? Second, to what extent did he appreciate the striking correlation between geographic isolation and the diversity of endemic finch forms and thus take steps to separate his collections according to the different islands he visited? Third, what aspects of Darwin's understanding of this avian group were retrospective, that is, developed after he had left the Galapagos and had returned to England? Given the fame of this episode in Darwin's life, there has been a surprising degree of misunderstanding and misinformation regarding these three questions. In fact, over the years Darwin's finches have become the focus for a considerable legend in the history of science, one that ranks alongside other famous stories that celebrate the great triumphs of modern science.

DARWIN IN THE GALAPAGOS

Morphology and Classification

It has frequently been asserted that Darwin's finches, along with certain other organisms from the Galapagos Archipelago, were what first alerted Darwin to the possibility that species might be mutable.[2] But as David Lack (1947:9) has

2. See notes 33, 54, and 64.

pointed out, Darwin did not even discuss the finches in the diary of his voyage on the *Beagle*, except for a single reference in passing; and his treatment of them in the first edition of his *Journal of Researches* (1839:461–462) was brief and matter of fact compared with the famous statement about them that he added to the 1845 edition. Given these facts, Lack concluded that Darwin's evolutionary understanding of the finches was largely retrospective. This interpretation is essentially correct, although Lack, who did not examine Darwin's unpublished scientific notes from the *Beagle* voyage, failed to appreciate the reasons for Darwin's gradual insight. There are two different versions of Darwin's voyage scientific notes that discuss his Galapagos finches. The first version, which forms part of his manuscript notes on zoology, was drafted in late October 1835, shortly after he left the Galapagos Archipelago on his way to Tahiti. The second and somewhat expanded version, which nevertheless follows the first in its general contents, was copied into a separate notebook for ornithological observations some nine months later.[3] Both accounts describe the tendency for the various finch species to feed together on the ground in the arid and sparsely vegetated lowlands of the islands. Darwin emphasized that insects were surprisingly scarce in the Galapagos and remarked that seeds, laid down in the loose volcanic soil after the annual rainy season, supplied the principal source of food for the birds. "Hence these Finches," he commented in his *Ornithological Notes*, "are in number of species & individuals far preponderant over any other family of birds" (1963[1836]:261). Darwin also noted the impossibility of distinguishing species on behavioral grounds, given the similar feeding habits of most of the birds. "There appears to be much difficulty," he acknowledged, "in ascertaining the species."[4]

Oddly, Darwin seems to have been more preoccupied with the unusual coloration of the finches than with the extreme variation in their beaks. "Amongst the species of this family," he wrote in this connection, "there reigns to me an inexplicable confusion" (1963[1836]:261). Some of his specimens were jet black, and others, by intermediate shades, passed into a brown or olive plumage. His collections, he believed, tended to show that the jet black coloration was peculiar to the old cock birds alone, but exceptions to this rule also seemed to exist. Finally, he noted almost as an afterthought that "a gradation in the form of the bill, appears to me to exist"

3. Toward the end of the voyage Darwin prepared a series of separate specimen catalogues for the use of the specialists who later took charge of his collections after the *Beagle's* return to England. Darwin's *Ornithological Notes* (1963[1836]) constitutes one of twelve such catalogues. On the dating of these catalogues, see Sulloway (1982a).

4. DAR 31.2: MS p. 340. All DAR numbers refer to the Darwin MSS, Cambridge University Library.

(1963[1836]:261).[5] In these voyage notes Darwin did not comment further on this problem of the bills, and he also offered no explanation as to why so many similar species of finches were to be found in the Galapagos.

The unusual nature of the bills in Darwin's finches has given rise to one of the more dramatic aspects of the legend that surrounds these birds. According to several commentators, the finches prompted Darwin and Captain Robert FitzRoy, while they were visiting the Galapagos, to fall into "one of their numerous disputes" (Grinnell 1974:260–261). FitzRoy, noting that the shape of the beaks varied slightly by island, supposedly concluded that each form was a separate species created for the particular island on which it was found. Darwin, on the other hand, is said to have thought the finches were derived from a mainland species and had become modified by their new surroundings. They were therefore, Darwin concluded, only varieties of a single species, a conclusion that FitzRoy considered "blasphemous rubbish."[6] Thus what most impressed Darwin, according to Grinnell (1974) and others, was the remarkable similarity and gradation in the characters of the various Geospizinae.[7]

The claim that Darwin and FitzRoy must have argued over the evolutionary implications of the Galapagos finches—although frequently presented with considerable conviction—is supported by little real evidence. There is, moreover, considerable evidence to the contrary. The supposed basis for this argument is a brief comment about the finches that FitzRoy made in his subsequently published *Narrative*. There he remarked: "All the small birds that live on these lava-covered islands have short beaks, very thick at the base, like that of a bull-finch. This appears to be one of those admirable provisions of Infinite Wisdom by which each created thing is adapted to the place for which it was intended" (1839:503). FitzRoy went on to say that such thick beaks were ideally suited for picking up insects or seeds from the hard lava and were also useful in crushing berries to obtain the moisture contained in them. He did not, however, comment about any variation in the beaks or about the geographical distribution of these species, and his account seems to apply exclusively to the large-billed forms of *Geospiza*.[8] In any event, FitzRoy was only saying what Darwin himself had noticed, namely, that "the greater number of birds haunt, and are adapted for, the dry & wretched looking thickets of the coast land" (1963[1836]:261).

5. Darwin did not mention the apparent gradation in the bills in his earlier account, written in October 1835.
6. See Moorehead (1969:205–206), whose words are quoted here; Barlow (1963:261n1); and Railing (1978).
7. See, for example, Bowman (1963:107), Ruse (1979:116), and Ospovat (1981:91).
8. Kottler (1978:283) has likewise made this point in criticizing Grinnell's (1974) account.

Had FitzRoy considered the issue of geographic variation among Darwin's finches, it is likely that he would have taken a very different stand from the one that legend has assigned him. From his *Narrative* it is clear he believed that "every animal varies more or less in outward form and appearance" owing to local differences in climate and geography (1839:253). Thus he insisted that such island variants as the differing foxes on East and West Falkland islands were only varieties, not separate species, caused by slightly different environments. So plastic did FitzRoy consider species in nature, and so critical was he of naturalists who repeatedly made local geographic races into separate forms, that he actually regarded the two Falkland foxes and the various Galapagos tortoises as mere varieties of species found elsewhere in the world (1839:250–254, 505). As for FitzRoy's religious fanaticism, commonly thought to have motivated his scientific debates with Darwin, FitzRoy himself later made it very clear in his *Narrative* that he did not undergo the religious conversion reflected in certain aspects of that work until after the *Beagle* had arrived back in England.[9]

To return to the related claim that Darwin considered the various finches to be mere varieties modified by their new environment, this assertion is contradicted by perhaps the most curious aspect of Darwin's voyage thoughts about these birds. I am referring to the individual entries by which Darwin recorded each numbered specimen in his voyage notes. From these entries it is clear that what initially impeded his understanding of the finches was not their extreme similarity but rather their apparent differences. In fact, Darwin evidently thought he was dealing with a highly diverse family of birds having at least three and perhaps four different subfamilies. He referred, for example, to the large-beaked birds as "Gross-beaks," to the smaller-beaked birds as "Fringilla," or true finches, and to the cactus finch as "Icterus" (a separate family of birds that now includes the orioles, meadowlarks, and blackbirds). Just how greatly Darwin was misled by certain of the Galapagos finches is poignantly illustrated by his misclassification of the warbler finch as a "wren," or warbler.[10] As for the remarkable woodpecker finch,

9. In the *Narrative* FitzRoy referred to "men who, like myself, formerly are willingly ignorant of the Bible" and admitted that he had previously known "so little of that [biblical] record... that I fancied some events there related might be mythological or fabulous. . . ." To these remarks he added: "While led away by sceptical ideas, and knowing extremely little of the Bible, one of my remarks to a friend [surely Darwin], on crossing vast plains composed of rolled stones bedded in diluvial detritus some hundred feet in depth, was 'this could never have been effected by a forty days' flood.' . . . I was quite willing to disbelieve what I thought to be the Mosaic account, upon the evidence of a hasty glance, though knowing next to nothing of the record I doubted..." (1839:657–659). See also Keynes (1979:6) on this point.
10. See Darwin's *Ornithological Notes*, where twenty-one specimens are classified "Fringilla," four "Icterus," four "Fringilla/Gross-beaks," one "Wren," and one without a name

thought by many to have stimulated Darwin's greatest evolutionary curiosity, this species was not even collected by Darwin; and its unusual tool-using behavior was not reported until 1919.[11] Darwin collected, in fact, only nine of the present thirteen species of "Darwin's finches." Of these, he properly identified as finches only six species—less than half the present total—placing them in two separate groups, large- and small-beaked Fringillidae.[12]

Darwin's difficulties in properly classifying his Galapagos finches during the *Beagle* voyage should by no means be taken as a sign that he was ornithologically inexperienced or inadept. Darwin was quite knowledgeable as a taxonomist, and he generally managed to classify his *Beagle* specimens under the appropriate family, genus, and sometimes even species using the published guides available to him on the voyage.[13] With the current triumph of Darwin's evolutionary views, however, it has become difficult for us to appreciate the confusion and puzzlement that such an unusual avian

(1963[1836]: 262–264). Similarly, in his voyage zoology notes he wrote that "far the preponderant number of individuals belongs to the Finches & the Gross-beaks" (DAR 31.2: MS pp. 340–341). These four separate designations are confirmed by Darwin's master catalogue of specimens, now at Down House. For the cactus finch, however, he wrote "Icterus (??)" after specimens 3320, 3321, 3322, and 3323, showing his obvious puzzlement over the whole problem of how to classify this divergent species. He also referred to this last species as the "Icterus like Finch" in his voyage zoology notes, but reiterated the "Icterus" classification nine months later in his *Ornithological Notes*. Darwin's use of these various designations was by no means an impressionistic or hasty manner of describing his specimens. Darwin was not in the habit of using ornithological terminology imprecisely in his voyage notes and catalogues. The term "Icterus," for example, is used throughout his *Ornithological Notes* to characterize many bona fide members of the Icteridae.

11. See Gifford 1919:256. The erroneous presumption that Darwin saw all the species of Darwin's finches, including the woodpecker finch, is endorsed by Peterson (1963:12), Huxley and Kettlewell (1965:136n44), Taylor and Weber (1968:877), Moorehead (1969:202), Thornton (1971:163), Thompson (1975:10–11), and Kimball (1975:434–435, 1978:587).

12. Although Gould indeed named thirteen species of Darwin's finches, four of these have since been recognized as variant forms of the other nine species. Thus Gould's *Geospiza strenua* is a subspecies of *G. magnirostris. G. dentirostris* and *G. dubia* are both examples of *G. fortis*. In addition, Gould's *Cactornis assimilis* is a subspecies of *G. scandens*. Of the nine true finch species that Darwin actually collected, two were misidentified by him as nonfinches, leaving only seven species that he might have distinguished in the field. I am assuming, however, that Darwin, like Gould, confused at least one large-billed specimen of the sharp-beaked ground finch (*G. nebulosa nebulosa*) with the cactus finch (*G. scandens*), because the requisite number of specimens of the latter species cannot otherwise be accounted for in his voyage catalogue. This leaves only six species apparently distinguished and recognized by Darwin as finches while on the *Beagle*.

13. On the voyage Darwin had with him Lesson's *Manuel d'ornithologie* (1828), the seventeen-volume *Dictionnaire classique d'histoire naturelle* (Bory de Saint-Vincent 1822–1831), Molina's *History of Chili* (1809), and various other books dealing with natural history, voyages, and travels.

group as the Geospizinae was capable of eliciting among nineteenth-century ornithologists.

What evidently misled Darwin most of all in his voyage understanding of these birds is the odd relationship that prevails between beak and plumage in the group. As David Lack (1947:12) has pointed out, closely related species of continental passerine birds are usually extremely similar in their beaks and other structural features, differing chiefly in their plumage. Most of Darwin's finches, on the other hand, are almost identical in plumage, whereas the beaks differ considerably between even the closest species. So anomalous is this condition that an ornithologist basing his classifications upon the customary relationship between beak and plumage would unhesitatingly place Darwin's finches in at least six or seven genera, and perhaps even several subfamilies (Lack 1947:14). This is precisely what Darwin did and is why, in part, he was so preoccupied with the confusing nature of the plumage in these birds.[14] John Gould, the eminent British ornithologist who later named and classified Darwin's finches, astutely recognized the misleading nature of these traditional characters; and he was subsequently able to persuade Darwin and others of the close affinities of the whole group (Gould 1837a, 1841). Thus it was not until the *Beagle* voyage was over that Darwin's finches actually became Darwin's *finches* in the sense that we now comprehend.

Geographic Distribution

Darwin's thoughts on the geographic distribution of his finches, and especially the nature of his labeling practices while he was in the Galapagos, have been the subject of much discussion. The importance of resolving these issues lies in ascertaining to what extent Darwin appreciated the highly endemic nature of each separate island's flora and fauna as he proceeded from island to island within the Galapagos.

In his *Journal of Researches* Darwin later reported that the possibility of the different islands possessing separate forms was first brought to his attention

14. Darwin was not alone in mistaking certain of the Galapagos finches for the forms they appear to mimick. Adolphe-Simon Néboux, who visited the Galapagos Islands in 1836 as surgeon of the French frigate *Vénus*, later described *Geospiza scandens* (the cactus finch) as a "Tisserin," or weaverbird (Néboux 1840). A case parallel to that of Darwin's and Néboux's confusion about the Galapagos finches may be seen in the initial efforts of ornithologists to classify the various species of Hawaiian honeycreepers, the other celebrated case of adaptive radiation among birds. Ornithologists at one time placed these birds in four families and eighteen genera before the evolutionary unity of the group, which is now recognized as a single family with only ten genera, was finally accepted. See Greenway 1964:374; Tyne and Berger 1976:545; and Gruson 1976:162.

by Nicholas O. Lawson, the vice-governor of the archipelago. Lawson, whom Darwin met on Charles Island, informed him that "the tortoises differed from the different islands, and that he could with certainty tell from which island any one was brought" (Darwin 1845:394). This discussion took place sometime between September 25 and 27, during the second of Darwin's five weeks in the archipelago.[15] "I did not for some time," Darwin commented, "pay sufficient attention to this statement, and I had already partially mingled together the collections from two of the islands. I never dreamed that islands, about fifty or sixty miles apart, and most of them in sight of each other, formed of precisely the same rocks, placed under a quite similar climate, rising to a nearly equal height, would have been differently tenanted. . . . But I ought, perhaps, to be thankful that I obtained sufficient material to establish this most remarkable fact in the distribution of organic beings" (1845:394).

Darwin did, fortunately, notice that the mockingbird he had collected on Charles Island differed from the form he had previously collected on Chatham Island. This discovery made him pay particular attention to their collection; and he subsequently made efforts to obtain, and to keep separate, specimens from the next two islands he visited (1841:63). These next two islands were Albemarle, where Darwin spent only part of a day, and James, where he spent a week. To Darwin's eyes, the mockingbird specimens from Chatham and Albemarle appeared to be the same, but those from James and especially Charles were noticeably different.[16] In his zoology notes Darwin commented about these specimens at the time: "This bird which is so closely allied to the Thenca of Chili (Callandra of B. Ayres) is singular from existing as varieties or distinct species in the different Isds.—I have four specimens from as many Isds—There will be found to be 2 or 3 varieties.—Each variety is constant in its own Island.—This is a parallel fact to the one mentioned about the

15. According to FitzRoy (1839:490), Lawson came on board the *Beagle* on September 25 and then escorted a party, including Darwin and FitzRoy, to the settlement in the highlands. Darwin spent four days on Charles Island, the last being September 27. See also Darwin's *Diary* (1933:336).

16. DAR 31.2: MS p. 342v. "The Thenca of Albermarle [*sic*] Island is the same as that of Chatham Isd—." Contrary to Darwin's voyage opinion, the mockingbirds from Albemarle (*Nesomimus parvulus*) and Chatham (*N. melanotis*) are now recognized as separate species by some ornithologists, whereas the James and Albemarle forms are both assigned to *N. parvulus*. Gould (1841:62–63), to confuse matters further, later synonymized the Chatham and James forms under the name *melanotis*, which merely goes to show that the Chatham, Albemarle, and James forms are all very similar in appearance and would be classified by many ornithologists as subspecies. The Charles Island form of the mockingbird (*N. trifasciatus*) is more noticeably distinct, but even this form would be ranked as a subspecies by some ornithologists. See Harris 1974:128; and Mayr and Greenway 1960:447–448.

Tortoises."[17] It was this singular fact in the distribution of the mockingbirds that subsequently prompted Darwin to write in his *Ornithological Notes*:

> When I recollect, the fact that from the form of the body, shape of scales & general size, the Spaniards can at once pronounce, from which Island any Tortoise may have been brought. When I see these islands in sight of each other, & possessed of but a scanty stock of animals, tenanted by these birds, but slightly differing in structure & filling the same place in Nature, I must suspect they are only varieties. The only fact of a similar kind of which I am aware, is the constant asserted difference—between the wolf-like Fox of East and West Falkland Islds.—If there is the slightest foundation for these remarks the zoology of Archipelagoes—will be well worth examining; for such facts [would *inserted*] undermine the stability of Species. (1963[1836]:262)

This famous statement, written approximately nine months after leaving the Galapagos Archipelago, is Darwin's first tentative admission of the possibility that species might be mutable.[18]

To what extent, then, did the finches help to reinforce this insight? According to Lack (1947:23), Darwin also began to separate the members of the finch tribe as a result of the vice-governor's remarks to him on Charles Island. Thereafter, Lack maintains, Darwin kept his ornithological collections from each island separate. Lack's assertion is based upon a detailed examination of Darwin's type specimens, many of which are labeled as coming from the last island Darwin visited, and upon the following statement made by Darwin in his *Journal of Researches*:

> Unfortunately most of the specimens of the finch tribe were mingled together; but I have strong reasons to suspect that some of the species of the sub-group Geospiza are confined to separate islands. If the different islands have their representatives of Geospiza, it may help to explain the singularly large number of the species of this sub-group in this one small archipelago, and as a probable consequence of their numbers, the perfectly graduated series in the size of their beaks. Two species of the sub-group Cactornis, and two of Camarhynchus, were procured in the archipelago; and of the numerous specimens of these two sub-groups shot by four collectors at James Island, all were found to belong to one species of each; whereas the numerous specimens shot either on Chatham or Charles Island (for the two sets were mingled together) all belonged to the two other species: hence we may feel almost sure that these islands possess their representative species of these two sub-groups. (1845:395)

17. DAR 31.2: MS pp. 341–342.
18. On the dating of Darwin's *Ornithothogical Notes*, see Sulloway (1982a).

Darwin's own testimony clearly implies that only the specimens from Chatham and Charles were mingled together, since he was later able to compare those specimens as a group with the specimens collected on James Island.

David Lack's insistence that Darwin began to separate and label his specimens by locality after leaving Charles Island is nevertheless called into question by the seemingly inaccurate nature of several of the island localities actually recorded by Darwin. Indeed, Darwin's type specimens have provided a considerable nightmare of taxonomic problems for subsequent ornithologists, based largely upon their controversial localities. Darwin claimed, for example, that specimens of a peculiarly large-beaked form of *Geospiza magnirostris* came from Chatham and Charles islands. But after more than a century of subsequent collecting without finding any such large-billed specimens, ornithologists found themselves faced with a puzzle. Either this form had become extinct on Chatham and Charles islands, where no *magnirostris* specimens (large or small) had ever been found by other expeditions; or else Darwin's specimens must have come from islands other than those indicated. Swarth (1931:147–149), noting that the largest bills among *G. magnirostris* are found in the northern part of the archipelago, including James Island, believed that Darwin's specimens came from that island. Although Darwin's specimens are still somewhat larger than the present James Island race of this species, Swarth concluded that some evolution in bill size must have occurred since Darwin's visit. Darwin also reported taking specimens of the smaller-billed *G. [magnirostris] strenua* on Chatham Island, and these specimens as well have generally been thought to have come from James Island (Fig. 3 [not reprinted]).[19]

David Lack, who at first agreed with the judgment of Swarth and others,[20] later changed his mind, given Darwin's testimony that only the specimens from the first two islands had been mingled together. Yet Lack himself distrusted other of Darwin's localities, including some involving specimens from the one island—James—where Lack claimed Darwin had kept his specimens separate. According to Lack (1945:14), one of Darwin's specimens of *Cactornis scandens*, labeled as coming from James Island, is actually an example of *Geospiza difficilis* (now *nebulosa*), the sharp-beaked ground finch, and belongs to a form that is not found on James Island today. So either measurable evolution has occurred in the size of the beak, or, more probably, the specimen came from Charles Island, where FitzRoy collected a very similar specimen of this now extinct island race. Altogether, there is serious doubt about the

19. See, for example, Rothschild and Hartert 1899:155; Swarth 1931:149; and Lack 1945:9. Similarly, Hellmayr has concluded: "There seems hardly any doubt that in the case of *G. strenua* and *G. magnirostris* the localities, as given . . . in the 'Zoology of the Beagle,' are altogether untrustworthy" (1938:130n3).

20. See Lack 1940:49; 1945:9–10.

accuracy of eight of the fifteen localities recorded on Darwin's Geospizinae type specimens.[21]

Not only is the accuracy of Darwin's localities in doubt, but so is the means by which Darwin might have recorded this information. From his voyage specimen catalogues and other scientific notes it is very difficult to see how he could have supplied as much information as he later did in this regard. His *Ornithological Notes*, for example, lists localities for only three of his thirty-one Geospizinae, namely, for three specimens of a very distinctive species (*Camarhynchus psittacula*) that he recalled having seen on only one island—James. Moreover, this information was apparently recorded to indicate the rarity of the species rather than its locality per se. For the same reason Darwin also noted such information for two other Galapagos birds.

Darwin is known, of course, to have used FitzRoy's collections after the voyage to supplement his own record of localities. But this source of information still does not account for the localities entered on Darwin's own type specimens. Presumably, Darwin might have recorded localities on his specimen tags rather than in his catalogues. For this reason ornithologists have repeatedly bemoaned the fact that no original labels in Darwin's or John Gould's hand have ever been found among Darwin's type specimens at the British Museum. In the nineteenth century it was the custom of the museum curators to throw away the original collector's labels and to replace them with neatly printed museum labels. Information thought worthy of preserving was transferred to the new labels. But much valuable information, such as the original collector's numbers, was inevitably lost. George Robert Gray, who assisted Darwin with the *Birds* volume of the *Zoology of the Voyage of H.M.S. Beagle* and who later received Darwin's types from the Zoological Society when it closed its museum, was a typical offender in this regard (Sharpe 1906:84–85).

The question of whether or not Darwin recorded island localities directly on the specimen tags is largely resolved, however, by the fortunate discovery of one (and probably the only surviving) original label for his ornithological specimens. Having vainly sought, like previous investigators, for original labels among Darwin's type specimens, it occurred to me to examine all those Darwin specimens at the British Museum (National History) that are not

---

21. These doubtful localities involve the following birds: two specimens of *Geospiza magnirostris* (British Museum registry nos. 1855.12.19.80 and 1855.12.19.113, labeled as coming from Chatham Island but thought to have come from James); two specimens of *G. parvula* (British Museum nos. 1855.12.19.167 and 1855.12.19.194, labeled as coming from Chatham Island but elsewhere assigned to James [Darwin 1841:102]); one specimen of *Cactornis scandens* (British Museum no. 1855.12.19.20, labeled as coming from James but assigned by Lack to an extinct race of *G. nebulosa* [formerly *difficilis*] on Charles or Chatham Island); and three specimens of *G. strenua* (British Museum nos. 1855.12.19.81, 1855.12.19.83, and 1855.12.19.114, labeled as coming from Chatham but thought to have come from James).

endemic to the Galapagos. One such specimen was at last found (*Dolichonyx oryzivorus*—the American bobolink), bearing what appears to be Darwin's original crude paper tag. Comparison of the specimen number (3374) with Darwin's manuscript catalogue shows that the number is indeed Darwin's and that it is inscribed in his own hand (Fig. 4 [not reprinted]).[22] On the reverse side of the tag the genus name, "Dolychonyx," is written in pencil, in an unidentified hand; and below it, in ink, the species name, "oryzivorus," appears, apparently in John Gould's hand. A second and smaller label, added when the specimen was presented to the Zoological Society in 1837, records Darwin's name, the date of accession, and, on the back, Darwin's original specimen number. The specimen was acquired by the British Museum in 1881, after Gould's death, along with many other birds from his huge personal collection. A third label (not shown) was attached to the specimen at this time.

Being a migrant species with an unusually wide range (from Canada to Chile), the bobolink is an occasional visitor to the Galapagos in the autumn of each year. Coincidentally, in its autumn plumage the bobolink is not unlike a Darwin's finch, although Darwin initially thought the bird was a pipit of very unusual structure.[23] When Gould first examined the bird in 1837, he thought it was a new species of finch. But he later discovered that it was an already described North American species and apparently decided to keep the specimen for his own collection.[24] This circumstance, together with the lack of scientific importance of the specimen, enabled its original Darwin and Zoological Society labels to survive.

What is particularly important about this specimen with regard to Darwin's labeling practices is that no island locality is recorded on either of the two earliest tags. Darwin did consider this information worth recording in his *Ornithological Notes*, however, since the bird had been encountered on one island only—James. Thus it appears that whatever island localities Darwin thought worth recording, such as those for three finch and four mockingbird

22. In addition, the paper is similar to that used by Darwin on the *Beagle* voyage. The registry number of this specimen at the British Museum is 1881.5.1.2394.

23. In his *Ornithological Notes* he wrote: "Anthus. was shot by Fuller on James Isd: it was the only one specimen seen during our whole residence. It is described as rising from the ground suddenly & again settling on the ground.—Showed in its flight long wings, like a Lark; uttered a peculiar cry.—Its structure appear[s] very interesting" (1963[1836]:265).

24. Whether Gould acquired the specimen in 1837, or whether he perhaps acquired it as late as 1855, when the Zoological Society closed its museum and sold all its ornithological specimens, is not known. Gould also possessed other Darwin type specimens. In 1857 he sold 251 ornithological specimens to the British Museum, including 2 specimens of *Geospiza* that once belonged to Darwin (reg. nos. 1857.11.28.247 and 1857.11.28.248). See "Zoological Accessions Aves 1854–1873," p. 64, and "Zoological Accessions Aves 1880–1884," p. 106; British Museum (Natural History), Sub-department of Ornithology, Tring.

specimens, were recorded in the master catalogue of specimens and in the *Ornithological Notes* rather than on the crude paper tags.[25]

In short, Darwin does not appear to have altered his collecting or labeling practices while he was in the Galapagos Archipelago. After he left Charles Island, his collecting procedures continued to reflect the typological and creationist assumptions he had brought with him to that archipelago. What localities he did record were noted as largely incidental information to remind himself later of scarce species or noteworthy habitats. He continued, moreover, to collect only a few specimens of each species; and he entirely failed to collect finches on the third island he visited—Albemarle—even though almost every finch within miles was gathered in front of him at a spring near Bank's Cove.[26] Darwin thereby passed up the chance of collecting an additional species, and two endemic subspecies, of Galapagos finches. Similarly, although Darwin (1844:98) asked his fellow shipmates to bring him geological specimens from all the larger islands he was personally unable to visit, he made no such request for zoological specimens. Even after leaving James Island and setting sail for Tahiti, Darwin apparently continued to treat the vice-governor's comment about the tortoises, and his own discovery with regard to the mockingbirds, as isolated anomalies. For if he had fully appreciated the revolutionary im-

25. This conclusion is confirmed by an analysis of the locality information published by Waterhouse (1845) in his paper on Darwin's Galapagos insects. Of twenty-nine species, fourteen have island localities and fifteen do not. Each of these fourteen localities is recorded as well in Darwin's specimen catalogue; and the island and habitat information given by Waterhouse corresponds exactly to Darwin's own wording in that catalogue. Thus only where this information was recorded in Darwin's notes was it preserved for later use. Darwin apparently recorded such information incidentally as part of the habitat description. For example, specimens 3363 and 3364 are followed by the comment: "Small insects, sweeping; high up, central parts of Charles Island" ("Printed Numbers 3345[–3907]," Down House). In his section on advice to collectors, which appeared only in the first edition of his *Journal of Researches* (1839:598–599), Darwin recommended that a number be placed on each specimen immediately after it was procured and that this number be entered in the specimen catalogue "during the very same minute" so that the locality would never be subject to doubt. If localities had been recorded on the numbered tags, this precaution would have been unnecessary. Finally, that none of Darwin's ornithological specimens had localities on the labels is reinforced by Gould's failure to provide any island designations for the Galapagos species he named in January and February of 1837. See "Zoological Society of London. Minutes of Scientific Meetings Oct. 1835 to Aug. 1840," pp. 120–121, 123–124, 129–130, 134; and Gould 1837a, b, c, d.

26. In his *Diary*, Darwin wrote in this connection: "'To our disappointment the little pits in the Sandstone contained scarcely a Gallon [of water] & that not good. It was however sufficient to draw together all the little birds in the country; Doves & Finches swarmed round its margin" (1933:338; entry for October 1, 1835). Similarly, FitzRoy commented: "Around this scanty spring draining continually through the rock, all the little birds of the island appeared to be collected, a pretty clear indication of there being then no other fresh-water within their reach . . ." (1839:495).

plications of these facts, he would never have allowed his *Beagle* shipmates to devour and discard all thirty adult tortoises brought on board ship as a source of fresh meat for the cruise across the Pacific (FitzRoy 1839:498).[27]

These conclusions regarding Darwin's collecting procedures during his Galapagos visit bring us back once again to the problem of his finches and their dubious localities. In particular, if Darwin recorded only three island localities for these birds in his scientific notes, how and when did he derive the many additional localities that are now to be found on his type specimens? To answer this question I must take up the topic of what happened to Darwin and his finches after they returned from the *Beagle* voyage.

DARWIN'S RETURN TO ENGLAND

The *Beagle* anchored in Falmouth, England, on October 2, 1836, after a voyage of nearly five years. During the next several months Darwin arranged for the disposal of his collections and began to prepare his *Journal* for publication. In mid-December he took up residence in Cambridge in order to look over all of his geological specimens. It was not until January 4, 1837, that he finally delivered his collection of birds and mammals to the Zoological Society in London.[28]

27. These tortoises, from Chatham Island, were brought on board the *Beagle* just five days before Darwin returned from James Island. FitzRoy had earlier embarked eighteen Chatham Island tortoises, and these were devoured as well. FitzRoy did, however, bring two Hood Island tortoises back to England ("Zoological Accessions 1837," p. 1; British Museum [Natural History], Mammals Library, London). Two other very small tortoises also survived the *Beagle* voyage—apparently brought home as pets (DAR 29.3:40, MS p. 7v). When Darwin finally realized the significance of having an expert taxonomist decide whether the reported differences between the tortoises were of specific distinction, these four tortoises were the only ones available. Although they were from three different islands (Hood, Charles, and James), they were all too young to be of value (Darwin 1839:465). Darwin also missed an opportunity to bring back an adult carapace of the unusual saddleback form of tortoise on Charles Island. According to FitzRoy (1839:492), numerous shells were lying around at the Charles Island settlement, where they were being used as flower pots. Within about ten years of Darwin's visit, the Charles Island tortoise was extinct. Zoologists had to wait nearly a century to find remains of this form in a lava cave (Broom 1929).

28. Several of his specimens, including his bobolink, still bear this date of accession on the labels. It seems likely that Darwin presented the specimens in person since he came to London from Cambridge that same day to deliver a paper before the Geological Society (Darwin 1837a). He also wrote a letter dated January 4 that was read that afternoon at a meeting of the Zoological Society Council. According to the minutes of that meeting, Darwin's letter "announced a present to the Society of his entire Collection of Mammalia and Birds made during His Majesty's Surveying Vessel Beagle. It was ordered that the best thanks of the Society be returned to Mr. Darwin for his liberal and valuable contribution to its preserved Collections: and that his wishes with respect to the disposal of the duplicate specimens in this Collection, and to the mounting and describing of the same be strictly complied with" (unpublished "Zoological Society Minutes of Council," 5:79–80).

For the next two months Darwin continued to reside in Cambridge, with the exception of a brief visit to London on February 18 to hear Charles Lyell's anniversary address to the Geological Society.[29] At this meeting Darwin learned about the latest taxonomic findings regarding his valuable collection of South American fossil Mammalia. Richard Owen, who had taken charge of these bones, had recently reported his preliminary results to Lyell and had given him permission to make them public at the anniversary meeting. In his address, Lyell (1837:511) emphasized that Darwin's fossils had confirmed a law previously deduced with regard to Australia, namely, the close relationship that prevails between the past and present Mammalia of large continents. With this confirmation of "the law of succession," Darwin had received a source of evidence that would shortly prove instrumental in his conversion of the theory of transmutation. But it was the case of the Galapagos birds that was to be the most decisive in this respect.

On March 6, having finished looking over his geological specimens in Cambridge, Darwin moved to London in order to be near the various specialists who were working on his zoological collections. His first meeting with John Gould, who had been busy naming Darwin's ornithological specimens over the previous two months,[30] took place between March 7 and 12. It was at this time that Darwin first learned the results of Gould's analysis of his Galapagos collections.[31] The Galapagos finches were not, as Darwin had previously thought, members of widely different genera or even families, but rather one peculiar group of thirteen species that Gould placed in one genus and three closely allied subgenera (1837a). Gould had astutely realized the basic peculiarity of these birds, namely, that "the bill appears to form only a secondary character." Furthermore, he had even got the warbler finch right.[32]

29. According to Wilson (1972:442n21) and Herbert (1974:248n99), Darwin did not attend this meeting, but his presence is recorded in the manuscript minutes of the society's meetings. See "Ordinary Minute Book," 8:219.

30. After receiving Darwin's specimens, Gould exhibited, discussed, and named portions of Darwin's collection at the next five consecutive meetings of the Zoological Society (January 10 and 24, February 14 and 28, and March 14). See Gould 1837a, b, c, d, and e.

31. For the dating of this meeting, and evidence that Darwin and Gould had not discussed the Galapagos specimens before this time, see Sulloway (1982a).

32. See "Zoological Society of London. Minutes of Scientific Meetings Oct. 1835 to Aug. 1840," p. 120; manuscript record of the meeting of January 10, 1837. At this meeting, Gould recognized only eleven or twelve species of finches in three genera (*Geospiza, Camarhynchus*, and *Cactornis*), apparently not at first realizing that the warbler finch (*Certhidea olivacea*) was one of the Geospizinae. As Gould continued to work his way through the rest of Darwin's collection, group by group, he soon realized his mistake, which he had probably corrected by the time Darwin moved to London in March. The discrepancy between the number of finches reported as being named by Gould on January 10 (twelve species in the Zoological

More important still for Darwin's evolutionary thinking, Gould (1837d) had declared that three of the four island forms of Galapagos mockingbird brought to England by Darwin were distinct species, a possibility that Darwin had already asserted "would undermine the stability of Species." For the Galapagos as a whole, Gould pronounced twenty-five of the twenty-six land birds as new and distinct forms, found nowhere else in the world. Even four of the eleven waders and waterbirds—a gull, a rail, a heron, and a turnstone—were considered new by Gould (Darwin 1839:461). Darwin was frankly stunned, not only by the realization that three separate species of mockingbirds indeed inhabited the different islands of the Galapagos, but also by the fact that most of these Galapagos species, even though new, were closely related to those found on the American continent.[33] It was these two conclusions, together with the findings about his fossils, that finally convinced him that species must be mutable and that subsequently prompted the famous entry in his private journal: "In July [1837] opened first note book on 'Transmutation of

---

Society's "Minutes" and eleven species in three contemporary newspaper accounts) is probably the result of Gould's subdivision of one species into two shortly after the January 10 meeting. For further information, see Sulloway (1982a). On May 10 Gould again brought Darwin's finches before the Zoological Society, naming fourteen species in four genera, including *Certhidea* (see the manuscript "Minutes," pp. 164–165). Gould's fourteenth species, *Geospiza incerta*, lived up to its name, for he subsequently synonymized it under one of the others. A curious remnant of this change of mind remains in the published *Proceedings of the Zoological Society*, for although it is said that fourteen species were named, only thirteen names and descriptions follow (Gould 1837a). Also of interest is the fact that the published *Proceedings* lists under the January 10, 1837, meeting the names and descriptions that were given only later by Gould at the May 10 meeting. Thus the published record, by transferring the events of May 10 back to January 10 and by deleting the earlier presentation, obscures the difficulties that Darwin's finches caused even such a celebrated ornithologist as John Gould.

33. It is often claimed that Darwin was impressed by the American character of his Galapagos finches (see, for example, Silverstein 1974:505; and Ruse 1979:164). But Darwin's finches played no role in this aspect of his evolutionary insight. Rather it was the mockingbirds, the flycatchers, the dove, and numerous other typically American species that established this generalization about the Galapagos avifauna. The finches, in contrast, were placed with the Fringillidae in the nineteenth century, and this family of birds was then believed to be worldwide. It is only in this century that the Fringillidae and Emberizidae, under which Darwin's finches are now classified, have been distinguished as families of Old and New World finchlike species, respectively. Although Darwin's finches have no close ancestor on the American continent today, some ornithologists believe they arose from a form related to the emberizine genus *Volatinia* (and several similar genera). These species are all seed-eating ground birds that range from the southern United States to northern Chile and Argentina (Paynter and Storer 1970:vii). Relying on osteological and other evidence, Steadman (in press) has argued that the Geospizinae evolved from *Volatinia jacarina*, the blue-black grassquit. He also contends that the Cocos Island finch and the Galapagos finches were established by two independent invasions of this species from Central and South America, respectively.

Species'—Had been greatly struck from about Month of previous March on character of S. American fossils—& species on Galapagos Archipelago. These facts origin (especially latter) of all my views" (de Beer 1959:7).

Reconstructing the Finch Localities

In the wake of Gould's taxonomic findings, many of them quite unexpected, Darwin soon realized that the enigma of the finches could largely be explained if they, like the mockingbirds, were confined to separate islands. He therefore began to solicit information from those shipmates on the *Beagle* who had made their own private ornithological collections and who, unlike himself, had fortunately kept accurate records of the islands from which they had procured their specimens. Captain FitzRoy's extensive collection, which had gone to the British Museum on February 21, 1837, offered relatively easy access, and Darwin later acknowledged his use of it in the *Zoology* (1841:99).[34] What Darwin did not say in the *Zoology*, however, was that he also employed two other shipmates' collections, including that of his own servant, in attempting to reconstruct these island localities. The first of these sources of information came from Harry Fuller, who had spent a week collecting with Darwin on James Island. Altogether Fuller collected eight specimens of *Geospiza*, one from Chatham Island and seven from James. The collection of Darwin's servant, Syms Covington, was somewhat smaller and included only four finches, one from Chatham Island and three from Charles Island.[35]

34. For the date of FitzRoy's presentation of specimens, which included 187 skins, see the manuscript catalogue "Zoological Accessions Aves, 1837–1851–3," pp. 7–15; British Museum (Natural History), Sub-department of Ornithology, Tring. FitzRoy presented one further specimen on March 15, 1837, an egg of *Rhea darwinii*. FitzRoy's Galapagos portion of the collection included 50 skins, 21 of them finches, all with an island locality. Some of these Galapagos specimens were retained by FitzRoy, however; and only 24 Galapagos skins, 13 of them finches, actually went to the British Museum. Because FitzRoy's specimens were all labeled by island, his collection establishes that *Geospiza nebulosa* (Gould 1837a), an extinct race of the sharp-beaked ground finch (*G. difficilis*), was once present on Charles Island. According to the international rules of nomenclature, the name *G. nebulosa* therefore has priority over *G. difficilis*, which was first applied by Sharpe (1888) to another subspecies of this species. Thus the name of this taxon will henceforth be *G. nebulosa*. For further details about FitzRoy's collection, see Sulloway (1982b).

35. I have recently succeeded in locating and identifying these twelve specimens, which are now at the University Museum of Zoology, Cambridge (Fuller's collection), and the British Museum (Natural History), Sub-department of Ornithology, Tring (Covington's collection). Although only two of the birds have island localities on their labels, I have been able to resupply this information for the other ten specimens based upon two independent manuscript sources. Of particular importance is the fact that Fuller and Covington

Records of Darwin's use of locality information from the collections of FitzRoy, Fuller, and Covington are among Darwin's manuscripts at Cambridge University Library (Figs. 5 and 6 [neither reprinted])[36] There are four such sheets, in Darwin's hand. Although none of the sheets [is] dated, indirect evidence indicates that Darwin lost little time after he became an evolutionist in trying to reconstruct the Galapagos finch localities. One of the four sheets, which bears an 1836 watermark (manufacturer unknown), comprises a series of questions about Galapagos specimen localities that Darwin evidently sent to FitzRoy, and that was answered by an unidentified amanuensis or clerk (Fig. 5). On this same sheet an amanuensis working for Darwin also asked from what island of the Falklands a specimen of fox had come. Darwin mentioned the results of this latter inquiry in his *Journal of Researches* (1839:250–251), which was already in press by mid-August 1837. Similarly, Darwin's statement in his *Journal* (1839:475) that he "very much" suspected that certain species of Galapagos finches were confined to separate islands corroborates the conclusion that he had already examined the various *Beagle* collections by the time his *Journal* went to press. Since Darwin had reached the Galapagos chapter of his *Journal* by late May or early June and since he had finished with the whole of the *Journal* by the end of June, his efforts to collate the various *Beagle* Geospizinae by locality probably date from June at the latest.[37]

---

collected specimens of the large-billed form of *Geospiza magnirostris* on Chatham and Charles islands, respectively. For further details about these collections and their history, see Sulloway (1982b). Ironically, that other shipmates on the *Beagle*, but not Darwin, recorded island localities for their birds marks Darwin as the only real scientist aboard that ship. For Darwin collected with a theory, however mistaken, in mind. The other shipmates were mere collectors, and their labeling practices reflect that fact.

36. See DAR 29.3:26, 28–30.

37. That Darwin's manuscript notes on this question were initially compiled in connection with the writing of his *Journal* is reinforced by another consideration. On the list of Covington's and Fuller's birds, which occupies one of the four sheets, Darwin mistakenly referred *Camarhynchus psittacula* to the genus *Geospiza* (Fig. 6). He also misspelled *psittacula* as *spittacula*. This same species name is misspelled and assigned to the genus *Geospiza* in a list of Galapagos species that Darwin compiled in the spring of 1837 during a meeting with John Gould (Sulloway 1982a). Darwin was not, therefore, entirely familiar in the spring of 1837 with the generic or specific names that Gould had just given these species. The use of erroneous generic and specific names on the locality list for Covington's and Fuller's birds suggests that these notes too were compiled about this time. The name *psittacula* was altered to *psittaculus* in the *Zoology* (1841:103), so these notes on Covington's and Fuller's specimens clearly predate that change. I would assign Darwin's two other sheets of notes on his Galapagos finch localities to late 1840, when he was working on the final installment of the ornithological portion of the *Zoology*. One sheet, which records all thirteen of FitzRoy's finch localities, may be dated by the use of the specific name *Camar[h]y[nchus] psittaculus*. The other, although it bears the name *psittacula*, is probably of the same date, since it contains a collated list of localities for all the *Beagle* collections as published in the *Zoology* (1841:100–106).

It was undoubtedly at this time, that is, sometime in the spring or early summer of 1837, that Darwin also tried to reconstruct the island localities of his *own* Galapagos specimens. For a few birds Darwin was able to infer from his notes or from memory that he had collected these specimens on only one island. This was the case, for example, for an owl, a swallow, a flycatcher, and for three finch specimens with a peculiar beak shaped like that of a parrot (*Camarhynchus psittacula*). In addition, from his *Beagle* shipmates Darwin apparently acquired several finch specimens that were lacking in his own collection, and at least one of these had a locality attached (Sulloway 1982b).

Unfortunately, certain of Darwin's attempts to reconstruct the island localities of his own specimens involved guesswork, and errors inevitably crept in. In his master catalogue of specimens, for example, he drew a line under the first eight Geospizinae and wrote "¿Chatham Isd??"[38] The reason Darwin surrounded this locality designation with three question marks is evident from the order of the catalogue entries as a whole. As may be seen from the number sequence assigned to his birds, Darwin ticketed, numbered, and catalogued the entire collection only after leaving the Galapagos Archipelago in late October 1835. Within the list of birds, the entries proceed topsy-turvy, with specimens from the different islands entered in no apparent order.[39] It is hardly surprising, then, that at least two of the eight specimens that Darwin later assigned to Chatham Island appear to have been mislabeled (Sulloway 1982b).

In the process of attempting to correlate the results from four different collections, Darwin inadvertently made other mistakes. In the *Zoology* (1841:101) he later gave the locality of *Geospiza fortis* as Charles and Chatham islands; but this was clearly an error, since the *Beagle* specimens all came from Charles

38. See "Printed Numbers 3345[–3907]," Down House, under specimen nos. 3312–3319. The catalogue is written in ink. The line under the first eight specimens and the comment "¿Chatham Isd??" were added later in pencil, almost certainly after Darwin's return to England.

39. Of those specimens for which island localities are listed (eighteen) or were later published by Darwin (two), or for which localities can be reconstructed on the basis of other evidence (nine), the sequence runs: James (3299); James (3303); James (3304); Charles (3306); Chatham (3307); Chatham (3308); Charles or James (3309); James (3310); the eight specimens of finch that Darwin later assigned to Chatham with three question marks (3312–3319); James (3330–3332); James (3340); Charles or James (3342–3344); Chatham (3345); Albemarle (3349); James (3350); James (3356); James (3362); and James (3374). I have deduced seven of these twenty-nine localities from information unknown to Darwin. *Certhidea olivacea* exhibits distinctive characteristics by island, and Darwin's specimens (3310 and 3340) definitely belong to the James Island form of this species. *Pyrocephalus dubius* (3345) is confined to Chatham Island, and hence Darwin's specimens of *P. nanus* (3309, 3342–3344), a form that replaces *dubius* elsewhere in the archipelago, must have come from either Charles or James Island. The localities of two other specimens (3299 and 3362) can be deduced from Darwin's statement that they came from a salt lagoon, which he visited on James Island. Darwin also visited a salt lagoon on Albemarle Island, but he does not appear to have collected at this site.

and James.[40] Further inaccuracies are associated with Darwin's claim about geographic representation among the various species of the Geospizinae. Eager to squeeze whatever evolutionary evidence he could from these finches, Darwin systematically collated the island localities of the four *Beagle* collections to see if any of the species represented one another on the different islands. In two genera, *Cactornis* and *Camarhynchus*, he claimed this to be the case. Of the numerous specimens shot by four collectors at James Island, he reported, all belonged to *Cactornis scandens* and *Camarhynchus psittacula*, whereas the specimens collected either on Chatham or Charles were those of *Cactornis assimilis* and *Camarhynchus crassirostris*. "Hence we may feel almost sure," he concluded, "that these islands possess their representative species of these two subgroups" (1845:395).

Darwin's analysis of these two genera was plagued by several errors. In actual fact, FitzRoy had collected a specimen of *Cactornis assimilis* on James, not Charles or Chatham Island, thus invalidating half of Darwin's claim. Furthermore, Darwin had not collected long enough on any of these islands to realize that the various finch species are by no means confined to single islands. *Camarhynchus crassirostris*, for example, is found not only on Charles Island, where Darwin believed his own specimens had probably been taken, but also on Chatham and James. Similarly, *Cactornis scandens* and *Camarhynchus psittacula* are not confined to James Island, as Darwin had thought, but are found on the other islands he visited. Thus Darwin's claim about geographic representation in this group of four species is not only wrong in every detail, but it is not even substantiated by the *Beagle*'s own collections. It is no wonder, then, that Darwin was so excited and relieved in 1845 by Joseph Hooker's rigorous demonstration of representation in his several hundred species of Galapagos plants. To Hooker he wrote in July of that year, "I cannot tell you how delighted and astonished I am at the results of your examination; how wonderfully they support my assertion on the differences in the animals of the different islands, about which I have always been fearful" (1877, 2:22). Darwin lost no time in adding Hooker's welcome results to his *Journal of Researches*, which he was then engaged in revising for the second edition.[41]

40. In his manuscript notes on the collections of FitzRoy, Fuller, and Covington, Darwin listed this locality correctly as "Charles [and] James Isd." See DAR 29.3:28. Nevertheless, because John Gould probably mistook at least one Chatham Island specimen of *Geospiza fortis* for that of *G. [magnirostris] strenua*, the actual locality for the *Beagle* collections of *G. fortis* should have been Chatham, Charles, and James islands. Similarly, *G. [magnirostris] strenua*, reported as coming from Chatham and James islands in the *Zoology* (1841:101), was in fact collected only on James Island. See Sulloway (1982b) for further discussion of Gould's classification mistakes.

41. It is ironic, and Darwin (1839:629) was the first to admit it, that his Galapagos plants proved so valuable precisely because he was least accomplished in that field of natural

Fortunately, the errors and uncertainties associated with Darwin's ornithological specimens did not affect the published results of the *Zoology of the Voyage of H.M.S. Beagle* that much. Of the seventeen type localities that Darwin published for his finches, fifteen were either provided or corroborated by the other shipmates' collections. Darwin himself, employing an educated guess, was able to supply localities for two additional species that only he had collected. In the end only two species of finches remained without any locality whatsoever.

Unfortunately, what later ornithologists generally failed to appreciate was that these published localities were not necessarily those of Darwin's own specimens. In fact, the largely borrowed nature of Darwin's published localities for his Galapagos finches has had one curious repercussion that has confused even further the localities of the *Beagle* type specimens. A number of originally unlabeled Darwin specimens appear to have acquired island localities later in a completely circular fashion, based upon the published information provided in the *Zoology of the Voyage of H.M.S. Beagle*. Curators at the British Museum apparently noticed that certain Galapagos species were indicated in the *Zoology* as coming from one island only. They therefore assumed that unlabeled Darwin specimens of these species must have come from those published localities. The specimens in question now carry these island localities on their labels; and in the British Museum's published list of type specimens there are notes to see the relevant pages of the *Zoology of the Voyage of H.M.S. Beagle*.[42] In certain instances (for example, in the case of Darwin's specimens of *Otus galapagoensis*, *Hirundo concolor*, and *Dolichonyx oryzivorus*), these derivative localities are indeed correct, since Darwin was the only person on the *Beagle* to collect these species, whose localities he was later able to recall. But this same process of circular relabeling is apparently what

history. For this reason he collected "blindly" from each island he visited, mistaking representative species for duplicate specimens. That he fortunately recorded the island localities of his plant specimens reflects the way in which they were collected. Plants must be placed in a plant press soon after collection, and the plants from a given island would all tend to be pressed together rather than intermixed with plants from a separate island. Similarly, Darwin recorded separate island localities for his saltwater fish because they had to be numbered and preserved in spirits of wine soon after being caught.

42. The following specimens at the British Museum (Natural History), Sub-department of Ornithology, Tring, appear to have acquired localities—either on the labels or in the published type specimen catalogue—by reference to the *Zoology*: *Camarhynchus psittacula* (reg. no. 1855.12.19.22); two specimens of *Cactornis scandens* (nos. 1855.12.19.20 and 1855.12.19.125); two specimens of *Geospiza parvula* (nos. 1855.12.19.167 and 1855.12.19.194); *Otus galapagoensis* = *Asio flammeus* (no. 1855.12.19.153); *Larus fuliginosa* (no. 1855.12.19.218); *Hirundo concolor* = *Progne modesta* (no. 1860.1.16.54); and *Dolichonyx oryzivorus* (no. 1881.5.1.2394). See Warren 1966:104, 108; Warren and Harrison 1971:127, 420, 448, 494; and Mayr and Greenway 1960:87.

accounts for at least four of Darwin's finches being given localities that do not necessarily belong to them.[43]

More ironically still, three of Captain FitzRoy's accurately labeled specimens have also suffered from this relabeling process, based once again upon Darwin's published testimony. In one instance FitzRoy's specimen of *Camarhynchus psittacula*, which was procured on James Island, was relabeled as coming from Charles Island. This error was precipitated by the loss of Darwin's three type specimens of *C. crassirostris*. *C. crassirostris and C. psittacula* are somewhat similar species. FitzRoy's slightly aberrant specimen of *psittacula*, which was later thought to be the missing type of *crassirostris*, was accordingly reassigned to that species. But the island locality now had to be altered as well to agree with Darwin's dubious but "official" information for the type of *C. crassirostris*![44] The classification error was eventually caught by Swarth (1931:208), but the specimen in question still bears two island localities. Similarly, two other FitzRoy specimens, one being the type of *Geospiza nebulosa*, were also relabeled incorrectly, owing once again to Darwin's published localities.[45]

In short, the published designations of the *Zoology* were seen by later ornithologists and museum curators as more definitive than the accurately labeled FitzRoy specimens that had largely supplied this information. Swarth (1931:11) actually dismissed FitzRoy's localities wholesale, assuming his speci-

43. Darwin's specimens of *Geospiza parvula* (nos. 1855.12.19.167 and 1855. 12.19.194) do not necessarily come from James Island, as the labels and Warren and Harrison (1971:420) have claimed. According to Lack (1945:14–15), one of the two Darwin specimens of *Cactornis scandens* (no. 1855.12.19.20), which are both labeled as coming from James Island, is actually a specimen of *G. difficilis* (now *nebulosa*) and belongs to the extinct Charles Island form of this species. Darwin was unable to supply the island locality for *C. assimilis*, which he probably did not distinguish from *C. scandens*, so it is unlikely that he was certain about the localities of any of his *scandens* specimens. Once again, see Warren and Harrison (1971:494) for the circular derivation of these *C. scandens* localities.

44. Both the reassignment of this specimen to *Camarhynchus crassirostris* and the change in its island locality were apparently done prior to Sharpe's (1888:16) catalogue of specimens at the British Museum.

45. The source of the first of these two errors began with Salvin's (1876:482) reassignment of the species *Geospiza nebulosa* to *G. fortis*. Since Darwin's specimens of *G. fortis* were supposed to have come from Chatham and Charles islands, and since the only extant specimen of *G. fortis* in the British Museum bears a Charles Island locality, subsequent ornithologists apparently assumed the FitzRoy specimen had come from the other published locality (e.g., Sharpe 1888:11). Later, the erroneous Chatham Island locality was crossed out and the Charles locality reinstated, possibly by Kinnear (see note 46), but the presence of two island localities on this specimen has proved confusing for subsequent ornithologists (e.g., Lack 1945:14–15).

The second incorrectly labeled FitzRoy specimen is the type of the Galapagos rail (*Zapornia spilonota* Gould = *Laterallus spilonotus*, British Museum reg. no. 1837.2.21.404). Rothschild and Hartert (1899:184–185), noting that Darwin (1839:459) had described seeing water rails on James Island, erroneously concluded that the bird was collected by him on that

mens could have come from practically anywhere in the archipelago.[46] David Lack (1945, 1947), although not going quite so far, assumed that all of FitzRoy's specimens were really Darwin's and that those specimens labeled as coming from either Chatham or Charles Island could have come from either locality. With all of these confusions about the localities of Darwin's and FitzRoy's specimens, it is little wonder that the *Beagle* types have proved so problematical to ornithologists over the last hundred years.

Darwin's Finches and Darwinian Theory, 1837–1859

The largely retrospective nature of Darwin's understanding of his Galapagos finches is apparent not only from his postvoyage attempts to reconstruct their island localities but also from his theoretical conceptions about these birds. Contrary to the legend, Darwin's finches do not appear to have inspired his earliest theoretical views on evolution, even after he finally became an evolutionist in 1837; rather it was his evolutionary views that allowed him, retrospectively, to understand the complex case of the finches.

Not only was this retrospective understanding surprisingly slow in coming; but it was far more limited than is generally assumed. The finches are not mentioned, for example, in any of the four notebooks on "Transmutation of Species" that Darwin commenced in July 1837 and kept until later 1839. Nor are they mentioned in the later portions of the *Red Notebook*, written between March and July 1837, which predate this series and which contain his earliest speculations on the transmutation of species.[47] Although Darwin frequently

---

island. FitzRoy, however, collected his specimen on Charles Island. In the *Zoology* (1841:132), Darwin gave only "Galapagos Archipelago" as the locality for this species. Swarth (1931: 53) and Warren (1966:279), following Rothschild and Hartert, have perpetuated the erroneous James Island locality for FitzRoy's specimen.

46. Swarth's erroneous conclusion was reinforced by the fact that some of FitzRoy's specimens have their localities recorded not on the specimen tags but only in the museum's "Zoological Accessions Aves 1837–1851–3" register. Norman B. Kinnear, who worked in the Bird Room of the British Museum (Natural History), nevertheless understood that the localities of Fitz-Roy's specimens had been recorded in this old register. Using this information, he inserted a number of footnotes into Swarth's (1931) monograph indicating the localities of various unlabeled FitzRoy specimens. Swarth, however, chose to disregard this information, arguing that "there have been so many chances for dissociation of specimens and data that my every instinct impels me to rely upon the evidence supplied by the specific or subspecific characters of the specimens rather than on what has been written about them" (1931:146n). Unfortunately, Swarth's ornithological intuitions were not as accurate as FitzRoy's recorded localities.

47. The *Red Notebook* has been transcribed with extensive editorial annotations by Herbert (1980), who supplies documentation for an approximate dating of this notebook. Elsewhere I provide a more precise dating of the evolutionary passages in the *Red Notebook* (Sulloway 1982a).

discussed in these notebooks the two subjects with which the finches are usually associated—speciation through geographic isolation and adaptive radiation into unfilled niches—he always cited examples other than the finches.[48]

In the first edition of his *Journal of Researches* (1839), Darwin said very little about the finches except to comment that certain subgenera of *Geospiza*
probably had their representative forms on different islands and that this circumstance would help to explain the "wide range of character" found in the
group. In the spring and early summer of 1837, when Darwin was preparing his
Journal for publication, he believed that species diverged primarily through
geographic isolation and ensuing adaptation to varying local circumstances.
Since he considered the different islands of the Galapagos to have identical
climates and geographic conditions, he apparently believed isolation alone
was the cause of the small differences that separate most representative
species in archipelagoes. But he did not address himself, either publicly or
privately, to the enigma of how such differences might arise under identical environmental conditions or how they could become as pronounced as
they are in some species of Darwin's finches. It should be pointed out that
Darwin's brief discussion of the finches in his *Journal* predates by more than
a year his discovery of the principle of natural selection. He therefore did not
have an adequate appreciation of how evolution, and particularly adaptation,
are effected through competition between life forms. Nor did he appreciate
that islands within an archipelago might differ biotically without differing
climatically or geographically.[49]

48. In the first of the four notebooks on transmutation of species (July 1837 to February
1838), Darwin's favorite examples of speciation through geographic isolation were the
Galapagos tortoises and mockingbirds (p. 7); the English and Irish hares (pp. 7, 221, 262);
and various other cases of representative forms in archipelagoes or on islands and nearby
continents (pp. 11, 31, 50, 69 [excised], 82, 138, 156, 166 [excised], 187 [excised], 221, 241, 249
[excised] [brackets in original]). Darwin broached the topics of divergence and adaptive
radiation in this first notebook in a number of different contexts: the prevalence of Edentata
in South America (pp. 13, 54, 106); the prevalence of marsupial types in Australia (pp. 14–15,
141); the tendency for every organic group to adapt some of its forms to air, land, and water
(pp. 23–24, 45–46, 263); and instances of species that have adopted new stations, often
evolving new structures and new behaviors, normally occupied by other, very different species (pp. 55–56 [excised], 137, 141, 144, 193). See de Beer 1960–1961; and de Beer, Rowlands,
and Skramovsky 1967.

49. Later, in the *Origin of Species*, Darwin reflected upon these conceptual difficulties: "But
this dissimilarity between the endemic inhabitants of the [Galapagos] islands may be used
as an argument against my views; for it may be asked, how has it happened in the several islands situated within sight of each other, having the same geological nature, the same height,
climate, &c., that many of the immigrants should have been differently modified, though
only in a small degree. This long appeared to me a great difficulty: but it arises in chief part
from the deeply-seated error of considering the physical conditions of a country as the most
important for its inhabitants; whereas it cannot, I think, be disputed that the nature of the

Even after he had hit upon the principle of natural selection and was writing about the finches in the *Zoology of the Voyage of H.M.S. Beagle* (1841:99–106), Darwin simply reiterated what he had already said in his *Journal* about the possibility that geographic representation contributed to the group's "fine gradation" of character. It was not until he drafted his Essay of 1844, in which he set down a 230-page outline of his theory of evolution by natural selection, that he finally set forth a theoretical model with sufficient sophistication to begin to deal with the enigma of the Galapagos finches. Imagining a volcanic island newly elevated from the ocean floor and far from any point of land, Darwin noted that the first colonists would rarely be completely adapted to the many vacant "stations" they encountered there. Not only would the physical conditions of the new and rugged volcanic environment differ from those in the colonists' homeland, but the absence of the colonists' usual competitors would further ensure altered conditions of existence. Each successive colonist would in turn contribute to "new and varying conditions" for the island biota as a whole (1909 [1844]:185). Hence natural selection would act continuously on the various colonists, Darwin concluded, to produce ever more adapted forms. If the island were turned into an archipelago by the continued action of subterranean forces, new opportunities for colonization and evolution would eventually give rise to representative species or races, "as is so wonderfully the case with the different islands of the Galapagos Archipelago" (1909 [1844]:187).

One of Darwin's novel insights in his Essay of 1844 was that no two islands in an archipelago that is continually stocked by random colonists would ever possess exactly the same inhabitants. It is this circumstance, he now appreciated, that causes differential evolution among the representative species of neighboring islands. Commenting in the related context of temporary archipelagoes created by the repeated elevation and subsidence of a continent, Darwin concluded that through evolution "the inhabitants of the most *dissimilar* stations . . . would be more closely allied than the inhabitants of two very *similar* stations on two of the main divisions of the world" (1909 [1844]:190). Although Darwin did not apply this Essay idea to the case of the Galapagos finches, he clearly had some such general concept in mind when, the following year, he added the famous remark to his *Journal of Researches* that "one might really fancy that from an original paucity of birds in this archipelago, one species [of finch] had been taken and modified for different ends" (1845:380).

In spite of these declarations in the Essay and in the 1845 edition of the *Journal*, Darwin had not yet fully grasped the notion of adaptive radiation. At this time he still did not understand why divergence *necessarily* takes place

---

other inhabitants, with which each has to compete, is at least as important, and generally a far more important element of success" (1859:400).

after a species multiplies itself through geographic isolation and then comes into secondary contact with its geographic representatives. It is not immediately clear, for example, why any divergence at all should occur among the species of a scantily populated archipelago. With few competitors or predators to challenge the colonists of an isolated island group, what real evolutionary pressure would there be for representative species, once formed, to evolve significant differences beyond simple reproductive isolation?

Looking back in his *Autobiography*, Darwin recalled that his failure to resolve this problem of divergence was the one major omission from his Essay of 1844. The solution, which finally came to him in the 1850s, was that "the more diversified the descendants from any one species become in structure, constitution, and habits, by so much will they be better enabled to seize on many and widely diversified places in the polity of nature, and so be enabled to increase in numbers" (1859:112).[50] In other words, natural selection favors the most divergent offspring of every species because divergence, by minimizing competition, increases the individual's chances for survival. With his principle of divergence, Darwin at last had an explanation for why adaptive radiation tends to occur in those cases, such as the Galapagos finches, where geographic isolation enables early colonists to exploit the many unoccupied stations of a new environment.

Surprising as it may seem, Darwin did not publish anything more about his famous finches after the brief and cryptic hint about them he had inserted into the second edition of his *Journal*. And publicly, at least, he never actually put his finches forward as evidence for the theory of evolution.[51] In the *Origin of Species* (1859) the Geospizinae go unmentioned, although the Galapagos Islands are employed on six different occasions to illustrate the general relation between the inhabitants of oceanic islands and those of the nearest continent, the phenomenon of representative species, and the absence of certain classes of organisms, such as mammals, from remote islands. The closest that Darwin came in these later years to discussing the origins of his Galapagos finches from an evolutionary point of view was in *Natural Selection*, the longer version of the *Origin* that was interrupted in 1858 by Alfred Russel Wallace's anticipation of the theory of natural selection. In that larger work Darwin contrasted the situation of Madeira, which annually receives stray birds from the neighboring continent and which possesses only one endemic species

50. On the dating of Darwin's insight into the principle of divergence, see Browne (1980), who argues that it occurred in 1857. Darwin himself dated this discovery to about 1852. Actually, the idea came to him in a series of stages between the late 1840s and 1857, being applied first to species and higher taxa with allopatric distributions, and being extended later to include sympatric divergence among varieties of the same species. See Ospovat 1981:170–190.
51. The first published evolutionary account of the Galapagos finches is apparently that of Salvin (1876).

among its twenty land birds, with the far more isolated state of the Galapagos, where twenty-five of the twenty-six land birds have reached endemic status. In Darwin's view, as few as eleven species originally colonized the Galapagos, and there they must have encountered a wide range of open places in the economy of nature:

> hence I suppose that nearly all the birds had to be modified, I may say improved by selection in order to fill as perfectly as possible their new places; some as Geospiza, probably the earliest colonists, having undergone far more change than other species; Geospiza now presenting a marvellous range of difference in their beaks, from that of a gross-beak to a wren; one sub-species of Geospiza mocking a starling, another a parrot in the form of their beaks. (1975[1856–1858]:257)

When he abstracted material from his "Big Book" for the *Origin of Species*, Darwin dropped the example of the finches from the corresponding discussion (1859:104–105, 390–391).

How is it that Darwin elected to omit from his *Origin of Species* what today is probably the most cited "textbook" example of the validity of his evolutionary views? The answer to this question is that Darwin clearly did not consider the case of the Galapagos finches to be in any way crucial to his argument. In this connection we must distinguish what we now know about Darwin's finches from what Darwin knew about them in 1859.

To establish a presumption that his Galapagos finches had indeed evolved such divergent forms through adaptive radiation, it was first necessary to show that the different shapes of their beaks were in some way effective in reducing competition. But Darwin lacked precisely this information. According to his own testimony, the several species of *Geospiza* were "indistinguishable from each other in their habits," feeding together on the ground in large irregular flocks (1841:99). These observations were not only incomplete but also incorrect. *Geospiza magnirostris*, the large ground finch, is actually a solitary species that rarely feeds on the ground with the other seed eaters. Moreover, all four species of ground finches have somewhat different diets; and one species, *G. nebulosa*, is restricted to the humid zone on the islands visited by Darwin. Similarly, Darwin erroneously believed that the habits of three other tree-dwelling species were identical to those of the genus *Geospiza*. But two of these species (*Camarhynchus parvulus* and *C. psittacula*) are insectivorous, and the third (*Platyspiza crassirostris*) has a purely vegetarian diet. To Darwin's eyes, only the cactus finch (*G. scandens*) seemed to be distinguishable by its habit of feeding upon the prickly-pear cactus. Thus Darwin failed to correlate feeding habits in the Galapagos finches with their diverse beaks, and partly for this reason most subsequent ornithologists thought that there was no relationship.

As for *Certhidea olivacea*, the warbler finch, there were frequent debates throughout the remainder of the nineteenth century about whether this species was really a finch at all; and Darwin himself entertained some doubts about the matter (1841:105). Most ornithologists actually rejected Gould's perceptive classification until after the turn of the century, when anatomical studies and observations of breeding behavior finally convinced them that *C. olivacea* was indeed one of the Geospizinae.[52] Similarly, some nineteenth-century ornithologists also doubted that the genus *Camarhynchus*, which includes five of the thirteen species of Galapagos finches, had the same evolutionary origins as the six species of *Geospiza*.[53] Thus as a case of divergence, or adaptive radiation, Darwin's finches were a speculative and problematical example at best, lacking proof on just those points that were crucial to the whole argument.

Above all, what now sets the Geospizinae apart as a convincing paradigm of evolution in action is the evidence associated with their geographic distribution and intraspecific variation. In 1859 Darwin had scant knowledge of the role geographic isolation had played in the evolution of the Geospizinae. Moreover, what evidence he did possess was unfortunately wrong. This fact became readily apparent after Dr. Habel visited the Galapagos in 1868 and brought back an extensive collection of Geospizinae described by Philip Sclater and Osbert Salvin (1870). In his 1876 monograph on the avifauna of the Galapagos, Salvin rather charitably commented that "Mr. Darwin's views as to the exceedingly restricted range of many of the species must be considerably modified" (1876:461). In fact none of the species of Geospizinae collected by Darwin have turned out, as he suggested, to be restricted to single islands. Only when ornithologists returned to the Galapagos Archipelago in a series of expeditions from the late 1880s to the 1930s, collecting numerous finch specimens from each of the islands and analyzing the statistical variations in

52. Salvin (1876:476), Sclater (1886:27–28), and Ridgway (1897:497) placed *Certhidea olivacea* with the Coerebidae. Rothschild and Hartert (1899:148) were less certain, placing this species either with the Mniotiltidae or the Coerebidae. Snodgrass and Heller (1904:234) preferred to classify it with the Mniotiltidae. On anatomical grounds, however, Snodgrass (1903), Sushkin (1925, 1929), and Lowe (1936) all recognized the close affinity between the warbler finch and the rest of Darwin's finches. This opinion, which was accepted and corroborated by Swarth (1931:138) and Lack (1947:13), is no longer questioned.
53. Species of the genus *Camarhynchus* differ from those of *Geospiza* not only in their beaks and diets but also in plumage. The males of *Camarhynchus*, in particular, are never fully black, developing that coloration only around the head and upper body. This difference in plumage is partly what prompted Salvin (1876:470) to doubt that *Camarhynchus* and *Geospiza* had the same evolutionary origins. Similarly, three decades earlier, Lafresnaye (1843) doubted the relatedness of these two genera, referring a species of *Camarhynchus* to the South American finch genus *Guiraca*. Lafresnaye's classification was followed by Prévost and des Murs (1855:209–212).

characters of the different species and subspecies, did it finally become possible to appreciate the evolutionary richness presented by this one group of birds. It was precisely this evolutionary richness that Darwin, with his limited number of specimens from only three islands, did not have at his disposal when he wrote the *Origin of Species*. Only the legend of Darwin's finches makes us think differently.

THE LEGEND OF DARWIN'S FINCHES

The legend of Darwin's finches encompasses two principal themes. The first involves the claim that the different forms of the finches, along with the tortoises and the mockingbirds, first convinced Darwin that species must be mutable while he was still in the Galapagos Archipelago. The legend's second theme holds that Darwin's observations on the finches inspired all his later theories by providing him with a decisive example of evolution in action. In particular, the finches are said to have elucidated the crucial roles of geographic isolation and adaptive radiation as mechanisms of evolutionary change.[54] "Probably no evidence was more important to his [evolutionary] thinking," writes one such spokesman for the legend, "than the example of his finches" (Kimball 1978:587). In the most extreme form of the myth, Darwin is said to have collected species and observed behavioral traits, such as the remarkable tool-using habit of the woodpecker finch, that were not even known in his own lifetime.[55]

As it turns out, Darwin made absolutely no effort while in the Galapagos

54. Many authorities have stressed the role of Darwin's finches in converting Darwin to the theory of evolution while he was still in the Galapagos. See, for example, Huxley 1954:6, 1960:9; Eibl-Eibesfeldt 1961:18; Peterson 1963:11–12; Darling and Darling 1963:34; Moorehead 1969:202; Grzimek 1973:359; Ohtey 1976:135; Dobzhansky et al. 1977:12; and Jensen et al. 1979:486. The following commentators, who do not date Darwin's conversion or who place it later than the actual Galapagos visit, still emphasize the critical role of the finches in that conversion: Swarth 1931:10; Wynne-Edwards 1947:687; Mayr 1947:217; Eiseley 1961:172–173; de Beer 1963:132; Moody 1970:303; Leigh 1971:136; Thornton 1971:12, 161–162; Grinnell 1974:259, 263; Gruber 1974:130; Dorst 1974, 2:252; Silverstein 1974:505; Thompson 1975:10; Kimball 1978:587; Freeman 1978:147; Ralling 1978; and Ruse 1979:164. Most of these authors, regardless of their dating of Darwin's conversion, argue that the finches provided Darwin with a decisive model for his general theory of evolution.

55. See, for example, Peterson 1963:12; Taylor and Weber 1968:877; Moorehead 1969:202; Kimball 1975:434–435, 1978:587; and Thompson 1975:10–11. Even Darwin scholars have occasionally implied that Darwin knew of certain evolutionary evidence, such as the correlation between beaks and the diverse feeding habits of the Galapagos finches, that dates from this century. See de Beer 1963:83; Huxley and Kettlewell 1965:44; and Gruber 1974:160. Similarly, Ruse (1979:164–165) implies that Darwin had a qualitative understanding of the relationship between isolation and endemism among Darwin's finches, but this was first documented by Lack (1947).

to separate his finches by island; and what locality information he later published, he reconstructed after his return to England, using other shipmates' carefully labeled collections. As for Darwin's supposed insight into evolution by adaptive radiation while he was still in the Galapagos, the more the various species of finch exhibited this remarkable phenomenon, the more Darwin mistook them at the time for the forms they were mimicking. Even after his return to England, when John Gould had clarified the affinities of this unusual avian group, Darwin was slow to understand how the Galapagos finches had evolved. In particular, he possessed only a limited and largely erroneous conception of both the feeding habits and the geographic distribution of these birds—information that was vital to a proper explanation of their evolution. Lastly, far from being crucial to his evolutionary argument, as the legend would have us believe, the finches were not even mentioned by Darwin in the *Origin of Species*.[56]

In spite of the legend's manifest contradictions with historical fact, it successfully holds sway today in the major textbooks of biology and ornithology, and is frequently encountered as well in the historical literature on Darwin. It has become, in fact, one of the most widely circulated legends in the history of the life sciences, ranking with the famous stories of Newton and the apple and of Galileo's experiments at the Leaning Tower of Pisa, as a classic textbook account of the origins of modern science.

To appreciate the growth of this pervasive legend, one must understand the tradition of ornithological research that the Galapagos Islands inspired in the post-*Origin* period. In his 1876 monograph on the Galapagos avifauna, Osbert Salvin was already calling that archipelago "classic ground" in the history of biology. It was here, he asserted, that Charles Darwin had made a series of insights and deductions, "the importance of which in their bearing upon the study of natural science has never been equaled" (1876:461). This proud and reverent attitude was echoed by most subsequent ornithologists working on the avifauna of the Galapagos, and Salvin's words were frequently quoted by fellow monographers to bolster their feeling of being on "classic" ground.[57]

Meanwhile, scientific expeditions continued to visit the Galapagos at increasingly regular intervals. Habel's visit in 1868 was followed by eight more expeditions during the remainder of the century, an average of one every four years. More important, with the triumph of Darwin's evolutionary views there had ensued a veritable revolution in collecting techniques. Whereas Darwin, in accordance with prevailing typological collecting procedures, had

56. Because it is so widely held that Darwin's finches led Darwin to develop the theories published in the *Origin of Species*, it has naturally been assumed by some authors that the finches were given a prominent place in that work. See, for example, Lack 1945:4; and Gillsäter 1968:85.

57. See, for example, Ridgway 1890:102n, 1897:459; and Rothschild and Hartert 1899:136.

brought home only 31 finches and 64 birds altogether from this archipelago, Habel collected 460 specimens in 1868, Georg Baur about 1,100 specimens in 1891, Charles Harris 3,075 specimens in 1897, and the California Academy of Sciences an astonishing 8,691 specimens in 1905–1906. Even before this last expedition, Walter Rothschild and Ernst Hartert could proclaim that more ornithological specimens had been collected from the Galapagos than had "ever been brought together from any area of similarly small dimensions" (1899:136). Darwin's finches, in turn, had become perhaps the best known avian group in the world.

With this spectacular growth in the number of collected specimens, there eventually came a similarly impressive advance in the biological understanding of Darwin's finches. It is now recognized that Darwin's original thirteen species constitute only nine present-day species. (John Gould, having insufficient material, split his "species" too finely.) Another four species collected after Darwin's visit have been recognized, bringing the present total, coincidentally, back to thirteen.[58] But it was only after half a century of debate about the status of numerous island subspecies, many of which were elevated to the rank of full species by their describers, that the present number of species was finally agreed upon.[59] Among these thirteen species, Lack (1969:254) has recognized thirty-five subspecies; and it is these subspecies, rather than the various species, as Darwin had claimed, that represent one another within the Galapagos group.[60]

58. *Camarhynchus pallidus*, the woodpecker finch, was first collected by Habel in 1868. *Geospiza difficilis* (now *G. nebulosa difficilis*), the sharp-beaked ground finch, was also first described from Habel's collection, although FitzRoy's specimen of *G. nebulosa* and one of Darwin's specimens are apparently earlier examples of this species (see Lack 1947:23; and Sulloway 1982b). *G. conirostris*, the large cactus finch, and *C. pauper*, the medium tree finch, were first collected in 1888 by the *Albatross* expedition. The last of the Galapagos finches—*C. heliobates*, the mangrove finch—was collected in 1899 by Heller and Snodgrass. The sole Cocos Island member of the Geospizinae (*Pinaroloxias inornata*) was collected by Richard Hinds in 1840 and was described by Gould (1843).

59. Sixty-seven different specific and subspecific names were at one time applied to the Geospizinae. Among the various authorities, Rothschild and Hartert (1899) recognized twenty-one species of Darwin's finches, Snodgrass and Heller (1904) twenty species, Swarth (1931) twenty-eight species, and Lowe (1936) thirty-seven species. It is to Lack (1945, 1947) that we owe the present reduction to thirteen species.

60. Representation is very infrequent at the species level among the Galapagos finches, since most of the "representatives" long ago spread to other islands. At least eight islands have 9 to 11 species each of Darwin's finches, and the average number of species for the sixteen main islands is 7.5 (Lack 1969:254). Only 2 of the 13 species (*Geospiza scandens* and *G. conirostris*) may be said to represent one another on different islands of the Galapagos. This instance was unknown to Darwin, because *G. conirostris* inhabits islands he did not visit, and was not collected until 1888.

This detailed understanding of geographic variation among the various subspecies of Darwin's finches has in turn resolved a long-standing debate over the evolutionary origins of the whole group. As late as the 1930s it was still believed by many ornithologists that Darwin's finches had arisen by some means other than geographic isolation. Percy Lowe (1936) insisted, for example, that the finches constituted "hybrid swarms" of just a few originally unhybridized species. Even Bernhard Rensch (1933), a champion of the theory of geographic speciation, was much puzzled as to how the Geospizinae might have evolved by this means. At this time only Erwin Stresemann (1936) defended the theory of geographic speciation in connection with Darwin's finches.

It was the subsequent researches of David Lack (1940, 1945, 1947) that put an end to these debates and finally turned Darwin's finches into a rigorous and paradigmatic demonstration of speciation through geographic isolation. Among other findings, Lack showed that the percentage of endemic subspecies on each island of the Galapagos is directly proportional to the degree of geographic isolation from the center of the archipelago (Fig. 7). This is why Cocos Island, which is isolated both from the Galapagos and from the mainland, has the highest level of endemism (100 percent) as well as only one species of finch. Isolation promotes endemism; but extreme isolation, by preventing recolonization, excludes the possibility of speciation and adaptive radiation. The Galapagos Archipelago, unlike Cocos Island, has provided just the right conditions for this radiation process.

Although most species of Darwin's finches have spread throughout the archipelago, thus obscuring the role that isolation has played in their evolutionary origins, there is one species that has apparently just commenced this speciation process. The large tree finch, *Camarhynchus psittacula*, exhibits four well-defined forms in the Galapagos Archipelago (Fig. 8 [see figure 23 in the Lack selection in chapter 9, this volume). Because two of these forms have coexisted in the past on Charles Island, and do not appear to have interbred, they are now classified as separate species (*C. psittacula* and *C. pauper*). With its small finch-like bill, *C. pauper* appears to be the earliest and most primitive form, which must have evolved on Charles Island. From there it evidently spread to the northwest and evolved into the closely related form *affinis*. Another form, *psittacula* (*sensu stricto*), has replaced it in the center of the archipelago. Still another form, *habeli*, is found on Abingdon and Bindloe islands to the north. Had not *C. psittacula* (*sensu stricto*) recolonized Charles Island to the south and remained separate, the four forms would be classified by most ornithologists as races of one species. *C. psittacula* and *C. pauper* therefore represent the earliest stages in the origins of a new species of Darwin's finch. *C. pauper* is, in fact, the only species of the Galapagos finches to be confined to one island (Lack 1947:126–128).

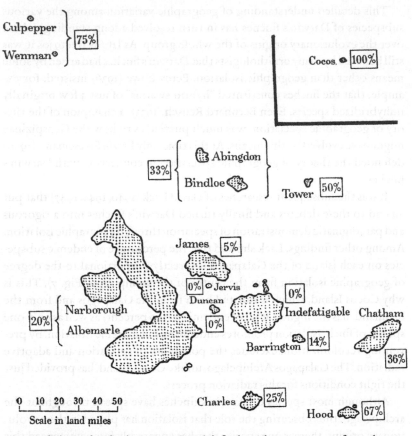

Fig. 7. Isolation and endemism among Darwin's finches. The percentages refer to the proportion of endemic species and subspecies on each island. (From Lack 1947:121.) I have slightly altered the percentages for Charles and Chatham islands to reflect the distributions of *Geospiza magnirostris magnirostris* and *G. nebulosa nebulosa*, as set forth by me elsewhere (1982b). [Figure supplied by Frank Sulloway.]

Perhaps the most remarkable discovery about Darwin's finches was first intimated by David Lack. Reversing his own previous opinion (1945) and rejecting the similar views of most earlier ornithologists, Lack argued that the differences in the beaks of the various finches were highly adaptive with regard to feeding. Previously, ornithologists, following Robert Snodgrass (1902:380–381), had maintained that beak size was not necessarily adaptive and that the tendency for certain of the larger species to feed on larger seeds was merely an incidental result of differences in the size of the beaks. The discovery that the different finch species often recognize one another by the size and shape of their bills reinforced this erroneous view that feeding habits were not a primary consideration in bill form.

Impressed by Gregory Gause's (1934:19–20) contention that no two species with similar ecologies can coexist in the same territory, Lack finally acceded to the conclusion that differences in the beaks must help to reduce competition for food resources. Reinterpreting the same data presented in his earlier publications, Lack was able to substantiate this hypothesis by pointing to the phenomenon of character displacement in Darwin's finches (1947:81–90). On islands where the large, medium, and small ground finches (*Geospiza magnirostris*, *G. fortis*, and *G. fuliginosa*) are found, bill measurements show distinct and confined ranges for the three species. But where one or more of these species is rare or absent, bill size becomes more variable and is always extended in the direction of the missing species. The same phenomenon is found in other species of Darwin's finches and is most dramatically illustrated on the smaller, outlying islands, where only three or four species occupy the niches shared by eight or ten species in the center of the archipelago. Thus the shape as well as the average size of the bill in most species of Darwin's finches is related to the nature of their competitors on each island, with natural selection tending to minimize competition by inducing character displacement and hence adaptive radiation.[61]

These findings on geographic isolation, speciation, and character displacement among the Geospizinae were first brought together in David Lack's celebrated book *Darwin's Finches* (1947). Altogether this work supplies abundant evidence for considering these birds, more than most other avian groups, as a classic paradigm of evolution and adaptive radiation in action. But Lack's *Darwin's Finches* was not just a milestone in the progress of Darwinism. It was also a crucial step in the evolution of the legend about Darwin and his finches. In fact, with the publication of Lack's book in 1947 the legend became fully established.

Three aspects of Lack's book, in particular, helped to crystallize the legend by blurring the crucial distinction between what was "Darwin's" in connection with his famous finches and what was not. First, as a sweeping testimonial to the validity of Darwinism, Lack's researches were closely associated with the triumph of the evolutionary synthesis in the late 1930s and early 1940s. In line with other biological research in this period, Lack showed that the evolutionary dynamics of the Galapagos finches agreed with a strictly Darwinian model of evolution incorporating genetic variation, geographic isolation, and natural selection as the principal mechanisms of evolutionary change. In this sense, then, *Darwin's Finches* was a return to Darwin's own version of evolutionary theory after nearly a century of disputes among rival

61. More recently, Robert Bowman (1961) has shown by detailed analysis of stomach contents that the various species of Darwin's finches indeed subsist upon different diets related to the size and shape of their beaks.

doctrines. Reflecting this triumph of Darwinian theory was the whole design of Lack's book, which included relevant quotations from Darwin's *Journal of Researches* and his *Origin of Species* at the heads of each chapter. It is hardly surprising, then, that many readers of *Darwin's Finches* tended to synonymize Darwin's understanding of his finches with the neo-Darwinian understanding of them.

The second aspect of Lack's book that spurred the growth of the legend was Lack's use of the term "Darwin's finches." Although Lack was not the first person to use this term, it was he who succeeded in popularizing it.[62] In one sense the term is felicitous, because not all the Geospizinae are confined to the Galapagos Islands, and thus the name "Galapagos finches" is inappropriate for the whole group. This was, in fact, the chief reason for Lack's use of the expression "Darwin's finches." But as the term became more popularly known through Lack's book, people tended to assume that these birds had been so named because, as one biologist put it, "they helped to persuade Darwin of the truth of evolution" and were crucial as well to his later theories.[63] Through this act of eponymy, Darwin was increasingly given credit after 1947 for finches he never saw and for observations and insights about them he never made. The coincidence that Darwin's Galapagos finches were described by John Gould as thirteen species, the same number that Lack himself recognized in the archipelago, greatly contributed to this additional source of the legend.

Finally, Lack encouraged the growth of the legend in a third way when he mistakenly insisted that Darwin had indeed separated his ornithological collections by island after leaving the second of the four islands he visited. This conclusion provided convincing evidence that Darwin, at the time, had suspected the evolutionary implications of his collections as a whole and that he had taken steps to correct his earlier oversights in this regard.[64] It mattered

62. Apparently the first person to use the term "Darwin's finches" was Lowe (1936:310).

63. See, for example, Leigh (1971:136), quoted in the text; Thornton (1971:162); and Moody, who writes: "No group of Galápagos animals is of more interest to students of evolution than are the birds, partly because of the role played by these birds in influencing the thinking of Darwin. He was particularly impressed by the varied adaptations exhibited by the unique finches of the archipelago. In commemoration of this fact, Dr. David Lack, has had the happy inspiration to term them 'Darwin's finches'" (1953:268).

64. Lack's erroneous conclusion that Darwin began to separate his ornithological collections by island while he was still in the Galapagos has been endorsed by Himmelfarb (1959:115), Eiseley (1961:171), Barlow (1967:12), Moorehead (1969:202), and Gruber (1974:130). Lack (1963) later reaffirmed this view in a brief article on Nicholas Lawson, the vice-governor on Charles Island who first told Darwin about the differences in the tortoises. Kottler (1978:282–283) has argued that Darwin labeled his ornithological collections by island from the very beginning, but this view clearly cannot be maintained in the light of the facts presented here.

little that David Lack had also debunked another aspect of the legend when, confronted by the historical evidence, he duly acknowledged the apparently slow growth of Darwin's thinking about the finches. Readers and reviewers of Lack's book, steeped in a Darwinian conception of the finches, naturally assumed that Darwin too had fully understood their evolutionary import and that he had merely chosen to bide his time before making his revolutionary views public. "One cannot help thinking," V. C. Wynne-Edwards commented in a typical review of Lack's classic treatise, "what a delight his book would have brought to Charles Darwin, who was so deeply stimulated by his own observations of the Galapagos finches during the voyage of the 'Beagle'" (1947:687). "This miniature example of evolution," Lois and Lewis Darling have similarly written of the Galapagos finches, "was as impressive when young Charles Darwin visited the Galápagos in 1835 as it is today" (1963:34). Even Lack himself seems to have become increasingly caught up by the legend he helped to create when he insisted that Darwin's finches had "provided one of the chief stimuli for their discoverer's theory of evolution" and had thereby "change[d] the course of human history."[65]

In the years since the publication of *Darwin's Finches*, the tremendous popular success of Lack's book has helped to make these birds famous far out of proportion to their actual role in furthering evolutionary theory. Certain other avian groups, such as the Hawaiian honeycreepers, as well as the insects of the Hawaiian Islands, offer even more dramatic examples of explosive evolutionary radiation. But these other cases, although well known to evolutionary biologists, have not permeated the popular biological literature as have Darwin's finches. These birds have become, in fact, the standard textbook example of the historical origins and factual basis of Darwin's theory of evolution. It is the textbooks, moreover, that have given fullest expression to the legend of Darwin's finches. By telescoping history around one dramatic moment of insight in the Galapagos Archipelago, the textbooks have developed the legend into a compelling and appropriately empirical account of the origins of modern evolutionary biology.[66] Through the legend, Darwin is continually celebrated as a scientific hero who single-handedly solved the biological riddle of the Sphinx when he recognized the different Galapagos finches for an extraordinary microcosmic example of evolution in action. In many ways it is perhaps asking too much to deny Darwin a share in the scientific triumph that his legendary finches have come to represent for Darwinism. Legends are, after all, to celebrate heroes; and there is something

65. See Lack 1964:178, 1953:67.
66. The role of textbooks in rewriting history to accord with a linear and strictly empiricist conception of science has been discussed by Kuhn (1970:137–143, 167) and Brush (1974). See also Sulloway 1979:420–422, 503–503.

definitely heroic—more so than even the legend has captured—about Darwin's scientific triumph based on only a fraction of the evidence we know today about the Galapagos Archipelago and its famous finches.

ACKNOWLEDGMENTS

I thank the following persons and institutions for their assistance in connection with this article: the Charles Darwin Research Station, Isla Santa Cruz, Galapagos Archipelago, where I was a visitor in 1968 and again in 1970; I. C. J. Galbraith; Peter J. Gautrey; David Kohn; Ernst Mayr; the Miller Institute for Basic Research in Science; Raymond A. Paynter, Jr.; David W. Snow; David Stanbury; and R. L. M. Warren.

REFERENCES

Barlow, Nora, ed. 1967. *Darwin and Henslow: The Growth of an Idea. Letters 1831–1860.* Berkeley: University of California Press.

Bory de Saint-Vincent, ed. 1822–1831. *Dictionnaire classique d'histoire naturelle.* 17 vols. Paris: Rey et Gravier.

Bowman, Robert I. 1961. "Morphological Differentiation and Adaptation in the Galapagos Finches." *Univ. Calif. Publ. Zool.,* 58:1–302.

—— 1963. "Evolutionary Patterns in Darwin's Finches." *Occ. Pap. Calif. Acad. Sci.,* no. 44:107–140.

Broom, R. 1929. "On the Extinct Galapagos Tortoise that Inhabited Charles Island." *Zoologica,* 9:313–320.

Browne, Janet. 1980. "Darwin's Botanical Arithmetic and the 'Principle of Divergence,' 1854–1858." *J. Hist. Biol.,* 13:53–89.

Brush, Stephen G. 1974. "Should the History of Science Be Rated X?" *Science, 183:*1164–1172.

Darling, Lois, and Lewis Darling. 1963. *Bird.* London: Methuen.

Darwin, Charles Robert. 1837a. "Observations of Proofs of Recent Elevation on the Coast of Chili, Made during the Survey of His Majesty's Ship Beagle, Commanded by Capt. FitzRoy, R. N." *Proc. Geol. Soc. London,* 2:446–449.

—— 1837b. "Remarks upon the Habits of the Genera *Geospiza, Camarhynchus, Cactornis,* and *Certhidea* of Gould." *Proc. Zool. Soc. London,* 5:49.

—— 1839. *Journal of Researches into the Geology and Natural History of the Various Countries Visited by H.M.S. Beagle, under the Command* of *Captain FitzRoy, R.N. from 1832 to 1936.* London: Henry Colburn.

——, ed. 1841. *The Zoology of the Voyage of H.M.S. Beagle, under the Command of Capt. Fitz-Roy, R.N., during the Years 1832–1836.* Part III: *Birds.* London: Smith, Elder.

—— 1844. *Geological Observations on the Volcanic Islands Visited during the Voyage of H.M.S. Beagle. . . .* London: Smith, Elder.

—— 1845. *Journal of Researches into the Natural History and Geology of* the *Countries Visited during the Voyage of H.M.S. Beagle round the World. . . .* 2nd ed. London: John Murray.

—— 1859. *On the Origin of Species by means of Natural Selection, or, The Preservation of Favoured Races in the Struggle for Life.* London. John Murray.

—— 1887. *The Life and Letters of Charles Darwin, Including an Autobiographical Chapter.* Edited by Francis Darwin. 3 vols. London: John Murray.

——— 1909. *The Foundations of the Origin of Species: Two Essays Written in 1842 and 1844.* Edited with an Introduction by Francis Darwin. Cambridge: Cambridge University Press.

——— 1933. *Charles Darwin's Diary of the Voyage of H.M.S. "Beagle."* Edited by Nora Barlow. Cambridge: Cambridge University Press.

——— 1958 [1876]. *Autobiography: With Original Omissions Restored.* Edited with Appendix and Notes by Nora Barlow. London: Collins.

——— 1963 [1836]. *Darwin's Ornithological Notes.* Edited with an Introduction, Notes, and Appendix by Nora Barlow. *Bull. Brit. Mus. (Nat. Hist.) Hist. Ser.,* 2, no. 7.

——— 1975 [1856–1858]. *Charles Darwin's Natural Selection: Being the Second Part of His Big Species Book Written from 1856 to 1858.* Edited by R. C. Stauffer. Cambridge: Cambridge University Press.

De Beer, Gavin, ed. 1959. *Darwin's Journal. Bull. Brit. Mus. (Nat. Hist.) Hist. Ser.,* 2, no. 1.

———, ed. 1960–1961. *Darwin's Notebooks on Transmutation of Species.* Parts I–V. *Bull. Brit. Mus. (Nat. Hist.) Hist. Ser.,* 2, nos. 2–6.

——— 1963. *Charles Darwin: Evolution by Natural Selection.* London: Thomas Nelson and Sons.

De Beer, Gavin, M. J. Rowlands, and B. M. Skramovsky, eds. 1967. *Darwin's Notebooks on Transmutation of Species.* Part VI: *Pages Excised by Darwin. Bull. Brit. Mus. (Nat. Hist.) Hist. Ser., 3,* no. 5.

Dobzhansky, Theodosius, Francisco J. Ayala, G. Ledyard Stebbins, and James W. Valentine. 1977. *Evolution.* San Francisco: W. H. Freeman.

Dorst, Jean. 1974. *The Life of Birds.* 2 vols. Translated by I. C. J. Galbraith. London: Weidenfeld and Nicolson.

Eibl-Eibesfeldt, Irenäus. 1961. *Galapagos: The Noah's Ark of the Pacific.* Translated by Alan Houghton Brodrick. Garden City, N.Y.: Doubleday.

Eiseley, Loren. 1961. *Darwin's Century: Evolution and the Men Who Discovered It.* Garden City, N.Y.: Anchor Books/Doubleday.

FitzRoy, Robert. 1839. *Narrative of the Surveying Voyages of His Majesty's Ships Adventure and Beagle, between the Years 1826 and 1836, Describing Their Examination of the Southern Shores of South America, and the Beagle's Circumnavigation of the Globe.* Vol. 2: *Proceedings of the Second Expedition, 1831–1836, under the Command of Captain Robert FitzRoy, R.N.* With *Appendix.* London: Henry Colburn.

Freeman, R. B. 1978. *Charles Darwin: A Companion.* Folkestone, Eng.: William Dawson & Sons.

Gause, Gregory Frantsevich. 1934. *The Struggle for Existence.* Baltimore: Williams & Wilkins.

Gifford, E. W. 1919. "Field Notes on the Land Birds of the Galapagos Islands and of Cocos Island, Costa Rica." *Proc. Calif. Acad. Sci.,* 2:189–258.

Gillsäter, Sven. 1968. *From Island to Island: Oases of the Animal World in the Western Hemisphere.* Translated by Joan Tate. London: George Allen and Unwin.

Gould, John. 1837a. "Remarks on a Group of Ground Finches from Mr. Darwin's Collection, with Characters of the New Species." *Proc. Zool. Soc. London,* 5:4–7.

——— 1837b. "Observations on the Raptorial Birds in Mr. Darwin's Collection, with Characters of the New Species." *Proc. Zool. Soc. London,* 5:9–11.

——— 1837c. "Exhibition of the Fissirostral Birds from Mr. Darwin's Collection, and Characters of the New Species." *Proc. Zool. Soc. London,* 5:22.

——— 1837d. "[Exhibition of] Three Species of the Genus *Orpheus,* from the Galapagos, in the Collection of Mr. Darwin." *Proc. Zool. Soc. London,* 5:27.

——— 1837e. "On a New *Rhea* (*Rhea Darwinii*) from Mr. Darwin's Collection." *Proc. Zool. Soc. London,* 5:35.

————— 1841. *The Zoology of the Voyage of H.M.S. Beagle, under the Command of Captain FitzRoy, R.N., during the Years 1832–1836.* Edited and superintended by Charles Darwin. Part III: *Birds.* London: Smith, Elder.

————— 1843. "On Nine New Birds Collected during the Voyage of H.M.S. 'Sulphur.'" *Proc. Zool. Soc. London,* 11:103–106.

Gray, George Robert. 1870. *Hand-List of Genera and Species of Birds, Distinguishing Those Contained in the British Museum.* Vol. 2: *Part II. Conirostres, Scansores, Columbae, and Gallinae.* London: Printed by Order of the Trustees.

Greenway, James Cowan. 1964. "Honeycreepers (2)." In *A New Dictionary of Birds,* ed. A. Landsborough Thomson, pp. 373–374. London: Thomas Nelson and Sons.

Grinnell, George. 1974. "The Rise and Fall of Darwin's First Theory of Transmutation." *J. Hist. Biol.,* 7:259–273.

Gruber, Howard E. 1974. *Darwin on Man: A Psychological Study of Scientific Creativity.* Together with *Darwin's Early and Unpublished Notebooks,* transcribed and annotated by Paul H. Barrett. New York: E. P. Dutton.

Gruson, Edward S. 1976. *Checklist of the Birds of the World: A Complete List of the Species, with Names, Authorities and Areas of Distribution.* London: William Collins & Sons.

Grzimek, Bernhard. 1973. *Grzimek's Animal Life Encyclopedia.* Vol. 9: *Birds.* New York: Van Nostrand Reinhold.

Harris, Michael. 1974. *A Field Guide to the Birds of the Galapagos.* London: William Collins & Sons.

Hellmayr, Charles E. 1938. *Catalogue of Birds of the Americas and the Adjacent Islands in Field Museum of Natural History.* Part XI: *Ploceidae—Catamblyrbynchidae—Fringillidae. Zoological Series, Field Museum of Natural History,* 13, Part XI.

Herbert, Sandra. 1974. "The Place of Man in the Development of Darwin's Theory of Transmutation. Part I. To July 1837." *J. Hist. Biol.,* 7:217–258.

—————, ed. 1980. *The Red Notebook of Charles Darwin.* With an Introduction and Notes. *Bull. Brit. Mus. (Nat. Hist.) Hist. Ser.,* 7. Published as a book by Cornell University Press, Ithaca, N.Y.

Himmelfarb, Gertrude. 1959. *Darwin and the Darwinian Revolution.* New York: Doubleday.

Huxley, Julian. 1954. "The Evolutionary Process." In *Evolution as a Process,* ed. Julian Huxley, A. C. Hardy, and E. B. Ford, pp. 1–23. London: George Allen & Unwin.

————— 1960. "The Emergence of Darwinism." In *Evolution after Darwin,* ed. Sol Tax, vol. 1: *The Evolution of Life: Its Origins, History, and Future,* pp. 1–21. Chicago: University of Chicago Press.

Huxley, Julian, and H. B. D. Kettlewell. 1965. *Charles Darwin and His World.* New York: Viking Press.

Jensen, William A., Bernd Heinrich, David B. Wake, Marvalee H. Wake, and Stephen L. Wolfe. 1979. *Biology.* Belmont, Calif.: Wadsworth.

Keynes, Richard Darwin. 1979. *The Beagle Record: Selections from the Original Pictorial Records and Written Accounts of the Voyage of H.M.S. Beagle.* Cambridge: Cambridge University Press.

Kimball, John W. 1975. *Man and Nature: Principles of Human and Environmental Biology.* Reading, Mass.: Addison-Wesley.

————— 1978. *Biology.* 4th ed. Reading, Mass.: Addison-Wesley.

Kottler, Malcolm J. 1978. "Charles Darwin's Biological Species Concept and Theory of Geographic Speciation: The Transmutation Notebooks." *Ann. Sci.* 35:275–297.

Kuhn, Thomas. 1970. *The Structure of Scientific Revolutions.* 2nd ed. Chicago: University of Chicago Press.

Lack, David. 1940. "The Galapagos Finches." *Bull. Brit. Orn. Club*, 60:46–50.

———— 1945. *The Galapagos Finches (Geospizinae): A Study in Variation. Occ. Pap. Calif. Acad. Sci.*, no. 31.

———— 1947. *Darwin's Finches: An Essay on the General Biological Theory of Evolution.* Cambridge: Cambridge University Press.

———— 1953. "Darwin's Finches." *Sci. Amer.*, April, pp. 66–72.

———— 1963. "Mr. Lawson of Charles." *Amer. Sci.*, 51:12–13.

———— 1964. "Darwin's Finches." In *A New Dictionary of Birds*, ed. A. Landsborough Thomson, pp. 178–179. London: Thomas Nelson and Sons.

———— 1969. "Subspecies and Sympatry in Darwin's Finches." *Evolution*, 23:252–263.

Lafresnaye, Frédéric de. 1843. "G. cendré. C. cinerea." *Magasin de Zoologie*, 2nd ser., 5:1–2.

Leigh, Egbert Giles, Jr. 1971. *Adaptation and Diversity: Natural History and the Mathematics of Evolution.* San Francisco: Freeman, Cooper.

Lesson, René-Primevère. 1828. *Manuel d'ornithologie ou description des genres et des principles des espéces d'oiseaux.* 2 vols. Paris: Roret.

Lowe, P. R. 1936. "The Finches of the Galapagos in Relation to Darwin's Conception of Species." *Ibis*, 13th ser., 6:310–321.

Lyell, Charles. 1837. "Address to the Geological Society, Delivered at the Anniversary, on the 17th of February, 1837, by *Charles Lyell*, Jun. Esq., President." *Proc. Geol. Soc. London*, 2:479–523.

Mayr, Ernst. 1947. Review of *The Galapagos Finches (Geospizinae)* and *Darwin's Finches*, by David Lack (1945, 1947). *Quart. Rev. Biol.*, 22:217.

Mayr, Ernst, and James C. Greenway, Jr., eds. 1960. *Check-List of Birds of the World: A Continuation of the Work of James L. Peters.* Vol. 9. Cambridge, Mass.: Museum of Comparative Zoology.

Molina, J. Ignatius. 1809. *The Geographical, Natural, and Civil History of Chili.* London: Longman, Hurst, Rees, and Orme.

Moody, Paul Amos. 1953. *Introduction to Evolution.* New York: Harper & Brothers.

———— 1970. *Introduction to Evolution.* 3rd ed. New York: Harper & Row.

Moorehead, Alan. 1969. *Darwin and the Beagle.* New York: Harper & Row.

Néboux, Adolphe-Simon. 1840. "Description d'oiseaux nouveaux recueillis pendant l'expédition de la Vénus." *Revue Zool.*, 3:289–291.

Olney, P. J. S. 1976. "The Policy of Keeping Birds in the Society's Collections 1826–1976." *Symposia Zool. Soc. London*, no. 40:133–145.

Ospovat, Dov. 1981. *The Development of Darwin's Theory: Natural History, Natural Theology, and Natural Selection, 1838–1859.* Cambridge: Cambridge University Press.

Paynter, Raymond A., Jr., and Robert W. Storer. 1970. *Check-List of Birds of the World: A Continuation of the Work of James L. Peters.* Edited by Raymond A. Paynter, Jr. (in consultation with Ernst Mayr). Vol. 13: *Emberizinae, Catamblyrhynchinae, Cardinalinae, Thraupinae, Tersininae.* Cambridge, Mass.: Museum of Comparative Zoology.

Peterson, Roger Tory. 1963. *The Birds.* Life Nature Library. New York: Time, Inc.

Prévost, Florent, and Oeilliet des Murs. 1855. "Oiseaux." In *Voyage autour du monde sur la frégate la Vénus, pendant les années 1836–1839*, by Abel du Petit-Thouars. Vol. 5, Part 1: *Zoologie: Mammifères, oiseaux, reptiles et poissons*, pp. 177–284. Paris: Gide et J. Baudry.

Ralling, Christopher, producer. 1978. *The Voyage of Charles Darwin.* A seven-part film depicting the life of Charles Darwin, written by Robert Reid. Part 6: *Suppose that All Animals and Plants Are Represented by a Tree and the Tree-like Branches.* British Broadcasting Corporation.

Rensch, Bernhard. 1933. "Zoologische Systematik und Artbildungsproblem." *Verb. dtsch. zool. Ges., 35 (Zool. Anz.*, suppl. 6):19–83.

Ridgway, Robert. 1890. "Scientific Results of Explorations by the U.S. Fish Commission Steamer Albatross. No. 1.—Birds Collected on the Galapagos Islands in 1888." *Proc. U.S. Nat. Mus.*, 12:101–139.

—— 1897. "Birds of the Galapagos Archipelago." *Proc. U.S. Nat. Mus.*, 19:495–670.

Rothschild, Walter, and Ernst Hartert. 1899. "A Review of the Ornithology of the Galapagos Islands, with Notes on the Webster-Harris Expedition." *Novit. Zool.*, 6:85–205.

—— 1902. "Further Notes on the Fauna of the Galapagos Islands." *Novit. Zool.*, 9:373–418.

Ruse, Michael. 1979. *The Darwinian Revolution: Science Red in Tooth and Claw*. Chicago: University of Chicago Press.

Salvin, Osbert. 1876. "On the Avifauna of the Galapagos Archipelago." *Trans. Zool. Soc. London*, 9:447–510.

Sclater, Philip Lutley. 1886. *Catalogue of the Passeriformes, or Perching Birds, in the Collection of the British Museum*. Vol. 11: *Fringilliformes: Part II*. London: Printed by Order of the Trustees.

Sclater, Philip Lutley, and Osbert Salvin. 1870. "Characters of New Species of Birds Collected by Dr. Habel in the Galapagos Islands." *Proc. Sci. Meetings Zool. Soc. London*, pp. 322–327.

Sharpe, R. Bowdler. 1888. *Catalogue of Birds in the British Museum*. Vol. 12: *Catalogue of the Passeriformes, or Perching Birds, in the Collection of the British Museum. Fringilliformes: Part III. Containing the Family Fringillidae*. London: Printed by Order of the Trustees.

—— 1906. "Birds." In *The History of the Collections Contained in the Natural History Departments of the British Museum*, ed. E. Ray Lankester, vol. 2: *Separate Historical Accounts of the Several Collections Included in the Department of Zoology*, pp. 79–515. London: William Clowes and Sons.

Silverstein, Alvin. 1974. *The Biological Sciences*. San Francisco: Holt, Rinehart, and Winston.

Snodgrass, Robert Evans. 1902. "The Relation of the Food to the Size and Shape of the Bill in the Galapagos Genus *Geospiza*." *Auk*, 19:367–381.

—— 1903. "Notes on the Anatomy of *Geospiza*, *Cocornis* and *Certhidea*." *Auk* 20:402–417.

Snodgrass, Robert Evans, and Edmund Heller. 1904. "Papers from the Hopkins-Stanford Galapagos Expedition, 1898–1899. XVI. Birds." *Proc. Wash. Acad. Sci.*, 5:231–372.

Steadman, David W. In press. "The Origin of Darwin's Finches." *Trans. San Diego Soc. Nat. Hist.*

Stresemann, Erwin. 1936. "Zur Frage der Artbildung in der Gattung *Geospiza*." *Orgaan der Club Van Nederlandische Vogelkundigen*, 9:13–21.

Sulloway, Frank J. 1979. *Freud, Biologist of the Mind: Beyond the Psychoanalyic Legend*. New York: Basic Books; London: Burnett Books/André Deutsch.

—— 1982a. "Darwin's Conversion: The *Beagle* Voyage and Its Aftermath." *J. Hist. Biol.*, 15 (in press).

—— 1982b. *The Beagle Collections of Darwin's Finches (Geospizinae)*. *Bull. Brit. Mus. (Nat. Hist.) Zool. Ser.* (in press).

Sushkin, Petr Petrovich. 1925. "The Evening Grosbeak (*Hesperiphona*), the Only American Genus of a Palaearctic Group." *Auk*, 42:256–261.

—— 1929. "On Some Peculiarities of Adaptive Radiation Presented by Insular Faunae." In *Verhandlungen des VI. internationalen Ornithologen-Kongresses in Kopenhagen 1926*, pp. 375–378. Berlin.

Swarth, Harry S. 1931. *The Avifauna of the Galapagos Islands*. *Occ. Pap. Calif. Acad. Sci.*, no. 18.

Taylor, William T., and Richard J. Weber. 1968. *General Biology*. 2nd ed. Princeton, N.J.: D. Van Nostrand.

Thompson, Richard F. 1975. *Introduction to Physiological Psychology*. New York: Harper & Row.

Thornton, Ian. 1971. *Darwin's Islands: A Natural History of the Galápagos*. Garden City, N.Y.: Natural History Press.

Tyne, Josselyn van, and Andrew J. Berger. 1976. *Fundamentals of Ornithology*. 2nd ed. New York: John Wiley & Sons.

Warren, Rachel L. M. 1966. *Type-Specimens of Birds in the British Museum (Natural History)*. Vol. 1: *Non-Passerines*. London: Trustees of the British Museum (Natural History).

Warren, Rachel L. M., and C. J. O. Harrison. 1971. *Type-Specimens of Birds in the British Museum (Natural History)*. Vol. 2: *Passerines*. London: Trustees of the British Museum (Natural History).

Waterhouse, George R. 1845. "Description of Coleopterous Insects Collected by Charles Darwin, Esq., in the Galapagos Islands." *Ann. Mag. Nat Hist.*, 16:19–41.

White, Michael J. D. 1978. *Modes of Speciation*. San Francisco: W. H. Freeman.

Wilson, Leonard G. 1972. *Charles Lyell*. Vol. 1: *The Years to 1841: The Revolution in Geology*. New Haven: Yale University Press.

Wynne-Edwards, V. C. 1947. Review of *Darwin's Finches*, by David Lack (1947). *Ibis*, 89:685–687.

Journal of the History of Biology 15, no. 1 (1982): 1–53. With kind permission from Springer Science and Business Media.

Place: Historical Expeditions to the Galápagos

It is true that most men who found themselves adrift in the Galápagos in early days were hungry and very thirsty. The life they looked after there was their own. From their perspective, and depending on the time of year, the Galápagos Archipelago had little to offer. It is hard to imagine a more forsaken landscape than that of Herman Melville's Galápagos. For this American novelist the islands embody changelessness and death. Even a forgotten cemetery would provide more comfort in its humanity, and desolate Greenland charms by comparison. Instead of the mellow contrast of seasons, Melville's Galápagos offers only a sliver of sweet cane or coconut shell washed up upon its clinkered, blasted shore from some place far away. Already so ruined, the place cannot be ruined further, and so remains changeless. Even the wind and tides have no direction.

Changeless. Can this possibly be the same place Annie Dillard describes, the place where life was born, ideas were born, and where life begets more and more life? For Melville the place is a hell or purgatory. For Dillard, the islands evoke no thoughts of death. She seems to challenge whether there is even such a thing in a world where lava becomes soft protein. "Geography is life's limiting factor," says Dillard. Life conforms to it. And so, for her, the Galápagos Archipelago is a place not of changeless devastation, but of constantly changing creation and re-creation. We are blown apart, evolve, then blown back together, she says. It sums up, as it were, our position in the universe—isolated in company, close enough, but not too close. We may be different species, but we still recognize each other. Geography, which determines the distance between us—past and present—allows for an appreciative cohabitation. The life Dillard looks after there is everything that breathes and wiggles or soaks up the sun.

Geography is crucial to evolutionary theory. It determines what happens where (or doesn't happen). These two interpretations of the same Pacific islands, written by nonscientists, challenge one to wonder: if the geography of place is so central to evolutionary theory, how does one's moral and practical outlook on that geography shape what ones sees there, what one looks for there, and what one concludes about nature there?

A passage from Darwin's *Journal of Researches* shows his own double mindedness about the Galápagos: "such wretched-looking little weeds would have better become an arctic, than an equatorial Flora. The thin woods, which cover the lower parts of all the islands, excepting where the lava has recently flowed, appear from a short distance quite leafless, like the deciduous trees of the northern hemisphere in winter. It was some time before I discovered, that not only almost every plant was in full leaf, but that the greater number were now in flower" (C. R. Darwin 1839, 454).

At first he saw nothing but dead or dormant scrub. But he looked again and realized the plants were in full flower. What caused that change of heart? Just a closer look, probably with a hand lens.

It is important that he carried a hand lens. Other travelers did not. Indeed, Darwin went there deliberately to collect, examine, catalogue, describe. By paying his own way on the voyage, he was even relieved from the general colonial mission to discover natural resources to exploit. He could pay close attention to the most unpromising of foliage, ride tortoises, and pull the tails of iguanas if he wanted; and he did. His job was to be companion to the captain and the collecting hands and feet and eyes for his Natural History mentors back home. That is what *he* went for. Melville was no ordinary whaler, but his perspective as a crew member would have shaped a far different impression than that of a captain's companion. A deadly, annihilating monotony would have struck him with much greater force than Darwin, who for company had his collections, books, servants, and officer friends.

Not only did Darwin see life where other early island visitors saw death by parching (both physical and spiritual), he saw hints of the animate in patterns he observed in the inanimate: a remarkable transference from geology to biology. He eventually came to read a history of life directly from the rocks. He applied Lyellian geological principles—that present geological formations are the result of observable, gradual processes such as erosion, uplifting, and volcanism—to the gradual changes that occur also in species themselves. It was the geological phenomena he saw or read about during his voyage—volcanoes causing islands to emerge, earthquakes raising coastlines discernibly, deep canyons etched out by rivers running below—that made his mind comfortable with the idea of gradual change over a very long time as the cause of extant topography. Further, he closely attended to the material relationship between geology and biology: whole islands, the coral atolls, formed by the growth of marine larvae with specific tolerances for depth and light; long-dead creatures themselves having turned to rock, unearthed with picks and mattocks along with the mineral samples they were embedded in. A history of geology quite literally revealed a history of biology in its matrix. In reference to claims that biogeographic patterns are determined by coincidental acts of the Creator, Darwin explicitly proclaims that, just as general laws apply to inorganic matter, so too should they apply to the organic: "it is absolutely opposed to every analogy, drawn from the laws imposed by the Creator on inorganic matter, that facts, when connected, should be considered as ultimate and not the direct consequences of more general laws" (C. R. Darwin 1909, 182).

Darwin said himself that the combination of the American fossils and the geographic distribution of Galápagos species led him to pursue the question of what exactly constitutes a species. His "horizontal history," mentioned in his notebook on transmutation (chapter 1), reflected a sort of sideways excavation, where distance, like depth in soil strata, reflects time. The degree of affinity between organisms was somehow related to the geographic distance between them—a major insight to be sure—and this affinity also reflected a historical relationship between them through inheritance—another great insight. Together, they led him to one of his most tantalizing statements about the Galápagos: "Hence, both in space and time, we seem to be brought somewhat near to that great fact—that mystery of mysteries—the first appearance of new beings on this earth" (C. R. Darwin 1845, 337).

Space and time together explain. Time is reflected in the vertical history of geology and the horizontal history of geographic distribution. In a sense, this perspective represents a corporealizing of the process of change itself. The substrate—rocks or flesh—is less important than the natural laws that determine how they change; both—rocks and flesh—are carved and shaped by a process of erosion, an elimination of their softer bits. It is the process itself that is the focus of inquiry. Along with simple sustenance for visiting whalers, the substrates provide evidence of a process at work. This process, of course, was none other than the process of creation. Darwin saw biology in geology, evolution in geography. Darwin's objectivity, his deliberate exclusion of a moral agenda and insistence on viewing the biological with the same eyes as the geological, had, nevertheless, major moral implications for his contemporary audience, and indeed for our own creationist contemporaries.

But everyone looks to nature for evidence. For Melville the Galápagos were a curse, for Dillard a blessing, for Darwin a deliberate neutrality. For other visitors, like the renaissance and eighteenth-century buccaneers, the archipelago provided a respite, where animals they could eat ate in turn, and then became either palatable or rank. Season also plays a part, no doubt, with a more foreboding landscape greeting visitors in the dry season than after a particularly lush El Niño. Each visitor brings his own eyes, and what he sees depends on why he is there and what he is looking for. The following readings show how a single place—the Galápagos islands—has been perceived and used over time by various people: from buccaneers, to whalers, and curious visitors wanting to see them for themselves.

READINGS

The first excerpt is by Fray Thomas, Bishop of Panama, who gives the very first written description of the Galápagos in 1535. The excerpt is taken from a very informative article by J. R. Slevin, who gives a detailed history of explorations to the Galápagos for various purposes and provides an excellent bibliography for the historical visits to the islands, from the sixteenth century to the 1950s when he wrote.

Next is an excerpt from a famous buccaneer of the English Royal Navy, William Dampier, who served in the late seventeenth century. He was an observant man, and contributed to the study of natural history by transmitting knowledge of direct use to mariners. He noted the geographic locations and habits of animals with admirable accuracy, but to different purpose than Darwin's. He explains that what an animal eats determines its taste; where it forages indicates where a sailor can find it. Knowing natural history can feed crew members and save their lives.

Last are the readings by Hermann Melville, who arrived as a whaler in 1856, and by Annie Dillard, who arrived as a journalist in 1975, during the peak of environmental activism. They present vastly different impressions from two very different times, with Darwin's theories and the Evolutionary Synthesis intervening.

Time Line of Visits to the Galápagos Archipelago

This is a summary of expeditions and visits to the Galápagos, based largely on Slevin's 1959 article. An asterisk indicates a scientific/biological expedition.

1475 (or thereabouts): Some islands of the Galápagos were discovered by Tupac Yupanqui (Inca king) and his 20,000 men on rafts. It is reported that he brought back the jawbone of a horse, which could have been introduced by unrecorded sailors. It is thought that horses were first introduced to South America by Hernando Cortés in 1519, however, casting some doubt on this story.

1535: Alternatively, the Galápagos were discovered by Fray Tomás de Berlanga, bishop of Panama, as he set off to Chile to arbitrate a dispute between Pizarro and his lieutenant Diego de Almagro. He drifted off course and found himself in the Galápagos. He composes the first written document to mention the islands. "God had showered stones."

1546: The Galápagos were visited by Diego Rivandeniera, a captain under Diego Centano who was waging war with Pizarro. Rivandeniera deserted Centano so as not to be captured by Pizarro, seized a ship, and went adrift to the Galápagos.

1570 (or thereabouts): The Galápagos first appeared on maps by Abraham Ortelius, published in Antwerp. They are referred to as "Insulae de los Galopegos."

1593: Richard Hawkins became the first English captain to visit the islands.

1684: The buccaneer captains Eaton and Cook, with Davis, Dampier, and Cowley as passengers, landed in Albany Bay. Cowley produced the first navigational chart of the archipelago. Accounts of this visit, written by Davis and Cowley, are in the British Museum. Davis reports they are called "Enchanted Islands," and describes the "brimstone" of Albemarle Island. He described several of the islands. The buccaneers revisited the islands repeatedly, leaving and taking stores and using them as hideouts.

1700: French navy captain de Beauchesne Gouin was sent to establish colonies in South America and used the Galápagos to restock his ship.

1708: English captain and buccaneer Woodes Rogers, with captain Stephen Courtney, visited the islands with instructions to fight the Spanish and French.

1720: Captain Dampier and Captain Clipperton replenished supplies in the islands, then headed off to plunder Spanish settlements and capture ships as prizes.

1788: The first whalers from Enderby and Sons, in England, began whaling in the South Pacific.

*1790: The first scientific mission was sent to the Galápagos by the Spanish, under the command of the Sicilian Alessandro Malaspina. All records of this voyage have been lost.

1791: New England whalers from Nantucket sent several boats to the South Pacific for whaling.

1793–1794: James Colnett, in the *Rattler*, was sent to investigate the whaling opportunities. He made extensive descriptions of the geography and resources of the islands and found many relics of the buccaneers. He described much of the flora and fauna, and his journals were read and cited by Darwin frequently.

*1795: Captain George Vancouver, in the *Discovery*, explored the islands. The cruise carried a botanist, Archibald Menzies. Menzies collected the first sample of *Scalesia*, the genus in the daisy family that diversified in the Galápagos, taking the form of shrubs and trees.

1801: Amasa Delano, an explorer, described the natural history of the islands. He mocked the pelicans, was spooked by the tortoises, first described the lava lizards, broke the necks of and ate the ring doves, and admired the endemic "diver" birds (blue-footed boobies). He mentioned many endemic species (although the finches did not appear in his lists). He appreciated the Galápagos as a scarce resource in a vast ocean:

> Were it not for the advantages which these islands afford to the American and English whaling ships, in getting a supply of fresh provisions when in these seas, I know not what they would do, especially in time of war, as they could not possibly attain any from the Spanish settlements. The advantages which ships derive from these islands are very important, in consequence of their lying so much in the way of those employed in the whaling business, who can always get a supply of fresh provisions here, and not be put much out of their course; without which they would not be able to keep to sea but for a few months, before their people would die with the scurvy.

1807: Patrick Watkins was marooned on Charles Island by a British ship and became the first permanent human inhabitant of the islands. He grew vegetables to sell to the sailors. After seven years, he stole a whaleboat from some crewmen who were seeking tortoises and water inland, and sailed with other men to Guayaquil, arriving *alone*.

1822: Captain Basil Hall visited in the H.M.S. *Conway* with the intention of measuring the compression of the earth at the equator. Disappointed with the accuracy of the measurements, he declared he could have made better measurements by staying at home.

1825: Lord Byron visited in the H.M.S. *Blonde*, carrying the bodies of the king and consort of Hawaii, who had visited London as guests of the British government, and died. He went to replenish water in the watering hole mentioned by Delano but found it dry.

*1825: David Douglas, a botanist, and John Scouler from Glasgow visited James Island with the Hudson Bay Company's brig *William and Ann*. They made the first extensive botanical collections of the islands in three days. Douglas made reference to the "smaller birds" of the island, possibly the finches: "Many of the smaller kinds perched on my hat, and even unconsciously settled upon the gun (the instrument of their destruction) which I carried on my shoulder."

*1829 (or thereabouts): Hugh Cuming, who lived in Chile, financed his own trip for collecting purposes. He built his own vessel, the *Discoverer*, and made plant collections. He presented his results to the Zoological Society of London in 1832.

1832: The Galápagos were annexed to the newly independent country of Ecuador. English names for the islands were replaced by Spanish names (but
of course the English still referred to them by their English names). General
José Villamil founded the first colony on Charles Island. The colony was used
to supply ships. The government of Ecuador also used the colony as a penal
colony, and Villamil and his people, finding their neighbors less than salubri-
ous, abandoned it to the convicts.

*1835: The H.M.S. *Beagle*, captained by Robert Fitzroy, landed with naturalist
Charles Darwin aboard. The visit lasted for five weeks, during which time
many geological measurements were made and natural-history collections
were secured. They visited Chatham, Charles, James, and Albemarle.

1838: Sir Edward Belcher of the Royal Navy sailed from Cocos Island to the
Galápagos in the H.M.S. *Sulphur* for exploratory purposes.

*1846: Thomas Edmonstone visited on the H.M.S. *Herald*, captained by Henry
Kellet and accompanied by H.M.S. *Pandora*. Edmonstone was sent with en-
treaties directly from Darwin to make separate collections on each island.
He did, but unfortunately died during the voyage, his collections becoming
irreparably disorganized as a consequence. The naturalist Berthold Seeman
was also on board, and observed on Charles Island no tortoises, but many
wild dogs, pigs, goats, and cattle.

1846: Henri Louns, compte de Gueydon, visited the islands in the French
brig-of-war, *Le Genie*, for five weeks for exploratory purposes. He affirmed
most of the accounts and measurements of the *Beagle* expedition.

1860 and thereafter: The whaling industry waned, in part due to damage to
the *Shenandoah* and the sinking of the "Stone Fleet" by Union soldiers in the
American Civil War.

*1868: Simeon Habel sailed from New York and made extensive collections of
birds, insects, reptiles, and mollusks, which he took to Vienna. These collections were analyzed by Osbert Salvin. Salvin published his account of the avifauna in 1876. He corresponded with Darwin about the distribution of finch
species among the islands. Darwin was surprised that the species were not
more restricted to their respective islands.

1869: Manuel Cobos started a settlement on Chatham Island to grow sug-
arcane. He apparently ruled cruelly, keeping its inhabitants as slaves. The
colony was named Progreso. Another colony was established nearby by
Señor Valdizán for the purpose of harvesting orchilla, a particular type of
lichen used for dyes. Cobos was hacked to death with machetes in 1904, and
Valdizán was likewise assassinated by the abused colonists.

1870–1913: Several British warships visited the islands for restocking, including Sir Arthur Farquar in the *Zealous* (1870), Frederick Henry Sterling in the *Triumph* (1880), and Lieutenants Wintour and Chadwick in the *Hyacinth* (1895). The *Hyacith* was attacked by the Spaniards, but the passengers survived to write about it.

*1873: Louis Agassiz, founding director of Harvard University's Museum of Comparative Zoology, organized an extensive collecting mission from the vessel, the *Hassler*. Dr. Franz Steindachner of the Royal Zoological Museum of Vienna and Count Louis François de Portales of the U.S. Coast Survey were also aboard. The visit lasted only nine days, but extensive collections were made. Agassiz, unconvinced by the finches, remained a believer in divine creation.

*1875: Theodore Wolf visited the islands on an Ecuadorian vessel and made extensive collections of plants. The specimens were destroyed in storage, however.

*1875: Commander W. E. Cookson visited in the H.M.S. *Peterel* and collected tortoises at the bequest of the British Museum curator, Albert Günther. They collected all the tortoises they could find on Abingdon because they feared they were becoming extinct there and wanted to collect as many as they could before they disappeared.

*1888: Z. L. Tanner of the U.S.F.S. *Albatross* collected tortoises, birds, and plants from several islands in the Galápagos.

*1891: George Baur visited many of the islands and made collections of several taxa, including tortoises. Baur wrote extensively about the geology and natural history of the Galápagos.

1893: Don Antonio Gil founded a small colony on Albemarle Island and named it Santo Thomas. The inhabitants raised cattle.

*1896: Robert Ridgway published an account of the birds of the Galápagos, based on collections made by the *Albatross* expedition.

*1897: Lord Walter Rothschild sponsored the Webster-Harris expedition, which left from San Francisco, apparently for the purposes of satisfying a tortoise fetish. They returned with sixty living tortoises, which Rothschild kept on his property.

*1898: Edmund Heller and Robert E. Snodgrass, from the Zoology Department of Stanford University, visited the islands in the *Julia E. Whalen*. They stayed for six months and made extensive herpetological and botanical collections. They also hunted the Galápagos fur seal, which was nearly extinct by then.

1900 (or thereabouts): Sixty Norwegians tried to colonize Charles Island, but were unsuccessful. They scattered to Chatham and Indefatigable islands instead and attempted to grow fruit trees.

*1901: Rollo Beck visited the islands and collected many tortoises, all of which went to Walter Rothschild's private collection.

1905: All but one British navy gunboat (the H.M.S. *Shearwater*) quit their patrols around the islands.

*1905: The California Academy of Sciences sent off a huge collecting and surveying expedition to the Galápagos. Rollo Beck headed the expedition in the *Academy*. After a full year, they returned with the largest collections of specimens ever obtained from there.

1914: Panama Canal opened.

*1923: The Harrison-Williams Galápagos Expedition visited, with William Beebe aboard. He visited again in 1925 on the *Arcturus* Oceanographic Expedition. Beebe wrote a popular account of the visit in *Galapagos—World's End*, which inspired many tourists to visit.

*1924: The Norwegian Zoological Expedition, led by Alf Wollebaek on the *Monsunen*, visited the islands, and several papers resulted. Also, the British scientist, James Hornell visited on the *St. George* and made extensive collections of birds, mammals, and marine life.

*1928–1935: Captain Alan Hancock made several trips to the Galápagos, bringing many students of natural history and evolutionary biology to them. The *Allan Hancock Pacific Expeditions* is a collection of papers resulting from these trips. They were published by the University of Southern California.

*1928, 1931–1932: William K. Vanderbilt collected specimens from the islands and published his results in the *Bulletin of the Vanderbilt Marine Museum* at Huntington, Long Island.

*1929: The Honorable Gifford Pinchot visited in the *Mary Pinchot* and published his results on insects, birds, and fishes in the United States National Museum papers.

*1929: The Cornelius Crane Pacific Expedition from the Field Museum of Natural History in Chicago spent ten days in the islands, with Karl P. Schmidt as head of the expedition. They travelled in the *Illyria*, and Sidney Shurecliffe wrote an account of the voyage in *Jungle Islands—The Illyria in the South Seas*.

*1930: The Vincent Astor Expedition explored mainly Indefatigable Island and brought back several tortoises.

*1931: Henry S. Swarth published his account *Avifauna of the Galapagos Islands*.

*1932: The Templeton-Crocker Expedition spent two months collecting insects, fishes, mollusks, fossils, and birds.

1934: The first airplane came to the Galápagos from the navy in Panama to evacuate a naturalist stricken with appendicitis.

*1936: After visiting the Galápagos and observing the finches, P. R. Lowe declared them a heterogeneous swarm of hybrids rather than distinct species.

*1936 and 1937: The Philadelphia Academy of Natural Sciences sponsored two expeditions: the Dennison Crockett Expedition and the George Vanderbilt South Pacific Expedition. Dillon Ripley published an account of the second expedition as *The Trial of the Money Bird*.

1940s: During World War II, the Americans were given permission to build a naval base on Baltra Island to monitor the Japanese in the Pacific.

*1940: David Lack spent four months observing finches on the Galápagos. He wrote the major work on the finches and established the finches as a model system for studying evolutionary biology.

1946: The United States left the naval base in Baltra. A penal colony was established on Isabela. It was disbanded in 1959.

*1947–1948: The Swedish Deep-Sea Expedition, led by Dr. Hans Petterson, sampled biota from great depths. They also collected plants that were studied by Carl Skottsberg.

*1952–1953: Dr. Bruce W. Halstead of the School of Tropical and Preventive Medicine at Loma Linda, CA, studied poisonous and venomous fishes in the Galápagos waters.

1959: Anniversary of Darwin's *Origin of Species*. The Ecuadorian government declared the Galápagos Islands as its first national park, and the Charles Darwin Foundation and its Charles Darwin Research Station was established.

*1960: R. I. Bowman studied the feeding behavior of the finches in order to challenge Lack's interpretation of the differentiation of the beaks.

*1970–present: Peter and Rosemary Grant began their decades-long studies of the finches. Shortly after, many other evolutionary biologists and ecologists conducted research, hosted by the Charles Darwin Research Station.

Letter to Emperor Carlos V of Spain (1535)

❋ FRAY TOMÁS

* Puerto Viejo, April 26, 1535

Sacred Imperial Catholic Majesty:

It seemed right to me to let Your Majesty know the progress of my trip from the time when I left Panama, which was on the twenty-third of February of the current year, until I arrived in this new town of Puerto Viejo.

The ship sailed with very good breezes for seven days and the pilot kept near land and we had a six-day calm; the currents were so strong and engulfed us in such a way that on Wednesday, the tenth of March, we sighted an island: and, as on board there was enough water for only two more days, they agreed to lower the life-boat and go on land for water and grass for the horses, and once out, they found nothing but seals and turtles, and such big tortoises, that each could carry a man on top of itself, and many iguanas that are like serpents.

On another day, we saw another island larger than the first, and with great sierras; and thinking that on account of its size and monstrous shape, there could not fail to be rivers and fruits, we went to it, because the distance around the first one was about four or five leagues and around the other, ten or twelve leagues, and at this juncture the water on the ship gave out and we were three days in reaching the island on account of the calms, during which all of us, as well as the horses, suffered great hardships.

The boat once anchored, we all went on land, and some were given charge of making a well, and others looking for water over the island: from the well there came out water saltier than that of the sea: on land they were not able to find even a drop of water for two days and with the thirst the people felt they resorted to a leaf of some thistles like prickly pears, and because they were somewhat juicy, although not very tasty, we began to eat of them and squeeze them to draw all the water from them, and drawn, it looked like slops, or lye, and they drank of it as if it were rose water.

On Passion Sunday, I had them bring on land the things necessary for saying Mass, and after it was said. I again sent the people in twos and threes, over different parts. The Lord deigned that they should find in a ravine among the rocks as much as a hogshead of water, and after they had drawn that, they found more and more. In time, eight hogsheads were filled and the barrels and the jugs that there were on the boat, but through the lack of water we lost one man and two days after we left the island we lost another: and ten horses died.

From this island we saw two others, one much larger than all, which was easily fifteen or twenty leagues around; the other was medium; I took the latitude to know where the islands were and they are between half a degree

and a degree and a half south latitude. On this second one, the same conditions prevailed as on the first: many seals, turtles, iguanas, tortoises, many
birds like those of Spain, but so silly they do not know how to flee, and many
were caught in the hand. The other two islands we did not touch; I do not
know their character. On this one, on the sands of the shore, there were small
stones, that we stepped on as we landed, and they were diamond-like stones
and others amber colored; but on this whole island, I do not think that there
is a place where one might sow a bushel of corn, because most of it is full of
very big stones, so much so, that it seems as though some time God had showered stones; and the earth that there is, is like dross, worthless, because it has
not the power of raising a little grass, but only some thistles, the leaf of which I
said we picked. Thinking that we were not more than twenty or thirty leagues
from this soil of Peru, we were satisfied with the water already mentioned,
although we might have filled more of our casks; but we set sail, and with
medium weather we sailed eleven days without sighting land, and the pilot
and the master of the ship came to me to ask me where we were and to tell me
there was only one hogshead of water on the ship. I tried to take the altitude
of the sun that day and found that we were three degrees south latitude, and
I realized that with the direction we were taking, we were becoming more
and more engulfed, that we were not even heading for land, because we were
sailing south; I had them tack on the other side, and the hogshead of water I
had divided as follows: half was given for the animals, and with the other half
a beverage was made which was put into the wine cask, for I held it as certain
that we could not be far from land, and we sailed for eight days, all of which
the hogshead of the beverage lasted, by giving a ration to each one with which
he was satisfied. And when the hogshead gave out and there was no relief for
us, we sighted land and we had calm for two days, during which we drank
only wine, but we took heart on sighting land. We entered the bay and river
of the Caraques on Friday, the ninth April, and we met there the people of the
galleon from Nicaragua, who had left Nicaragua eight months before, so we
considered our trip good in comparison with theirs.

The bay of the Caraques is at half a degree south latitude and on the maps
it is three degrees; from this bay to Puerto Viejo it is nine leagues along the
sea-coast; and the said bay is one of the most beautiful ports that there can be
in the world, and the boats can moor there, and they can sail up it three or four
leagues and they do not know whether any more. Commander Pedro de Alvarado landed here and destroyed a town of Indians that was there and frightened others; and it is a pity to see the havoc wherever he went with his men.

From this bay I landed with the passengers and we set out on foot, because
our animals were worn out from coming to this town from Puerto Viejo, and
walking, we came to a valley which is called Charapoto, which has a very good
river, where there are many Indians now peaceful, because Captain Francisco

Pizarro has behaved so well that he is at peace with about thirty chiefs. This captain and lieutenant governor is so well looked upon by them that they bring him food of corn and fish and venison, and whatever is necessary, and if by chance when they go to see some land they capture some Indians, they immediately return them to their native soil and they give them a cross, so that on account of it no Spaniard will do them any harm, and that any one who wants to come to see it should bring some sign and that way no harm will be done to them. If he learns that any gold or silver is taken from them, he immediately has it returned to them, and some of them have brought it to him and he tells them that he has not come for their gold or their silver, but rather so that they may know God and your Majesty, and that your Majesty will give them masters, who will have charge of teaching them the things of the Holy Catholic Faith, and that, on account of their solicitude, they must undertake to feed them. The keeping of this Captain seems to me very good for the serving of Our Lord and Your Majesty, and for relieving your royal conscience; and since he has a great thing to do, I have told him your Majesty's intention in this matter, and he is determined not to deviate from it very soon.

There are great gold mines, and here I have the information from those who were with Alvarado, that six leagues from this town there are very good gold mines. There is thought to be a bed of emeralds, because the Indians have them in their jewelry; ordinarily the said Indians have their touches and points and some of them of very great qualities. It is thought that before half a year a good part of this land will be peaceful, owing to the good treatment given them by the already mentioned captain and lieutenant; and your Majesty should support him in it, and it is necessary considering the importunities of the men that he has, because they follow with longing eyes every trace of gold that they see.

God willing, I shall leave for the town of San Miguel in four days. The whereabouts of the Governor, Don Francisco Pizarro, is not known at present. He is quite far from here, although some think and they said that he is coming to the Town of Truxillo, which is between San Miguel and Xauxa.

The Lord fill Your Sacred Majesty with holy love and grace for many years and with the conservation of your realms and an increase of other new ones as I hope. From this new town of Puerto Viejo, the twenty-sixth of April, in the year fifteen hundred and thirty-five. I am Your Sacred Imperial Catholic Majesty's most true servant and subject and perpetual Chaplain who kisses your royal feet and hands.

Fray Tomás eps. Locastelli Auril.

The account of Fray Tomás, bishop of Panama, excerpted from Joseph R. Slevin, "The Galapagos Islands: A History of Their Exploration," *Occasional Papers of the California Academy of Science* 25 (1959): 14–16.

From *A New Voyage round the World* (1683–1688)

※ WILLIAM DAMPIER

It was the 31st day of May when we first had sight of the islands Galapagos: some of them appeared on our weather bow, some on our lee bow, others right ahead. . . . in the evening the ship that I was in and Captain Eaton anchored on the east side of one of the eastermost islands, a mile from the shore, in sixteen fathom water, clean, white, hard sand.

The Galapagos Islands are a great number of uninhabited islands lying under and on both sides of the Equator. The eastermost of them are about 110 leagues from the Main. They are laid down in the longitude of 181, reaching to the westward as far as 176, therefore their longitude from England westward is about 68 degrees. But I believe our hydrographers do not place them far enough to the westward. The Spaniards who first discovered them, and in whose charts alone they are laid down, report them to be a great number stretching north-west from the Line, as far as 5 degrees north, but we saw not above 14 or 15. They are some of them 7 or 8 leagues long, and 3 or 4 broad. They are of a good height, most of them flat and even on the top; 4 or 5 of the eastermost are rocky, barren and hilly, producing neither tree, herb, nor grass, but a few dildoe-trees, except by the seaside. The dildoe-tree is a green prickly shrub that grows about 10 or 12 foot high, without either leaf or fruit. It is as big as a man's leg, from the root to the top, and it is full of sharp prickles growing in thick rows from top to bottom; this shrub is fit for no use, not so much as to burn. Close by the sea there grows in some places bushes of burton-wood, which is very good firing. This sort of wood grows in many places in the West Indies, especially in the Bay of Campeachy and the Samballoes. I did never see any in these seas but here. There is water on these barren islands in ponds and holes among the rocks. Some other of these islands are mostly plain and low, and the land more fertile, producing trees of divers sorts unknown to us. Some of the westermost of these islands are nine or ten leagues long and six or seven broad; the mould deep and black. These produce trees of great and tall bodies, especially mammee-trees, which grow here in great groves. In these large islands there are some pretty big rivers; and in many of the other lesser islands there are brooks of good water. The Spaniards when they first discovered these islands found multitudes of iguanas, and land-turtle or tortoise, and named them the Galapagos Islands. I do believe there is no place in the world that is so plentifully stored with those animals. The iguanas here are fat and large as any that I ever saw; they are so tame that a man may knock down twenty in an hour's time with a club. The land-turtle are here so numerous that 5 or 600 men might subsist on them alone for several months without any other sort of provision: they are extraordinary large and fat; and so sweet that no pullet eats more pleasantly. One of the largest of these creatures will weigh 150 or 200 weight, and some of them are 2 foot, or 2 foot 6 inches over the challapee or belly. I did never see any but at this place that will weigh above 30 pound weight. I have heard that at the isle of St. Lawrence or Madagascar, and at the English Forest, an island near it called also Don Mascarin and now possessed by the French, there are very large ones, but whether so big, fat, and sweet as these, I know not. There are 3 or 4 sorts of these creatures in the West Indies. One is called by the Spaniards hecatee; these live most in fresh-water ponds, and seldom come on land. They weigh about 10 or 15 pound; they have small legs and flat feet, and small long necks. Another sort is called tenapen; these are a great deal less than the hecatee; the shell on their backs is all carved naturally, finely wrought, and well clouded: the backs of these are rounder than those before mentioned; they are otherwise much of the same form: these delight to live in wet swampy places, or on the land near such places. Both these sorts are very good meat. They are in great plenty on the isles of Pines near Cuba: there the Spanish hunters when they meet them in the woods bring them home to their huts, and mark them by notching their shells, then let them go; this they do to have them at hand, for they never ramble far from thence. When these hunters return to Cuba, after about a month or six weeks' stay, they carry with them 3 or 400 or more of these creatures to sell; for they are very good meat, and every man knows his own by their marks. These tortoise in the Galapagos are more like the hecatee except that, as I said before, they are much bigger; and they have very long small necks and little heads. There are some green snakes on these islands, but no other land animal that I did ever see. There are great plenty of turtle-doves so tame that a man may kill 5 or 6 dozen in a forenoon with a stick. They are somewhat less than a pigeon, and are very good meat, and commonly fat.

There are good wide channels between these islands fit for ships to pass, and in some places shoal water where there grows plenty of turtle-grass; therefore these islands are plentifully stored with sea-turtle of that sort which is called the green turtle. I have hitherto deferred the description of these creatures therefore I shall give it here.

SEA-TURTLE, THEIR SEVERAL KINDS.

There are 4 sorts of sea-turtle, namely, the trunk-turtle, the loggerhead, the hawksbill, and the green turtle. The trunk-turtle is commonly bigger than the other, their backs are higher and rounder, and their flesh rank and not wholesome. The loggerhead is so called because it has a great head, much bigger than the other sorts; their flesh is likewise very rank, and seldom eaten but in case of necessity: they feed on moss that grows about rocks. The hawksbill-turtle is the least kind, they are so called because their mouths are long and small, somewhat resembling the bill of a hawk: on the backs of these hawks-bill turtle grows that shell which is so much esteemed for making cabinets, combs, and other things. The largest of them may have 3 pound and a half of shell; I have taken some that have had 3 pound 10 ounces: but they commonly have a pound and a half or two pound; some not so much. These are but ordinary food, but generally sweeter than the loggerhead: yet these hawksbills in some places are unwholesome, causing them that eat them to purge and vomit excessively, especially those between the Samballoes and Portobello. We meet with other fish in the West Indies of the same malignant nature: but I shall describe them in the Appendix. These hawksbill-turtles are better or worse according to their feeding. In some places they feed on grass, as the green tortoise also does; in other places they keep among rocks and feed on moss or seaweeds; but these are not so sweet as those that eat grass, neither is their shell so clear; for they are commonly overgrown with barnacles which spoil the shell; and their flesh is commonly yellow, especially the fat.

Hawksbill-turtle are in many places of the West Indies: they have islands and places peculiar to themselves where they lay their eggs, and seldom come among any other turtle. These and all other turtle lay eggs in the sand; their time of laying is in May, June, July. Some begin sooner, some later. They lay 3 times in a season, and at each time 80 or 90 eggs. Their eggs are as big as a hen's egg, and very round, covered only with a white tough skin. There are some bays on the north side of Jamaica where these hawksbills resort to lay. In the Bay of Honduras are islands which they likewise make their breeding-places, and many places along all the coast on the Main of the West Indies from Trinidad de La Vera Cruz in the Bay of Nova Hispania. When a sea-turtle turns out of the sea to lay she is at least an hour before she returns again, for she is to go above high-water mark, and if it be low-water when she comes ashore, she must rest once or twice, being heavy, before she comes to the place where she lays. When she has found a place for her purpose she makes a great hole with her fins in the sand, wherein she lays her eggs, then covers them 2 foot deep with the same sand which she threw out of the hole, and so returns. Sometimes they come up the night before they intend to lay, and take a view of the place, and so having made a tour, or semicircular march, they return to the sea again, and they never fail to come ashore the next night to lay near that place. All sorts of turtle use the same methods in laying. I knew a man in Jamaica that made 8 pound Sterling of the shell of these hawksbill turtle which he got in one season and in one small bay, not half a mile long. The manner of taking them is to watch the bay by walking from one part to the other all night, making no noise, nor keeping any sort of light. When the turtle comes ashore the man that watches for them turns them on their backs, then hauls them above high-water mark, and leaves them till the morning. A large green turtle, with her weight and struggling, will puzzle 2 men to turn her. The hawksbill-turtle are not only found in the West Indies but on the coast of Guinea, and in the East Indies. I never saw any in the South Seas.

The green turtle are so called because their shell is greener than any other. It is very thin and clear and better clouded than the hawksbill; but it is used only for inlays, being extraordinary thin. These turtles are generally larger than the hawksbill; one will weigh 2 or 3 hundred pound. Their backs are flatter than the hawksbill, their heads round and small. Green turtle are the sweetest of all the kinds: but there are degrees of them both in respect to their flesh and their bigness. I have observed that at Blanco in the West Indies the green turtle (which is the only kind there) are larger than any other in the North Seas. There they will commonly weigh 280 or 300 pound: their fat is yellow, and the lean white, and their flesh extraordinary sweet. At Boca Toro, west of Portobello, they are not so large, their flesh not so white, nor the fat so yellow. Those in the Bay of Honduras and Campeachy are somewhat smaller still; their fat is green, and the lean of a darker colour than those at Boca Toro. I heard of a monstrous green turtle once taken at Port Royal in the Bay of Campeachy that was four foot deep from the back to the belly, and the belly six foot broad; Captain Roch's son, of about nine or ten years of age, went in it as in a boat on board his father's ship, about a quarter of a mile from the shore. The leaves of fat afforded eight gallons of oil. The turtle that live among the keys or small islands on the south side of Cuba are a mixed sort, some bigger, some less; and so their flesh is of a mixed colour, some green, some dark, some yellowish. With these Port Royal in Jamaica is constantly supplied by sloops that come hither with nets to take them. They carry them alive to Jamaica where the turtles have wires made with stakes in the sea to preserve them alive; and the market is every day plentifully stored with turtle, it being the common food there, chiefly for the ordinary sort of people.

Green turtle live on grass which grows in the sea in 3, 4, 5, or 6 fathom water, at most of the places before mentioned. This grass is different from manatee-grass, for that is a small blade; but this a quarter of an inch broad and six inches long. The turtle of these islands Galapagos are a sort of a bastard green turtle; for their shell is thicker than other green turtle in the West or East Indies, and their flesh is not so sweet. They are larger than any other green turtle; for it is common for these to be two or three foot deep, and their callapees or bellies five foot wide: but there are other green turtle in the South Seas that are not so big as the smallest hawksbill. These are seen at the island Plata, and other places thereabouts: they feed on moss and are very rank but fat.

Both these sorts are different from any others, for both he's and she's come ashore in the daytime and lie in the sun; but in other places none but the she's go ashore, and that in the night only to lay their eggs. The best feeding for turtle in the South Seas is among these Galapagos Islands, for here is plenty of grass.

There is another sort of green turtle in the South Seas which are but small, yet pretty sweet: these lie westward on the coast of Mexico. One thing is very strange and remarkable in these creatures; that at the breeding time they leave for two or three months their common haunts, where they feed most of the year, and resort to other places only to lay their eggs: and it is not thought that they eat anything during this season: so that both he's and she's grow very lean; but the he's to that degree that none will eat them. The most remarkable places that I did ever hear of for their breeding is at an island in the West Indies called Caymans, and the isle Ascension in the Western Ocean: and when the breeding time is past there are none remaining. Doubtless they swim some hundreds of leagues to come to those two places: for it has been often observed that at Cayman, at the breeding time, there are found all those sort of turtle before described. The South Keys of Cuba are above 40 leagues from thence, which is the nearest place that these creatures can come from; and it is most certain that there could not live so many there as come here in one season.

Those that go to lay at Ascension must needs travel much farther; for there is no land nearer it than 300 leagues: and it is certain that these creatures live always near the shore. In the South Sea likewise the Galapagos is the place where they live the biggest part of the year; yet they go from thence at their season over to the Main to lay their eggs; which is 100 leagues the nearest place. Although multitudes of these turtles go from their common places of feeding and abode to those laying-places, yet they do not all go: and at the time when the turtle resort to these places to lay their eggs they are accompanied with abundance of fish, especially sharks; the places which the turtle then leave being at that time destitute of fish, which follow the turtle.

When the she's go thus to their places to lay the male accompany them, and never leave them till they return: both male and female are fat the beginning of the season; but before they return the male, as I said, are so lean that they are not fit to eat, but the female are good to the very last; yet not so fat as at the beginning of the season. It is reported of these creatures that they are nine days engendering, and in the water, the male on the female's back. It is observable that the male, while engendering, do not easily forsake their female: for I have gone and taken hold of the male when engendering: and a very bad striker may strike them then, for the male is not shy at all: but the female, seeing a boat when they rise to blow, would make her escape, but that the male grasps her with his two fore fins, and holds her fast. When they are thus coupled it is best to strike the female first, then you are sure of the male also. These creatures are thought to live to a great age; and it is observed by the Jamaica turtlers that they are many years before they come to their full growth....

As soon as we came to an anchor, we made a tent ashore for Captain Cook who was sick. Here we found the sea-turtle lying ashore on the sand; this is not customary in the West Indies. We turned them on their backs that they might not get away. The next day more came up, when we found it to be their custom to lie in the sun: so we never took care to turn them afterwards; but sent ashore the cook every morning, who killed as many as served for the day. This custom we observed all the time we lay here, feeding sometimes on land-turtle, sometimes on sea-turtle, there being plenty of either sort. Captain Davis came hither again a second time; and then he went to other islands on the west side of these. There he found such plenty of land-turtle that he and his men ate nothing else for three months that he stayed there. They were so fat that he saved sixty jars of oil out of those that he spent: this oil served instead of butter to eat with doughboys or dumplings, in his return out of these seas. He found very convenient places to careen, and good channels between the islands; and very good anchoring in many places. There he found also plenty of brooks of good fresh water, and firewood enough, there being plenty of trees fit for many uses. Captain Harris, one that we shall speak of hereafter, came thither likewise, and found some islands that had plenty of mammee-trees, and pretty large rivers. The sea about these islands is plentifully stored with fish such as are at Juan Fernandez. They are both large and fat and as plentiful here as at Juan Fernandez. Here are particularly abundance of sharks. The north part of this second isle we anchored at lies 28 minutes north of the Equator. I took the height of the sun with an astrolabe. These isles of the Galapagos have plenty of salt. We stayed here but 12 days in which time we put ashore 5000 packs of flour for a reserve if we should have occasion of any before we left these seas. Here one of our Indian prisoners informed us that he was born at Realejo, and that he would engage to carry us thither. He being examined of the strength and riches of it satisfied the company so well that they were resolved to go thither.

Having thus concluded; the 12th of June we sailed from hence, designing to touch at the island Cocos, as well to put ashore some flour there as to see the island, because it was in our way to Ria Lexa.

Sixth edition. From *Journals of Voyages Conducted 1683–1688*, reprinted in *The Voyages of Captain William Dampier* (London: E. Grant Richards, 1906), 75–82.

From "The Encantadas" (1856)

❊ HERMAN MELVILLE

THE ENCANTADAS OR ENCHANTED ISLES

Sketch First

THE ISLES AT LARGE
That may not be, said then the ferryman,
Least we unweeting hap to be fordonne;

> For those same islands seeming now and than,
> Are not firme land, nor any certein wonne,
> But stragling plots which to and fro do ronne
> In the wide waters; therefore are they hight
> The Wandering Islands: Therefore do them shonne;
> For they have oft drawne many a wandring wight
> Into most deadly daunger and distressed plight;
> For whosoever once hath fastened
> His foot thereon, may never it recure,
> But wandreth evermore uncertain and unsure.
>
> Darke, dolefull, dreary, like a greedy grave,
> That still for carrion carcases doth crave:
> On top whereof ay dwelt the ghastly owle,
> Shrieking his balefull note, which ever drave
> Far from that haunt all other chearefull fowlle,
> And all about it wandring ghostes did wayle and howle.

Take five-and-twenty heaps of cinders dumped here and there in an outside city lot, imagine some of them magnified into mountains, and the vacant lot the sea; and you will have a fit idea of the general aspect of the Encantadas, or Enchanted Isles. A group rather of extinct volcanoes than of isles; looking much as the world at large might, after a penal conflagration.

It is to be doubted whether any spot on earth can, in desolateness, furnish a parallel to this group. Abandoned cemeteries of long ago, old cities by piecemeal tumbling to their ruin, these are melancholy enough; but, like all else which has but once been associated with humanity, they still awaken in us some thoughts of sympathy, however sad. Hence, even the Dead Sea, along with whatever other emotions it may at times inspire, does not fail to touch in the pilgrim some of his less unpleasurable feelings.

And as for solitariness; the great forests of the north, the expanses of unnavigated waters, the Greenland ice-fields, are the profoundest of solitudes to a human observer; still the magic of their changeable tides and seasons mitigates their terror; because, though unvisited by men, those forests are visited by the May; the remotest seas reflect familiar stars even as Lake Erie does; and in the clear air of a fine Polar day, the irradiated, azure ice shows beautifully as malachite.

But the special curse, as one may call it, of the Encantadas, that which exalts them in desolation above Idumea and the Pole, is, that to them change never comes; neither the change of seasons nor of sorrows. Cut by the Equator, they know not autumn, and they know not spring; while, already reduced to the lees of fire, ruin itself can work little more upon them. The showers refresh the deserts; but in these isles, rain never falls. Like split Syrian gourds left withering in the sun, they are cracked by an everlasting drought beneath a torrid sky. "Have mercy upon me," the wailing spirit of the Encantadas seems to cry, "and send Lazarus that he may dip the tip of his finger in water and cool my tongue, for I am tormented in this flame."

Another feature in these isles is their emphatic uninhabitableness. It is deemed a fit type of all-forsaken overthrow that the jackal should den in the wastes of weedy Babylon; but the Encantadas refuse to harbor even the outcasts of the beasts. Man and wolf alike disown them. Little but reptile life is here found: tortoises, lizards, immense spiders, snakes, and that strangest anomaly of outlandish nature, the *iguana*. No voice, no low, no howl is heard; the chief sound of life here is a hiss.

On most of the isles where vegetation is found at all, it is more ungrateful than the blankness of Atacama. Tangled thickets of wiry bushes, without fruit and without a name, springing up among deep fissures of calcined rock and treacherously masking them; or a parched growth of distorted cactus trees.

In many places the coast is rock-bound, or, more properly, clinker-bound; tumbled masses of blackish or greenish stuff like the dross of an iron-furnace, forming dark clefts and caves here and there, into which a ceaseless sea pours a fury of foam, overhanging them with a swirl of gray, haggard mist, amidst which sail screaming flights of unearthly birds heightening the dismal din. However calm the sea without, there is no rest for these swells and those rocks; they lash and are lashed, even when the outer ocean is most at peace with itself. On the oppressive, clouded days, such as are peculiar to this part of the watery Equator, the dark, vitrified masses, many of which raise themselves among white whirlpools and breakers in detached and perilous places off the shore, present a most Plutonian sight. In no world but a fallen one could such lands exist.

Those parts of the strand free from the marks of fire, stretch away in wide level beaches of multitudinous dead shells, with here and there decayed bits of sugar-cane, bamboos, and cocoa-nuts, washed upon this other and darker world from the charming palm isles to the westward and southward; all the way from Paradise to Tartarus; while mixed with the relics of distant beauty you will sometimes see fragments of charred wood and moldering ribs of wrecks. Neither will anyone be surprised at meeting these last, after observing the conflicting currents which eddy throughout nearly all the wide channels of the entire group. The capriciousness of the tides of air sympathizes with those of the sea. Nowhere is the wind so light, baffling, and every way unreliable, and so given to perplexing calms, as at the Encantadas. Nigh a month has been spent by a ship going from one isle to another, though but ninety miles between; for owing to the force of the current, the boats employed to tow barely suffice to keep the craft from sweeping upon the cliffs, but do nothing towards accelerating her voyage. Sometimes it is impossible for a vessel from afar to fetch up with the group itself, unless large allowances for prospective lee-way have been made ere its coming in sight. And yet, at other times, there is a mysterious indraft, which irresistibly draws a passing vessel among the isles, though not bound to them.

True, at one period, as to some extent at the present day, large fleets of whalemen cruised for spermaceti upon what some seamen call the Enchanted Ground. But this, as in due place will be described, was off the great outer isle of Albemarle, away from the intricacies of the smaller isles, where there is plenty of sea-room; and hence, to that vicinity, the above remarks do not altogether apply; though even there the current runs at times with singular force, shifting, too, with as singular a caprice.

Indeed, there are seasons when currents quite unaccountable prevail for a great distance round about the total group, and are so strong and irregular as to change a vessel's course against the helm, though sailing at the rate of four or five miles the hour. The difference in the reckonings of navigators produced by these causes, along with the light and variable winds, long nourished a persuasion, that there existed two distinct clusters of isles in the parallel of the Encantadas, about a hundred leagues apart. Such was the idea of their earlier visitors, the Buccaneers; and as late as 1750, the charts of that part of the Pacific accorded with the strange delusion. And this apparent fleetingness and unreality of the locality of the isles was most probably one reason for the Spaniards calling them the Encantada, or Enchanted Group.

But not uninfluenced by their character, as they now confessedly exist, the modern voyager will be inclined to fancy that the bestowal of this name might have in part originated in that air of spell-bound desertness which so significantly invests the isles. Nothing can better suggest the aspect of once living things malignly crumbled from ruddiness into ashes. Apples of Sodom, after touching, seem these isles.

However wavering their place may seem by reason of the currents, they themselves, at least to one upon the shore, appear invariably the same: fixed, cast, glued into the very body of cadaverous death.

Nor would the appellation, enchanted, seem misapplied in still another sense. For concerning the peculiar reptile inhabitant of these wilds—whose presence gives the group its second Spanish name, Galapagos—concerning the tortoises found here, most mariners have long cherished a superstition not more frightful than grotesque. They earnestly believe that all wicked sea-officers, more especially commodores and captains, are at death (and, in some cases, before death) transformed into tortoises; thenceforth dwelling upon these hot aridities, sole solitary lords of Asphaltum.

Doubtless, so quaintly dolorous a thought was originally inspired by the woe-begone landscape itself; but more particularly, perhaps, by the tortoises. For, apart from their strictly physical features, there is something strangely self-condemned in the appearance of these creatures. Lasting sorrow and penal hopelessness are in no animal form so suppliantly expressed as in theirs; while the thought of their wonderful longevity does not fail to enhance the impression.

Nor even at the risk of meriting the charge of absurdly believing in enchantments can I restrain the admission that sometimes, even now, when leaving the crowded city to wander out July and August among the Adirondack Mountains, far from the influences of towns and proportionally nigh to the mysterious ones of nature; when at such times I sit me down in the mossy head of some deep-wooded gorge, surrounded by prostrate trunks of blasted pines, and recall, as in a dream, my other and far-distant rovings in the baked heart of the charmed isles, and remember the sudden glimpses of dusky shells, and long languid necks protruded from the leafless thickets; and again have beheld the vitreous inland rocks worn down and grooved into deep ruts by ages and ages of the slow draggings of tortoises in quest of pools of scanty water; I can hardly resist the feeling that in my time I have indeed slept upon evilly enchanted ground.

Nay, such is the vividness of my memory, or the magic of my fancy, that I know not whether I am not the occasional victim of optical delusion concerning the Galapagos. For, often in scenes of social merriment, and especially at revels held by candle-light in old-fashioned mansions, so that shadows are thrown into the further recesses of an angular and spacious room, making them put on a look of haunted undergrowth of lonely woods, I have drawn the attention of my comrades by my fixed gaze and sudden change of air, as I have seemed to see, slowly emerging from those imagined solitudes, and heavily crawling along the floor, the ghost of a gigantic tortoise, with "Memento * * * * *" burning in live letters upon his back.

From *Great Short Works of Herman Melville* (New York: Harper and Row Publishers, 1968), 98–103.

From *Innocence in the Galapagos: Voyage to a Time Machine in the Pacific* (1975)

✳ ANNIE DILLARD

I

First there was nothing, and although you know with your reason that nothing is nothing, it is easier to visualize it as a limitless slosh of sea—say, the Pacific. Then energy contracted into matter, and although you know that even an invisible gas is matter, it is easier to visualize it as a massive squeeze of volcanic lava spattered inchoate from the secret pit of the ocean and hardening mute and intractable on nothing's lapping shore—like a series of islands, an archipelago. Like: the Galápagos. Then a softer strain of matter began to twitch. It was a kind of shaped water; it flowed, hardening here and there at its tips. There were blue-green algae; there were tortoises; there were men.

The ice rolled up, the ice rolled back, and I knelt on a plain of lava boulders in the islands called Galápagos, stroking a giant tortoise's neck. The tortoise closed its eyes and stretched its neck to its greatest height and vulnerability. I rubbed that neck, and when I pulled away my hand, my palm was green with a slick of single-celled algae. I stared at the algae, and at the tortoise, the way you stare at any life on a lava flow, and thought, Well—here we all are.

Being here is being here on the rocks. These Galapagonian rocks, one of them seventy-five miles long, dried under the equatorial sun between 500 and 600 miles west of the South American continent, at the latitude of the Republic of Ecuador, to which they belong.

There is a way a small island rises from the ocean affronting all reason. It is a chunk of chaos pounded into visibility *ex nihilo*, here rough, here smooth, shaped just so by a matrix of physical necessities too weird to contemplate, here instead of there, here instead of not at all. It is a fantastic utterance, as though I were to open my mouth and emit a French horn, or a vase, or a knob of tellurium. It smacks of folly, of first causes.

I think of the island called Daphnecita, little Daphne, on which I never set foot. It's in half of my few photographs, though, because it obsessed me: a dome of gray lava like a pitted loaf, the size of the Plaza Hotel, glazed with guano and crawling with red-orange crabs. Sometimes I attributed to this island's cliff face a surly, infantile consciousness, as though it were sulking in the silent moment after it had just shouted, to the sea and the sky, I didn't ask to be born. Or sometimes it aged to a raging adolescent, a kid who's just learned that the game is fixed, demanding, What did you have me for, if you're just going to push me around? Daphnecita: again, a wise old island, mute, leading the life of pure creaturehood open to any antelope or saint. After you've blown the ocean sky-high, what's there to say? What if we the people had sense or grace to live as cooled islands in an archipelago live, with dignity, passion, and no comment?

A Bosch Landscape

It is worth flying to Guayaquil, Ecuador, and then to Baltra in the Galápagos to see only the rocks. But these rocks are animal gardens. They are home to a Hieronymus Bosch assortment of windblown, stowaway, castaway, flotsam and shipwrecked creatures. Most exist nowhere else on earth. These reptiles and insects, small mammals and birds evolved unmolested on the various islands on which they were cast into unique species adapted to the boulder-wrecked shores, the cactus deserts of the lowlands, or the elevated jungles of the large island's interiors. You come for the animals. You come to see the curious shapes soft proteins can take, to impress yourself with their reality, and to greet them.

You walk among clattering four-foot marine iguanas heaped on the shore lava, and on each other, like slag. You swim with penguins; you watch flight-less cormorants dance beside you, ignoring you, waving the black nubs of their useless wings. Here are nesting blue-footed boobies, real birds with real feathers, whose legs and feet are nevertheless patently fake, manufactured by Mattel. The tortoises you touch are as big as stoves. The enormous land iguanas at your feet change color in the sunlight, from gold to blotchy red as you watch.

There is always some creature going about its beautiful business. I missed the boat back to my ship, and was left behind momentarily on uninhabited South Plaza Island because I was watching the Audubon's shearwaters. These dark pelagic birds flick along pleated seas in stitching flocks, flailing their wings rapidly—because if they don't, they'll stall. A shearwater must fly fast, or not at all. Consequently it has evolved two nice behaviors which serve to bring it into its nest alive. The nest is a shearwater-sized hole in the lava cliff. The shearwater circles over the water, ranging out from the nest a quarter of a mile, and veers gradually toward the cliff, making passes at its nest. If the flight angle is precisely right, the bird will fold its wings at the hole's entrance and stall directly onto its floor. The angle is perhaps seldom right, however; one shearwater I watched made a dozen suicidal looking passes before it van-ished into a chink. The other, alternative, behavior is spectacular. It involves choosing the nest hole in a site below a prominent rock with a downward-angled face. The shearwater comes careering in full tilt, claps its wings, stalls itself into the rock, and the rock, acting as a backboard, banks it home.

The animals are tame. They have not been persecuted, and show no fear of man. You pass among them as though you were wind, spindrift, sunlight, leaves. The songbirds are tame. On Hood Island I sat beside a nesting waved albatross while a mockingbird scratched in my hair, another mockingbird jabbed at my fingernail, and a third mockingbird made an exquisite progres-sion of pokes at my bare feet up the long series of eyelets in my basketball shoes. The marine iguanas are tame. One settler, Carl Angermeyer, built his house on the site of a marine iguana colony. The gray iguanas, instead of moving out, moved up on the roof, which is corrugated steel. Twice daily on the patio, Angermeyer feeds them a mixture of boiled rice and tuna fish from a plastic basin. Their names are all, unaccountably, Annie. Angermeyer beats on the basin with a long-handled spoon, calling "Here AnnieAnnie AnnieAnnie"—and the spiny reptiles, fifty or sixty strong, click along the steel roof, finger their way down the lava boulder and mortar walls, and swarm round his bare legs to elbow into the basin and be elbowed out again smeared with a mash of boiled rice on their bellies and on their protuberant, black, plated lips.

The wild hawk is tame. The Galápagos hawk is related to North America's Swainson's hawk; I have read that, if you take pains, you can walk up and pat it. I never tried. We people don't walk up and pat each other; enough is enough. The animals' critical distance and mine tended to coincide, so we could enjoy an easy sociability without threat of violence or unwonted intimacy. The hawk, which is not notably sociable, nevertheless endures even a blundering approach, and is apparently as content to perch on a scrub tree at your shoulder as anyplace else.

In the Galápagos, even the flies are tame. Although most of the land is Ecuadorian national park, and as such rigidly protected, I confess I gave the evolutionary ball an offsides shove by dispatching every fly that bit me, marveling the while at its pristine ignorance, its blithe failure to register a flight trigger at the sweep of my descending hand—an insouciance that was almost, but not quite, disarming. After you kill a fly, you pick it up and feed it to a lava lizard, a bright-throated four-inch creature that scavenges everywhere in the arid lowlands. And you walk on, passing among the innocent mobs on every rock hillside; or you sit, and they come to you.

We are strangers and sojourners, soft dots on the rocks. You have walked along the strand and seen where birds have landed, walked, and flown; their tracks begin in sand, and go, and suddenly end. Our tracks do that: but we go down. And stay down. While we're here, during the seasons our tents are pitched in the light, we pass among each other crying "greetings" in a thousand tongues, and "welcome," and "goodbye." Inhabitants of uncrowded colonies tend to offer the stranger famously warm hospitality—and such are the Galápagos sea lions. Theirs is the greeting the first creatures must have given Adam—a hero's welcome, a universal and undeserved huzzah. Go, and be greeted by sea lions.

I was sitting under a ledge of pewter cloud with Soames Summerhays, the ship's naturalist, on a sand beach under cliffs on uninhabited Hood Island. The white beach was a havoc of lava boulders black as clinkers, sleek with spray, and lambent as brass in the sinking sun. To our left a dozen sea lions were body-surfing in the long green combers that rose, translucent, half a mile offshore. When the combers broke, the shoreline boulders rolled. I could feel the roar in the rough rock on which I sat; I could hear the grate inside each long backsweeping sea, the rumble of a rolled million rocks muffled in splashes and the seethe before the next wave's heave.

To our right, a sea lion slipped from the ocean. It was a young bull; in another few years he would be more dangerous, bellowing at intruders and biting off great dirty chunks of the ones he catches. Now this young bull, which weighed maybe 120 pounds, sprawled silhouetted in the late light, slick as a drop of quicksilver, his glistening whiskers radii of gold like any crown. He hauled his packed bulk toward us up the long beach; he flung himself with an enormous surge of fur-clad muscle onto the boulder where I sat. "Soames," I said—very quietly—"he's here because *we're* here, isn't he?" The naturalist nodded. I felt water drip on my elbow behind me, then the fragile scrape of whiskers, and finally the wet warmth and weight of a muzzle, as the creature settled to sleep on my arm. I was catching on to sea lions.

Walk into the water. Instantly sea lions surround you, even if none has been in sight. To say that they come to play with you is not especially anthropomorphic. Animals play. The bull sea lions are off patrolling their territorial shores; these are the cows and young, which range freely. A five-foot sea lion peers intently into your face, then urges her muzzle gently against your underwater mask and searches your eyes without blinking. Next she rolls upside down and slides along the length of your floating body, rolls again, and casts a long glance back at your eyes. You are, I believe, supposed to follow, and think up something clever in return. You can play games with sea lions in the water using shells, or bits of leaf, if you are willing. You can spin on your vertical axis, and a sea lion will swim circles around you, keeping his face always six inches from yours, as though he were tethered. You can make a game of touching their back flippers, say, and the sea lions will understand at once; somersaulting conveniently before your clumsy hands, they will give you an excellent field of back flippers.

And when you leave the water, they follow. They porpoise to the shore, popping their heads up when they lose you and casting about, then speeding to your side and emitting a choked series of vocal notes. If you won't relent, they disappear, barking; but if you sit on the beach with so much as a foot in the water, two or three will station with you, floating on their backs and saying, Urr.

Pioneers

Few people come to the Galápagos. Buccaneers used to anchor in the bays to avoid pursuit, to rest, and to lighter on fresh water. The world's whaling ships stopped here as well, to glut their holds with fresh meat in the form of giant tortoises. The whalers used to let the tortoises bang around on deck for a few days to empty their guts; then they stacked them below on their backs to live—if you call that living—without food or water for a year.

Early inhabitants in the islands were a desiccated assortment of grouches, cranks, and ship's deserters. These hardies shot, enslaved, and poisoned each other off, leaving behind a fecund gang of feral goats, cats, dogs, and pigs whose descendants skulk in the sloping jungles and take their tortoise hatchlings neat. Now scientists at the Charles Darwin Research Station, on the island of Santa Cruz, rear the tortoise hatchlings for several years until their shells are tough enough to resist the crunch; then they release them in the wilds of their respective islands. Some few thousand people live on three of the islands; settlers from Ecuador, Norway, Germany, and France make a livestock or pineapple living from the rich volcanic soils. The settlers themselves seem to embody a high degree of courteous and conscious humanity, perhaps because of their relative isolation.

On the island of Santa Cruz, eleven fellow passengers and I climb into an open truck and bump for an hour up the Galápagos's only long road to visit Alf Kastdalen. Where the road's ascent ends, native villagers leave their muddy soccer game to provide horses, burros, and mules which bear us lurching up a jungle path to a mountain clearing, to the isolate Kastdalen farm. Alf Kastdalen came to the islands as a child with his immigrant parents from Norway. Now a broad, blond man in his late forties, he lives with his mother and his Ecuadorian wife and their children in a solitary house of finished timbers imported from the mainland, on 400 acres he claimed from the jungle by hand. He raises cattle. He walks us round part of his farm, smiling expansively and meeting our chatter with a willing, open gaze and kind words. The pasture looks like any pasture—but the rocks under the grass are round lava, the copses are a tangle of thorny bamboo and bromeliads, and the bordering trees dripping in epiphytes are breadfruit, papaya, avocado, and orange.

Kastdalen's house is heaped with books in three languages. He knows animal husbandry; he also knows botany and zoology. He feeds us soup, chicken worth chewing for, green naranjilla juice, noodles, pork in big chunks, marinated mixed vegetables, rice, and bowl after bowl of bright mixed fruits.

And his white-haired Norwegian mother sees us off; our beasts are ready. We will ride down the mud forest track to the truck at the Ecuadorian settlement, down the long, long road to the boat, and across the bay to the ship. I lean down to catch her words. She is gazing at me with enormous warmth. "Your hair," she says softly. I am blonde. Adios.

II

Charles Darwin came to the Galápagos in 1835, on the *Beagle*; he was twenty-six. He threw the marine iguanas as far as he could into the water; he rode the tortoises and sampled their meat. He noticed that the tortoise's carapaces varied wildly from island to island; so also did the forms of various mockingbirds. He made collections. Nine years later he wrote in a letter, "I am almost convinced (quite contrary to the opinion I started with) that species are not (it is like confessing a murder) immutable." In 1859 he published *On the Origin of Species*, and in 1871 *The Descent of Man*. It is fashionable now to disparage Darwin's originality; not even the surliest of his detractors, however, faults his painstaking methods or denies his impact.

Darwinism today is more properly called neo-Darwinism. It is organic evolutionary theory informed by the spate of new data from modern genetics, molecular biology, paleobiology—from the new wave of the biologic revolution which spread after Darwin's announcement like a tsunami. The data are not all in. Crucial first appearances of major invertebrate groups are missing from the fossil record—but these early forms, sometimes modified larvae, tended to be fragile either by virtue of their actual malleability or by virtue of their scarcity and rapid variation into "hardened," successful forms. Lack of proof in this direction doesn't worry scientists. What neo-Darwinism seriously lacks, however, is a precise description of the actual mechanism of mutation in the chromosomal nucleotides.

In the larger sense, neo-Darwinism also lacks, for many, sheer plausibility. The triplet splendors of random mutation, natural selection, and Mendelian inheritance are neither energies nor gods; the words merely describe a gibbering tumult of materials. Many things are unexplained, many discrepancies unaccounted for. Appending a very modified neo-Lamarckism to Darwinism would solve many problems—and create new ones. Neo-Lamarckism holds, without any proof, that certain useful acquired characteristics may be inherited. Read C. H. Waddington, *The Strategy of the Genes*, and Arthur Koestler, *The Ghost in the Machine*. The Lamarckism/Darwinism issue is not only complex, hinging perhaps on whether DNA can be copied from RNA, but also emotionally and politically hot. The upshot of it all is that while a form of Lamarckism recently held sway in Russia, neo-Darwinism is supreme in the West, and its basic assumptions, though variously modified, are not overthrown.

Fundamentalist Christians, of course, still reject Darwinism because it conflicts with the creation account in Genesis. Fundamentalist Christians have a very bad press. Ill-feeling surfaces when, from time to time in small Southern towns they object again to the public schools' teaching evolutionary theory. Tragically, these people feel they have to make a choice between the Bible and modern science. They live and work in the same world we do, and know the derision they face from people whose areas of ignorance are perhaps different, who dismantled their mangers when they moved to town and threw out the baby with the straw.

Even less appealing in their response to the new evolutionary picture were, and are, the social Darwinists. Social Darwinists seized Herbert Spencer's phrase "the survival of the fittest," applied it to capitalism, and used it to sanction ruthless and corrupt business practices. A social Darwinist is unlikely to identify himself by the term; social Darwinism is, as the saying goes, not a religion but a way of life. A modern social Darwinist wrote the slogan "If you're so smart, why ain't you rich?" The notion still obtains, I believe, wherever people seek power: that the race is to the swift, that everybody is *in* the race, with varying and merited degrees of success or failure, and that reward is its own virtue.

Philosophy reacted to Darwin with unaccustomed good cheer. William Paley's fixed and harmonious universe was gone, and with it its meticulous watchmaker god. Nobody mourned. Instead, philosophy shrugged and turned its attention from first and final causes to analysis of certain values here in time. "Faith in progress," the man-in-the-street philosophy, collapsed in two world wars. Philosophers were more guarded; pragmatically, they held a very refined "faith in process"—which, it would seem, could hardly lose. Christian thinkers, too, outside of Fundamentalism, examined with fresh eyes the world's burgeoning change. Some Protestants, taking their cue from Whitehead, posited a dynamic god who lives alongside the universe, himself charged and changed by the process of becoming. The Catholic Pierre Teilhard de Chardin, a paleontologist, examined the evolution of species itself, and discovered in that flow a surge toward complexity and consciousness, a free ascent capped with man and propelled from within and attracted from without by god, the holy freedom and awareness that is creation's beginning and end. And so forth. Like tortoises, like languages, ideas evolve. And they evolve, as Arthur Koestler suggests, not from hardened final forms, but from the softest plasmic germs in a cell's heart, in the nub of a word's root, in the supple flux of an open mind.

Darwin gave us time. Before Darwin (and Huxley, Wallace, et al.) there was in the nineteenth century what must have been a fairly unsettling period in which people knew about fossils of extinct species, but did not yet know about organic evolution. They thought the fossils were litter from a series of past creations. At any rate, for many, this creation, the world as we know it, had begun in 4004 B.C., a date set by the Irish Archbishop James Ussher in the seventeenth century. We were all crouched in a small room against the comforting back wall, awaiting the millennium which had been gathering impetus since Adam and Eve. Up there was a universe, and down here would be a small strip of man come and gone, created, taught, redeemed, and gathered up in a bright twinkling, like a sprinkling of confetti torn from colored papers, tossed from windows, and swept from the streets by morning.

The Darwinian revolution knocked out the back wall, revealing eerie lighted landscapes as far back as we can see. Almost at once, Albert Einstein and astronomers with reflector telescopes and radio telescopes knocked out the other walls and the ceiling, leaving us sunlit, exposed, and drifting—leaving us puckers, albeit evolving puckers, on the inbound curve of space-time.

III

It all began in the Galápagos, with these finches. The finches in the Galápagos are called Darwin's finches; they are everywhere in the islands, sparrowlike, and almost identical but for their differing beaks. At first Darwin scarcely noticed their importance. But by 1839, when he revised his journal of the *Beagle* voyage, he added a crucial sentence about the finches' beaks: "Seeing this gradation and diversity of structure in one small, intimately related group of birds, one might really fancy that from an original paucity of birds in this archipelago, one species had been taken and modified for different ends." And so it was.

The finches come when called. I don't know why it works, but it does. Scientists in the Galápagos have passed down the call: you say psssssh psssssh psssssh psssssh psssssh until you run out of breath; then you say it again until the island runs out of birds. You stand on a flat of sand by a shallow lagoon rimmed in mangrove thickets and call the birds right out of the sky. It works anywhere, from island to island.

Once, on the island of James, I was standing propped against a leafless palo santo tree on a semiarid inland slope when the naturalist called the birds.

From other leafless palo santo trees flew the yellow warblers, speckling the air with bright bounced sun. Gray mockingbirds came running. And from the green prickly pear cactus, from the thorny acacias, sere grasses, bracken, and manzanilla, from the loose black lava, the bare dust, the fern-hung mouths of caverns or the tops of sunlit logs—came the finches. They fell in from every direction like colored bits in a turning kaleidoscope. They circled and homed to a vortex, like a whirlwind of chips, like draining water. The tree on which I leaned was the vortex. A dry series of puffs hit my cheeks. Then a rough pulse from the tree's thin trunk met my hand and rang up my arm—and another, and another. The tree trunk agitated against my palm like a captured cricket: I looked up. The lighting birds were rocking the tree. It was an appearing act: before there were barren branches; now there were birds like leaves.

Darwin's finches are not brightly colored; they are black, gray, brown, or faintly olive. Their names are even duller: the large ground finch, the medium ground finch, the small ground finch; the large insectivorous tree finch; the vegetarian tree finch; the cactus ground finch, and so forth. But the beaks are interesting, and the beaks' origins even more so.

Some wield chunky parrot beaks modified for cracking seeds. Some have slender warbler beaks, short for nabbing insects, long for probing plants. One sports the long chisel beak of a woodpecker; it bores wood for insect grubs and often uses a twig or cactus spine, like a pickle fork, when the grub won't dislodge. They have all evolved, fanwise, from one ancestral population. The finches evolved in isolation. So did everything else on earth. With the finches, you can see how it happened. The Galápagos Islands are near enough to the mainland that some strays could hazard there; they are far enough away that those strays could evolve in isolation from parent species. And the separate islands are near enough to each other for further dispersal, further isolation, and the eventual reassembling of distinct species. (In other words, finches blew to the Galápagos, blew to various islands, evolved into differing species, and blew back together again.) The tree finches and the ground finches, the woodpecker finch and the warbler finch, veered into being on isolated rocks. The witless green sea shaped those beaks as surely as it shaped the beaches. Now on the finches in the palo santo tree you see adaptive radiation's results, a fluorescent splay of horn. It is as though an archipelago were an arpeggio, a rapid series of distinct but related notes. If the Galápagos had been one unified island, there would be one dull note, one super-dull finch.

IV

Now let me carry matters to an imaginary, and impossible, extreme. If the earth were one unified island, a smooth ball, we would all be one species, a tremulous muck. The fact is that when you get down to this business of species formation, you eventually hit some form of reproductive isolation. Cells tend to fuse. Cells tend to engulf each other; primitive creatures tend to move in on each other and on us, to colonize, aggregate, blur. (Within species, individuals have evolved immune reactions, which help preserve individual integrity; you might reject my liver—or, someday, my brain.) As much of the world's energy seems to be devoted to keeping us apart as was directed to bringing us here in the first place. All sorts of different creatures can mate and produce fertile offspring: two species of snapdragon, for instance, or mallard and pintail ducks. But they don't. When you scratch the varying behaviors and conditions behind reproductive isolation, you find, ultimately, geographical isolation. Once the isolation has occurred, of course, forms harden out, enforcing reproductive isolation, so that snapdragons will never mate with pintail ducks.

Geography is the key, the crucial accident of birth. A piece of protein could be a snail, a sea lion, or a systems analyst, but it had to start somewhere. This is not science; it is metaphor. And the landscape in which the protein "starts" shapes its end as surely as bowls shape water.

We have all, as it were, blown back together like the finches, and it's hard to imagine the isolation from parent species in which we evolved. The frail beginnings of great phyla are lost in the crushed histories of cells. Now we see the embellishments of random chromosomal mutations selected by natural selection and preserved in geographically isolate gene pools as faits accomplis, as the differentiated fringe of brittle knobs that is life as we know it. The process is still going on, but there is no turning back. It happened, in the cells; geographical determination is not the cow-caught-in-a-crevice business I make it seem. I'm dealing in imagery, working toward a picture.

Geography is life's limiting factor. Speciation—life itself—is ultimately a matter of warm and cool currents, rich and bare soils, deserts and forests, fresh and salt waters, deltas and jungles and plains. Species arise in isolation. A plaster cast is as intricate as its mold; life is a gloss on geography. And if you dig your fists into the earth and crumble geography, you strike geology. Climate is the wind of the mineral earth's rondure, tilt, and orbit modified by local geological conditions. The Pacific Ocean, the Negev Desert, and the rain forest of Brazil are local geological conditions. So are the slow carp pools and splashing trout riffles of any backyard creek. It is all, God help us, a matter of rocks.

The rocks shape life like hands around swelling dough. In Virginia, the salamanders vary from mountain ridge to mountain ridge; so do the fiddle tunes the old men play. These are not merely anomalous details. This is what life is all about: salamanders, fiddle tunes, you and me and things, the split and burr of it all, the fizz into particulars. No mountains and one salamander, one fiddle tune, would be a lesser world. No continents, no fiddlers. The earth, without form, is void.

A Bright Snarl

The mountains are time's machines; in effect, they roll out protoplasm like printer's rollers pressing out news. But life is already part of the landscape, a limiting factor in space; life, too, shapes life. Geology's rocks and climate have already become Brazil's rain forest, yielding shocking bright birds. To say that all life is an interconnected membrane, a weft of linkages like chain mail, is truism. But in this case, too, the Galápagos Islands afford a clear picture.

On Santa Cruz Island, for instance, the saddleback carapaces of tortoises enable them to stretch high and reach the succulent pads of prickly pear cactus. But the prickly pear cactus on that island, and on other tortoise islands, has evolved a tall treelike habit; those lower pads get harder to come by. Without limiting factors, the two populations could stretch right into the stratosphere.

Ça va. It goes on everywhere, tit for tat, action and reaction, triggers and inhibitors ascending in a spiral like spatting butterflies. Within life, we are pushing each other around. How many animal forms have evolved just so because there are, for instance, trees? We pass the nitrogen around, and other vital gases; we feed and nest, plucking this and that and planting seeds. The protoplasm responds, nudged and nudging, bearing the news.

And the rocks themselves shall be moved. The rocks themselves are not pure necessity, given, like vast, complex molds around which the rest of us swirl. They heave to their own necessities, to stirrings and prickings from within and without.

The mountains are no more fixed than the stars. Granite, for example, contains much oxygen and is relatively light. It "floats." When granite forms under the earth's crust, great chunks of it bob up, I read somewhere, like dumplings. The continents themselves are beautiful pea-green boats. The Galápagos archipelago as a whole is surfing toward Ecuador; South America is sliding toward the Galápagos; North America, too, is sailing westward. We're on floating islands, shaky ground.

So the rocks shape life, and then life shapes life, and the rocks are moving. The completed picture needs one more element: life shapes the rocks.

Life is more than a live green scum on a dead pool, a shimmering scurf like slime mold on rock. Look at the planet. Everywhere freedom twines its way around necessity, inventing new strings of occasions, lassoing time and putting it through its varied and spirited paces. Everywhere live things lash at the rocks. Softness is vulnerable, but it has a will; tube worms bore and coral atolls rise. Lichens in delicate lobes are chewing the granite mountains; forests in serried ranks trammel the hills. Man has more freedom than other live things; anti-entropically, he batters a bigger dent in the given, damming the rivers, planting the plains, drawing in his mind's eye dotted lines between the stars.

The old ark's a moverin'. Each live thing wags its home waters, rumples the turf, rearranges the air. The rocks press out protoplasm; the protoplasm pummels the rocks. It could be that this is the one world, and that world a bright snarl.

Like boys on dolphins, the continents ride their crustal plates. New lands shoulder up from the waves, and old lands buckle under. The very landscapes heave; change burgeons into change. Gray granite bobs up, red clay compresses, yellow sandstone tilts, surging in forests, incised by streams. The mountains tremble, the ice rasps back and forth, and the protoplasm furls in shock waves, up the rock valleys and down, ramifying possibilities, riddling the mountains. Life and the rocks, like spirit and matter, are a fringed matrix, lapped and lapping, clasping and held. It is like hand washing hand. It is like hand washing hand and the whole tumult hurled. The planet spins, rapt inside its intricate mists. The galaxy is a flung thing, loose in the night, and our solar system is one of its many dotted campfires ringed with tossed rocks. What shall we sing?

What shall we sing, while the fire burns down? We can sing only specifics, time's rambling tune, the places we have seen, the faces we have known. I will sing you the Galápagos Islands, the sea lions soft on the docks. It's all still happening there, in real light, the cool currents upwelling, the finches falling on the wind, the shearwaters looping the waves. I could go back, or I could go on; or I could sit down, like Kubla Khan:

> Weave a circle round him thrice,
>   And close your eyes with holy dread,
>   For he on honey-dew hath fed,
>   And drunk the milk of Paradise.

Harper's Magazine, May 1975, 74, 76–79. This essay was also published in Dillard's sole book of essays, *Teaching a Stone to Talk: Expeditions and Encounters* (New York: Harper and Row, 1982). Reprinted with permission from Annie Dillard.

Land: A Thousand Accidents

The geographic distribution of taxa gave Darwin his most convincing evidence for the transmutation of species. How the land itself changes over geological time, then, becomes directly relevant to interpreting the pace and mechanisms of evolutionary change. Specifically, if the correspondence between geographic distance and genealogical affinity among organisms was one of Darwin's major lines of evidence for evolutionary change, then understanding that distance—what effects isolation to varying degrees and over what lengths of time—becomes crucial to interpreting the evolutionary process. It becomes important to understand paths of geographic connectedness and modes of overcoming geographic isolation.

In the time of Lyell and Darwin, geologists were providing abundant evidence for the subsidence and uplifting of land, processes that altered connections between geographic regions considerably. A new land bridge could allow once distinct terrestrial biota to mix while isolating marine biota on either side. Uplifting could create impassable mountain ranges; subsidence could strand offshore islands or even continents. Continental drift had not yet entered the picture, being first mentioned in 1908 by Frank Bursley Taylor, and it would not become firmly established as fact until the mid-twentieth century.

The other mode of connection between biota of different places is through the mobility of the organisms themselves: dispersal. Dispersal is capricious, however. Anyone who has ever watched a milkweed feather drift in a breeze knows this: no telling exactly where it will land. Can such an unpredictable process determine the major patterns of distribution of organisms on earth?

Often with charming attention to unusual detail, Darwin attended closely to the ways organisms dispersed. He examined mud on the toes of birds to look for seeds and tiny mollusks. He noted rafts of vegetation in the sea and recorded strange visits of beetles and dust descending from the sky onto the *Beagle's* deck as the ship sailed in mid-ocean. He was interested in how things moved around, far and near, but especially far. He saw that improbable migrations were possible.

Dispersal also played an important role in discourse on creation in the nineteenth century. Noachian naturalists held that divine origination first populated the earth, and that after a global deluge, the world was repopulated from a single location. By the nineteenth century, however, many naturalists had been persuaded that creation occurred repeatedly throughout the history of the earth. In fact, most of the leading scientists of Darwin's time argued for multiple sequential episodes of creation. With the discovery of the fossilized remains of creatures no longer extant on earth, and the corresponding lack of extant organisms preserved in very ancient fossil beds, the phenomenon of extinction and apparent replacement with different biota profoundly challenged the view of a single creation event.

Many nineteenth-century adherents of divine origination conceded that not only did creation occur repeatedly over time, but each species was also created in a particular location to which it would be particularly well suited. In fact, the Galápagos finches themselves were cited as evidence for this, as Fitzroy states: "All the small birds that live on these lava-covered islands have short beaks, very thick at the base, like that of a bull-finch. This appears to be one of those admirable provisions of Infinite Wisdom by which each created thing is adapted to the place for which it was intended" (Fitzroy 1839, 2:503).

An influential paleontologist and comparative anatomist from Paris, Georges Cuvier, interpreted the fossil record to represent several cataclysmic extinction events followed by new rounds of creation during which entire populations were created at once throughout the natural distribution of a given species. Challenging the views of "catastrophists" such as Cuvier, Lyell interpreted the fossil record to reflect a more occasional process of creation, whereby species were created now and then, and now and then went extinct. As the land changed, so too did the biota. According to Lyell, each species' origination was localized and produced only a few individuals. These individuals propagated and migrated outward from their place of origin. "Each species may have had its origin in a single pair, and species may have been created in succession at such times and in such places as to enable them to multiply and endure for an appointed period, and occupy an appointed space on the globe" (Lyell 1830–1833, 2:124).

Before his commitment to evolutionary thinking, Darwin followed Lyell in this conjecture. Referring to land iguanas in the Galápagos, Darwin said, "It appears as if this species had been created in the centre of the Archipelago, and thence had been dispersed only to a certain distance" (C. R. Darwin 1839, 469). Compared to the catastrophists' more generous episodes of creation, during which whole regions were populated by entire populations at once, Lyellian creation seems miserly, producing just a few new beings who had to hang on, multiply, and hoof it.

Amid these debates, scholars of natural history began asking how many centers of origination existed, given the great diversity of life that was being discovered. The search for centers of creation, led by Philip Lutley Sclater, was a common motive for biogeographers in Darwin's time. Were whole regions given their full complement of biota simultaneously, or was each organism created and distributed independently? Did creation occur in just a few locations from which the organisms then dispersed to all suitable locations, or did each bit of land and sea have its own private creation? Organisms were not distributed globally, as Darwin and others noted. Otherwise, why have coypus in South America as opposed to beavers as in North America—both aquatic mammals inhabiting very similar ecologies—or chinchillas as opposed to picas—both alpine? Darwin made many such observations of distantly related taxa that were adapted to similar habitats in geographically distant locations. One center of creation certainly wasn't adequate to explain the distribution of species; nor, for Darwin and some of his contemporaries, were just a very small number.

Prevailing views in Darwin's time allowed that creation did occur in a number of distinct locations, and some dispersal among or within locations was undeniable. But what was the balance between creation and dispersal? Time plays a role. Clearly, if land itself is newly created, as in volcanic oceanic islands, and if unique species inhabit these islands, they must have been created after the land itself was created. Does creation occur on each rocky island just after it emerges? Is it, then, a constant process? Darwin wrote in Notebook B on the transmutation of species: "The question if creative power acted at Galapagos it so acted that birds with plumage & tone of voice purely American, North & South; geographical divisions are arbitrary and not permanent. This might be made very strong, if we believe the Creator created by any laws, which I think is shown by the very facts of the Zoological character of these islands—so permanent a breath cannot reside in space before island existed.—Such an influence must exist in such spots. We know birds do arrive, & seeds" (C. R. Darwin 1837–1838, 98). The creative power could not have existed on the islands before the islands themselves existed. And yet, birds do arrive, and seeds.

Darwin saw evidence of dispersal from the American mainland to the Galápagos. By the second edition of his *Journal of Researches* Darwin regarded the Galápagos Archipelago as "a satellite attached to America, whence it has derived a few stray colonists," not just "a little world within itself," as the first edition claimed. The affinity between Galápagos and American fauna is material, a direct migration of fauna from one location to the other. Yet he also saw that the dispersed species are modified from the original, and that the modification must have occurred after the geologically new islands had been formed. That modification, to Darwin, was creation itself—the origin of new beings. Creation had become an ever-present process; and that process was "transmutation."

Dispersal is the mechanism for the affinity between geographically distant taxa, and for Darwin it argued against divine and independent creation in situ. It took sophisticated observational skills to notice the affinity between organisms isolated on oceanic islands and those on the nearest mainland. It also took considerable imagination to suppose that those mainland organisms could have made the implausible journey to such an isolated place. Yet if one could not bring himself to believe in the dispersal step, how much more implausible then was the subsequent step of modification. Even among Darwin's contemporaries, scientists predisposed toward an evolutionary perspective sometimes had more difficulty with the notion of unlikely dispersal than with transmutation itself. The reading by George Baur provided in this chapter shows one such example.

Geological context surfaces in the ensuing controversy over the evolution of the Galápagos finches, namely on the questions of how the Galápagos Archipelago was formed and where the finches came from in the first place. It may surprise us today to learn that not everyone always agreed that the Galápagos are volcanic oceanic islands, never connected to any mainland. George Baur, for instance, argued strongly that they were continental islands, previously connected to South America, with indirect linkage to North America. This theory fit with his perception of the slow pace of evolution. If the islands were oceanic islands of volcanic origin, he reasoned, they must have only recently emerged from the ocean since the craters are still smoking away, and evolution must occur at an absurdly fast clip. Alternatively, if the islands are just the tips of much older land, gradually becoming fragmented over a much longer period of time, then the process of evolution would have more time. Moreover, if the organisms were already distributed upon the islands at their inception, then the additional time it would take for the organisms to make their way there would not be required. It is this second factor that presents the biggest impediment to Baur's acceptance of the Galápagos as oceanic islands: unlikely dispersal.

Unable to believe that the monstrously large island tortoises could have possibly crossed ocean waters, he argued that the "harmonious" distribution of taxa among all the islands indicates that there was a time when the creatures could wander freely from "island" to "island," and that this must have happened before they were, in fact, islands. That is, the mixing of biota occurred before the islands were geographically isolated from each other. Moreover, he argued that the biota were "harmonious" with that of the mainland. The differences among the taxa within the archipelago accumulated only after the land submerged to create isolated islands. The idea that dispersal was likely enough for original colonization, but uncommon enough to enforce isolation once colonists arrived on the islands, was just too improbable for him. It would, he says, require a thousand accidents. It is hard not to sympathize with him on that point.

But, given the incontrovertible evidence that the Galápagos are in fact volcanic oceanic islands and have always been so, and that there is precious little possibility of them being proven otherwise, we have come to terms with the improbable balance of dispersal and isolation that creates diversity on the Galápagos, and other archipelagos throughout the globe. Acceding to Darwin yet again: "There seems to be no more design in the variability of organic beings and in the action of natural selection than in the course which the wind blows" (C. R. Darwin, 1958/1888, 87).

READINGS

The first paper in this chapter is by the paleontologist and comparative anatomist George Baur. He argues for the continental origin of the Galápagos Archipelago, with the improbability of dispersal being able to account for the distribution of species as his primary evidence. Broadly put, his scenario is that organisms were more or less evenly distributed between the mainland and the Galápagos before the ocean rose and isolated the archipelago. As the waters continued to rise, the islands one by one became isolated from each other. Dispersal among them must have been minimal, since nearly every species was seen to be restricted to a single island. With isolation, the species diverged from one another because of the slightly different conditions on each island. He was unusual in arguing for interisland environmental differences as the cause of divergence (whether through adaptation or plasticity is not specified), as will be further discussed in chapter 5.

Following is an excerpt from the ornithologist Osbert Salvin, who studied the geographic distribution of the finches across multiple islands. Salvin unambiguously supports Darwin's thesis of dispersal from the mainland to the archipelago, and interprets the affinities between Galápagos and mainland taxa to suggest (if somewhat approximately) the time since dispersal and isolation. Unlike Baur, Salvin saw evidence for differences among the islands in species identities, but he saw a distinct propensity for dispersal among islands as well. Indeed he surprised Darwin by noting that the finch species were not, in contrast with the mockingbirds, restricted to one island each (see chapters 1 and 4). Apparently the birds could flit around a bit. It was this observation that challenged Darwin and all subsequent finch biologists to wonder how the species are kept distinct (if in fact they are).

Finally, a modern account of the historical geology of the Galápagos shows evidence for the repeated volcanic creations of separate islands as the Nazca plate moved over a hotspot. This account dredges up a surprise: while the present islands are 3 million years old or less, the authors discovered drowned islands at least 9 million years old, and evidence of volcanic island activity 20 million years old or more. While the original avian colonists still had to make the journey out to the hotspot, and still had to make the jump as the islands sank and scooted closer to the mainland, at least they had more time to do so.

The Galápagos Islands (1892)

❊ GEORGE BAUR

It was by accident that my attention was directed to a study of the Galápagos Islands. On the 9th of January, 1889, when assistant of Prof. O. C. Marsh of New Haven, a big land-tortoise from the miocene of Nebraska was unpacked at the Yale University Museum. This tortoise resembled very much the gigantic forms of the Galápagos Islands. The question at once arose: How did these large *land*-animals come to the *islands*? In the evening of the same day I wrote in my diary: "What is the origin of the Galápagos fauna? It is not introduced, but left there: the Galápagos originated through subsidence of a larger area of land; they do not represent oceanic islands, as generally believed, but are continental islands."

From this date I began to study the different works and notes, which had been written on the islands, becoming more and more convinced that my opinion was correct, and that an examination of the group would be of the greatest interest, not only in the question of the origin of continents and oceans but also in that of the origin of species. No other group of islands afforded such a splendid opportunity for the examination of these questions. They had never been inhabited by man before their discovery by the Spaniards in the sixteenth century; the first small colony was established on Charles Island in 1832, but does not exist to-day. Now a settlement is found only on Chatham Island. It was to be expected, therefore, that nearly all the islands presented their original condition, only influenced in a small degree by man. After I had resigned my position with Professor Marsh, in January, 1890, it occurred to me that it might, perhaps, be possible to bring together the necessary funds for an expedition to the islands. In February, I worked out a plan for an expedition, which was presented, through Prof. v. Kupffer, to the Royal Academy of Berlin. The matter was discussed by the Academy but it was concluded that the high sum of 20,000 marks, which I had considered necessary for a complete biological and physiographical survey of the islands would probably not be in accordance with the results obtained. After this the matter was laid before various institutions in this country, but with the same negative result. During this time, I published two papers on the Galápagos; one about the gigantic tortoises, and one about the variation of the lizard Tropidurus. It was in the latter paper that I expressed, for the first time in print, my opinion of the continental origin of these islands.

After I had been appointed to Clark University, I took up the matter again, more and more convinced of its great importance. On the 10th of December, 1890, a paper was read before the Biological Club of the University: "Ideas on the origin of the Galápagos Islands and the origin of species." A trial to interest Clark University failed, however. Later on I spoke about the importance of the expedition at Boston, New York and Princeton. Everywhere I found great interest, but it seemed impossible to bring together the necessary funds. At last I sent my paper to the printers. At that moment, Mr. Stephen Salisbury came forward and offered a sum which seemed large enough, with other amounts contributed by the Elizabeth Thompson Fund of Boston and my friend Prof. H. F. Osborn, to secure the success of an expedition. In Mr. C. F. Adams of Champaign, Ill., who had great experience through his collecting trips in Borneo and New Zealand, I found a most useful companion. We left New York on the *City of Para*, on which steamer, through the great courtesy of Mr. George J. Gould, we had received free passage. We arrived at Panama, May 9th, and at Guayaquil, May 13th, but it was not before June 1st that we could leave Guayaquil for the islands on a small sloop. Chatham, the most eastern of the islands and the only one inhabited, was reached in the evening of June 9th. Mr. Manuel Cobos has established there an extensive sugar plantation with great success. Besides coffee, many other tropical fruits are cultivated. Wild cattle exist there in abundance as in some of the other islands. We remained at Chatham, making extensive collections, until one of Mr. Cobos's sloops arrived from Guayaquil. This was engaged, and on June 27th, we left Chatham to visit the other islands. The rent we had to pay for the sloop was higher than anticipated; and I have again to acknowledge the liberality of Mr. Salisbury and Mr. Gould, without which the successful accomplishment of the expedition would have been impossible. During the two months following, all the islands south of the equator with the exception of Narborough, were visited. It was intended on the second trip to examine the other islands, but unfortunately this plan could not be carried out completely. When we reached Chatham, I found news from home necessitating an immediate return. Therefore only Tower, Bindloe and Abingdon were visited. Wenman and Culpepper, two small, rocky islands to the northwest were not touched at. Notwithstanding the programme could not be followed entirely, the expedition proved to be a great success. The collections made are the most extensive. I may mention for instance, that on Albemarle, where so far only four species of birds had been collected, more than forty were obtained. Animals which had not been found since Darwin's visit in 1835 were again secured. A peculiar gull which had been considered exceedingly rare, only five specimens being in existence in all the museums of the world, was found to be quite common, and to show a very much more extensive distribution than was supposed. Of the gigantic tortoises, a large collection was made, notwithstanding the many hardships which were experienced. Some of these tortoises had a weight of more than four hundred pounds: one of them is the largest ever carried from the islands, so far as I know, the carapace having a length of four feet in straight line.

The collections and observations made on the islands seem to prove without doubt, that the opinion of the continental origin of the islands is the correct one. These volcanic islands are nothing but the tops of volcanic mountains of a greater area of land, which has sunken below the level of the ocean. This is proved by the absolutely harmonious distribution of the organisms. We do not find the same animals on the islands, but nearly every island has its own races. This important fact was for the first time noted by David Porter, who pointed out that the different islands contain different races of the tortoise. This view was fully supported by Darwin, who states that the inhabitants of Charles Island could tell from the aspect of the tortoise from which particular island it came. The same is true for many of the land birds, for the lizards, the land shells, and for some of the insects.

Now let us suppose for a moment, that the opinion generally believed to-day, that the Galápagos are oceanic islands lifted out of the ocean, is correct. In this case there must have been a time when not a single organism existed on the islands. Only by accidental introduction from some other part of the earth could the islands be populated; but on such a supposition we are absolutely unable to explain the harmonious distribution, we cannot explain why every, or nearly every, island has its peculiar race or species, not represented on any other island. If some animals could be carried over hundreds of miles to the islands, why are they not carried from one island to the other? But besides that, how could we make plain the presence of such peculiar forms as the gigantic land-tortoises for instance? According to the elevation theory, we can only think of an accidental importation of these tortoises by some current, because they are unable to swim. After the islands had been elevated out of the sea, it happened once, by a peculiar accident, that a land-tortoise was carried over. Alone it could not propagate. This was only possible after a similar accident imported another specimen of *the same species*, of *the other sex*, to *the same island*. Or we could imagine that at the same time animals of both sexes were thus accidentally introduced. By this we could at least explain the population of a single island. But how did all the other islands become populated? To explain this we would have to invoke a thousand accidents.

The most simple solution is given by the theory of subsidence, however. All the islands were formerly connected with each other, forming a single large island; subsidence kept on and the single island was divided up into several islands. Every island developed, in the course of long periods, its peculiar races, because the conditions on these different islands were not absolutely identical.

That it has been made probable, that the Galápagos are of continental origin, I consider one of the most important results of the expedition. If the Galápagos originated through subsidence, we can believe the same of the Sandwich Islands, which also show harmony in the distribution of their organisms. It is not at all improbable that formerly large continental areas spread where we find to-day the Pacific Ocean; that an Atlantis, a Lemuria, so often demurred at, existed after all. New, extensive and methodical explorations of the different groups of islands in the Pacific, Atlantic and Indian Oceans, which have to be made, will be able to decide this interesting question.

Another great result will, I feel certain, come to light after the collections have been fully worked up. The change of the species can be followed, stage by stage, on the different islands; so far as I can anticipate, it will be shown that variation goes on in definite lines determined by the surroundings; that the surroundings and time are the most important and principal factors of variation, and that *natural selection* plays only a secondary role, and very often none at all.

Proceedings of the American Antiquarian Society, at the Annual Meeting October 21, 1891, pp. 3–8.

From *On the Avifauna of the Galapagos Archipelago* (1876)

✴ OSBERT SALVIN

CONCLUDING REMARKS

Before concluding this paper a few conjectural remarks on the process by which these islands have become tenanted with bird-life may not be out of place.

Considering their purely volcanic nature, it cannot reasonably be doubted that these islands have always been islands since they emerged from the sea. Such is Mr. Darwin's view; and it is fully indorsed by Dr. Hooker and others. The birds that are now found, being related to American birds, must have emigrated thence and become modified by the different circumstances with which they became surrounded. The oldest immigrants seem to be indicated by their generic difference from their continental allies, the more modern comers by their merely specific distinctness, and the most recent by their identity with birds now found on the adjoining continent. On this view the islands were first taken possession of by individuals of the parent stock of *Certhidea* and *Conirostrum*, *Geospiza* and *Guiraca*, *Camarhynchus* and *Neorhynchus*. Then came perhaps the ancestors of *Buteo*; after these followed those of *Mimus*, *Pyrocephalus*, and *Myiarchus*; *Strix* and *Asio*, *Zenaida*, *Larus*, and *Spheniscus*. Then those of *Dendroeca*, *Progne*, *Butorides*, *Nycticorax* and *Porzana*, and, finally,

Dolichonyx orizivora, Ardea herodias and the Ducks, Flamingo, Gannets, Plovers, and Sandpipers, though of these last a constant stream of immigrants may have been maintained from the earliest times. It must be remembered, however, that no precise order of immigration can be absolutely laid down, even approximately; for one term in the proposition is an absolutely unknown quantity. We know nothing of the rate of change that has taken place in any one species. Outward circumstances may have acted upon one species so as to leave it little changed in a given time, whilst in the same time another species may have assumed distinctive generic characters. Viewing the very peculiar physical characters possessed by these islands when contrasted with the neighbouring American shores, it would seem reasonable to consider that the rate of change demanded of an immigrant species would be high; consequently the origin of the islands need not be dated back to a more distant period than seems indicated by their volcanic origin. But I am writing of the Birds alone; other forms of life found in these islands present far more complex problems for solution, into which I am not prepared to enter.

Transactions of the Zoological Society of London 10 (1876): 447–510. Excerpts from pp. 509–510.

Drowned Islands Downstream from the Galapagos Hotspot Imply Extended Speciation Times (1992)

❊ DAVID M. CHRISTIE, ROBERT A. DUNCAN, ALEXANDER R. MCBIRNEY, MARK A. RICHARDS, WILLIAM M. WHITE, KAREN S. HARPP, AND CHRISTOPHER G. FOX

Abstract.—The volcanic islands of the Galapagos archipelago are the most recent products of a long-lived mantle hotspot.[1,2] Little is known, however, of the submarine Galapagos platform on which the islands are built, or of the Cocos and Carnegie submarine ridges produced by past motion of the Cocos and Nazca plates across the hotspot.[3,4] In 1990 we surveyed selected areas around the Galapagos platform and as far east as 85° 30′ W on the Carnegie ridge, where we dredged abundant well-rounded basalt cobbles from a small seamount with a terraced summit region. Cobbles were also dredged from several other seamounts. We interpret these features, especially the presence of cobbles, as evidence for erosion near sea level and conclude that these seamounts were volcanic islands before subsiding to their present depths. Radiometric ages for these drowned islands range from 5 to 9 Myr, consistent with predicted plate motions. They indicate that the time available for speciation of Galapagos organisms is much longer than the age range of the existing islands.

The oldest known and best documented of the drowned islands, referred to here as the 85° 40′ W seamount, seems to have formed over the hotspot

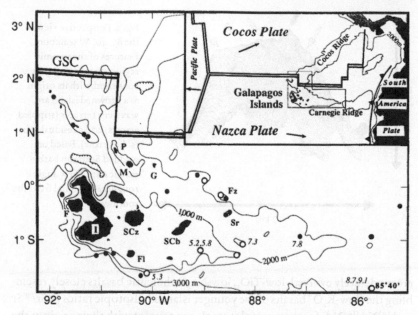

Fig. 1. The generalized bathymetry of the Galapagos area. The Galapagos platform is the broad, generally flat region bounded by the 1,000-m contour. Open circles, dredge sites believed to have sampled drowned islands; closed circles, dredge sites that recovered pillow basalts from seamounts that may or may not have been islands but yielded no evidence of subaerial processes. Numbers are $^{40}Ar/^{39}Ar$ radiometric ages. Island and seamount names are abbreviated. F, Fernandina; Fl, Floreana; Fz, Fitzroy seamount; G, Genovesa; I, Isabela; M, Marchena; P, Pinta; SCb, San Cristóbal; SCz, Santa Cruz; Sr, Sunray seamount. GSC is the Galapagos spreading centre.

at least 9 Myr ago. This small seamount is the middle of three similar east-west aligned edifices, previously surveyed along 2° S latitude on the southern flank of the Carnegie ridge (ref. 5, and P. Lonsdale, unpublished data from the Ceres 04 expedition). The basal diameter of this seamount is ~10 km, roughly the size of the smaller present-day islands such as Isla Pinzon. It rises steeply to a generally flat summit region at 2,000 m depth with a small peak towards its northwest edge (Fig. 2). The summit morphology can be interpreted in terms of a number of wave-cut terraces at different depths and a residual spire. Much stronger evidence for wave action is provided by the recovery of abundant, highly rounded basalt pebbles and cobbles in a dredge haul from the upper north slope. All the cobbles recovered are composed of aphyric basalt with greenish-brown discoloration due to low-temperature seawater alteration. Many have large, irregularly shaped, interconnected vesicles of a type that is common in subaerial flows but which we have not observed in lavas erupted in deep water. The four analyzed samples from this seamount

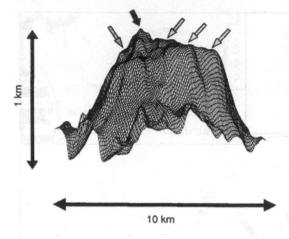

1 km

10 km

Fig. 2. Perspective view of the 85° 40′ W seamount. Features of the summit region that we interpret as consistent with its origin as a drowned island are wave-cut terraces (stippled arrows) and a residual spire (solid arrow). Based on gridded Seabeam bathymetric data. View angle is roughly horizontal from the southwest.

are moderately evolved, low-TiO_2, low-Na_2O tholeiitic basalts closely resembling the "low-K_2O" basalts of the younger islands.[6,7] Isotopic ratios ($^{87}Sr/^{86}Sr$ and $^{143}Nd/^{144}Nd$) for these samples are also consistent with their origin at the Galapagos hotspot.

Basalt pebbles and cobbles were also recovered at a number of other sample sites (Table 1 and Fig. 1). A second large, previously unmapped seamount to the north of Sunray seamount has been named "Fitzroy" after the captain of HMS *Beagle*. It yielded well-rounded cobbles, although it lacks morphologic features consistent with sea-level erosion, perhaps indicating more rapid subsidence. In addition, three sample sites on smaller edifices, which were not sufficiently well surveyed for morphologic studies, yielded rounded pebbles of various sizes. In all cases, chemical compositions of basalts lie within the range of samples from the Galapagos Islands. These occurrences imply the existence of a number of other drowned islands on the eastern end of the Galapagos platform and western Carnegie ridge.

The production of volcanic ridges and associated seamount chains as lithospheric plates move over stationary volcanic hotspots is a well-documented phenomenon.[8–10] In the case of the Galapagos hotspot, the Cocos and Carnegie ridges record the motions of the Cocos and Nazca plates, respectively, away from a hotspot that was located beneath the Cocos-Nazca spreading centre until ~5 Myr ago.[3] At that time, the spreading centre moved north, away from the hotspot, severing the connection between Cocos ridge and the hotspot. In the present Galapagos archipelago, volcanoes have been active on almost all the islands during the past 100,000 years, but the maximum measured ages increase systematically eastward from the youngest volcano, Isla Fernandina (Fig. 3). This island age progression is consistent

TABLE 1. Locations of dated samples and others indicative of emergent conditions

Dredge	Latitude	Longitude	Depth (m)	Age/Sample no.*	Notes
1	2° 00.5′ S	85° 39.7′ W	2,500–2,200	8.7 ± 0.4/PL02-1-12 9.1 ± 0.4/PL02-1-46	Small flat-top terraced seamount. Well-rounded vesicular and massive basalt pebbles and cobbles.
4	0° 55.9′ S	87° 07.0′ W	1,800–1,300	7.8 ± 1.2/PL02-4-19	Small seamount. Angular basalt pieces only.
5	1° 17.3′ S	88° 21.4′ W	875–575		Small seamount. Cobbles of vesicular basalt.
8	1° 07.0′ S	88° 15.8′ W	1,000–900	7.3 ± 1.5/PL02-8-1	Small steep seamount. Two cobbles vesicular basalt.
11	0° 12.0′ S	88° 40.5′ W	1,300–700		Large seamount (Fitzroy). Cobbles vesicular and massive basalt.
13	0° 05.8′ N	89° 02.9′ W	1,400–700		Composite seamount. Two orange weathered finely vesicular basalt cobbles in addition to pillow basalt fragments.
17	1° 11.4′ S	89° 06.6′ W	650–300	5.8 ± 0.8/PL02-17-4 5.3 ± 0.4/PL02-17-1	East-west ridge. Well-rounded basalt pebbles. Branching coral.
20	1° 37.7′ S	90° 10.7′ W	1,050–300		Small seamount. Basalt pebbles (1.3 cm).

*40Ar/39Ar plateau ages.

with predicted eastward Nazca plate motion at 40–50 mm yr^{-1} over the past 3 Myr.[11,12] Earlier than this, the Nazca plate may have moved rather faster (~75 mm yr^{-1}) over the hotspot[13] (Fig. 3). The dated cobbles from the 85° 40' W seamount are ~2 Myr younger than the maximum age (~11 Myr) predicted for this plate velocity, well within the lifespan of many of the present islands. Particularly when we consider that the cobbles probably represent the younger products of this volcano, the measured ages clearly support the idea that this seamount was a volcanic island that first became active over the Galapagos hotspot ~10–11 Myr ago. Dated samples from four other Carnegie ridge seamounts (Table 1, Fig. 3) define a clear age-distance trend consistent with prolonged hotspot activity.

Although the existence of drowned islands along Carnegie ridge is a predictable consequence of plate motion over the Galapagos hotspot,[2] the actual

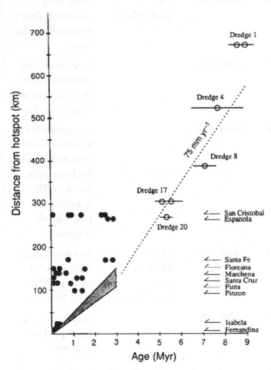

Fig. 3. Radiometric ages against distance from the centre of the Galapagos hotspot, beneath Isla Fernandina. Filled circles are K/Ar ages from island samples, open circles are ^{40}Ar/^{39}Ar ages from dredged seamount samples (1 σ error bars). Reported ages (Table 1) are weighted means of concordant ages from gas fractions released at sequential heating steps. Predicted velocities for the Nazca plate over the hotspot[11,12] fall within the shaded region for the past 3 Myr, whereas a faster motion (dotted line) is required for earlier times.[13]

documentation of their existence may help to resolve a controversy among evolutionary biologists. Some have argued that the geological youth of the present islands ($\leq$3 Myr) requires that all adaptive radiation has occurred within that period.[14] Others have maintained, on the basis of molecular studies of proteins, that much longer periods are required for the divergence of some stocks, most notably the iguanas, which may have required as much as 15–20 Myr for divergence from a common gene pool.[15]

Our geological observations and radiometric data indicate that there have been islands present over the Galapagos hotspot for at least 9 Myr and probably much longer. The geometry of the Cocos and Carnegie ridges is consistent with hotspot volcanism in the Galapagos region for the past 15–20 Myr, and large portions of the Caribbean plate (of age 80–90 Myr) have been linked to the initiation of Galapagos volcanism through compositional stud-ies[16] and plate reconstruction.[13] We consider it likely that islands have ex-isted through the entire 80–90 Myr history of hotspot activity.

ACKNOWLEDGMENTS

We thank the government of Ecuador, the Parque Nacional Galapagos and the Charles Darwin Research Station for permission to carry out the field work, the officers and crew of the RV *Thomas Washington* for cooperation and the NSF for funding.

Nature 335 (1992): 246–248. Reprinted with permission from Macmillan Publishers, Ltd. (Nature, copyright 1992).

NOTES

1. Morgan, W. J. *Nature* 230, 42–43 (1971).

2. Morgan, W. J. *Am. Assoc. Petrol. Geol. Bull.* 56, 203–213 (1972).

3. Hey, R. N. *Geol. Soc. Am. Bull.* 88, 1404–1420 (1977).

4. Cox., A. in *Patterns of Evolution in Galapagos Organisms* (eds Bowman, R. I. & Leviton, A. E.) 11–23 (American Association for the Advancement of Science, Washington DC, 1983).

5. Malfait, B. T. thesis. Oregon State Univ. (1975).

6. Geist. D. J., McBirney, A. R. & Duncan, R. A. *Geol. Soc. Am. Bull.* 97, 555–566 (1986).

7. Baitis, H. W. & Swanson, F. J. *Nature* 259, 195–197 (1976).

8. Dalrymple, G. B., Lanphere, M. A. & Clague, D. A. in *Init. Rep. Deep-Sea Drilling Project 55* (eds Jackson, E. D. *et al.*) 659–676 (US Government Printing Office, Washington, 1981).

9. O'Connor, J. M. & Duncan, R. A. *J. geophys. Res.* 95, 17475–17502 (1990).

10. Duncan, R. A. & Clague, D. A. in *The Ocean Basins and Margins 7 A* (eds Nairn, A. E. M., Stehli, F. G. & Uyeda, S.) 89–121 (Plenum, New York, 1985).

11. Minster, J. B. & Jordan. T. H. *J. geophys. Res.* 83, 5331–5354 (1978).

12. Gripp, A. E. & Gordon, R. G. *Geophys. Res. Lett.* 17, 1109–1112 (1990).

13. Duncan, R. A. & Hargraves, R. B. in *The Caribbean–South American Plate Boundary and Regional Tectonics, Geol. Soc. Am. Mem. 162* (eds Bonini, W. E., Hargraves, R. B. & Shagam, R.) 81–94 (1985).

14. Hickman, C. S. & Lipps, J. H. *Science 227,* 1578–1580 (1985).

15. Wiles, J. S. & Sarich, V. M. in *Patterns of Evolution in Galapagos Organisms* (eds Bowman, R. I. & Leviton, A. E.) 177–186 (American Association for the Advancement of Science, Washington DC, 1983).

16. Sen, G., Hickey-Vargas, R., Waggoner, D. G. & Maurasse, F. *Earth planet. Sci. Lett. 87,* 423–437 (1988).

A Confusion of Finches

Pedigree: from Old French, *pé de grue*, or the foot of a crane. This etymology derives from the manner of depicting genealogical relationships within a branching lineage, representing a common ancestor and his descendents, and incidentally resembling the footprint of a bird. For breeders and social commentators alike, the emphasis of "pedigree" is on the continuity, the common and defining quality, passed from one generation to the next: a name perhaps, or wealth, a strength or weakness of character, a proven and stable quality. A "good pedigree" implies something predictably good about the descendent. Yet the symbol and the word itself is the branch—the divergence from a straight line.

The art of taxonomy is the art of analyzing both affinities and differences of organisms—of inferring shared ancestry and interpreting divergence. The confusing history of the taxonomic and genealogical classification of Darwin's finches illustrates just how important taxonomy was to the development of evolutionary theory. First, grouping entities into intuitive categories forced people to analyze the undeniable affinities and differences among organisms. Darwin and many others of his time eventually interpreted these affinities in terms of shared ancestry.

If ancestry accounted for the similarities, then how taxa acquired their differences also required explanation. And if those differences are large, they reveal something important about the impressive efficacy of the process of modification. Imagine the shock when a group of birds, looking as different from each other as warblers, blackbirds, and finches, were all classified as a single new supergenus of closely related finch species. Features, including beak shape, varied within this single group to a degree comparable to the variation seen across families in other bird groups. A single founding stock could apparently undergo enormous change.

This view, naturally, did not go unchallenged. Considerable controversy centered on just how closely related members of the group of finches are, and how many different founding ancestors colonized the islands. Taxonomy therefore was not only a practice of classification and organization of diversity, but an analysis of processes through which differences among taxa arose and were maintained.

The debates regarding finch taxonomy were in the form of the reclassification of some of the subgenera of the Geospizinae subfamily into a number of distinct genera. Salvin even classified the genus *Camarhynchus* into a different subfamily; as such, if the subgenera were actually different subfamilies they would not have had to diverge from their common ancestor to such an outrageous degree, since they likely came from different founding ancestors. Even more controversial was the warbler finch, which fluttered around between families for more than five decades, adopted regularly by both warblers and bananaquits. Even Darwin questioned its assignment into the subfamily Geospizinae, given how different it appears from the ground finches and given that it looks so very much like a warbler, down to the way it flicks it wings.

Taxonomy also forced people to question what makes up a meaningful biological unit. What, in effect, is a species? Much hangs on Darwin's distinction between "only varieties" and species. Why?

In Darwin's time and before, the word *species* meant a single created entity. As such, it was a loaded term. The number of species reflected the number of events of creation. Varieties were only reflections of how environmental circumstances caused deviations from the created form. In examining the Galápagos species, Darwin saw the affinities among Galápagos and American taxa as evidence of migration from continental America, and subsequent change from the original founding form. If that change resulted in new species, as Gould's classification asserted, then that change itself is none other than the origin of species and a form of creation. In that sense, the classification of the endemic species into new species, as opposed to varieties of American taxa, established independently the progression from migration to speciation. Darwin discovered the affinities through migration; Gould discovered the speciation.

Moreover, if divergence and speciation occurred not only between America and the isolated Galápagos, but within the archipelago itself, then creation/speciation occurred with surprising frequency—certainly favoring the Lyellian scenario over the episodic creation of the catastrophists. And because it seemed to occur so frequently and ubiquitously, it was, most likely, a constant process, operating in the past and present. That process of change, as opposed to creation through discreet events, is what Darwin called the origin of species through transmutation.

The debate as to whether species are real, discreet biological units has persisted long past the scientific debates over creation, however. This is somewhat perplexing, especially to those of us studying plants, in which species boundaries are notoriously murky. In fact, the finches are a lot like plants that way.

There is uncertainty, persistent to this day, about the distinctions that define varietal (or subspecific) differences and specific differences. The prevailing view holds that the latter differences are permanent whereas the former are differences that erode due to interbreeding or even simple environmental change. Conceptually, then, a species is an evolutionarily independent unit; changes in one such entity cannot be passed to other similar entities. For that reason, differences between entities cannot erode, and diversity is preserved. This is an extremely important concept.

In regard to the finches, enormous effort has been devoted to trying to identify these discreet units. At first, a unit was defined as discreet if it was morphologically distinct. If one could divide the birds into two separate groups according to a given set of characters, say size and shape, and then find a conspicuous difference between the groups, even if there were some minor overlap, then each group was often considered to be a distinct species. With a few collections from a few islands this was a relatively easy task. But as more collections were obtained from more islands, those overlaps tended to increase, and the question of how much overlap was permissible in the definition of a separate species became increasingly important. The fact that a type, which might be recognizable on one island, differed discernibly from island to island led to a proliferation of species names at the time those collections were becoming available, especially after the *Albatross* expedition in 1888.

Later, taxonomists questioned whether morphological distinctiveness alone was adequate to define species. Ernst Mayr published his "biological species concept" in 1942, which defined species according to whether they exchanged genetic material. Since then, species came to be considered units that are evolutionarily, that is genetically, distinct, not just morphologically so. A fundamental conceptual shift had occurred: species came to be defined by the population genetic process that produces them, and the process as much as the outcome became the subject of study. David Lack published his taxonomic treatment of Darwin's finches in 1945 and 1947 and applied Mayr's definition when naming only thirteen species (but thirty-five subspecies). He also had access to a huge number of collections and saw more continuous variation where previous taxonomists saw separate groups.

The labile classification of the finches shows what a problem it can be to define and recognize evolutionarily independent entities. To some evolutionary biologists, getting it right—finding the meaningfully discreet units— becomes an irresistible challenge. Others accept that species boundaries are sometimes trespassed, and they are not so bothered by blurry plant and finch species. It seems that boundaries among evolving entities are neither impregnable, giving them total evolutionary independence, nor wholly penetrable, giving panmixia and total blending. There is everything in between, ranging from minor differentiation among populations of a fairly homogeneous species to small degrees of genetic exchange between highly differentiated groups—even between what are often called species. Sure, plenty of species out there, from an evolutionarily standpoint, are completely isolated from each other, crooning, "with me, it's all or nothin'," but there is also a lot of border crossing. If species are no longer considered to be the outcome of a single creation event, the unassailability of the distinctions is no longer so crucial. Indeed, if speciation is an ongoing process, we would expect the full range of outcomes.

READINGS

The finches have had their genealogical history rewritten over and over as new evidence emerged. It is still not known for certain where they came from initially or who might have been their founding ancestor. Indeed, theories of who their progenitors were ranged from a virtual international embassy of bird diversity to one lonely collection of about thirty lost souls blown to sea during a rare, ill-fated outing. The following readings recount the changing classification of the finches over time in attempts to infer ancestral relationships with founder species, ancestral relationships among the finches, and processes of divergence.

A casual classification of the finches began with Darwin's original notes (as in chapter 1), implying three completely separate lineages of finches, blackbirds, and a warbler (a.k.a. "wren"). John Gould consolidated them all into a single supergenus (with four subgenera) and thirteen species. Presented first are passages from *The Zoology of the H.M.S.* Beagle, which contains Gould's classification and Darwin's notes, as well as a set of beautiful plates.

The number of genera fluctuated over time. Salvin, included next, made *Camarhynchus* a distinct subfamily: *Camarhynchus*, he says "is not to be traced to the same origin as *Geospiza* and *Cactornis*" but shares a common ancestor with *Spermophila*. Ridgway subsequently assigned *Certhidea* to a different family and split *Camarhynchus* into three subgenera: *Camarhynchus*, *Cactospiza*, and *Platyspiza*. He also assigned *Cactornis*, previously a distinct genus according to Salvin, as a subgenus of *Geospiza*. Following this trend, the number of genera grew, reflecting in part the hypothesis of a greater number of colonizing ancestors.

The number of species also fluctuated immensely. Beginning with Gould's thirteen species, more species were added as more collections were made. By the time Heller and Snodgrass finished their collections in 1898, all the species had been collected (even if all the subspecies had not). The main cause of variation in the numbers of species that were named thereafter depended primarily on how the samples were designated, not on how many samples existed. From Salvin onwards, the species lists of Galápagos finches were presented separately for each island. At first, as more collections were made from more islands, more species were added. Salvin, presented here, followed Gould but added a few new collections.

Ridgway, presented next, tended to classify what are today recognized as different subspecies into different species, especially if found on different islands. This incidentally fulfilled Darwin's prophesy that each species would be restricted to a single island, more than did Salvin's classification. With Ridgway, the number exploded to forty-nine species in two families, three distinct genera, and five subgenera. However, he admitted to his tendency to add species: "Whenever there seemed to be a well-defined average difference between specimens from different islands, I have not hesitated to separate them as local forms. No other course, indeed, is practicable; for were 'lumping' once begun there could be no end to it" (468).

Later, Snodgrass and Heller (1904) consolidated several of Ridgway's species into subspecies. Swarth (1931) generally followed them, but added several more species to *Certhidea*. Lowe, like Ridgway, was ambivalent about whether to classify entities as species, and his paper is presented here. Despite referring to Swarth's thirty-seven species and subspecies all as species, in the end, he declared them "one heterogeneous swarm" originating from a few different founders. He is puzzled by the causes of the variation, dismissing adaptation and invoking isolation and hybridization as the source.

Today following Lack, whose chapter is included here, we are back to thirteen species on the Galápagos, in three or four genera, depending on whether *Cactospiza* is lumped with *Camarhynchus*, plus the lonely Cocos Island finch. The thirteen species are not all the same as Gould's original classification. Only nine of those are still recognized, and the other four have been designated subspecies or varieties. This taxonomic treatment is supported by more recent genetic analyses, with one frequently cited study being that of Petren, Grant, and Grant (1999), from which a key figure has been redrawn and included here.

This present classification is not without controversy. In particular, Robert Zink keeps the controversy alive by arguing in 2002 (echoing Lowe) that two of the primary genera in fact are monospecific species of great behavioral and morphological diversity (not unlike the Cocos Island finch, incidentally). This is a conspicuous challenge to much current research on the finches; instead of attempting to infer the ancient patterns of colonization, divergence, and occasional mixing through hybridization, the question becomes how to explain the generation and maintenance of such diversity *within* a species—a process that would have repeated itself on island after island.

TABLE 1. Classification of Darwin's finches over time from Gould to the present

First collected and described	Present species	Darwin 1835	Gould 1841	Sclater and Salvin 1870	Lack 1945
Beagle 1835 Described by Gould 1837	*Geospiza magnirostris* (large ground finch)	Grossbeak	*Geospiza magnirostris* + **Geospiza strenua**	*Geospiza magnirostris* + *Geospiza strenua*	*Geospiza magnirostris*
Beagle 1835 Described by Gould 1837	*Geospiza fortis* (medium ground finch)	Grossbeak	*Geospiza fortis* + **Geospiza dentirostris** + **Geospiza dubia** (white bill)	*Geospiza fortis* + *Geospiza dentirostris* + *Geospiza dubia*	*Geospiza fortis*
Beagle 1835 Described by Gould 1837	*Geospiza fuliginosa* (small ground finch)	Fringilla	*Geospiza fuliginosa*	*Geospiza fuliginosa*	*Geospiza fuliginosa*
Beagle 1835 Described by Gould 1837	*Geospiza scandens* (cactus ground finch)	Icterus	*Cactornis scandens* (from James) + **Cactornis assimilis** (from Chatham or Charles)	*Cactornis scandens* + *Cactornis assimilis* + *Cactornis abingdoni*	*Geospiza scandens*
Albatross 1888 Described by Ridgway 1890	*Geospiza conirostris* (large cactus ground finch)	X	X	X	*Geospiza conirostris*

Albatross 1888 Described by Sharpe 1888 Earlier possible sample from *Beagle* described by Gould 1837	*Geospiza difficilis* (sharp-beaked finch)	Fringilla	*Geospiza nebulosa* (sample of extinct *G. difficilis?*)	*Geospiza nebulosa*	*Geospiza difficilis*
Habel 1868 Described by Sclater and Salvin 1870	*Cactospiza pallidus* (woodpecker finch)	X	X	*Cactornis pallida*	*Camarybnchus pallidus*
Heller and Snodgrass 1898 Described by Heller and Snodgrass 1901	*Cactospiza heliobates* (mangrove finch)	X	X	X	*Camarybnchus beliobates*
Beagle 1835 Described by Gould 1837	*Camarybnchus parvulus* (small tree finch)	Fringilla	*Geospiza parvula*	*Geospiza parvula* + **Camarhynchus prosthemelas**	*Camarybnchus parvulus*

TABLE 1. (Continued)

First collected and described	Present species	Darwin 1835	Gould 1841	Sclater and Salvin 1870	Lack 1945
Albatross 1888 Described by Ridgway 1890	*Camarhynchus pauper* (medium tree finch)	X	X	X	*Camarhynchus pauper*
Beagle 1835 Described by Gould 1837	*Camarhynchus psittacula* (large tree finch)	Fringilla	*Camarhynchus psittacula*	*Camarhynchus psittacula* + **Camarhynchus habeli**	*Camarhynchus psittacula*
Beagle 1835 Described by Gould 1837 Designated *Platyspiza* by Ridgway in 1896	*Platyspiza crassirostris* (vegetarian finch)	Fringilla	***Camarhynchus crassirostris***	*Camarhynchus variegatus**	*Platyspiza crassirostris*
Beagle 1835 Described by Gould 1837	*Certhidea olivacea* (warbler finch)	Wren	*Certhidea olivacea*	*Certhidea olivacea* + **Certhidea fusca**	*Certhidea olivacea*
Richard Hinds 1840 Described by Gould 1843	*Pinaroloxias inornata* (Cocos finch)	X	*Pinaroloxias inornata*		*Pinaroloxias inornata*

Notes: X indicates that the species had not yet been collected. Boldface indicates an addition of a specimen or a change in designation from the previous classification. The "wren" would be called a warbler in North America. + indicates splitting compared to previous classification.

*Identified from R. W. R. J. Dekker and C. Quaisser, *Type Specimens of Birds in the National Museum of Natural History, Leiden. Part 3. Passerines: Pachycephalidae—Corvidae (Peters's Sequence)*, Nationaal Natuurhistorisch Museum Naturalis, Technical Bulletin 9 (Leiden: Nationaal Natuurhistorisch Museum, 2006).

TABLE 2. Lability in genus designations of Darwin's finches over time

Genus or subgenus	Present	Gould 1841	Sclater and Salvin 1870	Ridgway 1896	Snodgrass and Heller 1904[a]	Swarth 1931	Lack 1945
Geospiza	Genus 6 species	Genus 8 species	Genus Follows Gould; 8 species	Genus 15 species	Genus 7 species (16 species + subspecies)	Genus 9 species (15 species + subspecies)	Genus 6 species (16 species + subspecies)
Cactornis	Pooled with Geospiza	Subgenus Geospiza but with distinct genus name 2 species	Follows Gould, added 2 species 4 species	Subgenus within Geospiza; named Geospiza 8 species	Subgenus within Geospiza 2 species (6 species + subspecies)	Pooled with Geospiza	Pooled with Geospiza
Cactospiza	Genus 2 species	Not collected	Designated Cactornis, but the two newly collected species designated subgenus Cactospiza by Ridgway	Subgenus pooled with Camarhynchus; named Camarhynchus 2 species	Subgenus within Geospiza; named Geospiza 2 species (no additional subspecies)	Genus 3 species (5 species + subspecies)	Pooled with Camarhynchus

Notes: The column heads indicate the authority and date of the classification scheme; columns indicate the number of species classified for each genus (genus indicated in left-most column). All species had been collected by the time of Ridgway's classification, so from that point onward, the differences in the numbers of species and genera indicate differences in placement rather than differences in the collections available. Gould classified the finches within the family Coccothraustinae. By Ridgway's treatment, they were classified within the family Fringillidae. Swarth promoted them to the status of a new family, Lowe demoting them back to the Fringillidae. Lack adhered, referring to them as a subfamily within the Fringillidae. Now they are classified within the Emberizidae.

[a]Closely following Rothschild and Hartert.

From *The Zoology of the H.M.S.* Beagle (1841)

✳ CHARLES R. DARWIN

Editor's note: Latin descriptions have been omitted.

FAMILY: COCCOTHRAUSTINAE

Genus: *Geospiza*. Gould

This singular genus appears to be confined to the islands of the Galapagos Archipelago. It is very numerous, both in individuals and in species, so that it forms the most striking feature in their ornithology. The characters of the species of Geospiza, as well as of the following allied subgenera, run closely into each other in a most remarkable manner.

In my *Journal of Researches*, p. 475, I have given my reasons for believing that in some cases the separate islands possess their own representatives of the different species, and this almost necessarily would cause a fine gradation in their characters. Unfortunately I did not suspect this fact until it was too late to distinguish the specimens from the different islands of the group; but from the collection made for Captain FitzRoy, I have been able in some small measure to rectify this omission.

In each species of these genera a perfect gradation in colouring might, I think, be formed from one jet black to another pale brown. My observations showed that the former were invariably the males; but Mr. Bynoe, the surgeon of the *Beagle*, who opened many specimens, assured me that he found two quite black specimens of one of the smaller species of Geospiza, which certainly were females: this, however, undoubtedly is an exception to the general fact; and is analogous to those cases, which Mr. Blyth has recorded of female linnets and some other birds, in a state of high constitutional vigour, assuming the brighter plumage of the male. The jet black birds, in cases where there could be no doubt in regard to the species, were in singularly few proportional numbers to the brown ones: I can only account for this by the supposition that the intense black colour is attained only by three-year-old birds. I may here mention, that the time of year (beginning of October) in which my collection was made, probably corresponds, as far as the purposes of incubation are concerned, with our autumn. The several species of Geospiza are undistinguishable from each other in habits; they often form, together with the species of the following subgenera, and likewise with doves, large irregular flocks. They frequent the rocky and extremely arid parts of the land sparingly covered with almost naked bushes, near the coasts; for here they find, by scratching in the cindery soil with their powerful beaks and claws, the seeds of grasses and other plants, which rapidly spring up during the short rainy season, and as rapidly disappear. They often eat small portions of the succulent leaves of the *Opuntia Galapageia*, probably for the sake of the moisture contained in them: in this dry climate the birds suffer much from the want of water, and these finches, as well as others, daily crowd round the small and scanty wells, which are found on some of the islands. I seldom, however, saw these birds in the upper and damp region, which supports a thriving vegetation; excepting on the cleared and cultivated fields near the houses in Charles Island, where, as I was informed by the colonists, they do much injury by digging up roots and seeds from a depth of even six inches.

1. Geospiza magnirostris. *Gould*

PLATE XXXVI

Long. tot. 6 unc.; *alae*, 3 1/2; *caudae*, 2; *tarsi*, 1; *rostri*, 7/8; *alt. rostri*, 1.

Sooty black; with the vent cinereous white, the bill black, washed with brownish, and the feet black.

Female, or young male: Deep fuscous, with each feather margined with olive, the abdomen much paler, with the under tail-coverts cinereous white, the feet and bill like those of the male.

Habitat, Galapagos Archipelago. (Charles and Chatham Islands)

I have strong reasons for believing this species is not found in James Island. Mr. Gould considers the *G. magnirostris* as the type of the genus.

Plate XXXVI *Geospiza magnirostris*

Plate XXXVII Geospiza strenua

2. Geospiza strenua. *Gould*

PLATE XXXVII

Long. tot. 5 1/2 unc.; *alae*, 3; *caudae*, 1 3/8; *tarsi*, 3/4; *rostri*, 5/8; *alt. rostri*, 3/8.

Sooty black, with the under tail coverts white; the bill brown, tinged with black, and the feet black.

Female: Upper part of the body fuscous, with the margins of each feather, except those of the wings and tail, pale cinereous-olive; the throat and breast/fuscous: the abdomen, sides, and under tail-coverts pale cinereous-fuscous; the bill brownish.

Habitat, Galapagos Archipelago (James' and Chatham Islands).

3. Geospiza fortis. *Gould*

PLATE XXXVIII

Long. tot. 4 3/4 unc.; *alae*, 3; *caudae*, 1 1/2; *tarsi*, 10/12; *rostri*, 7/12.

Deep sooty black; with the under tail-coverts and the bill reddish brown tinged with black; the feet black.

Female (or young male): The body above, breast and throat, deep fuscous, with each feather margined with cinereous-olive: the abdomen, and under tail-coverts pale cinereous-brown; the bill reddish fuscous, with the apex yellowish, and the feet like those in the male.

Habitat, Galapagos Archipelago (Charles and Chatham Islands).

4. Geospiza nebulosa. *Gould*

Long. tot. 5 unc.; *alae*, 2 3/4; *caudae*, 1 3/4; *tarsi*, 3/4; *rostri*, 5/8; *alt. rostri*, 1/2.

Male: Upper part of the head and body blackish fuscous, with each feather margined

Plate XXXVIII Geospiza fortis

with cinerous olive; the body beneath paler, with the lowest part of the abdomen and under tail-coverts ashy; the bill and feet deep fuscous.

Habitat, Galapagos Archipelago (Charles Island).

5. Geospiza fuliginosa. *Gould*

Long. tot. 4 1/2 unc.; *alae*, 2 1/2; *caudae*, 1 5/8; *tarsi*, 3/4; *rostri*, 1 1/2; *alt. rostri*, 3/8.

Deep sooty black, with the under tail coverts white; the bill fuscous, and the feet blackish fuscous.

Female: Upper part of the body; the wings and tail deep fuscous, with each feather margined with ashy ferrugineous; beneath the body cinereous, with each feather towards the middle darker; the bill brown and the feet blackish brown.

Habitat. Galapagos Archipelago (Chatham and James' Island).

6. Geospiza dentirostris. *Gould*

Long. tot. 4 3/4 unc.; *alae*, 2 3/8; *caudae*, 1 3/4; *rostri*, 1/2; *alt. rostri*, 3/8.

The margin of the upper mandible produced into a tooth; the vertex and above the body fuscous, with each feather towards the middle darker; the margins of the secondaries and wing coverts straw colour; the throat and breast pale brown, darker towards the middle of each feather; the sides and under tail-coverts cinereous white; the bill rufous fuscous, and the feet obscure lead colour.

Habitat, Galapagos Archipelago.

Mr Gould considered this specimen a female, from the appearance of its plumage; but from dissection, I thought it was a male.

Plate XXXIX Geospiza parvula

7. Geospiza parvula. *Gould*

PLATE XXXIX

Long. tot. 4 unc.; *alae*, 2 3/8; *caudae*, 1 1/2; *tarsi*, 3/4; *rostri*, 3/8; *alt. rostri*, 5/16.

The head, throat, and back, sooty black; the lower part of the back cinereous olive; the tail and wings blackish brown, margined with cinereous; the sides olive with fuscous spots; the abdomen and under tail-coverts white; the bill and feet blackish brown.

Female: The upper surface cinereous brown; the throat, breast, abdomen, and the under tail coverts, pale cinereous tinged with straw colour.

Habitat, Galapagos Archipelago (James' Island).

8. Geospiza dubia. *Gould*

Long. tot. 3 3/8 unc.; *alae*, 2 3/4; *caudae*, 1 5/8; *tarsi*, 7/8; *rostri*, 5/8; *alt. rostri*, 3/8.

Upper surface fuscous, with each feather margined with cinereous olive; the streak above the eye, cheeks, throat, and beneath the body, cinereous olive, with the middle of each feather fuscous; the wings and tail brown, with each feather margined with cinereous ash; the bill white, and the feet obscure fuscous.

Habitat, Galapagos Archipelago (Chatham Island).

Plate XL Camarhynchus psittaculus

Sub-genus: *Camarhynchus*. Gould

Camarhynchus psittaculus is the typical species.

1. Camarhynchus psittaculus. *Gould*

PLATE XL

Long. tot. 4 3/4 unc.; *alae*, 2 3/4; *caudae*, 1 3/4; *tarsi*, 7/8; *rostri*, 1/2; *alt. rostri*, 1/2.

The upper part of the head and body fuscous; the wings and tail darker; the throat, and beneath the body cinereous white, tinged with straw-colour; the bill pale yellowish fuscous, and the feet fuscous.

Habitat, Galapagos Archipelago (James' Island).

The species of *Camarhynchus* do not differ in habits from those of *Geospiza*; and the *C. psittaculus* might often be seen mingled in considerable numbers in the same flock with the latter. Mr. Bynoe procured a blackish specimen, which, doubtless, was an old male; I saw several somewhat dusky, especially about the head.

2. Camarhynchus crassirostris. *Gould*

PLATE XLI

Long. tot. 5 1/2 unc.; *alae*, 3 3/4; *caudae*, 2; *tarsi*, 1 1/8; *rostri*, 1/2; *alt. rostri*, 1/2.

Upper part of the body deep brown, with each feather margined with cinereous olive; the throat and breast cinereous olive, with the middle of each feather darker; the abdomen, sides, and under tail coverts cinereous tinged with straw colour.

Plate XLI Camarhynchus crassirostris

Habitat, Galapagos Archipelago (Charles Island?)

I am nearly certain that this specimen is not found in James Island. I believe it came from Charles Island, and probably there replaces the *C. psittaculus* of James Island. I obtained three specimens, one male, and two females; from the analogy of so many species in this group, I do not doubt the old male would be black.

Sub-genus: *Cactornis*. Gould

Cactornis scandens is the typical species.

1. Cactornis scandens. *Gould*

PLATE XLII

Long. tot. 5 unc.; *rostri*, 3/4; *alae*, 2 5/8; *caudae*, 1 3/4; *tarsi* 3/4.

Deep sooty black, with the under tail-coverts white; the bill and feet blackish-brown.

Female: Upper surface of the body, throat and breast intensely brown, with the margins of each feather pale; the abdomen and the under tail coverts cinereous, tinged with straw-colour; the bill pale fuscous, and the feet blackish fuscous.

Habitat: Galapagos Archipelago (James' Island).

The species of this sub-genus alone can be distinguished in habits from the several foregoing ones belonging to *Geospiza* and *Camarhynchus*. Their most frequent resort is the *Opuntia Galapageia*, about the fleshy leaves of which they hop and climb, even with their back downwards, whilst feeding with their sharp beaks, both on the fruit and the flowers. Often, however, they alight on the ground, and mingled with the flock of the above mentioned species, they search for seeds in the parched volcanic soil. The extreme scarceness of the jet-black specimens, which I mentioned under the head of the genus *Geospiza*, is well exemplified in

Plate XLII Cactornis scandens

the case of the *C. scandens*, for although I daily saw many brown-coloured ones (and two collectors were looking out for them), only one, besides that which is figured, was procured, and I did not see a second.

2. Cactornis assimilis. *Gould*

PLATE XLIII

Long. tot. 5 1/2 unc.; *rostri*, 3/4; *alae*, 2 3/4; *caudae*, 1 3/4; *tarsi*, 3/4.

Upper surface of the body sooty black, margined with cinereous, as well as the throat and abdomen; the bill pale rufous brown; the feet blackish brown.

Habitat, Galapagos Archipelago.

I do not know from which island of the group this species was procured; almost certainly not from James Island. Analogy would in this case, as in that of *Camarhynchus crassirostris*, lead to the belief that the old male would be jet black. By a mistake this bird has been figured standing on the *Opuntia Darwinii*, a plant from Patagonia, instead of the *O. Galapageia*. I may here mention that a third and well characterized species of *Cactornis* has lately been sent by Captain Belcher, R.N. to the Zoological Society; as Captain Belcher visited Cocos Island, which is the nearest land to the Galapagos Archipelago, being less than 400 miles distant, it is very probable that the species came thence.

Sub-genus: *Certhidea*. Gould

Of the foregoing sub-genera, *Geospiza*, *Camarhynchus* and *Cactornis* belong to one type, but with regard to *Certhidea*, although Mr Gould confidently believes it should also be referred to the same division, yet as in its slighter form and weaker bill, it has so much the appearance of a member of the Sylviadae,

Plate XLIII Cactornis assimilis

Plate XLIV Certhidea olivacea

he would by no means insist upon the above view being adopted, until the matter shall have been more fully investigated.

Certhidea olivacea. *Gould*

PLATE XLIV

Long. tot. 4 unc.; *rostri*, 1/2; *alae*, 2; *caudae*, 1 1/2; *tarsi*, 3/4.

Upper part of the head, body, wings and tail, olivaceous brown; the throat, and beneath the body, cinereous; the bill and feet pale brown.

Habitat, Galapagos Archipelago (Chatham and James Islands).

I believe my specimens, which include both sexes, were procured from Chatham and James Islands; it is certainly found at the latter.

Part 3, *Birds*, described by J. Gould, and with notes by Charles R. Darwin (New York: New York University Press, 1841), 146–165.

From *On the Avifauna of the Galapagos* (1876)

❋ OSBERT SALVIN

V. ON THE VARIATION OF THE SPECIES IN CERTAIN GENERA, AND THE CONSEQUENT DIFFICULTY IN DEFINING SPECIFIC LIMITS

The acquisition of a large series of specimens of the different so-called species of *Geospiza* renders the question as to the definiteness of the points of specific distinction between them more difficult of solution than ever. Distinctions are plainly enough to be seen between such birds as *G. magnirostris* and *G. parvula*, where great disparity in size is obvious. But these differences are gradually erased by almost insensible steps by the interposition of *G. strenua*, *G. fortis*, and *G. fuliginosa*. The series before me includes specimens that can almost as well be referred to either of two contiguous species, so that their position can only be determined by assigning to each species what must be called arbitrary standards of measurements on the bill alone. The question follows, Do these birds, in their natural relations to one another, keep themselves to groups of individuals dependent upon the size of their bills? The answer is to be found rather by field-observation than by measuring skins. Were different species, though closely allied, found to inhabit different islands, the case would be much simplified. But what do we find? Charles Island possesses three out of eight so-called species, Chatham Island has not less than seven, James Island four, Indefatigable Island three, Bindloe three, and Abingdon four. Thus we see that in Chatham Island every gradation in size is represented.

Then as to coloration. The assumption of the black plumage by the old males seems to be a slow process, the fully black feathering being only arrived at in probably the third year; and it is probable, from the comparatively scanty numbers of the black individuals, that the cock birds breed in the mottled plumage long before the fully black dress is assumed. Observations on paired and nesting birds would here be invaluable.

It would seem, then, that with these singular birds the sexual selection displayed amongst them is such that it is almost a matter of indifference whether the cock birds are mottled or black, and also that the nature of their food and the general conditions of life are such that birds with huge bills as large as a Grosbeak's, as well as those with bills no bigger than a Bunting's, can equally find sustenance, variation as regards the length of wing, tail, and tarsi being equally unimportant. The members of this genus present a field where natural selection has acted with far less rigidity than is usually observable.

The gap between *Geospiza* and *Cactornis* is fairly defined—not but what we see in some specimens of the former more elongated bills than in others, showing a tendency in *Geospiza* to develop in the direction of *Cactornis*. But it would seem that the connecting links are gone; hence our ability to define the differences between these genera.

Camarbynchus belongs to a somewhat different type; and I am inclined to believe that it is not to be traced to the same origin as *Geospiza* and *Cactornis*, but to a common ancestor with, perhaps, *Spermophila*, the bill of the somewhat abnormal ally of that genus (*Neorbynchus nasesus*), recalling to some extent the peculiarities of the bill of *Camarbynchus*. So much for structural characters. In coloration *Camarbynchus* resembles both *Geospiza* and *Cactornis*; but the males of none of the species are so black as those of the adults of the other genera.

It seems worthy of notice that, though the different species of Finches are not restricted to any one particular island, there appears to be a prevalence of some one species in each. Dr. Habel's collection shows that in Indefatigable Island two dominant species occur—*Geospiza fortis* and *G. fuliginosa*. In Bindloe Island *G. strenua* abounds, whilst in Abingdon we find *G. parvula*. It may thus be argued that the tendency of each of these islands is to produce the form most prevalent in it; but, on the other hand, I am not able to say what attention Dr. Habel devoted to each species.

Transactions of the Zoological Society of London 10 (1876): 447–510. Excerpt from pp. 469–470.

From *Birds of the Galapagos Archipelago* (1896)

✳ ROBERT RIDGWAY

Editor's note: Table on p. 472: This table shows a list of species, classified by the island from which it was obtained. The table also indicates the various collections Ridgway used for his taxonomic treatment.

Figures from pp. 498, 511, 512, 546, 547, 548: These figures show the generic and subgeneric designations of species according to Ridgway's classification. The maps indicate the islands on which the species are found. Ridgway meticulously indicated the geographic location of all his taxonomic groups, reflecting his concern with interpreting classification in terms of biogeography. This habit also reveals the challenge of working with the new collections from so many more of the islands than were previously available.

Proceedings of the U.S. National Museum 19 (1896): 459–670.

Names of species.	Darwin, 1835.	Néboux, 1836–1839.	Kinberg, 1852.	Kellett and Wood, ——.	Habel, 1868.	Cookson, 1875.	Markham, 1880.	Jones, 1884.	"Albatross," 1888.	Townsend, 1891.	Barr and Adams, 1891.
1. *Nesomimus macdonaldi*					[x]				x		x
2. *Dendroica aureola*					[x]						x
3. *Certhidea cinerascens*									x		x
4. *Geospiza conirostris*									x		x
5. *Geospiza media*									x		
6. *Geospiza fuliginosa*									x		x
7. *Myiarchus magnirostris*									x		x
8. *Asio galapagoensis*					[x]						x
9. *Buteo galapagoensis*					[x]						
10. *Sula nebouxii*					[x]						
11. *Butorides plumbeus*									x		x
12. *Nyctanassa violacea*									x		x
13. *Poecilonetta galapagensis*					[x]						x
14. *Nesopelia galapagoensis*					[x]				x		
15. *Haematopus galapagensis*											x
16. *Arenaria interpres*									x		
17. *Heteractitis incanus*									x		
18. *Creagrus furcatus*											x
19. *Anous galapagensis*									x		x
20. *Diomedea exulans*[a]					[x?]						
21. *Diomedea migripes??*					[x]						
Total by each collector	0	0	0	0	[9]	0	0	0	12	0	13

[a]Wolf, Ein Besuch aus den Galápagos-Inseln, 1879, p. 13. Perhaps one of the "two kinds of Albatrosses" seen at Hood Island by Dr. Habel (see Trans. Zool. Soc., IX, Pt. IX, 1876, pp. 453, 459) was this species.

Names of species.	Darwin, 1835.	Néboux, 1836–1839.	Kinberg, 1852.	Kellett and Wood, ——.	Habel, 1868.	Cookson, 1875.	Markham, 1880.	Jones, 1884.	"Albatross," 1888.	Townsend, 1891.	Burr and Adams, 1891.
1. Nesomimus adamsi	x	x							x	x	x
2. Dendroica aureola	x?	x						x	x	x	x
3. Certhidea luteola	x								x	x	x
4. ?Progne modesta?ª											
5. Geospiza magnirostris	x										
6. Geospiza strenua	x										
7. Geospiza dubia	x	x							x	x	x
8. Geospiza fuliginosa	x	x							x	x	x
9. Geospiza parvula		x								x	
10. ?Geospiza dentirostris ᵇ											
11. Geospiza fatigata ??					z				x		
12. Camarhynchus variegatus									x	x	x
13. Camarhynchus salvini			x						x	x	x
14. Myiarchus magnirostris	x	x							x	x	x
15. Pyrocephalus dubius	x?								x	x	x
16. Coccyzus melanocoryphus									x		x
17. Buteo galapagoensis	x?							x			x
18. Fregata aquila											x
19. Pelecanus californicus									x		x
20. Sula nebouxii									x		x
21. Butorides plumbeus								x			
22. Nyctanassa violacea											x

ªSee Sharpe, Cat. Birds Brit. Mus., X, 1885, p. 176.

ᵇFide Sharpe, Cat. Birds Brit. Mus., XII, p. 12.

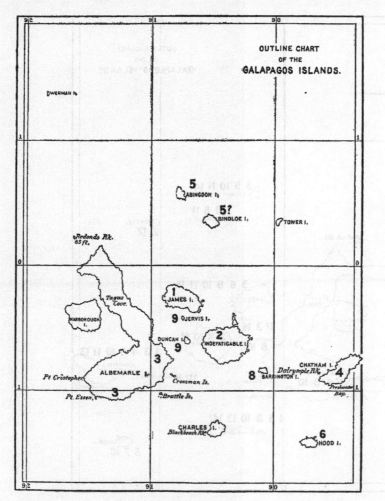

Ascertained range of the genus *Certhidea*, Gould.

1. Certhidea olivacea, Gould. *2. Certhidea salvini*, Ridgway. *3. Certhidea albemarlei*, Ridgway.
4. Certhidea luteola, Ridgway. *5. Certhidea fusca*, Sclater and Salvin. *6. Certhidea cinerascens*,
Ridgway. *7. Certhidea mentalis*, Ridgway. *8. Certhidea bifasciata*, Ridgway. *9.* (Undetermined
form.)

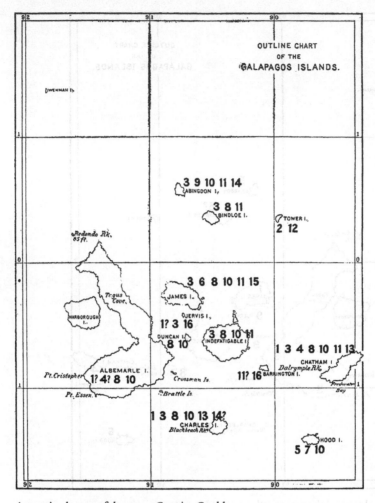

Ascertained range of the genus *Geospiza*, Gould.

a. Subgenus *Geospiza*. 1. *Geospiza magnirostris*, Gould. 2. *Geospiza pachyrhyncha*, Ridgway. 3. *Geospiza strenua*, Gould. 4. *Geospiza dubia*, Gould. 5. *Geospiza conirostris*, Ridgway. 6. *Geospiza bauri*, Ridgway. 7. *Geospiza media*, Ridgway. 8. *Geospiza fortis*, Gould. 9. *Geospiza fratercula*, Ridgway. 10. *Geospiza fuliginosa*, Gould. 11. *Geospiza parvula*, Gould. 12. *Geospiza acutirostris*, Ridgway. 13. *Geospiza dentirostris*, Gould. 14. *Geospiza difficilis*, Sharpe. 15. *Geospiza debilirostris*, Ridgway. 16. (Undetermined form.)

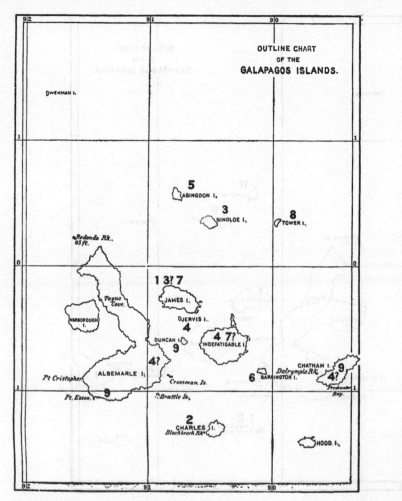

Ascertained range of the genus Geospiza, Gould.
b. Subgenus *Cactornis*, Gould 1. *Geospiza scandens* (Gould). 2. *Geospiza intermedia*, Ridgway. 3. *Geospiza assimilis* (Gould). 4. *Geospiza fatigata*, Ridgway. 5. *Geospiza abingdoni* (Sclater and Salvin). 6. *Geospiza barringtoni*, Ridgway. 7. *Geospiza brevirostris*, Ridgway. 8. *Geospiza propinqua*, Ridgway. 9. (Undetermined form.)

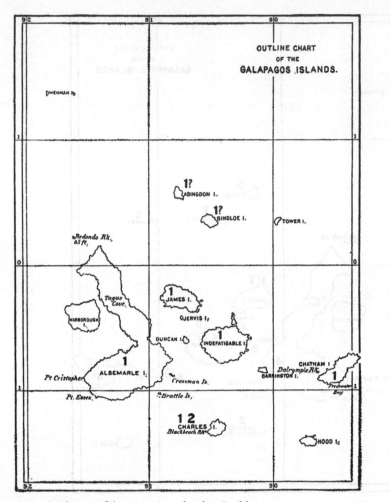

Ascertained range of the genus *Camarhynchus*, Gould.

a. Subgenus *Platyspiza*, Ridgway. 1. *Camarhynchus variegatus*, Sclater and Salvin. 2. *Camarhynchus crassirostris*, Gould.

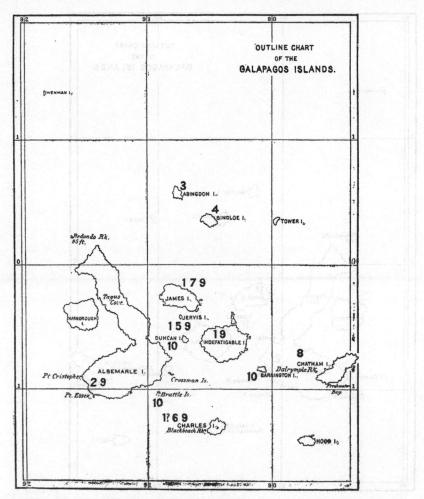

Ascertained range of the genus *Camarhynchus*, Gould.
b. Subgenus *Camarhynchus*, Gould. 1. *Camarhynchus psittaculus*, Gould. 2. *Camarhynchus affinis*, Ridgway. 3. *Camarhynchus habeli*, Sclater and Salvin. 4. *Camarhynchus bindloei*, Ridgway. 5. *Camarhynchus compressirostris*, Ridgway. 6. *Camarhynchus pauper*, Ridgway. 7. *Camarhynchus incertus*, Ridgway. 8. *Camarhynchus salvini*, Ridgway. 9. *Camarhynchus prosthemelas*, Sclater and Salvin. 10. (Undetermined form.)

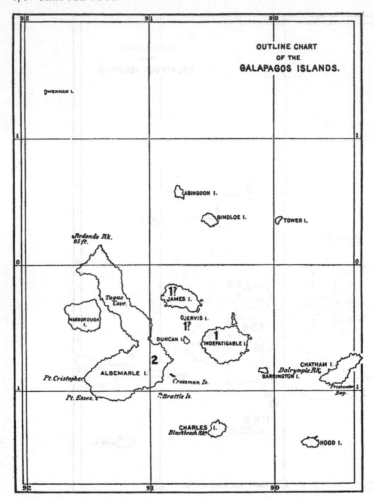

Ascertained range of the genus *Camarhynchus*, Gould.

c. Subgenus *Cactospiza*, Ridgway. 1. *Camarhynchus pallidus* (Sclater and Salvin). 2. *Camarhynchus productus*, Ridgway.

The Finches of the Galapagos in Relation to Darwin's Conception of Species (1936)

❊ PERCY R. LOWE

I feel it was an honour to have been asked to say something about the birds of the Galapagos in connection with the centenary of Darwin's visit.

Instead of taking this invitation too literally, I propose to confine myself to a consideration of one group of birds only, viz., Darwin's Finches, or the Geospizids.

I do so because I think it is true to say that there is no group of birds in the whole world which has more right to occupy the attention of zoologists at the present moment; for the problem presented by the very extraordinary diversity seen within this group of Finches appears to me to be a problem of first-class biological importance.

We know that it was the diversity presented by these Finches, as well as the Mocking-birds, tortoises, and plants, which started Darwin down that brilliant corridor of thought which led to his conception of the origin of species. But the facts which Darwin thought he had gathered as the result of his collection of Finches came very far short of the truth.

Speaking of the diversity of the flora and fauna on the different islands in his "Journal of Researches" (vol. iii. pp. 474–475, ed. 1839), he says: "Unfortunately I was not aware of these facts (of diversity) till my collection was nearly completed. It never occurred to me that the productions of islands only a few miles apart, and placed under the same physical conditions, would be dissimilar. I, therefore, did not attempt to make a series of specimens from the separate islands."

Apparently, however, he did realize that several of the islands produced their own form of Mocking-bird, for, strangely enough, he immediately goes on to say: "I ascertained (and have brought home with me the specimens) that one species of Mocking Bird (*Nesomimus*) is exclusively found in Charles Island, a second on Albemarle, a third common to James and Chatham. I examined many specimens in the different islands and in each the respective kind was alone present."

Going on to consider the Finches, he says: "I very much suspect that certain members of the series (Finches, p. 475) are confined to different islands: therefore if the collection had been made on any one island it would not have presented so perfect a gradation. It is clear that if several islands have *each their peculiar species* (italics mine) of the same genera, when these are placed together they will have a wide range of character."

It is clear then that after Gould had worked out Darwin's collection of Finches from the Galapagos, and after Darwin had realized that "a nearly perfect gradation of structure in this one group can be traced in the form of the beak from one exceeding in dimensions that of the largest grosbeak to another hardly differing from that of a warbler" (Journ. Researches, p. 462), he still thought that the diversity and distribution of the Finches among the islands was on all fours with that of the Mocking-birds: that is to say, he thought that the various islands he had visited had their own peculiar Finch.

Five species of these Mocking-birds, with six subspecies, have since Darwin's day been recognized, and each form is restricted to its own island. Moreover, the group is obviously very closely related indeed to the mainland genera of Mimidae.

On the other hand, the Geospizids are entirely peculiar to the Galapagos, and they represent A HETEROGENEOUS SWARM whose diversity has been the despair of systematists, and whose distribution among the islands is completely abnormal as compared with the generality of island groups. So far as I am aware nothing exactly comparable with this distribution can be found among present existing birds, except that in the case of the Drepanidae of Hawaii one finds an amazing diversity in the shape of the bill and in other characters among the genera and species of this family. But here we have obvious isolating factors connected with physical and physiological conditions which, as well shall presently see, are apparently absent in the Finches.

Yet even in the case of the Drepanidae it seems to me inconceivable that it was the food-factor which produced the diversity in the bills and other characters; but rather that "nature," or the germ-plasm, turned out a diversity of bill- and colour-forms which either fitted some particular food or other environmental factor, *or did not persist.*

I have also called attention (Ibis, 1923, pp. 521–523, and Bull. B. O. C. vol. i. 1930, p. 27) to the remarkable fact that a similar form of diversity, but on a very much smaller scale, so far as examples are known, occurs on Nightingale Island in the Tristan d'Acunha, group in the case of the Finch genus *Nesospiza* (text-fig. 1).

There are also one or two other groups of birds, such as the Silver Pheasants (*Gennaeus*), whose genetics have been studied by Ghigi of Bologna, and the diversity seen in these groups presents features which may perhaps come near to the case of Darwin's Finches, except that in their case isolating factors are present; but these I cannot stop to consider here.

What then *are* the facts in the case of the Finches of the Galapagos?

Text-figure 1. To show diversity in size and shapes of bill in Finches of the genus *Nesospiza*, taken on Nightingale Island, South Atlantic. (a) *Nesospiza acunhae.* (b) *Nesospiza wilkinsi.*

SOME GENERAL FACTS

Darwin's Finches form by far the most striking, by far the most interesting, and by far the most dominant group of land-birds on the Galapagos. Of the eighty-nine breeding species which have been recorded from the islands, thirty-seven or more belong to this group of Finches.

So far as the Finches are concerned, the Galapagos may be divided into nine rather large islands occupying a more or less central position and eleven quite small ones (text-fig. 2), some of which, like Culpepper and Wenman, are situated at a considerable distance (70 miles) from the central group, and others in their more or less near vicinity.

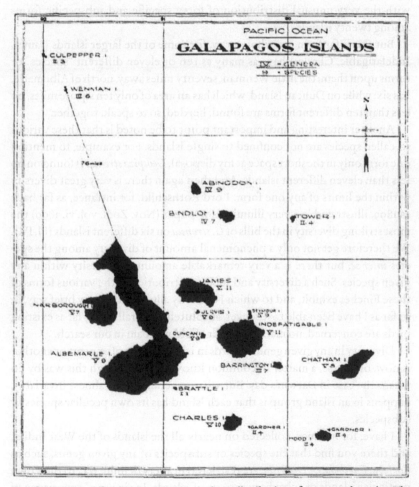

Text-figure 2. Map of Galapagos Islands, to show distribution of genera and species. The number of genera represented on each island is indicated in roman, the number of species and subspecies in arabic, figures.

On every one of the twenty islands, large or small, Geospizids are found.

Mr. Harry Swarth ("Occasional Papers, California Academy of Sciences," xviii. 1931), the latest monographer of the birds of the Galapagos, tells us that no less than sixty-seven specific and subspecific names have been bestowed on the Finches which inhabit these twenty islands, and in his opinion forty of them are valid.

Mr. Swarth has examined an immense amount of material, comprising practically all the collections in America, in the British Museum, in the Paris Museum, in Lord Rothschild's collection, and others on the Continent, including all the types.

If, then, we accept Swarth's probably not too drastic rejection of twenty-seven specific names as being synonyms of one sort of another, we are faced with the very unusual distribution of forty specific and subspecific forms among twenty islands.

But this is far from the whole story, for some of the larger islands (James, Indefatigable, Charles) have as many as ten or eleven different "species" or forms upon them; the little Wenman, seventy miles away, north of Albemarle, has six; while on Duncan Island, which has an area of only ten square miles, no less than ten different forms are found, herded, so to speak, together.

Another interesting and important point to be noted is that these various so-called species are not confined to single islands. For example, to mention one form only in the short space at my disposal, *Geospiza strenua* is found on no less than eleven different islands. And then again there is very great diversity within the limits of any one form. Lord Rothschild, for instance, as far back as 1899, illustrated in a very illuminating paper (Nov. Zool. vol. vi. 1899) the most striking diversity in the bills of *G. strenua* on six different islands (Pl. IV.). We therefore get not only a phenomenal amount of diversity among the species *inter se*, but there is a very remarkable amount of diversity within any given species. Such a diversity and such a distribution as the various forms of these Finches exhibit, and to which I can only allude in this very brief way, is, so far as I have been able to discover, absolutely unparalleled as far as existing birds are concerned, no matter how far afield we roam in our search.

Diversity in any given genus of birds in island groups of one sort or another is now, of course, a matter of common knowledge, although this was by no means the case in Darwin's day. But what generally, if not almost invariably, happens in an island group is that each island has its own peculiar species or subspecies.

I have, for instance, collected on nearly all the islands of the West Indies, and there you find that the species or subspecies of any given genus, such as *Coereba*, *Euetheia*, or *Loxigilla*, are restricted in a very definite way to their own particular islands or subsidiary groups of islands. In six years' experience of winter collecting I do not remember ever finding a species or subspecies on any island other than its own proper home, even though most of the islands are in sight of their immediate neighbours. I know of only one exception in the West Indies to this rule, and that is on Cuba, a large island between 800 and 900 miles long. I still remember my delight and surprise in coming across the *two* species of *Territristis*, which seem to be restricted to different ends of that island. But, as a matter of fact, we need not have gone so far afield as the West Indies to find examples of the normal mode of distribution in island groups; for as I have already said, of the ten or eleven species or subspecies of Mocking-bird found in the Galapagos no two different forms are found upon any one island.

SOME DETAILS OF DIVERSITY

Coming down to details, I must now refer, very briefly, to some of the forms the diversity takes in the case of these Finches. It shows itself in two conspicuous ways—in the form, shape, and size of the bill, as also in a corresponding degree in the size and general shape of the body, and in general coloration or colour-pattern.

You have already noted on the screen the extraordinary range of the *diversity in the bills*. In very large series the intergradation is almost complete, and the birds themselves may vary in size and form from that of a good big plump Grosbeak, with an enormously heavy bill, to that of a warblerlike form (*Certhidea*) with a relatively longer and quite slender bill. Indeed, the last form has been referred to the American Warblers (Mniotiltidae), and not to the Finches. My anatomical investigations have, however, so far as I am personally concerned, satisfied me that it is a generalized or unspecialized form of Finch. The contrast between the two extremes is phenomenal.

The diversity and intergradation in coloration is also extremely puzzling. It has been almost the despair of systematists. Colour-pattern is vague, and depends apparently upon the dose of some melanistic or pigmentary factor; and the amount of this depends again to a large extent upon age and sex. The rule is for the female to be streaked, dusky, or yellowish. The so-called genera have a greater or lesser dose of black; for instance, in *Geospiza* you get an all black condition in the fully mature male. In *Platyspiza* and *Camarhynchus* a black-headed condition only. In *Cactornis* males and females are alike unpigmented. In *Certhidea* you get a pale yellowish bird with a chestnut throat. A chestnut throat crops up in the almost black West Indian Finch genus, *Pyrrhulagra*. Swarth has called attention to the interesting fact that a different percentage of males in the fully mature plumage occurs upon different islands. Males in the most fully mature, saturated or perfect plumage are very rare in all species. It seems, therefore, difficult to resist the conclusion that the differentiation by systematists of the many species or subspecies has been to a large extent based on or influenced by the degree of black pigment saturation. There seems, in fact, to be a progression from a plain buffy-yellow through streaked and black-headed phases to all black. Then again, as regards one group known as *Geospiza*, three "modes" in actual size of the bird stand out conspicuously. There is a large size (*G. magnirostris*), a medium size (*G. fortis*), and a small size (*G. fuliginosa*). In other respects these three forms are all alike, and all occur together on most of the islands in mixed flocks feeding together. Swarth has called attention to the fact that the large-billed *G. magnirostris* diminishes in size of bill from north to south. In *G. fortis* and *G. fuliginosa*, on the other hand, the size diminishes from south to north. *Platyspiza* ranges almost unchanged throughout the islands.

Another interesting phenomenon exhibited by these Finches is that the coloration in general points to a definite tendency to an inhibition of development leading to a permanent condition of *juvenility*, a tendency seen also in other bird-groups in the same islands. Finally, I might add that I found melanism a very striking feature among the small Finches of the genus *Euetheia*, which I found extremely common on an extremely arid, cactus-grown, and isolated island named Blanquilla, a hundred miles from the coast of Venezuela. The physical conditions on Blanquilla must be remarkably similar to those obtaining in the Galapagos. Some of the Finches I shot there are so similar in appearance to certain species of Darwin's Finches that it would take an expert to distinguish them.

When we go back to the time when North and South America were unconnected, and the Caribbean Sea was in all probability no more than a large inlet of the Pacific, bounded to the north and east by a continuous range of high mountains, now depressed and forming the present-day chain of the Greater and Lesser Antilles, these small West Indian and continental Finches are bound to raise the question as to whether they may not be connected directly or indirectly with the origin of Darwin's Finches.

GENERIC DISTINCTIONS

A word, too, must be said in regard to the degree of generic distinctions. Few authors seem to agree on the point. Rothschild and Hartert group the entire community of these Finches under one single genus, *Geospiza*. At the opposite extreme Swarth, who has probably had infinitely more material under his observation than anyone else, considers there are six, including *Certhidia* [*sic*].

I have lately had the privilege of making an anatomical study of some valuable material in spirit collected by Dr. Gregory Bateson and some from Tring, and I should be much more inclined to agree with Rothschild and Hartert. The more material that will be collected for anatomical study the more likely it would seem that Rothschild and Hartert will be proved to be right, so complete is the intergradation; but if ever generic distinction was justified as a *purely practical convenience* for systematists it seems to be so here. If we are trying to get at the truth of what these Finches really are, I think generic splitting probably handicaps us, but splitting always has the advantage of demonstrating diversity, and lumping the disadvantage of causing it to be lost sight of.

From my anatomical investigations I find as many or more differences in form and shape of the bill and skull within the genus *Geospiza* than, say, between the genera *Cactornis* and *Geospiza* (Pl. III.).

I may add here that Swarth has endeavoured to raise the Geospizids to the status of a family. From my anatomical studies I can find no justification for such a proceeding. I find, on the contrary, that they belong most definitely

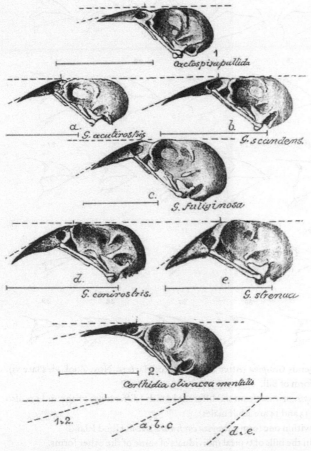

Plate III. Skulls of species of genera *Geospiza*, *Cactospiza* and *Certhidea* to show as much diversity *within* the genus *Geospiza* as exists between it and *Cactospiza* or *Certhidea*. The dotted lines below represent different angles of bill depression in different "genera" or groups of species.

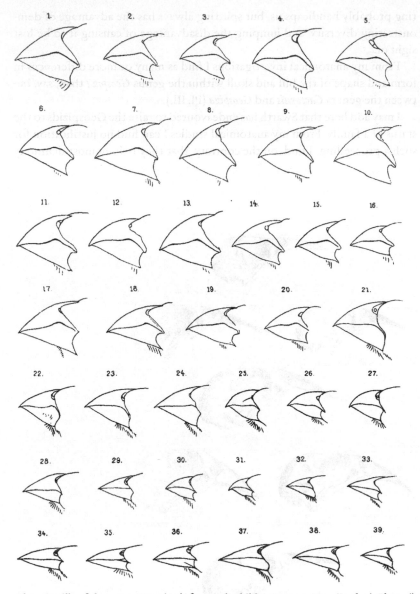

Plate IV. Bills of the genus *Geospiza* (After Rothschild & Hartert, Nov. Zool. vi, Plate vi). To show Diversity in form of Bill.

Figs. 1–16. Bills of *Geospiza strenua* from six different islands (differences in age and sex also shown: Nos. 1, 2, 7, 10, 13 and 14 are adult males).

Figs. 17–20. Diversity within one form *(Geospiza conirostris)* from Hood Island.

Figs. 21–39. Diversity in the bills of typical individuals of some of the other forms.

to the Fringilline division of the Fringillidae. The mere presence of palato-maxillaries (not to be confounded with maxillo-palatines) alongside the pre-palatine bars of the palate is almost enough in itself to settle this point.

ENVIRONMENTAL FACTORS

The correlation of food and other environmental factors with the extraordinary diversity obtaining in these Finches has failed to throw light on the problem. The proventriculus and gizzard show a lack of specialization throughout the whole group. Their smallness and simplicity in the large-billed forms is very noticeable, as I have proved by dissection. Large-billed and small-billed forms all seem to be alike in regard to the simplicity of the structure and small size of the stomach relative to the bulk of the bird. I found no differentiation in the stomachs of birds with big finch-like bills or in those with warbler-like bills. Beebe, from personal observation, has stated that birds utterly dissimilar in relative proportions of bills were feeding upon identical food.

As regards physical and other conditions in the various islands presenting factors which might have contributed to the diversity exhibited, I only have time to say that a study of all that has been written fails to reveal any; and all writers seem to be in agreement on this point.

Isolating factors of whatever kind seem, in fact, so far as one can gather, to be absent, unless a want of synchronization in the breeding periods of the various groups of Finches is subsequently found to obtain. It is true that Swarth thought that some groups of Finches were more restricted to small trees and bushes than were others which have been known as Ground-Finches, but this seems to be a very vague form of isolation, and requires special and long-continued observation in the field. Moreover, all accounts agree in stating that birds of very diverse structure are found living together. There seem to be no differential environmental factors, which could be regarded as having a survival value, which could obviously count one way or the other; and, therefore, no scope for Natural Selection.

EVIDENCE OF HYBRIDIZATION

Very little work in the field seems to have been done in respect of the question whether hybridization occurs or not. Beebe, on Daphne Island, had evidence of the mating of birds of considerable difference in size of body, wing, and bill. The male, he says, was breeding in partly immature plumage, the female feeding a young bird, not more than two weeks out of the nest, and at the same time completing a nest which already contained a new-laid egg. Assuredly, he writes, a mad country for birds. Beebe also found that 50 per cent. of the eggs of these Finches were quite infertile. His observations extended to four species. If this infertility is found to be constant, it is a strong point in favour of hybridization. Gregory Bateson collected a jar full of birds in spirit in which he had detected evidence of hybridization.

THE NATURE OF THESE FINCH GROUPS

How then are we to account for this extraordinary *mixed swarm or crowd* (*Plethogam*), for it is nothing else, of Finches which exhibit such a bewildering diversity, intergradation, and distribution.

In an address as Chairman before the British Ornithologists' Club (December 1929), I applied the terms *mictogone* (μηξις, a mixture; γονη, offspring) or *plethogam* (marriage-crowd) to such an assemblage, and compared the condition to that which we see in New Zealand in the case of the extinct Moas, in which group thirty-eight species, referable to seven genera, have been described, ranging from giants like *Dinornis maximus* to Dodo-like dwarfs such as *Anomalopteryx didiformis*. All this heterogeneous swarm of Moas, which exhibited diversity in many different respects, was herded together into an area one-eighth the size of Great Britain and Ireland, though doubtless their original area of distribution was much larger. Bearing these and other facts in mind, it is difficult to resist the conflation that in the Finches of the Galapagos we are faced with a *swarm of hybridization segregates* which remind us strangely of the "plant swarms" described by Cockayne and Lotsy in New Zealand forest as the result of natural crossings.

I think it was William Bateson who always maintained that the Finches of the Galapagos could only be explained on the assumption that they were segregates of a cross between ancestral forms distributed over a large insulated area which was subsequently broken up by subsidencies or upthrusts leading to the present disposition of the islands. But I do not think even Bateson realised the hotch-potch swarm of finch-forms which these islands exhibit.

Then again this swarm of diverse forms has been ascribed to the fact that the natural tendency to vary, resident in all organisms, has been uncontrolled by any selective action. *But why should this occur in the case of the Finches and not with other resident bird-groups on the island*, such, for instance, as the Mocking-birds?

Are we then to conclude that there is probably little or no real difference in the origin of the extraordinary diversity exemplified in the Drepanidae of Hawaii as compared with that of the Finches of the Galapagos, except that in the case of the Drepanidae there has been far more scope of isolating factors, so that we get a far greater number of pure-line products?

But it is not for me to try and explain the problem of these Finches before an audience of the kind assembled here. I have simply brought before you a few facts solely in the hope that properly qualified investigators, properly equipped, may be sent to the Galapagos with the sole object of studying on the spot and for a sufficiently long period, by means of *actual breeding experiments*, the genetics of this very interesting group of birds. For, as Lord Rothschild has said: "If the collections (of skins) which have already been made are not sufficient to throw light upon the problem *no collections will ever do so.*"

The problem is only likely to be solved by experiments and observations on the spot, and that ought to be comparatively easy, for these Finches are so tame that they can almost be picked off the bushes.

Ibis 6 (2) (1936): 310–321. Reprinted with permission from Wiley-Blackwell.

From *Darwin's Finches* (1947)

❋ DAVID LACK

* Chapter II: Classification

Something more is included in our classification than mere resemblance.... Propinquity of descent— the only known cause of the similarity of organic beings,—is the bond, hidden as it is by various degrees of modification, which is partially revealed to us.... Thus the grand fact in natural history of the subordination of group under group, which, from its familiarity does not sufficiently strike us, is in my judgement explained.

Charles Darwin: *The Origin of Species*, Ch. XIII*

DIFFICULTIES

Before any discussion of them is possible, Darwin's finches must be named, although this in itself commits one to a partial interpretation of the problems of their evolution. Fortunately, there is no longer any serious dispute as to the correct naming of the various forms. At one time matters were otherwise, and authorities differed considerably as to which specimens should be ascribed to different species, which were island forms of the same species, and which were merely varieties.

These differences of opinion arose from the inadequacy of the collected material, but they were considerably accentuated by the unfamiliar nature of the variations found among the finches, which often ran contrary to the experience of museum systematists accustomed to working with European or North American birds. In continental passerine birds, closely related species tend to differ from each other chiefly in plumage, and they are usually simi-

---

* To provide a succinct quotation, I have taken the last sentence of the above from a paragraph which precedes the rest. But there is no distortion of Darwin's meaning.

lar in beak and other structural characters. Differences in beak more usually characterize the broader units, the different genera. In Darwin's finches, on the other hand, closely related species usually differ markedly in beak, but little, if at all, in plumage, and plumage differences are chiefly important in distinguishing the different genera.

A further unusual feature in Darwin's finches is that some of the species are highly variable. For example, individuals of the ground-finch *Geospiza fortis* are so variable in beak that they were for a long time considered to belong to at least two, and by some authorities to three or more, separate species. Extensive collections were needed to establish that in fact only one exceptionally variable form was involved. Had only two specimens of this species been known, one at the minimum of size and the other at the maximum, systematists might well have placed them in separate genera.

ENGLISH NAMES

The species of Darwin's finches are most clearly and simply referred to by their scientific Latin names, but to help the reader to picture the birds, I have given each species a brief descriptive English designation. English names have been invented for some of the species by previous systematists, and Hellmayr (1938) has produced a complete list. But Hellmayr's names tend to be mere translations of the Latin ones, and are not sufficiently descriptive to help the general reader, while a few are actually inappropriate. Because of these objections, and because Hellmayr's names are mostly of recent origin and are not yet sanctioned by custom, I considered it justifiable to replace them by a new and more appropriate set. This change should not cause confusion, as each species is always referred to by its scientific Latin name, and the preceding English name should be regarded merely as a supplementary description. There are not, of course, any local or traditional names for the finches in the Galapagos.

The genera, species and island forms of Darwin's finches are briefly summarized in Tables II, III and IV which follow later in this chapter. In addition, the species are illustrated in Fig. 3, also in the Frontispiece and Plates III, IV and V [plates IV and V are not reprinted]. It is hoped that these illustrations, the three tables and the English designations of the birds will keep the characteristics of the various forms sufficiently in readers' minds for the argument of later chapters to be clear.

THE SUBFAMILY

On anatomical grounds, Snodgrass (1903), Sushkin (1925, 1929) and Lowe (1936) are agreed that the varied forms comprising Darwin's finches are all closely related to each other, including the peculiar warbler-finch *Certhidea*,

Plate III (i) *Geospiza magnirostris* (ii) *Camarhynchus parvulus* (iii) *Camarhynchus pallidus* with its stick (iv) *Camarhynchus crassirostris* (v) The Indefatigable mockingbird (vi) The Hood mockingbird (i) *by H. Gibb.* (ii), (iii), (v) *and* (vi) *by R. Leacock,* (iv) *by D. Lack.*

which superficially looks much more like a warbler than a finch. *Certhidea* was placed with the other finches when first described by Gould in 1837, though Darwin wondered whether this was correct. Later workers considered that the bird was not a finch at all, and placed it in various families, latterly with the American warblers (Mniotiltidae). However, the findings of Snodgrass have now been generally accepted, and our own field study has revealed strong similarities in breeding and other behaviour, confirming the close relationship of all the forms with each other, including *Certhidea*.

The whole group may be termed the Geospizinae, a subfamily of the finches (Fringillidae), to which Sushkin and Lowe showed that they are related. For simplicity they are referred to in this book as Darwin's finches. The term "Galapagos finch" is less satisfactory, since one species, namely *Pinaroloxias inornata*, occurs not in the Galapagos, but on Cocos Island, 600 miles to the north-east. Similarities in plumage and anatomy show that *Pinaroloxias* is undoubtedly one of the Geospizinae. With this exception, Darwin's finches are confined to the Galapagos.

All of Darwin's finches are greyish brown, short-tailed birds, with fluffy rump feathers. In some species the two sexes are alike in plumage, in others the males are distinguished by a varying amount of black feathering, while the male *Certhidea* has an orange-tawny throat. All the species build roofed nests, large for the size of the bird, and lay white eggs spotted with pink, four to a clutch. All are territorial and monogamous, and in all species courtship includes display with nest material and the feeding of the female by the male. There are many other similarities in breeding behaviour and many resemblances in internal anatomy. They vary in size from a small warbler to a very large sparrow, and their feeding habits are exceedingly diverse, as are their beaks, which range from delicate and thin to stout and huge.

GENERA

Swarth (1931), the latest authority, divided Darwin's finches into six genera, while a seventh has been used by some other workers. The beak differences between these seven subgroups are so marked that, if they were continental passerine birds, they would unhesitatingly be classified as separate genera. However, in most other respects, including plumage and breeding habits, some of these subgroups are closely similar to each other. Further, since Darwin's finches are in this book divided into only fourteen species, the employment of six or seven generic names seems excessive. For these reasons only four genera are used here, as set out in Table II.

DETERMINATION OF SPECIES AND SUBSPECIES

The scientific classification of birds, started by Aristotle and continued after a long interval by Gesner, Ray, Linnaeus and many others, was in the first place a tabulation of existing knowledge regarding the birds of their own countries. The species concept of the early naturalists, though at first implicit rather than explicit, was more or less sound, because it was based on familiar animals. It was early appreciated that birds consist of separate kinds or species, that the members of each kind have in common certain characteristics

TABLE II. Genera of Darwin's finches

Genus	Designation	Full male plumage	Beak	Chief food	Number of species
Geospiza	Ground-finch	Wholly black	Finch-like (one longer)	Seeds (cactus in one case)	6
Camarhynchus	Tree-finch	Partly black or without black	Thick, shape variable	Insects (fruit in one case)	6
Certhidea	Warbler-finch	Orange-tawny throat	Slender, warbler-like	Small insects	1
Pinaroloxias	Cocos-finch	Wholly black	Slender decurved	Small insects	1

Note: As here used, *Geospiza* includes the species *scandens*, sometimes placed in a separate genus *Cactornis*, while *Camarhynchus* includes the three genera *Platyspiza*, *Camarhynchus* and *Cactospiza* of Swarth (1931).

in which they differ from all other kinds, and that the individuals of one kind do not usually breed with those of any other.

Later, when travellers brought back bird specimens from distant parts, naturalists began to give names to birds which they had not studied in their natural haunts, a practice which has continued up to the present time. Darwin's finches, for example, have been named and described in seven big systematic papers, those of Gould (1841), Salvin (1876), Ridgway (1897), Rothschild and Hartert (1899, 1902), Snodgrass and Heller (1904) and Swarth (1931), while these and other authors, including Hellmayr (1938), have also contributed shorter papers and notes. But of these men, only Snodgrass and Heller had seen alive the birds of which they wrote, though Swarth was able to visit the Galapagos some years after his paper was published.

The naming of the first bird species was based on a wide experience of living birds in their natural homes. Obviously the determination of species from specimens brought by others from strange places is a much more superficial procedure. However, the great majority of the birds of the world have been named solely from the appearance of their skins, the latter preferably being collected as a series, together with information as to age, sex and locality. Despite the superficiality of this information, later field study has nearly always confirmed the diagnosis made by the skilled museum systematist, and has shown that the species which he has named are in fact separate breeding-units. It has sometimes been claimed that the museum worker is engaged in compiling a purely artificial catalogue of the animal kingdom. But, in birds at least, this statement is untrue, and the discontinuities which separate names imply exist in nature. Even in the exceptionally difficult group of Darwin's finches, our field observations have shown that no fundamental alterations are required in the classification by Swarth (1931) based purely on museum specimens, and the only changes introduced here have been with the object of reducing the number of essential names to a minimum. Owing to the comparative ease with which bird species can be distinguished, and helped by the fact that there are a moderately large but not an excessive number of species, avian systematics is in a far more advanced state than that of any other class of animals. As a result, birds provide particularly good material for studying the origin of species.

Widely distributed bird species tend to be subdivided into geographical races or subspecies. The term subspecies has been used rather loosely in some other groups of animals, so it may be emphasized that in birds it is synonymous with a geographical race. In continental birds, there is usually no difficulty in deciding whether two different forms should be classified as subspecies or species. Subspecies, as the term implies, differ from each other to a smaller extent than do full species, the differences chiefly involving shade of plumage and size. But a more important criterion is that of geographical distribution. Subspecies of the same species always breed in separate geographical regions, and where their respective breeding zones adjoin, they often interbreed freely and intergrade in characters. On the other hand, two forms which breed in the same region without normally interbreeding are always classified as separate species, however similar they may be to each other in appearance.

Since, as discussed later, new species evolve from subspecies, there is no absolute division between the two categories, and the decision in border-line cases is inevitably arbitrary. Difficulty occurs chiefly in regard to related forms which occupy separate geographical regions, like subspecies, but which differ from each other more markedly than is usual among races of the same species. This happens not uncommonly among related island forms, and as these are isolated from each other geographically, it is often impossible to know whether or not they would freely interbreed with each other if they met. In general, it is more useful to emphasize the affinities rather than the distinctiveness of such island forms. Hence in Darwin's finches I have wherever possible treated them as subspecies of one species rather than as separate species. This also has the advantage of reducing the number of essential names. On the other hand, Swarth treated many of these distinctive island forms as separate species. On such a point, nomenclature is arbitrary.

SPECIES OF DARWIN'S FINCHES

With the above points in mind, Darwin's finches have in this book been divided into fourteen species. The characteristics of these birds are considered in detail in later chapters. The chief differences concern the size and shape of the beak, and at this stage a sufficient summary is provided by Table III, together with Fig. 3, as set out overleaf.

TABLE III. Species of Darwin's finches

Scientific name	Descriptive designation
1. *Geospiza magnirostris* Gould	Large ground-finch
2. *Geospiza fortis* Gould	Medium ground-finch
3. *Geospiza fuliginosa* Gould	Small ground-finch
4. *Geospiza difficilis* Sharpe	Sharp-beaked ground-finch
5. *Geospiza scandens* (Gould)	Cactus ground-finch
6. *Geospiza conirostris* Ridgway	Large cactus ground-finch
7. *Camarhynchus crassirostris* Gould	Vegetarian tree-finch
8. *Camarhynchus psittacula* Gould	Large insectivorous tree-finch
9. *Camarhynchus pauper* Ridgway	Large insectivorous tree-finch on Charles
10. *Camarhynchus parvulus* (Gould)	Small insectivorous tree-finch
11. *Camarhynchus pallidus* (Sclater and Salvin)	Woodpecker-finch
12. *Camarhynchus heliobates* (Snodgrass and Heller)	Mangrove-finch
13. *Certhidea olivacea* Gould	Warbler-finch
14. *Pinaroloxias inornata* (Gould)	Cocos-finch

Notes: (i) *Geospiza conirostris* has obvious affinities with *G. scandens* and replaces it geographically, but it is so distinctive that it is given a separate specific name. It should perhaps be reckoned as part of the *G. scandens* superspecies, but this is not certain.

(ii) The distinctive island forms related to *Camarhynchus psittacula* have hitherto been regarded as separate species, but they are here united under *C. psittacula*, except for *C. pauper*, which, though no more distinctive than the rest, must be treated as a separate species, as it occurs together with *C. psittacula* on Charles without intermingling. The significance of this case is discussed later.

(iii) Since the genera *Certhidea* and *Pinaroloxias* each include only one species, there is no need to give both generic and specific names in the text, and the latter are normally omitted. But the names of the other species are required.

(iv) In *Camarhynchus*, confusion is possible as regards the gender of the specific names. The International Rules of Nomenclature state that where the specific name is an adjective it should follow the gender of the generic name, but where it is a noun in apposition it should keep its own gender. In 1837 Gould named *Camarhynchus psittacula*; *psittacula* is to be regarded as a noun in apposition, so Gould's alteration to *psittaculus* in 1841 has not been adopted. But when the species originally named *Geospiza parvula* and *Cactornis pallida* are, as here, transferred to the genus *Camarhynchus*, they become *parvulus* and *pallidus* respectively, as the latter are adjectives. There are an increasing number of systematists who feel that the termination of an adjectival specific name should not be changed in this way, particularly as so many modern generic names are indeterminate as to gender. This view has much to recommend it, but as the international rules of nomenclature are explicit on this point, and have not, as yet, been changed, they are followed in this book.

Fig. 3. Darwin's finches; male and female of each species. Numbers refer to Table III. About 2/7 life-size.

ISLAND SUBSPECIES

Most species of Darwin's finches occur on a number of islands. In some cases the island populations differ sufficiently to justify division into subspecies, in other cases the differences are less marked, and in yet others they are barely perceptible. Provided the degree of difference is described, the question of which island forms should be named is unimportant and inevitably somewhat arbitrary. In this book I have treated a form as a separate race or subspecies when I could distinguish at least 75 per cent of the available specimens from the specimens of any other form. As a result, I have used fewer subspecific names than did Swarth (1931). Incidentally, the general reader need not trouble to remember the subspecific names, since for most purposes each form is sufficiently indicated by its specific name together with the name of the island or islands on which it occurs. The distribution of the various species and island forms is shown in Table IV.

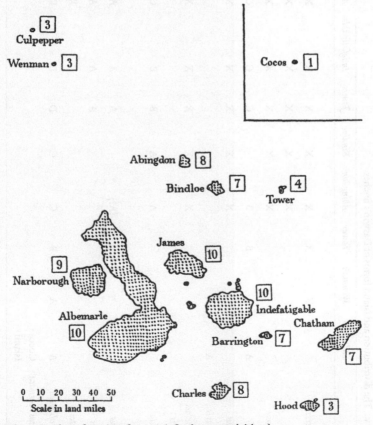

Fig. 4. Number of species of Darwin's finches on each island.

TABLE IV. The distribution and island forms of Darwin's finches

Species	Culpepper	Wenman	Tower	Abingdon	Bindloe	James	Indefatigable	Albemarle	Narborough	Barrington	Chatham	Hood	Charles
Geospiza magnirostris		X	X	X	X	X	X	X	X	X	X		
G. fortis				X	X	X	X	X	X	X	X	X	X
G. fuliginosa				X	X	X	X	X	X	X	X	X	X
G. difficilis	A	A	B	B	C	C	C	(C?)	(B?)				
G. scandens				X	X	X	X	X		X	X		X
G. conirostris	A		B									C	
Camarhynchus crassirostris				X	X	X	X	X	X	X	X		X
C. psittacula				A	A	B	B	C	C	B	B		B
C. pauper													X
C. parvulus				(A?)		A	A	A		A	B		A
C. pallidus						A	A	A			B		(A?)
C. heliobates								X	X				
Certhidea olivacea	A	A	B	C	C	D	D	D	D	E	F	G	H
Pinaroloxias inornata	Cocos Island												

Notes: (i) "X" denotes a resident breeding population, and the other letters a division into island forms as follows:

Geospiza difficilis: A = septentrionalis Rothschild and Hartert; B = difficilis Sharpe; C = debilirostris Ridgway.

Geospiza conirostris: A = darwini Rothschild and Hartert; B = propinqua Ridgway; C = conirostris Ridgway.

Camarynchus psittacula: A = habeli Sclater and Salvin; B = psittacula Gould; C = affinis Ridgway.

Camarhynchus parvulus: A = parvulus (Gould); B = salvini Ridgway.

Camarhynchus pallidus: A = pallidus (Sclater and Salvin); B = striatipectus (Swarth).

Certhidea olivacea: A = becki Rothschild; B = mentalis Ridgway; C = fusca Sclater and Salvin; D = olivacea Gould; E = bifasciata Ridgway; F = luteola Ridgway; G = cinerascens Ridgway; H = ridgwayi Rothschild and Hartert.

(ii) One adult Geospiza difficilis debilirostris collected on South Albemarle, two adults probably referable to G. difficilis difficilis from Narborough, three Camarhynchus parvulus from Abingdon and one C. pallidus from Charles are very probably representatives of all breeding populations on these islands, but the numbers collected are too small for this to be considered certain, so they are listed with a query in the table.

(iii) Geospiza scandens defies consistent nomenclatural treatment; the forms on the adjacent islands of James and Bindloe are completely separable, so should receive separate subspecific names, but the gap between them is bridged by the populations on islands south of both of them. Hence all the island forms have hesitatingly been united under one name.

(iv) If Darwin's large specimens of Geospiza magnirostris are accepted as belonging to an extinct form from Charles, then the specimens of this species from all other islands ought probably to be referred to another subspecies, G. magnirostris strenua Gould.

So much collecting has been carried out that, apart from the doubtful cases in note (ii), Table IV can probably be taken as a complete list of the finch species breeding on each Galapagos island. In addition there are a few cases, not listed in the table, in which finches have wandered from one island to another without becoming established. The Academy expedition caught a juvenile of the cactus ground-finch *Geospiza scandens* at sea 20 miles south of Indefatigable, and have four records of the medium ground-finch *G. fortis* seen a few miles offshore. The same expedition took fifteen juvenile specimens of *G. fortis* on Hood. No other collectors have found this species on Hood, and we saw none there, despite careful search, so the Academy birds were presumably stragglers. The same probably applies to one specimen of *G. fortis*, four of the small ground-finch *G. fuliginosa*, and one of the small insectivorous tree-finch *Camarhynchus parvulus*, taken by the Academy expedition on Wenman, as all were juveniles, and these species have not otherwise been found there. Wenman is so small that, if these species had been resident, they could scarcely have been overlooked by other collectors. Another presumed straggler in the Academy collection is a typical specimen of the large ground-finch *Geospiza magnirostris* from Charles, where no other recent expedition has found this form. The occurrence of these stragglers, nearly all juvenile individuals, is of importance when considering the degree of isolation of island forms.[1]

EXTINCT FORMS

Nearly all the finches collected by Charles Darwin are similar in appearance to those taken by later collectors, but there are two forms which have not been recorded since 1835. First, there are three male and four female specimens obviously referable to the large ground-finch *Geospiza magnirostris*, but which are considerably larger than any collected since. They are labelled as coming from Charles, but Swarth (1931) thought that this was probably a mistake, and concluded that the form of this species must have changed since the time of Darwin's visit. However, while it would be pleasing to demonstrate measurable evolution on the basis of specimens collected by Darwin, it seems far more probable that these large birds represent an extinct subspecies of *G. magnirostris* from Charles, where the bird no longer resides. Darwin collected typical present-day *G. magnirostris* on James, and expressly noted that the larger form did not occur there. Further, though Darwin unfortunately mingled

---

1. There are some further specimens of Darwin's finches, collected by other expeditions, which certainly do not represent breeding populations on the islands where they were found, and so may either be stragglers or are incorrectly labelled. These are one adult *Geospiza fortis* labelled as from Tower, one *G. scandens* of the Bindloe type from James, one *G. c. conirostris* from Gardner near Charles (which suggests an error for Gardner near Hood, where this form occurs), and two *Camarhynchus psittacula* from Chatham.

some of the specimens from different islands, his remarks (1839) make it clear that this applied only to specimens from the first two islands which he visited, namely, Chatham and Charles. When on Charles, Darwin came to realize that specimens from different islands might differ in appearance, and thereafter he kept the collections from each island separate. Hence specimens labeled as coming from either Chatham or Charles were presumably collected on either one or the other. Darwin's specimens most probably came from Charles, since he spent only a few hours ashore on Chatham, but several days on Charles. Further, Charles is a small island and was intensively settled just before Darwin's visit. Probably as an indirect result of this settlement, the endemic mockingbird *Nesomimus* has become extinct on Charles, and the same factor possibly brought about the extinction of the Charles form of *Geospiza magnirostris*.

Also among the *Beagle* specimens are two which in my opinion belong to an unknown form related to the sharp-beaked ground-finch *G. difficilis*. They have a similar shape of beak, though the beak is larger. Gould named one of these specimens *G. nebulosa*, and, I think mistakenly, referred a second rather similar specimen to another species. The one named *G. nebulosa* is labelled as coming from Charles, so it may well represent an extinct form of *G. difficilis* from this island, as the latter species does not occur there at the present time. It may be added that Darwin also collected a typical specimen of *G. difficilis*, presumably on James, as it belongs to the James form of this species.

INTERMEDIATE SPECIMENS

Rumour has it that the gardens of natural history museums are used for surreptitious burial of those intermediate forms between species which might disturb the orderly classifications of the taxonomist. Actually, specimens intermediate between two species are extremely rare in birds, and almost every specimen can be immediately assigned to a known species, except in a few cases where an obvious hybrid is involved. But in this respect Darwin's finches are exceptional, and specimens intermediate between two species, though rare, are less rare than usual. As discussed in Chapter X, some of these intermediates are probably freaks, and others are possibly of hybrid origin. Their existence has added to the difficulties of the systematist in correctly determining the species of Darwin's finches. . . .

* Chapter XI: An Evolutionary Tree

On the principle of the multiplication and gradual divergence in character of the species descended from a common parent, together with their retention by inheritance of some characters in common, we can understand the excessively complex and radiating affinities by which all the members of the same family or higher group are connected together.

Charles Darwin: *The Origin of Species*, Ch. XIII

THE ORIGINAL STOCK

From the seeds planted by Darwin, a forest of evolutionary trees came to adorn the text-books of zoology. Their cultivation is now somewhat out of fashion, but since the central theme of this book is that Darwin's finches evolved from a common stock, it is necessary to suggest the steps by which this could have come about, though in this matter no finality of judgment is possible. The difficulties in reconstructing the course of evolution are great, since resemblances between existing species may be due to close relationship, but may also be brought about by parallel evolution, or by the chance retention of the same primitive features. Further, island populations are often small, so that forms can become modified or extinct very quickly. Nor are there any fossil specimens to help fill the gaps in the living record.

Previous chapters have shown so many resemblances between the various species of Darwin's finches that it is scarcely necessary to reaffirm that they are all related to each other. All recent writers are agreed on this, Swarth (1931) from a study of skins, Snodgrass (1903), Sushkin (1925, 1929) and Lowe (1036) on anatomical grounds, and the present writer from a study of their breeding behaviour. But though all the species show marked similarities to each other, they do not show a close resemblance to any particular species of finch on the South or Central American mainland. Either the mainland ancestor has become extinct, or Darwin's finches have diverged from it so far that their close relationship is no longer apparent. Sushkin and Lowe have established that Darwin's finches are derived from the fringilline subfamily of the finches, but it is difficult to restrict their point of origin more closely.

The evidence suggests that the following are primitive geospizine features: black plumage in the male, streaked underparts and a rufous wing-bar in the female, a heavy finch-like beak, a diet of seeds, and a habitat in the arid lowland zone. One of Darwin's finches, namely, the sharp-beaked ground-finch *Geospiza difficilis*, possesses all these characteristics. The differentiation of this species into three well-marked races also suggests that it is a long-established form, while its irregular distribution on the outlying Galapagos islands suggests that it is in process of elimination by the small ground-finch *G. fuliginosa*, which was presumably evolved later.

The Culpepper and Wenman form of *G. difficilis* has an olive tinge to the upper parts, a buff tinge to the underparts and a rufous wing-bar, characters which are shared by two other species, namely, the peculiar Culpepper ground-finch *G. conirostris darwini* and the Cocos-finch *Pinaroloxias*. Such a discontinuous distribution of characters suggests that they are primitive. Hence *Geospiza difficilis septentrionalis* of Culpepper and Wenman can probably be regarded as the least modified living representative of Darwin's finches.

The small ground-finch *G. fuliginosa* shows many of the features which are

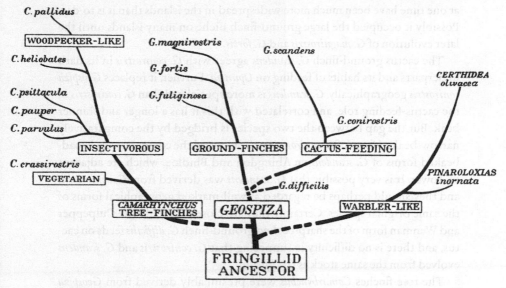

Fig. 21. Suggested evolutionary tree of Darwin's finches.

presumed to be primitive in Darwin's finches, but it lacks the rufous wing-bar. It may well have evolved from the sharp-beaked ground-finch *G. difficilis*, or from an ancestral form of the latter. Indeed, the Tower form of *G. difficilis* looks so like *G. fuliginosa* that, had *G. difficilis* occurred only on Tower (where *G. fuliginosa* is absent), the two would probably have been regarded as geographical races of the same species. It is even possible that *G. fuliginosa* was derived from *G. difficilis* via the Tower form of the latter, but I would not press this point, as I now consider that some of their resemblances are due to parallel evolution.

The medium ground-finch *G. fortis* differs from the small *G. fuliginosa* solely in size and proportions, and extreme individuals of the two species approach each other closely. Hence *G. fortis* may well have evolved from the smaller species. Similarly, the large ground-finch *G. magnirostris* differs from the medium *G. fortis* solely in size and proportions, and very possibly evolved from it.

The relationship of the large cactus ground-finch *G. conirostris* is more difficult to determine. The discontinuous distribution of this bird solely on the three outlying islands of Culpepper, Tower and Hood, together with its differentiation into well-marked races, indicate that it is a long-established species. This is corroborated by the fact that the Culpepper form *G. c. darwini* possesses primitive plumage features found otherwise only in the Cocos-finch *Pinaroloxias* and in the Culpepper and Wenman form of the sharp-beaked ground-finch *Geospiza difficilis*. Probably *G. conirostris* was derived from an ancestral form of *G. difficilis*, and its present distribution suggests that it may at one time have been much more widespread in the islands than it is to-day. Possibly it occupied the large ground-finch niche on many islands until the later evolution of G. *magnirostris* and G. *fortis*.

The cactus ground-finch G. *scandens* agrees with G. *conirostris* in its dark underparts and its habits of feeding on *Opuntia*. Further, it replaces *Geospiza conirostris* geographically. G. *scandens* is more specialized than G. *conirostris* in the cactus-feeding role, and correlated with this it has a longer and thinner beak. But the gap between the two species is bridged by the comparatively narrow-beaked form of G. *conirostris* on Tower and the comparatively broad-beaked forms of G. *scandens* on Abingdon and Bindloe, which are adjacent to Tower. It is very possible that G. *scandens* was derived from G. *conirostris*, and they should perhaps be regarded as well-marked geographical forms of the same original species. Certainly they are closely related. The Culpepper and Wenman form of the sharp-beaked ground-finch G. *difficilis* feeds on cactus, and there is no difficulty in supposing that G. *conirostris* and G. *scandens* evolved from the same stock as G. *difficilis*.

The tree-finches *Camarhynchus* were presumably derived from *Geospiza* stock by partial loss of black plumage in the male, reduced streaking in the female, and some modification in beak and feeding habits. The *Camarhynchus* line early diverged into two, the insectivorous forms on the one hand and the vegetarian tree-finch C. *crassirostris* on the other. The latter species has a distinctive song and a somewhat distinctive beak, but resemblances in plumage and beak indicate relationship with the insectivorous forms.

The insectivorous tree-finches have diverged further, into the small C. *parvulus* and the large C. *psittacula*. These two species differ only in size, and one presumably evolved from the other. The evolution of the pronounced island forms of C. *psittacula*, and the occurrence of two similar species C. *psittacula* and C. *pauper* together on Charles, is discussed later. Another insectivorous species, the woodpecker-finch C. *pallidus*, is very similar to the other species in song, plumage, feeding habits and beak, and obviously evolved from the same stock, but it has become more specialized for tree-climbing, wood-boring, and probing into cracks. Finally, the mangrove-finch C. *heliobates* shows considerable resemblance in beak to C. *pallidus* and is presumably closely related to it. Its darker plumage suggests that it may be more primitive than C. *pallidus*; it is not known whether it possesses the latter's specialized feeding habits.

An offshoot in a different direction is the Cocos-finch *Pinaroloxias inornata*. This bird has a slender beak like that of the warbler-finch *Certhidea*, but in plumage it shows marked affinities with the sharp-beaked ground-finch *Geospiza difficilis*, and particularly with the Culpepper and Wenman form, G. *d. septentrionalis*. The males of *Pinaroloxias* and *Geospiza d. septentrionalis* agree in having wholly black plumage save for rufous or buff under-tail coverts, and the females agree in having buff-tinged and much streaked under-parts, a rufous wing-bar and an olive tinge to the upper parts. The marked specialization of its beak indicates that *Pinaroloxias* is an early offshoot from the geospizine stock, and the several resemblances in plumage to *Geospiza difficilis* are presumably primitive characters which it has retained.

Finally, the warbler-finch *Certhidea olivacea* is so distinctive that it obviously diverged from the main geospizine stock very early. That it is a long-established form is also suggested by its differentiation into well-marked island forms. It agrees with the other species of Darwin's finches in internal anatomy, in display and nesting habits, and in the possession (by some island forms) of a rufous wing-bar. Its song is more distinctive than that of other species, but shows similarities to that of the sharp-beaked ground-finch *Geospiza difficilis* on Tower. *Certhidea* differs from *Geospiza* in having unstreaked underparts, but juveniles of the Charles form of *Certhidea* are heavily streaked, while streaking has been lost in some species of *Camarhynchus*. *Certhidea* is also peculiar in possessing an orange-tawny throat patch in the male, but this occurs sporadically in other forms. The freak variants "*Camarhynchus conjunctus*" and "*C. aureus*" are a further link between *Certhidea* and *Camarhynchus*, but they are perhaps hybrids, and their existence does not necessarily imply that *Certhidea* was derived from *C. parvulus*. The most peculiar feature of *Certhidea* is its slender warbler-like beak, but one other species, namely the Cocos-finch *Pinaroloxias*, has a slender beak. This may be due to parallel evolution, but it also seems possible that *Certhidea* and *Pinaroloxias* were derived from a common stock, and that the further modifications in *Certhidea*, particularly in plumage, appeared later.

The foregoing discussion may not be correct in every detail, but its purpose is to show the marked similarities between all of Darwin's finches, and that the derivation of these birds from a common finch-like stock presents no particular difficulties. The precise steps of this evolution cannot be known with certainty.

Gloucester, MA: Peter Smith Publisher, 1968. Excerpts from chapter 2, pp. 12–23; chapter 11, pp. 100–105.

From *Comparative Landscape Genetics and Adaptive Radiation of Darwin's Finches: The Role of Peripheral Isolation* (2005)

❋ KENNETH PETREN, PETER R. GRANT, B. ROSEMARY GRANT, AND LUKAS F. KELLER

Editor's note: A potential alternate topology joins the two Certhidea *species, making them the sister genus to the other taxa. Hypothesized hybridization between* Certhidea fusca *and the Cocos Island finch,* Pinaroloxias inornata, *could account for the observa-*

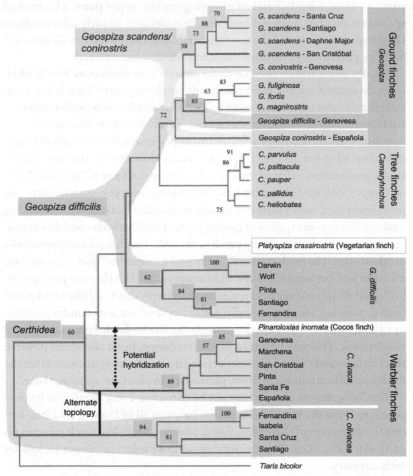

Evolutionary relationships among populations and species of Darwin's finches. Microsatellite genotypes at 16 loci were obtained from 1428 birds from 74 populations, and Ds and UPGMA [unweighted pair group method with arithmetic mean, a clustering algorithm] was used to produce the topology shown. Shaded areas connect populations from each of the three main groups: warbler finches (*Certhidea fusca* and *Certhidea olivacea*), sharp-beaked ground finches (*Geospiza difficilis*); and cactus finches (*Geospiza scandens* and *Geospiza conirostris*). Numbers indicate support for nodes based on bootstraps.

tion that Certhidea fusca *appears more closely related to* Pinaroloxias inornata *than it does to* Certhidea olivacea.

   The topology implies that the founding finches were most like the warbler finches, and that ancestors of the warbler finches founded both the Cocos Island finches and the other Galapagos finch genera. Most Geospiza difficilis *appear to be basal to the remaining finches, with the vegetarian finch diverging from the tree finches and ground finches before those two clades diverged from each other. The tree finches appear to be a natural grouping, as do the remaining ground/cactus finches. While the three ground finch species, Geospiza fuliginosa, G. fortis, and G. magnirostris, appear to be a natural group of closely related species, the group of cactus finches—G. conirostris and G. scandens—do not. Statistical support for some of these divergences is not very strong, indicating that additional studies would be useful.*

Figure excerpted and modified from *Molecular Ecology* 14 (2005): 2943–2957.

* PART 2 *
Adaptation and the Evolution of Diversity

If all the Galápagos finches live in essentially the same general environment, as Darwin initially thought, why are they so variable? This is a smaller version of the question asked by the eminent ecologist G. Evelyn Hutchinson in 1959, "Why are there so many kinds of animals?" If natural selection is actually so important, why isn't there just one terrific species on earth? What processes account for the proliferation of diversity in nature, what processes maintain that diversity, and what processes actually favor diversity?

The question addressed in chapter 5 is whether variation in nature has any adaptive significance at all. One possibility is that finch morphology is so variable precisely because the variation is adaptively irrelevant; all variants could be equally adapted, so none of them is eliminated. The variation could be preserved by isolation alone. This hypothesis was the prevalent one for many years.

Over time, detailed fieldwork challenged that hypothesis, as an association between bill morphology and diet was revealed. Bill characteristics affect feeding abilities—an essential skill in an environment with limited food—so they are likely to be subject to natural selection. An important controversy here, explored in chapter 6, is whether bill traits diverge due to adaptation to fixed resource differences among the islands, or whether they are constantly diverging in response to the ever-changing competitive environment imposed by other evolving species. In other words, is being different from other individuals a by-product or a target of natural selection?

David Lack provided the first evidence that bill morphology diverged among species in response to the presence of other species, but he did not actually document the process of natural selection. Chapter 7 presents studies that directly measure the strength of natural selection and identify constraints on the directional evolution of traits in natural environments.

Bill morphology influences survival, but reproduction is another important component of fitness in natural populations. An important constituent of reproductive success is mating success. Mating success depends on interactions among males as well as interactions between males and females—another instance where selection is imposed by biotic factors that are potentially evolving. Chapter 8 presents studies of sexual selection on male traits operating through female mate choice. It discusses how natural selection, via both viability and sexual selection combined, can contribute to the reproductive isolation between species.

* 5 *
What Matters? Variation and Adaptation

The majority of researchers who encountered the Galápagos finches before David Lack's highly influential book, *Darwin's Finches*, found no evidence that differences in their bill morphology was in any way adaptive. Several of them, including Darwin, Lowe, and Lack himself, believed that the islands in the Galápagos Archipelago were essentially similar in climate as well as topography, and found it difficult to guess why such diversity in bill form would be present in such a homogeneous landscape. In fact, the reason Darwin did not keep his collections from the various islands separate was that it did not occur to him that the specimens would differ among the islands. People like Louis Agassiz would use just this sort of unexplained variation as evidence for divine origination. Agassiz asserted that pronounced differences among taxa in a homogeneous environment could not be explained by evolution through natural selection but rather by the fact that the creator values diversity.

To add to the confusion, Darwin wrote that the finches were feeding in common mixed flocks, and this was also noted by Lowe and others. Apparently, all the finches were eating the same food, which excluded the most obvious adaptive explanation for different bill types. Lowe (chapter 4) saw the extreme variation as evidence for relaxed natural selection. His logic was that, without natural selection, there is no reason why a new form can't persist, provided it remains isolated enough to avoid being blended and diluted to oblivion. Islands conveniently supply such isolation. The continuous variation in bill shape and size, he hypothesized, was due to the hybridization of previously isolated and divergent forms.

The unaddressed question was how the forms become divergent in the first place if natural selection has no role. The development of the concept of random genetic drift by Sewall Wright in the early twentieth century provided the necessary framework for interpreting the divergence of traits that have no adaptive significance. In populations of finite size (that is, all real populations), some proportion of individuals could be eliminated from the population or prevented from mating by pure chance, and in the case of the finches, some perish regardless of the shape of their bill. Even if exactly the same number of individuals with large and small bills existed at the beginning of a season, slightly more survivors with larger bills than smaller bills could remain at the end of the season simply by chance. After all, a process that eliminated *exactly* the same number of each bill type each season wouldn't be random. Over time, the proportion of large bills to small bills will fluctuate and may drift more toward one type than another, and, eventually, one type could disappear altogether. If two populations drift randomly in different directions, and if they don't mix, they will become significantly different from one another eventually. Random genetic drift became a plausible mechanism whereby isolation alone could lead to divergent populations. And if processes other than natural selection (or divine creation) can cause isolated populations to diverge from each other, variation may have no adaptive significance at all; it may be merely an evolutionary frivolity caused by drift.

The importance of this line of reasoning cannot be overstated. Random drift has become the null hypothesis for evolutionary studies, the basis for the "Neutral Theory" of evolution, which describes expected evolutionary changes as a result of everything except natural selection. It is absolutely necessary to know how something is expected to evolve without natural selection in order to be able to recognize natural selection when it is operating.

Contrarian as usual, George Baur (chapter 3) argued that the islands do differ environmentally, and that the environmental differences induced morphological differences among the species located on different islands: "so far as I can anticipate, it will be shown that variation goes on in definite lines determined by the surroundings; and that the surroundings and time are the most important and principle factors of variation." He goes on to say that "natural selection plays only a secondary role, and very frequently none at all" (Baur 1890–1891, 418). Reasonably disagreeing about the ecological homogeneity across the archipelago, he nevertheless dismissed natural selection as a mechanism for maintaining differences among species. Baur seems to argue that environmental effects directly induced neutral morphological changes.

Two questions emerge from these readings: First, are the islands truly ecologically homogeneous, or do they differ in ecologically relevant factors, such as food availability? Second, is the great diversity of bill form in the finches the result of the total absence of natural selection acting on them?

READINGS

Snodgrass presents the first direct analysis of the relationship between bill morphology and diet in the finches. His conclusion, surprising today, is that no correlation exists between the food they eat and the size and shape of their bills. By examining the stomach contents of samples collected during the rainy season (between December and June), he found that a single species had a broad diet, and that many species that differed in bill morphology ate the same species of seeds. While not all finch species ate exactly the same seeds throughout their range, the examples in which the species, especially those co-occurring on the same island, actually ate different seeds were very few. His was not the final word, obviously, and he thoughtfully noted some of the limitations of his data set. For example, not all species were examined from all localities during the same season, with some species from one island examined in one month and another species from a different island examined in a different month, making their diets during any particular season and location difficult to compare. Snodgrass also acknowledged that he did not have evidence to determine the relative abundances of the different seed types. Finally, he obtained all the collections during a time of year in which seed supplies were comparatively abundant, a point that future researchers would note.

Importantly, Snodgrass also argued that the various islands differed in the availability of particular seeds, at least over time and in some cases from island to island. Rather than being a homogeneous landscape with a uniform climate, the archipelago comprises islands that differ from one another in ecologically significant ways. He stated clearly that the diet of the finches depends on the seeds available to them. This left an opening for future researchers to interpret differences among the islands' finches in terms of adaptation to different ecological conditions.

Equally surprising, Lack himself, who later became notable for establishing the adaptive significance of bill variation in Darwin's finches, first argued that the variation in bill morphology was due merely to isolation and genetic drift. Like many of his predecessors, he cited the apparent ecological similarity of the islands, and supported the adequacy of random genetic drift ("the Sewall Wright effect") as an explanation for differences among species and subspecies, especially since so little correspondence between bill shape and diet existed. He suggested that the primary biological significance of bill size and shape is that the birds might use these characteristics to recognize members of their own species, thus keeping the drifted pools distinct. Echoing previous voices, Lack argued that the depauperate island home of the finches, and in particular the lack of food competitors, provides an environment of relaxed natural selection.

The Relation of the Food to the Size and Shape of the Bill in the Galapagos Genus Geospiza (1902)

✳ ROBERT E. SNODGRASS

The Fringillid genus *Geospiza*[1] of the Galapagos Archipelago contains about thirty-four species and varieties. Four subgenera may be distinguished on a color basis, but the specific and varietal character are almost entirely in the shape and size of the bill. The bill being the feeding organ, it is most natural to look first for the cause of its variation in a variation of the character of the food.

Geospiza heliobates feeds entirely on insects. But it inhabits exclusively the "mangrove swamps" where there is nothing but insect food available. The other species are all seed-eaters, although they occasionally pick up a few ants and other small insects. The seeds that they eat are mostly small and they are usually swallowed whole, being found in this condition in the crop. Large seeds when eaten are broken into pieces by the beak before being swallowed, generally only fragments of such are to be found in the stomach. The birds feed a great deal upon the ground, picking up seeds that have fallen from the bushes, and at the same time taking in with the food a considerable amount of gravel.

With a view of determining whether there is any corresponding variation between the bills and the food, Mr. Edmund Heller and the writer, during 1898 and 1899, preserved the stomachs of two hundred and nine specimens of *Geospiza*. These represent *G. pachyrhyncha, G. strenua, G. conirostris, G. fortis fortis, G. fortis platyrhyncha, G. fuliginosa, G. scandens, G. scandens fatigata, G. scandens rothschildi, G. affinis, G. crassirostris, G. prosthemelas, G. heliobates.* The specimens were collected from the islands of Albemarle, Narborough, James, Seymour, Duncan, Charles, Hood, Barrington, Tower and Bindloe. The dates run from December till June, inclusive.

Comparison has been made of the food of individuals of the same species at different places, and of the food of different species at the same and at different places. The results are somewhat conflicting. In any case one would require a great amount of evidence to come to any definite conclusions. Then, too, there is always a doubt created by the fact that the specimens were not taken on the different islands during the same months, and by the fact that the

1. The name *Geospiza* is here used in the same sense as used by Rothschild and Hartert (Novit. Zool., VI, 1899), i.e., to include all of the related Galapagos genera of other authors, such as *Platyspiza, Camarhynchus, Geospiza* and *Cactornis*. Such a group is certainly a natural one; and in it lines of division are difficult to draw. Ridgway recognizes three genera: *Platyspiza, Camarhynchus* and *Geospiza*. The names of species are according to the synonymy in a paper yet to be published by Mr. Edmund Heller and the writer.

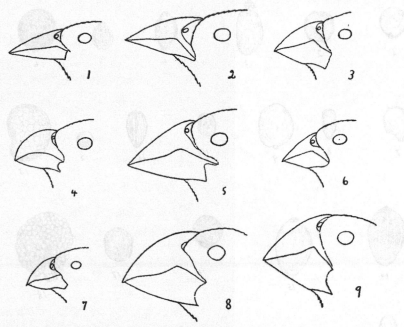

Bills of *Geospiza*. Natural Size. Fig. 1. *Geospiza scandens scandens*, James Island, from Ridgway. Fig. 2. *G. scandens rothschildi*, Bindloe Island. Fig. 3. *G. fortis fortis*, Albemarle Island, from Ridgway. Fig. 4. *G. crassirostris*, from Ridgway after Gould. Fig. 5. *G. conirostris conirostris*, Hood Island, from Ridgway. Fig. 6. *G. fuliginosa parvula*, Tagus Cove, Albemarle Island. Fig. 7. *G. prosthemelas prosthemelas*, Albemarle Island. Fig. 8. *G. strenua*, Albemarle Island, from Rothschild. Fig. 9. *G. pachyrhyncha*, Tower Island, from Ridgway.

seasons vary considerably at different localities. What might appear to be evidence of a difference in food habit between a species on one island and a different one on another island, might be nothing more than a seasonal change of diet due to different plants being in seed at the two times. However, a few conclusions may be positively deduced, the results being sufficient to warrant the discussion.

The detailed records of the two hundred and nine stomachs are omitted. The data obtained are given in the following table, and the seeds are illustrated on Plates XII and XIII. The seeds have not been identified, but the names are not necessary. They are drawn to show their relative sizes, and are referred to in the succeeding discussion by their numbers on the plates. Figures 1 to 44, inclusive, except figure 42, are magnified 64 times. The others are magnified only half as much. The stomachs of Mockingbirds (*Nesomimus*) from eight islands have been examined in the same way. The records of these are given at the end of the table, and the seeds are figured on the plate along with the *Geospiza* seeds. The stomachs of all contained a total of sixty kinds of seeds. Seeds nos. 59 and 60 are not figured.

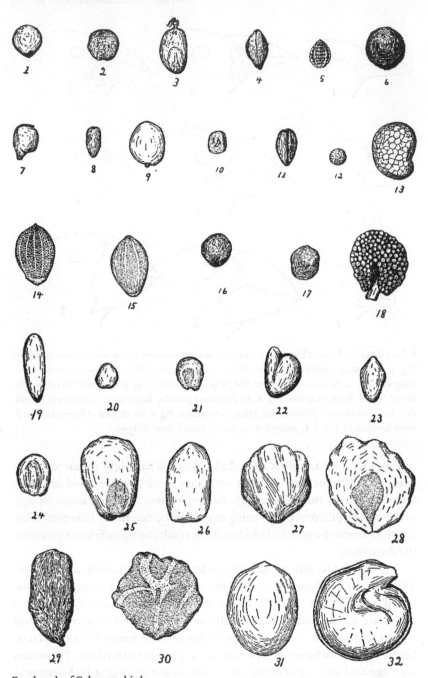

Food seeds of Galapagos birds.

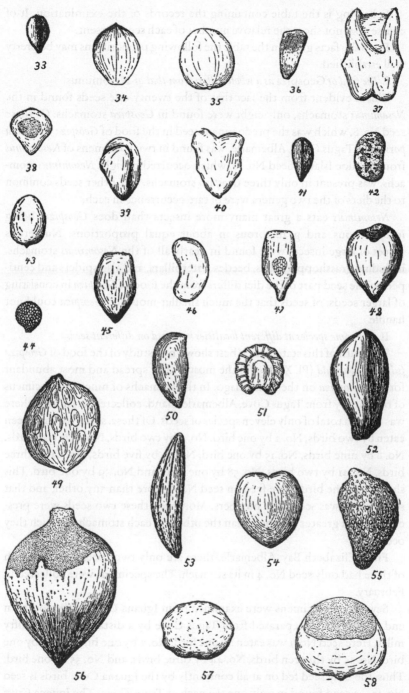

Food seeds of Galapagos birds.

Following is the table containing the records of the examination. It of course does not show the relative number of each seed present.

From the facts given in the table the following propositions may be pretty well established.

I. *The food of* Geospiza *as a whole differs from that of* Nesomimus.

This is evident from the fact that of the twenty-one seeds found in the *Nesomimus* stomachs, only eight were found in *Geospiza* stomachs. Of these seed No. 8, which was the predominant seed in the food of *Geospiza fuliginosa parvula* at Tagus Cove, Albemarle, was found in two specimens of *Nesomimus* from Bindloe Island. Seed No. 18, which occurred in eight *Nesomimus* stomachs, was present in only three *Geospiza* stomachs. The other seeds common to the diets of the two genera were of rare occurrence in each.

Nesomimus eats a great many more insects than does *Geospiza*, being insectivorous and granivorous in about equal proportions. Numerous pieces of large insects were found in nearly all of the *Nesomimus* stomachs, including grasshoppers, flies, beetles, caterpillars, and also spiders and centipedes. The seed part of the diet differs from the food of *Geospiza* in consisting of larger seeds, of seeds that the much smaller-mouthed *Geospiza* could not handle.

II. *The same species at different localities may feed on different seeds.*

The truth of this statement is best shown by a study of the food of *Geospiza fuliginosa parvula* (Pl. XI, Fig. 6) the most widely spread and most abundant form of *Geospiza* on the archipelago. In the stomachs of nineteen specimens of this variety from Tagus Cove, Albemarle Island, collected in January, there was found a total of only eleven species of seeds. Of these, seed No. 1 had been eaten by two birds, No. 2 by one bird, No. 4 by two birds, No. 6 by two birds, No. 8 by nine birds, No. 15 by one bird, No. 22 by five birds, No. 24 by three birds, No. 41 by two birds, No. 58 by one bird, and No. 59 by one bird. This shows that the birds here feed on seed No. 8 more than any other, and that seed No. 22 was second in numbers. Moreover, these two seeds were present in much greater numbers than the others in each stomach in which they occurred.

From Elizabeth Bay, Albemarle, there are only two specimens and each of these had only seed No. 4 in its stomach. The specimens were collected in February.

Seventeen specimens were examined from Iguana Cove at the southern end of Albemarle, separated from Tagus Cove by a distance of about fifty miles. Here seed No. 1 was eaten by one bird, No. 2 by one bird, No. 4 by one bird, No. 15 by thirteen birds, No. 44 by three birds, and No. 55 by one bird. Thus, the only seed fed on at all constantly by the Iguana Cove birds is seed No. 15—a seed found in only one stomach at Tagus Cove. The Iguana Cove specimens were collected in December.

TABLE OF FOOD SEEDS.

The numbers at the head of the columns refer to the seeds as figured and numbered in Plates XII and XIII.

SPECIES OF BIRDS.	5	8	10	11	13	14	15	17	18	22	23	26	28	29	44	46	57	59	LOCALITY.	DATE.
Geospiza pachyrhyncha (Pl. XI, Fig. 9).																	×		Tower	June
"																	×		"	"
"																	×		"	"
"																	×		"	"
"																	×		"	"
"																	×		"	"
"																	×		"	"
"				×						×		×××							Narborough	Jan.
"									××			×××	×						James	April
"						×	×				×	×	×			×			"	"
G. strenua (Pl. XI, Fig. 8).			×××				×					××	×	×		×	×	××	Bindloe	June
"	×		×									×	×		×				Hood	May
"			×	×		×××××	××××××	××××××				×	×						"	"
G. conirostris conirostris (Pl. XI, Fig. 5).					×	×××××	×××××						×						"	"
"													×						"	"
"																			"	"

TABLE OF FOOD SEEDS (CONTINUED).

SPECIES OF BIRDS.	2	3	4	5	8	9	15	21	22	23	24	28	33	35	38	40	41	48	57	59	LOCALITY.	DATE.
G. fortis fortis (Pl. XI, Fig. 3).					×							×									Tagus Cove	Jan.
"										×		×?									"	"
"		× × × ×																			"	"
"			×						×												"	"
"						×			×												"	"
"	×								×												"	"
"								×													"	"
"							× ×						×	× × ×							Narborough	"
"								×	×						×						James	"
"														×			× ×				"	"
G. fortis platyrhyn- cha.					× × × × × × ×				× × × ×				×		× ×	×	× ×	×	×	×	Iguana Cove	April
G. fuliginosa parvula (Pl. XI, Fig. 6).			× : ×		× ×						× × : ×										Tagus Cove	Jan.

TABLE OF FOOD SEEDS (CONTINUED).

SPECIES OF BIRDS.	LOCALITY.	DATE.	59	57	55	45	44	41	26	22	21	19	18	17	15	14	11	10	8	6	4	2	1
G. fuliginosa parvula (Pl. XI. Fig. 6).	Tagus Cove	Jan.																		××			
	"	"						×															×
	"	"																					×
	Elizabeth Bay	Feb.					×								×								
	Iguana Cove	Dec.			×										×					×		×	
	"	"													×××××								
	"	"													×								×
	"	"					××								×××××××					×	××		
	Narborough	"								××			××						×××		××		
	James	April	×					×							×								
	Seymour	"													×								
	"	"		×××											××××××	×××××××××							

TABLE OF FOOD SEEDS (CONTINUED).

SPECIES OF BIRDS.	LOCALITY.	DATE.	1	2	4	6	8	10	11	14	15	17	18	19	21	22	26	41	44	45	55	57	59
G. fuliginosa parvula (Pl. XI, Fig. 6). "	Seymour " " " " " " " " " " " " " Duncan " Barrington " " " " " "	April " " " " " " " " " " " " " May " " " " " " " " "						×		×	×													
G. scandens scandens (Pl. XI, Fig. 1). " " " " " " "	James " " " " "	April " " "		×				×		×	×		×		×	×	×	×				×	×
G. scandens fatigata " " " " "	Seymour " " " "	April " " " "	×					×		×	×											×	×

TABLE OF FOOD SEEDS (CONTINUED).

DATE.	LOCALITY.	60	57	55	46	44	41	40	39	36	31	25	20	16	15	14	12	11	10	4	2	SPECIES OF BIRDS.
April	Seymour		××												××××××	×××××			××××			G. scandens fatigata.
"	"			×									×		×××××××	××××			×			"
"	"											×		××		×××××××			×			"
May	Barrington		×							×	×				×				×			"
June	Bindloe				×		×								×		×	×				G. scandens rothschildi (Pl. XI, Fig. 2)
Dec.	Iguana Cove					×	×××××							××				×			G. affinis
Jan.	Narborough	×					××		×	×				×			×					G. crassirostris (Pl. XI, Fig. 4).
"	Albemarle														×			×				G. prosthemelas prosthemelas (Pl. XI, Fig. 7).
April	James						×							××			×	×	×	××		"

TABLE OF FOOD SEEDS (CONTINUED).

SPECIES OF BIRDS.	LOCALITY.	DATE.	6	7	8	13	15	18	19	25	30	32	33	37	41	45	49	50	52	53	54	55	56	57	58
Nesomimus melanotis parvulus	Tagus Cove	Jan.																		× ×	×				
	Elizabeth Bay	Feb.						×						× ×							× × × × × × × × × × × × ×				
	Narborough	Jan.	×								×		×				× ×								
N. melanotis melanotis	James	April						× ×		×															
N. melanotis dterythrus	Seymour															×			× ×						× ×
N. macdonaldi	Hood	May			× × × × ×									×							×			× × × ×
N. barringtoni	Barrington								×			×													×
N. personatus bauri	Tower	June		×	×																		× × × × × × × ×	×	
N. personatus bindloei	Bindloe				×	×					× × × ×							× ×						× × ×

Since we do not know what species of plants the different seeds belong to, we cannot say whether the differences in the food of the birds at Tagus Cove and Iguana Cove is due to a difference in the floras of the two localities, to a difference in the time of ripening of the seeds, or to a difference in the preferences of the birds with regard to the seeds at the two places. Since, however, seed No. 15 was found in abundance in the stomachs of birds taken on James, Seymour, Duncan and Barrington Islands in April and May, it would appear that the seeds should be ripe at Tagus Cove in January if they are ripe at Iguana Cove in December. That the plant occurs at Tagus Cove is shown by the fact that the seeds were found here in one stomach. The entire diet of the Tagus Cove birds consisted of seeds Nos. 1, 2, 4, 6, 8, 15, 22, 24, 41, 58 and 59; that of the Iguana Cove birds of seeds Nos. 1, 2, 4, 15, 44 and 55. Of the thirteen kinds of seeds only four are common to both sets. Hence, there is most evidently a difference in the food of the individuals at the two places, at approximately the same time of the year. It is, perhaps, most probably that this difference is due to the same seeds not being available in the same relative numbers at the two places.

Of five specimens from Narborough Island, taken in December, three had in their stomachs only seed No. 8, the other two contained only seed No. 22. These thus fed on the principal part of the diet of the Tagus Cove birds.

On James, Seymour, Duncan and Barrington Islands *Geospiza fuliginosa parvula* feeds almost exclusively on seeds Nos. 14 and 15. Specimens of one or both of these seeds were found in the stomachs of all the thirty-six birds examined, except in one from James and one from Barrington. Next to these, seeds Nos. 10 and 57 were most abundant, each being represented in ten stomachs. These four seeds were also by far the most abundant wherever they were found. The rest of the diet consisted of seed No. 2, found in two Seymour birds and in one Duncan bird; No. 11, found in one Seymour bird; No. 17, found in one Seymour bird; No. 18, found in one James bird; No. 20 found in one Duncan bird; and nos. 41 and 59, found in one James bird. These specimens were all collected in April and May.

The facts just detailed certainly show that the individuals of *Geospiza fuliginosa parvula* living on Narborough and at Tagus Cove, Albemarle, during December and January, have a different diet from those individuals living at Iguana Cove, Albemarle, in December, and on James, Seymour, Duncan and Barrington in April and May. The proof, from these facts, of proposition I, however, is somewhat invalided by the consideration that seeds Nos. 14 and 15 may ripen at Tagus Cove and on Narborough later than January. But seeds Nos. 8 and 22 were not found in any stomachs except in those of birds taken at Tagus Cove. We can see, at least, that the diet of birds depends on the local prevalence of certain seeds; and that, where the floras, differ, the foods of a species may differ.

III. *Different species at the same locality may feed on the same kinds of seeds.*

This proposition is much easier to prove than the last. Compare, for example, the food of *Geospiza fuliginosa parvula* (Pl. XI, Fig. 6) and of *G. scandens fatigata* on Seymour and Barrington Islands. As has already been shown, the food of the former species consists almost wholly of seeds Nos. 10, 14, 15 and 57, Nos. 14 and 15 being in the majority. An examination of the table will show that the food of *G. scandens fatigata* on the two islands is practically identical with that of *G. fuliginosa parvula*, consisting mainly of seeds Nos. 14 and 15, with Nos. 10 and 57 second in numbers.

The case of these two species, then, proves that species differing much in the size and shape of the bill (Pl. XI, Figs. 1 and 6) may have absolutely the same food habits. We have not, however, the material at hand to justify the statement of this as a general truth. We cannot show how far it actually holds true of other species on the archipelago. The similarity in the food of these two common forms on Seymour and Barrington Islands is so striking, however, that one is almost forced to the conclusion that all the species of *Geospiza* eat simply whatever seeds are accessible to them.

IV. *Different species at different localities may feed on the same kinds of seeds.*

The truth of the proposition may be seen by a comparison of the food of *Geospiza conirostris conirostris* (Pl. XI, Fig. 5) on Hood, *G. scandens fatigata* on Seymour and Barrington, and *G. fuliginosa parvula* at Iguana Cove, Albemarle, and on Seymour and Barrington. The largest part of the food of *G. conirostris conirostris* in May consisted of seeds Nos. 14 and 15. Next in numbers were seeds Nos. 10 and 17. All but two of the thirteen birds had eaten No. 14, and all but one No. 15, while Nos. 10 and 17 were each represented in six stomachs. Seed No. 5 was found in one bird, No. 23 in one bird, No. 26 in five birds, No. 28 in three birds, No. 29 in one bird, and No. 57 in one bird.

Hence, the food of *G. conirostris* on Hood Island is in the main the same as that of *G. fuliginosa parvula* at Iguana Cove, Albemarle, and on Seymour and Barrington Islands, and is also the same as that of *G. scandens fatigata* on Seymour and Barrington. There are thus three species of *Geospiza* with very different bills (Pl. XI, Figs. 1, 5 and 6), living at three localities, whose food is almost identical at approximately the same time of the year.

V. *Different species at the same or at different localities may feed on different seeds.*

If the size and shape of the bill is dependent on the character of the food, this proposition should be a general truth. However, the material under consideration affords only a few instances of it.

Geospiza pachyrhyncha (Pl. XI, Fig. 9) is peculiar to Tower Island. The stomachs of seven specimens taken in June contained only seed No. 57. We have no data to show what the food of other species on Tower consists of. Vegetation is extremely scant on the island, and all the birds may be forced to eat the same seed.

A specimen of *Geospiza strenua* (Pl. XI, Fig. 8) taken in January on Narborough had only seed No. 22 in its stomach. Five specimens taken in April on James Island had fed as follows: in one stomach were seeds Nos. 11, 26 and 28; in two others seeds Nos. 18 and 26; in another seeds Nos. 57 and 59; in the fifth seeds Nos. 14, 15, 18 and 59. Of two taken in June on Bindloe one had in its stomach only seed No. 46, the other only seed No. 44. These very scant data would seem to indicate that *Geospiza strenua* uses but little selection in the choice of its food. Altogether it has been found to eat seeds Nos. 11, 14, 15, 18, 22, 26, 28, 44, 46, 57 and 59. The James specimens alone had eaten seeds Nos. 11, 14, 15, 18, 26, 28, 57 and 59. This list is somewhat different from the diet of six specimens of *G. scandens fatigata* taken at the same time on James. The stomachs of these birds gave the following: seed No. 2 in one bird, No. 14 in two birds, No. 15 in three birds, No. 18 in one bird, No. 21 in one bird, No. 26 in one bird, No. 41 in four birds, No. 59 in one bird. The species of seeds forming the list in the two cases are almost the same, the main difference is in the proportions of the seeds present. It is a question whether the evidence in this case should not be given to proposition III. It, however, shows the weakness of the proof on which proposition V could be stated as a general fact.

A good example of the proposition under consideration may be derived from a comparison of the food of *Geospiza fortis* (Pl. XI, Fig. 3) at Tagus Cove, Albemarle, with that of *G. fuliginosa parvula* at the same locality. As has already been shown, the food of nineteen Tagus Cove specimens of the latter species was as follows: seed No. 1 had been eaten by two birds, seed No. 2 by one bird, seed No. 4 by two birds, seed No. 6 by two birds, seed No. 8 by nine birds, seed No. 15 by one bird, seed No. 22 by five birds, seed No. 24 by three birds, seed No. 41 by two birds, seed No. 48 by one bird. The table shows the following composition of the food of thirteen Tagus Cove specimens of *Geospiza fortis* taken also during January. Seed No. 2 had been eaten by one bird, seed No. 3 by four birds, seed No. 4 by one bird, seed No. 8 by one bird, seed No. 9 by one bird, seed No. 22 by five birds, seed No. 28 by two birds, seed No. 33 by one bird, seed No. 35 by three birds. The only important difference in these two cases is the predominance of seed No. 8 in the food of *G. fuliginosa* and its scarcity in that of *G. fortis*. These two species have somewhat similarly shaped bills (Pl. XI, Figs. 6 and 3), but that of *G. fortis* is the heavier.

Two specimens of *G. fortis platyrhyncha* from Iguana Cove, Albemarle, taken in December, had eaten only seeds Nos. 38, 40 and 48, seeds not found in the stomachs of any of the Tagus Cove, *G. fortis*, nor in any of the Iguana Cove specimens of *G. fuliginosa*.

Four specimens of the *Geospiza fortis* on James Island, which does not differ from the *G. fortis* of Tagus Cove, had eaten as follows, in April. Seed No. 15 occurred in two stomachs, seed No. 21 in two, seed No. 41 in two, seed No. 57 in one, and seed No. 59 in one. It will be observed that there is no seed common to the three sets, *i.e.*, in the food of the James Island, Iguana Cove, and Tagus Cove specimens of *Geospiza fortis*. The case of the James Island and Tagus Cove specimens belongs to proposition II; the *G. fortis platyrhyncha* differing from *G. fortis fortis* at two other localities belongs to proposition V. It is important to note that the food of all the individuals at any locality does not differ as a whole from that of the others, more than may the food of two individuals at the same locality.

Perhaps the best case under proposition V can be made out from a study of the food of *Geospiza crassirostris* (Pl. XI, Fig. 4) and of *G. fuliginosa parvula* at Iguana Cove, Albemarle. The food of five specimens of the former species, taken the last of December, consisted entirely of seeds Nos. 39 and 40, the former found in only one stomach, the latter in all. As has before been shown, *G. fuliginosa parvula* at Iguana Cove feeds almost entirely on seed No. 15, seeds Nos. 39 and 40 not being found in any of the stomachs.

VI. *Birds with small bills eat only small seeds; birds with large bills eat both small and large seeds.*

Geospiza fuliginosa (Pl. XI, Fig. 6) eats few seeds larger than Nos. 14 and 15. The only larger one found in their stomachs is No. 57, but this is a thin, flat seed, and is nearly always broken into small pieces before being swallowed. In the stomachs of *G. strenua*, one of the large-billed species (Pl. XI, Fig. 8), there occurred, besides numerous small seeds, such larger ones as Nos. 18, 26, 28, 45 and 58. In the stomachs of *G. conirostris* (Pl. XI, Fig. 5) most of these same larger seeds were found and also No. 29. *G. fortis* (Pl. XI, Fig. 3) eats such moderately large seeds as Nos. 35, 40, and 48 together with larger ones such as Nos. 28 and 57. An examination of the table will show, however, that the larger-billed species by no means confine themselves to large seeds. It appears most probable that they eat the larger seeds simply because their large bills makes it possible for them to eat a greater variety of seeds. There is no evidence that they show a special preference for large seeds.

The foregoing six propositions are about the only conclusions that we can deduce from a study of the material in hand. It is evident that if these propositions were stated severally as general facts they would be mutually conflicting. Each is true only in some cases.

If it be assumed that the various sizes and shapes of bills amongst the Geospizae have been developed as adaptations to differences in food habit, then it must be shown that the different species of the genus feed on different species of seeds. This cannot be done. We can prove definitely that some species with very different bills feed on exactly the same kinds of seeds. On the other hand some of the evidence seems to indicate that some of the species and subspecies do have different food habits. We cannot say, however, that these differences of diet are not forced upon different species as a result of their living in different localities. Especially is this probable since, in some cases, we find that individuals of the same species living at different localities feed on different seeds. This is due evidently to flora differences between the two regions. The evidence, then, seems to be in favor of the general conclusion that *there is no correlation between the food and the size and shape of the bill*. If this is true, then we must look elsewhere for an explanation of the variation of the *Geospiza* bill.

Auk 19 (1902): 367–381.

Evolution of the Galapagos Finches (1940)

✳ DAVID LACK

INTRODUCTION

The land faunas of oceanic islands have always excited considerable evolutionary speculation, and, starting with the "Origin of Species," the Geospizinae, the endemic Galapagos finches, have probably featured in as many evolutionary discussions as any group of animals. They differ from almost all other land birds of oceanic islands in that there is more than one species on each island. Further, some of the species seem to grade into each other, and others are linked by freak specimens. Some workers have supposed that some quite peculiar method of evolution must have been involved.

Assisted by grants from the Royal and Zoological Societies of London, I spent December 1938–April 1939 on the Galapagos investigating the breeding behaviour and ecology of Geospizinae, following this with a statistical study of museum material. Another object, to interbreed the different species, failed on the islands, but is being continued with captive birds brought to the California Academy of Sciences, San Francisco. The following is a brief summary of some general evolutionary problems. Full results will be published later[1] together with a list of the many persons who helped the expedition, the museum study, and in discussion.

SUBSPECIES

There are ten main species, each found on most of the islands. Some are divided into island subspecies, which differ mainly in beak size and minor plumage characters. The various Galapagos Islands present similar conditions for Geospizinae, and no species show regular or parallel trends of variation on different islands. Hence I consider that isolation of small island populations has led to non-adaptive differentiation in the manner postulated by Sewall

Wright.[2] That the two main factors in the formation of island subspecies are (1) degree of isolation (2) size of population, is indicated by the fact that small isolated islands have a higher proportion of endemic forms (also a smaller total of species) than have either small central islands or large isolated islands. Large isolated islands have a higher proportion of endemic forms than the central islands, large or small. Another point, the two most variable species, *Geospiza magnirostris* and *G. fortis*, are significantly less variable in three measured characters (namely, wing, culmen from nostril and depth of beak) on small isolated islands than they are on large central islands. (This relation did not hold in four other less variable species.)

SPECIES

Geospiza magnirostris (large), *G. fortis* (medium size) and *G. fuliginosa* (small) are three closely related species which do not differ at all in plumage but solely in size and relative size of beak, and to some extent in song. The beak is the chief specific character, the large species having relatively larger beaks. There is no allometric relation between beak and body size within each species, but the ratio is too subject to individual variation to form a reliable specific character. These three species occur together in the same habitat on many islands, the food and feeding habits of the first two seem identical, and of *G. fuliginosa* closely similar, and they use similar nest sites. On some, but not all, islands *G. fortis* is so variable that the smallest individuals appear more like large *G. fuliginosa* than like the largest individuals of their own species, and the latter appear more like small *G. magnirostris* than like the smallest individuals of their own species. (There is no evidence for selective mating by body and beak size within *G. fortis*.) Yet normally the three species keep distinct, and on most islands there is no evidence for hybridization, though this cannot be completely ruled out. There is a small overlap between the species in all characters, but nearly all specimens can be safely identified. However, a few specimens are intermediate in several characters together and cannot be safely allocated.

Field study showed no adaptive significance for the inter-specific beak differences *except* that they are the major (but probably not the sole) factor in specific recognition. A male usually attacks a rival by gripping its beak; frequently a male started to chase a member of a different species from behind, but usually, though not always, it stopped on coming round in front where its beak was visible. Experiments with wild birds which attacked and courted stuffed specimens confirmed the use of the beak in specific recognition.

Camarhynchus psittacula (large) and *C. parvulus* (small) occur together on most islands and present a similar problem to the *Geospiza* species, while on one island occurs a third species, *C. pauper*, differing from *C. psittacula* solely in a smaller depth of beak and in minor plumage details. In general, the closely related species of Geospizinae are not usually separated by habitat, nest site, breeding season or any other geological factor. Geneticists seem agreed that isolation promotes inter-sterility, and I consider that these species have evolved via geographically isolated subspecies which later met on the same island and did not interbreed. The species differ from each other in the same characters which distinguish island subspecies, that is, in beak and sometimes in minor plumage characters. When beak differences assist in specific recognition, they will be intensified by selection quite apart from their relation to feeding habits, since hybridization is at a selective disadvantage. It is extremely difficult to see any other significance for the specific beak differences, or to understand why, otherwise, they should persist. But the highly variable *G. fortis* still presents great difficulties in interpretation.

In two cases, the origin of a species via a geographical form seems clear. The *Camarhynchus psittacula* group is represented by one form on each main island except on Charles where two occur, *C. psittacula* and *C. pauper*. *C. pauper* was probably derived from the *psittacula* group via the form *affinis* on Albemarle

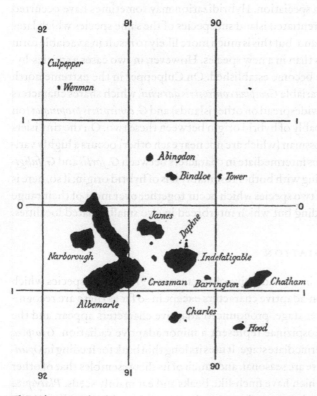

The Galapagos Islands.

to the west, as they are rather similar. *C. psittacula* on Charles is almost indistinguishable from *C. psittacula* on Indefatigable to the north, and probably colonized Charles from Indefatigable more recently, met *C. pauper* but kept distinct. Again, *G. fuliginosa* occurs on all the main Galapagos Islands except Tower in the north-east, where a rather similar species *G. difficilis* (formerly *acutirostris*) is found, which could have been derived from *G. fuliginosa* from Chatham to the south, and would probably have been classified as a subspecies of *fuliginosa*, but that it also occurs on Abingdon, to the west of Tower, together with a small form of *G. fuliginosa*, from which it keeps segregated.

Rensch[3] considers that, in general, most bird species have evolved via geographically isolated subspecies.

HYBRIDIZATION

Lowe[4] regarded hybridization as the main factor in species-formation in the Geospizinae. The amount of hybridization has been greatly exaggerated. We saw no certain cases in the field. The occasional freak and intermediate specimens are not necessarily hybrids, and in any event have probably played little, if any, part in speciation. Hybridization may sometimes have occurred between two differentiated island subspecies of the same species which later met on the same island, but this is much more likely to result in a variable form of the same species than in a new species. However, in two cases a species-hybrid seems to have become established. On Culpepper in the extreme north occurs the highly variable *Geospiza conirostris darwini*, which shares characters of *G. magnirostris* (widespread on other islands) and *G. conirostris propinqua* (on Tower) and is probably of hybrid origin between these two. On the tiny islets of Daphne and Crossman (which are not near each other) occurs a highly variable *Geospiza* species intermediate in characters between *G. fortis* and *G. fuliginosa*, and overlapping with both. Presumably it is of hybrid origin; if so, here is the unusual case of two species which occur together over most of their range without interbreeding but which interbreed in two small isolated localities.

ADAPTIVE RADIATION

Island forms differ in non-adaptive characters and give rise to species which differ mainly in non-adaptive characters except in so far as these are recognitional. But, at a later stage, pronounced adaptive characters appear, and the main genera of Geospizinae represent a minor adaptive radiation. *Geospiza scandens* is at an intermediate stage. It uses its long thin beak for feeding in *Opuntia* flowers, but these are seasonal, and much of its diet resembles that of other *Geospiza* species, which have finch-like beaks and eat mainly seeds. *Platyspiza crassirostris* eats mainly leaves. *Camarhynchus* species have beaks rather similar to *Platyspiza*, but eat mainly insects. *Cactospiza pallida*, clearly derived from *Camarhynchus*, has evolved in the direction of a woodpecker in both habits and beak. It also has a unique habit of holding a small twig or *Opuntia* spine lengthwise in its beak and probing insects out of cracks in trees, dropping this tool to seize the insect as it emerges. *Certhidea olivacea* is like a warbler in both beak and feeding habits. Despite their dissimilar feeding habits and appearance, all these forms have extremely similar breeding behaviour, which confirms previous anatomical findings as to their close relationship to each other.

EVOLUTIONARY FACTORS

The three main factors in the evolution of the Geospizinae have probably been: (1) The almost complete absence of food competitors. A few other land birds now frequent the islands, but all except *Nesomimus* are so little differentiated from mainland forms that they probably colonized long after the ancestral Geospizid arrived. (2) The almost complete absence of predators. Worthington[5] shows the marked inhibitory influence of predators on adaptive radiation.

These two factors result from the extreme isolation of the Galapagos and the difficulty of colonization from birds. Both must diminish the intensity of selection.

(3) Equally important, the opportunities for temporary isolation of different forms provided by the existence of a number of islands. One species of Geospizid, *Pinaroloxias inornata*, occurs outside the Galapagos, on the extremely isolated island of Cocos, and is so differentiated that is has clearly been isolated a long time, but there is only one island and still only one species of Geospizid. In general, no bird groups have evolved similarly to the Geospizinae on solitary islands, however isolated, but two parallel cases occur in isolated archipelagos. The two species of endemic finch, *Nesospiza*, in the Tristan da Cunha Islands differ primarily in beak,[6] so here is Galapagos in miniature, and at the other extreme is the marvelous adaptive radiation of the Drepanididae in the Hawaiian group.[7]

SECONDARY SEXUAL PLUMAGE

In *Geospiza* species the adult male plumage is black, but individuals frequently breed in the grey-brown juvenal plumage, or in partly black and partly juvenal plumage. In *Platyspiza* and *Camarhynchus*, the "full" male plumage is partly black, and many individuals breed in juvenal plumage, the percentage greatly varying on different islands. In *Cactospiza*, the male is normally coloured like the juvenal, but one had a black head. This tendency to lose the male secondary sexual plumage occurs in many other land birds of oceanic island. Such plumage is usually considered to have evolved through purely intra-specific selection, through threat and courtship display, and one would suppose threat and courtship were as essential on oceanic islands as elsewhere (Geospizinae display vigorously). But there is a further function of such plumage which has relation to other species, namely, enabling the female to recognize and pair up with a male of her own species, hybridization being at a selective disadvantage. The few land birds which have colonized oceanic islands are normally removed from all related species which the female might possibly confuse with her own, so this function disappears. The opposite of this process is seen in some continental Gallinaceous birds, and some birds of paradise, in which the sexes meet only for copulation and there is not the complex interrelated pairing behaviour typical of birds in which both sexes raise the young. Probably correlated with this, hybridization is relatively common, and at the same time there is an extreme development of male secondary sexual plumage, each species being strikingly distinct.

To sum up, while the Geospizinae present certain unusual features, there is no need to postulate some quite peculiar evolutionary agency. As on other oceanic islands, the almost complete absence of food competitors and predators has decreased the intensity of selection, so that peculiar types or habits have a greater chance of persisting. The existence of a number of islands has promoted non-adaptive differentiation of island subspecies, from which species have been evolved when two such forms have later met in the same area and kept distinct. The genera show a minor adaptive radiation. The lost of male secondary sexual plumage is correlated with its ceasing to function in specific recognition.

NOTES

1. Lack, D., "The Galapagos Finches (Geospizinae): a Study in Variation," *Proc. Cal. Acad. Sci.* (in the press).

2. Wright, Sewall, "Evolution in Mendelian Populations," *Genetics, 16*, 97–159 (1931).

3. Rensch, B., "Zoologische Systematik und Artbildungsproblem," *Verb. deutsch. Zool. Ges., 35*, 19–83 (Zool. Anzeiger 6 suppl.) (1933).

4. Lowe, P. R., "The Finches of the Galapagos in Relation to Darwin's Conception of Species," *Ibis*, 310–321 (1936).

5. Worthington, E. B., "On the Evolution of Fish in the Great Lakes of Africa," *Int. Rev. Hydrobiol. Hydrogr., 35*, 304–317 (1937).

6. Lowe, P. R., "Notes on some Land Birds of the Tristan da Cunha Group collected by the Quest Expedition," *Ibis*, 519–523 (1923).

7. Perkins, R. C. L., "Fauna Hawaiiensis," vol. 1, Pt. 4; Aves 368–466 (1903).

Nature 146 (1940): 324–327. Reprinted with permission from Macmillan Publishers Ltd. (Nature, copyright 1940).

Diversity as Adaptation?

After further analysis, Lack completely changed his mind. Having first seen the variation in finch bill morphology as evidence of relaxed competition and relaxed natural selection, he took a closer look, indeed a very close, careful and insightful look, and concluded that, in fact, competition for food *is* the primary cause of the differentiation of beak morphology among the finches.[1] Citing a lengthy, and previously neglected, work of E. W. Gifford (1919), and completely contradicting conclusions in his previous publications, Lack stated that the differences in beaks are correlated with marked differences in feeding habits. Gifford, by the way, was the first to observe the woodpecker finch probing for insects with a twig, but, a bit timid perhaps, he kept the sighting to himself for several years for fear of disbelief and ridicule. Unlike Snodgrass, Gifford visited the archipelago in September and October, probably before the rains, when food resources were likely to be scarcer. The reading selection here, from Lack's major book, shows his complete about-face over the adaptive significance of variation in the finch beaks.

The other important influence on Lack, aside from Gifford and his own observations in the field, was the seminal work of Georgii F. Gause. Gause (1934) demonstrated that the coexistence of two species that use exactly the same resources was a mathematical impossibility, since one species will inevitably exclude the other through even a slight competitive superiority. If species are able to change their resource use, however, competitive "nudging" might permit them to avoid total exclusion. That is, species might be able to nudge competitors from one resource to another, so that, eventually, they use different resources and can thereby coexist. This is called "character displacement," since characters—resource use, or the traits associated with it, such as bill shape and size—have evolved and produced differences between species in locations where they coexist. If two species impose natural selection on each other through competition, one would expect them to differ from each

1. It is rumored that the paper arguing for random drift as the explanation for the variation in finch morphology, presented in chapter 5, was held up in press due to World War II, and that two papers, arguing opposing views, were in press at the same time (or very close to it).

other more when they co-occur than when they occur alone. This is the pattern Lack looked for. This is the pattern he found.

It was Lack's detailed biogeographic analysis of variation within and between species of Darwin's finches that led to his controversial hypothesis of character displacement as a mechanism that creates ecological and morphological diversity. According to Lack, it was the competitive interactions between species in sympatry, where species co-occur, that caused adaptively significant divergence among closely related species. Similar evidence for character displacement bubbled out of the evolutionary literature for decades after Lack.

In contrast, other students of the finches, such as Robert Bowman, emphasized the ecological variation among the islands—especially in terms of the food available to the finches. They argued that the morphological variation within and between the finch species distributed across the islands is likely caused not by competition with other finches but by adaptation to different ecological conditions in allopatry (i.e., geographic isolation), particularly to different edible plant species on each island. Taking this alternative hypothesis very seriously, Bowman made extensive, meticulous observations on variation in both food availability among islands and in what foods the different finch species consumed. Precisely this sort of intensive field study was necessary to test these basic evolutionary and ecological hypotheses. From his perspective, competition was not the primary factor influencing morphological evolution in the finches; instead it was adaptation to enduring environmental (especially plant community) differences on the different islands.

Joseph H. Connell adopted this point of view as a legitimate alternative to coevolutionary divergence through character displacement, and published a prescriptive paper in 1980 in which he challenged ecologists to employ experimental methods in addition to purely observational approaches to test ecological theory. One cannot conclude that competition caused divergence unless one can experimentally demonstrate a negative effect of one species on another. However, ecologists argued, if the species have already evolved to avoid competition, then they will not at the present time negatively affect each other through competition. The divergence may be evidence of competition that existed in the past, but exists no longer. This argument Connell mocked, calling it the invocation of "the ghost of competition past." To assume that observed differences are due to past competition, without being able to demonstrate adverse effects of competition, is unsatisfactory. To rigorously test the hypothesis of competitive exclusion and character displacement, it is necessary to employ experimental manipulations that create competition—manipulations such as transplant experiments in which members of species that live without competition in a part of their range are transferred to a location in which they are forced to compete with another species.

The challenges of Connell and others gave a considerable boost to experimental ecology.

The debates over these two alternatives were intense. In the case of Darwin's finches, the central questions were: Do *stable* environmental differences among islands cause *stable* differences in the abundance of food resources to which allopatric populations of finches adapt? Or does competition for resources in sympatry cause food availability to vary across islands, due to *variation* in the abundance of competitors? These questions begged other, larger questions that persist in contemporary debate: Is evolution a response to a fixed or a moving (evolving) target? Specifically regarding the importance of competition in evolutionary processes, is being different from others per se adaptive? If so, the competitive environment is expected to continually evolve as formerly uncommon variants become more common.

Not just competition, but all sorts of biotic interactions among species— predation, parasitism, symbioses, toxicity—create challenges for adaptation in other species. This major insight was elaborated upon in Leigh Van Valen's classic paper articulating his "Red Queen Hypothesis" of evolution (published on page 1, volume 1 of Van Valen's homemade journal, *Evolutionary Theory*, hand typed and mimeographed for distribution). Just as the Red Queen of *Alice in Wonderland* had to run to stay in the same place, so too do species have to evolve in an ever-changing environment just to avoid extinction. The environment changes constantly because other species evolve. Indeed, other species *are* the environment. Evolution itself creates novel and frequently localized environmental challenges to which species must continually adapt, which in turn causes further change in environmental conditions for other species. The result is continuous evolutionary change and the proliferation of diversity.[2]

Of the possible biotic interactions, competition has acquired a special status for explaining the proliferation of diversity. To differ from one's competitors is a direct advantage if it enables the new use, or the more efficient use, of an underused resource. This dynamic is a form of "frequency-dependent selection," in which the adaptive value of a particular trait depends on the frequency of individuals that share that trait; in this case, the more uncommon

2. A more recent paper by several evolutionary ecologists suggests that the difference in the intensity of biotic interactions between temperate regions and the tropics contributes to the well-documented difference in species diversity between the two regions—one of the most consistent biogeographic patterns on earth, and one still without a definitive explanation. Temperate taxa, which are less diverse and for which biotic interactions are thought to be less of a challenge than predictable climatic challenges, can attain adaptive optima relatively easily and across a wide geographic range; tropical taxa, in contrast, may need to keep evolving and diversifying in the constantly evolving landscape of intense and localized biotic interactions.

the trait is, the more adaptive it is. If being different from others is itself adaptive, then that fact alone goes a long way toward explaining the ubiquity of variation.

READINGS

First presented are excerpts from D. Lack's classic 1947 book, *Darwin's Finches*, in which he presents his evidence for character displacement in the form of the geographic distribution of bill morphologies and an analysis of the overlap in bill traits among co-occurring species. The excerpt from his chapter 3 describes the general habitat in which different finch species are found, emphasizing the similarity of habitat shared by closely related species and alluding to the principle of competitive exclusion. Next are his chapters that interpret variation in bill morphology. His chapter 6 discusses the relationship between bill traits and diet, comparing distantly and closely related species. Chapter 7 of his book describes the pronounced variation in bill form within species, and chapter 8 examines and interprets the variation found within species in light of competition between species; bill shape differs within a species according to what other species occur with it. So compelling were the patterns he presented that a generation of evolutionary ecologists followed his lead, seeking similar patterns in other taxa and augmenting distributional data with further observations on resource use and experiments that manipulated resources and putative competitors.

Bowman's 1961 work challenges Lack's interpretation of competitive exclusion in Darwin's finches. Following the reasoning of R. E. Snodgrass (and G. Baur) regarding the possibility of environmental differences among the islands due to differences in plant abundance (chapter 5), Bowman systematically quantified this variation by examining the stomach contents of a great number of birds and comparing the diets of the birds on different islands and of different species. The resulting work is quite lengthy, and only excerpts are provided here. His work is important because he definitively broke with the assumption of ecological homogeneity across the archipelago, and applied patient and thorough observational methods to quantifying the ecological variation among islands—particularly with regard to the plant life as the food resource for the finches.

But Lack's interpretation found new support from the long-term studies initiated by Peter and Rosemary Grant and their associates. The third paper, by Dolph Schluter and colleagues, gives further evidence of competitive character displacement in at least one species. The work combines an analysis of interspecific overlap in bill morphology with an analysis of feeding behavior and interisland variation in resource abundance. While the research was still observational as opposed to experimental (it is difficult to conduct experimental manipulations on protected species), their evidence generally supports character displacement. Bill morphology differed more between species when they occurred together in sympatry than when they occurred alone in allopatry, and the difference in bill size was shown *not* to be due exclusively to differences in overall seed resources between islands of sympatry versus allopatry. Moreover, they documented that birds in allopatry that eat the seeds typically eaten by the other species have higher survival; that is, their competitor apparently consumes resources that would otherwise be beneficial to them. Through this and similar studies, bill morphology in Darwin's finches remains one of the most convincing examples of character displacement to this day.

From *Darwin's Finches* (1947)

✳ DAVID LACK

Editor's note: Only tables and figures that are part of the chapters included here are reprinted. Tables and figures from other chapters are not reprinted here, despite references to them.

* Chapter VI: Beak Differences and Food

The most curious fact is the perfect gradation in the size of the beaks in the different species of Geospiza, from one as large as that of a hawfinch to that of a chaffinch, and (if Mr Gould is right in including his sub-group, Certhidea, in the main group), even to that of a warbler. . . . The beak of Cactornis is somewhat like that of a starling; and that of the fourth sub-group, Camarhynchus, is slightly parrot-shaped.

Charles Darwin: *The Voyage of the "Beagle,"* Ch. XVII.

SUBGENERA

The chief way in which the various species of Darwin's finches differ from each other is in their beaks. Indeed, the beak differences are so pronounced that systematists have at various times used as many as seven different generic names for the birds. In this book the genera are reduced to four, but it is convenient to retain the other generic names as subgenera, since they emphasize the adaptive radiation of the finches, as set out in Table X. Table X refers only to the species on the central Galapagos islands, the forms on outlying islands being considered later.

The observations of Gifford (1919) and ourselves show that the marked beak differences between the subgenera of Darwin's finches are correlated with marked differences in feeding habits.

TABLE X. Adaptive radiation of central Galapagos islands

Subgenus	Species	Beak	Chief food	Comments
1. *Geospiza*	*magnirostris* *fortis* *fuliginosa* *difficilis*	Heavy, finch-like	Seeds	Four ground-finches, three in coastal zone differing in size, one in humid zone
2. *Cactornis*	*scandens*	Long, decurved	*Opuntia*	Cactus feeding
3. *Platyspiza*	*crassirostris*	Thick, short and somewhat decurved	Buds, leaves, fruit	Primarily vegetarian
4. *Camarhynchus*	*psittacula* *pauper* *parvulus*	Like the last	Insects	Two species (three on Charles), differing in size
5. *Cactospiza*	*pallidus* *heliobates*	Stout, straight	Insects, especially from wood	Two species, differing in habitat
6. *Certhidea*	*olivacea*	Slender	Small insects	Warbler-finch

Notes: (i) *Cactornis* is part of the genus *Geospiza*, and *Platyspiza* and *Cactospiza* are part of the genus *Camarhynchus*.

(ii) The seventh subgroup is *Pinaroloxias* of Cocos, with a slender beak and insectivorous habits. There is only one species.

(i) Subgenus *Geospiza*. The heavy finch-like beaks of the large, medium and small ground-finches *Geospiza magnirostris*, *G. fortis* and *G. fuliginosa* suggest a diet of seeds, and this is the chief food of these birds, particularly outside the breeding season. In addition, they eat various fruits, also flowers, buds, young leaves and large caterpillars at the seasons when these are particularly abundant. Small insects and spiders are taken occasionally.

Few observations are available on the food of the fourth species, the sharp-beaked ground-finch *G. difficilis*. The race *G. d. debilirostris* of the central Galapagos islands feeds on the ground in the humid forest, scratching about in the leaves (Gifford, 1919), thus occupying the niche filled in English woods by the blackbird *Turdus merula*. The beak of *Geospiza d. debilirostris* fits this habit, since it is rather more elongated than that of typical ground-finches such as *G. fortis* and *G. fuliginosa*.

(ii) Subgenus *Cactornis*. The long, somewhat decurved beak and the split tongue of the cactus ground-finch *Geospiza scandens* suggest a flower-probing and nectar-feeding habit, and the flowers of the prickly pear *Opuntia* are a staple food of this species. It also feeds regularly on the soft pulp of the prickly pear, and on all the types of food, including seeds, listed above as taken by the other ground-finches. Hence, though partly specialized in feeding habits, it has not completely departed from the usual ground-finch diet, and, probably correlated with this, its beak is thicker than that of typical flower-eating birds. Gifford (1919) recorded that on Charles *Geospiza scandens* has taken to feeding on the fruits of the introduced oranges and tropical plums, an interesting modification of its natural feeding habits.

(iii) Subgenus *Platyspiza*. The vegetarian tree-finch *Camarhynchus crassirostris* has a short and thick, somewhat decurved and slightly parrot-like beak, rather similar in appearance to that of other birds which, like it, feed primarily on leaves, buds, blossoms and fruits. This species rarely takes insects, and was not seen to eat grain, but it takes some types of seeds. It feeds mainly in the trees, but comes to the ground to take young leaves of herbaceous plants and fallen fruit. Its actions are leisurely, like those of many other bud- and fruit-eating birds.

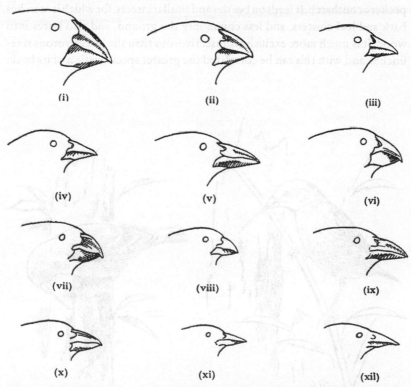

Fig. 9. Beak differences in Darwin's finches on central islands. 2/3 natural size (after Swarth). (i) *Geospiza magnirostris* (ii) *Geospiza fortis* (iii) *Geospiza fuliginosa* (iv) *Geospiza difficilis debilirostris* (v) *Geospiza scandens* (vi) *Camarhynchus crassirostris* (vii) *Camarhynchus psittacula* (viii) *Camarhynchus parvulus* (ix) *Camarhynchus pallidus* (x) *Camarhynchus heliobates* (xi) *Certhidea olivacea* (xii) *Pinaroloxias inornata*

(iv) Subgenus *Camarhynchus*. The insectivorous tree-finches *Camarhynchus psittacula*, *C. pauper* and *C. parvulus* have beaks very similar in shape to that of the vegetarian tree-finch *C. crassirostris*. This resemblance is presumably due to close relationship rather than to adaptive modification, since the food of the three former species is quite different from that of *C. crassirostris*, consisting mainly of beetles and similar insects, for which the birds examine the twigs, bark and leaf clusters, and also excavate shallow holes in soft wood. They feed chiefly in the trees and are agile in their movements, rather like tits, sometimes turning almost upside down in their search for food. They feed on the ground at times. While moderately small insects form their main food, they also eat nectar, buds, young leaves and large caterpillars at the seasons when these are abundant, and occasionally take grain.

(v) Subgenus *Cactospiza*. The woodpecker-finch *Camarhynchus pallidus* has a stout, straight beak, with obvious affinities to that of the insectivorous tree-finches, but more elongated, and modified in the direction of that of a woodpecker or nuthatch. It feeds on beetles and similar insects, for which it searches bark and leaf clusters, and less commonly the ground, and also bores into wood. It is much more exclusively insectivorous than the insectivorous tree-finches, and with this can be correlated the greater specialization of its beak.

Fig. 10. Camarhynchus pallidus and its stick. (Drawn by Roland Green from photographs by R. Leacock.)

C. pallidus further resembles a woodpecker in that it climbs up and down vertical trunks and branches. It is the only one of Darwin's finches to do this. It also possesses a remarkable, indeed a unique, habit. When a woodpecker has excavated in a branch for an insect, it inserts its long tongue into the crack to get the insect out. *C. pallidus* lacks the long tongue, but achieves the same result in a different way. Having excavated, it picks up a cactus spine or twig, one or two inches long, and holding it lengthwise in its beak, pokes it up the crack, dropping the twig to seize the insect as it emerges. In the arid zone the bird uses one of the rigid spines of the prickly pear *Opuntia*, but in the humid zone, where there is no *Opuntia*, it breaks off a small twig of suitable length from a tree or bush. It has been seen to reject a twig if it proved too short or too pliable. Sometimes the bird carries a spine or twig about with it, poking it into cracks and crannies as it searches one tree after another. This remarkable habit, first reported by Gifford (1919) and fully confirmed first by W. H. Thompson and later by the writer, is one of the few recorded uses of tools in birds. The nearest parallel is the use of fruits by the bower-bird *Ptilonorhynchus violaceus* for staining the stems in its bower (Gilbert, 1939).

The mangrove-finch *Camarhynchus heliobates* has a beak similar to that of *C. pallidus* and, like the latter, feeds almost exclusively on insects (Snodgrass, 1902a). Its feeding methods have not been recorded.

(vi) Genus *Certhidea*. The beak, general habits and appearance of the warbler-finch *Certhidea* are so like those of a warbler that it was for a long time considered to be one. Like a warbler, it searches the leaves, twigs and ground vegetation for small insects, and sometimes catches insects in the air. Nectar from flowers and young green leaves are also taken in season, but small soft insects form the great majority of its diet. The Hood form of this species also feeds commonly on small marine organisms collected at low tide below high-water mark, but this habit has rarely been observed on other islands, so provides an example of a racial difference in feeding habits.

(vii) Genus *Pinaroloxias*. The Cocos-finch *Pinaroloxias* is said to feed predominantly on insects and has a slender beak similar to that of *Certhidea*, except that it is a little longer and somewhat decurved. *Pinaroloxias* feeds both on the ground and in the trees, and has an unusually long, grooved and bifid tongue (Snodgrass, 1903; Gifford, 1919).

To summarize, the beak differences between most of the genera and subgenera of Darwin's finches are clearly correlated with differences in feeding methods. This is well borne out by the heavy, finch-like beak of the seed-eating *Geospiza*, the long beak of the flower-probing *Cactornis*, the somewhat parrot-like beak of the leaf-, bud- and fruit-eating *Platyspiza*, the woodpecker-like beak of the woodboring *Cactospiza*, and the warbler-like beaks of the insect-eating *Certhidea* and *Pinaroloxias*. Only in one group, namely, the insectivorous tree-finches of the subgenus *Camarhynchus* (*sens.*

strict.), is the beak not particularly suggestive of the feeding habits; these birds, though feeding primarily on insects, may be regarded as moderately unspecialized in both diet and beak.

CLOSELY RELATED SPECIES

While the beak differences between most of the subgenera of Darwin's finches are clearly adapted to differences in feeding methods, the same does not seem to hold for the beak differences between closely related species. Here the differences are chiefly in the size rather than the shape of the beak, and the species concerned have almost identical feeding methods. For instance, the large ground-finch *Geospiza magnirostris* has a huge beak, the medium *G. fortis* a fairly large one, and the small *G. fuliginosa* a smaller one. But all three species feed in similar places and have similar feeding methods, while their foods, listed in the preceding section, are of the same general nature. Similarly, the large insectivorous tree-finch *Camarhynchus psittacula* has a much larger beak than the small *C. parvulus*, but it occupies the same habitat and has almost identical feeding methods. The parallel with *Geospiza* is completed on Charles, where there occurs a third species, *Camarhynchus pauper*, intermediate in size and size of beak between the other two.

The significance of these marked beak differences between species otherwise similar has excited speculation from all who have discussed Darwin's finches. Our field observations confirm the similarity in feeding methods of the three species of *Geospiza* and that they often feed on the same things. However, we observed that the large, hard fruits of the manzanilla *Hippomane mancinella* are taken freely by the large *Geospiza magnirostris* and the medium *G. fortis*, but rarely if at all by the small *G. fuliginosa*; on the other hand, certain small seeds, notably of grasses, form a staple food of the small *G. fuliginosa*, but are taken less commonly by the medium *G. fortis*, and rarely if at all by the large *G. magnirostris*. Similarly, Snodgrass (1902a) showed from an analysis of stomach contents that the foods of the various species of *Geospiza* are often identical. However, the large *G. magnirostris* eats certain large seeds not taken by the small *G. fuliginosa*, while a series of the medium *G. fortis* and the small *G. fuliginosa* collected at the same time and place on North Albemarle were eating partly the same but partly different foods. Snodgrass was chiefly impressed by the similarity in the foods of the different species, but care is needed in interpreting his data, as too few stomach contents were collected to be truly representative of birds with such varied diets. In one case his results are definitely misleading, since they suggest that the small ground-finch *G. fuliginosa* and the cactus ground-finch *G. scandens* eat largely the same types of food. This happens to be true in the period immediately following the breeding season, when Snodgrass collected his specimens, for then *G. scandens*, like *G. fuliginosa*, eats many small seeds. But in the breeding season the diets of these two species are mainly different, *G. scandens* depending largely on prickly pear, as already noted.

Snodgrass concluded that the beak differences between the species of *Geospiza* are not of adaptive significance in regard to food. The larger species tend to eat rather larger seeds, but this he considered to be an incidental result of the difference in the size of their beaks. This conclusion was accepted by Gifford (1919), Gulick (1932), Swarth (1934) and formerly by myself (Lack, 1945). Moreover, the discovery mentioned in the last chapter, that the beak differences serve as recognition marks, provided a quite different reason for their existence, and thus strengthened the view that any associated differences in diet are purely incidental and of no particular importance.

My views have now completely changed, through appreciating the force of Gause's contention that two species with similar ecology cannot live in the same region (Gause, 1934). This is a simple consequence of natural selection. If two species of birds occur together in the same habitat in the same region, eat the same types of food and have the same other ecological requirements, then they should compete with each other, and since the chance of their being equally well adapted is negligible, one of them should eliminate the other completely. Nevertheless, three species of ground-finch live together in the same habitat on the same Galapagos islands, and this also applies to two species of insectivorous tree-finch. There must be some factor which prevents these species from effectively competing.

The above considerations led me to make a general survey of the ecology of passerine birds (Lack, 1944a). This has shown that, while most closely related species occupy separate habitats or regions, those that occur together in the same habitat tend to differ from each other in feeding habits and frequently also in size, including size of beak. In a number of the latter cases it is known that the beak difference is associated with a difference in diet, and this correlation seems likely to be general, since it is difficult to see how otherwise such species could avoid competing. It seems particularly significant that when two closely related species differ from each other in size and size of beak they often live in the same habitat, while such marked size differences are unusual in closely related species which occupy different habitats.

The general survey indicates that when two closely related species live in the same habitat, they do not usually take completely different foods. It is apparently sufficient if some of their foods are different, but the extent to which their diets can overlap is not known. In the case of the three species of *Geospiza*, there are similarities, but also established differences, in their diets, and though further evidence is much needed, it is provisionally concluded here that, so far from being unimportant and purely incidental, these food differences are essential to the survival of the three species in the same habitat. Further food analyses might most profitably be made in the latter part of the dry season, as this is the period when food is likely to be least varied and least abundant, and therefore most likely to limit the population density of the birds.

The food of the insectivorous tree-finches has not been analysed, but I suggest that, in this case also, the larger species probably tends to eat larger insects and the smaller species smaller insects, and that as a result the two species to some extent share out, instead of competing for, the available food supply. This view receives confirmation from evidence of a different nature. On Chatham, the large *Camarhynchus psittacula* does not occur, so it might be expected that a greater range of foods would thereby become available for the small species *C. parvulus*. It is therefore significant that the Chatham form of. *C. parvulus* has a larger beak than any other island form of this species. Indeed the beak overlaps in measurements with that of *C. psittacula* on other islands. This is just what would be expected if the beak difference between the two species is an adaptation for taking foods of different size, since on islands where *C. psittacula* is absent, but not elsewhere, unusually large individuals of *C. parvulus* will have survival value.

There is a parallel in the ground-finches, since the large *Geospiza magnirostris* is absent from Chatham and Charles, and on both these islands the medium *G. fortis* reaches a much greater maximum size than it does on the northern Galapagos islands where *G. magnirostris* is common. Indeed, some of the *G. fortis* from Chatham and Charles are as large as small specimens of *G. magnirostris* from the northern islands. This point is discussed further in Chapter VIII.

To sum up, though the three species of *Geospiza* have similar feeding methods, I consider that the marked difference in the size of their beaks is an adaptation for taking foods of different size and that, superficial appearances to the contrary, these three species are food specialists to an extent sufficient to enable them to live in the same habitat without effectively competing. A similar view is advocated for the difference in size of beak between *Camarhynchus psittacula* and *C. parvulus*. However, more detailed analysis of the foods taken by these species is extremely desirable, particularly in the case of *Camarhynchus*.

...

CONCLUSION

On the central Galapagos islands there are six subgenera of Darwin's finches, and five of the six show marked beak differences from each other, which are correlated with marked differences in feeding methods. Two of these subgenera, namely *Geospiza* and *Camarhynchus*, are further divided into species which differ primarily in size of beak, and these I consider to be adapted for taking partly different foods, though of the same general nature. Some of the finches are absent from outlying Galapagos islands; their food niches may then be filled by different species, or one form may take foods which on the central islands are divided between two species; in both cases there are corresponding beak modifications. To conclude, in Darwin's finches all the main beak differences between the species may be regarded as adaptations to differences in diet.

* Chapter VII: Size Differences between Island Forms

I have not as yet noticed by far the most remarkable feature in the natural history of this archipelago; it is, that the different islands to a considerable extent are inhabited by a different set of beings. . . . I never dreamed that islands about fifty or sixty miles apart, and most of them in sight of each other, formed of precisely the same rocks, placed under a quite similar climate, rising to a nearly equal height, would have been differently tenanted. . . . It is the circumstance, that several of the islands possess their own species of the tortoise, mocking-thrush, finches, and numerous plants, these species having the same general habits, occupying analogous situations, and obviously filling the same place. In the natural economy of the archipelago, that strikes me with wonder. It may be suspected that some of these representative species, at least in the case of the tortoise and of some of the birds, may hereafter prove to be only well-marked races; but this would be of equally great interest to the philosophical naturalist.

Charles Darwin: *The Voyage of the "Beagle,"* Ch. XVII

MEASUREMENTS

Several cases of striking beak differences between island forms of the same species were given in the later sections of the previous chapter. Many other island forms show less pronounced beak differences, and some of them also differ in size of body. These differences are for the most part average ones, and are best demonstrated by actual measurements. For this purpose two beak measurements have been taken, one of the culmen (upper edge) from the nostril to the tip of the beak, and the other of depth of the beak at its base. In addition, the wing was measured from the carpal joint (the main angle) to the tip of the longest primary. In Darwin's finches this latter measurement provides a more reliable indication of general body size than does the total length of the body measured from beak to tail.

. . .

Because of the above differences [in size between males and females], it is necessary to consider the two sexes separately when comparing the different forms of Darwin's finches. In this book, only the measurements of the males have been tabulated. Those of the females show precisely similar racial and specific variations, so that there is no need to include them as well; they are published elsewhere (Lack, 1945). When considering wing-length, it is also necessary to separate black or partly black males from those in plumage of

juvenile type, but the differences in beak measurements between these different types of males are so small that they can be neglected.

In considering racial differences in the average size of beak and wing,
it should be pointed out that in many of Darwin's finches these characters
do not vary altogether independently of each other. In nearly all the species, those individuals with longer beaks tend also to have deeper beaks, the
correlation being particularly marked in the ground-finches *Geospiza fortis*,
G. fuliginosa and *G. conirostris*, but altogether absent in the woodpecker-finch
Camarhynchus pallidus. A similar correlation is found in the house sparrow
Passer domesticus (Lack, 1940a), but is absent in most, though not all, forms of
the North American *Junco* (Miller, 1941). No other birds appear to have been
studied in this respect.

In five of Darwin's finches, namely, the ground-finches *Geospiza magnirostris*, *G. fortis*, *G. fuliginosa* and *G. conirostris* and the tree-finch *Camarhynchus
crassirostris*, the individuals with longer wings tend also to have larger beaks,
though the correlation is not so marked as that between length and depth of
beak. Hence in these species, and particularly in the medium ground-finch
Geospiza fortis, general body-size presumably has some influence on the size of
the beak, though other beak variations are independent of body-size. Neither
the house sparrow nor most forms of the junco show a correlation between
size of beak and length of wing, though this is found in a few forms of the
junco. Correlation coefficients for these characters are set out in Table XXII
[not reprinted].

DIFFERENCES BETWEEN ISLAND FORMS

For reference purposes, the average measurements of beak and wing are set
out in Table XXIII [not reprinted] for every species of Darwin's finch on every
island from which sufficient specimens have been collected. The three following tables give the limits of measurement of the various forms, and standard deviations are given in Tables XXIX and XXX [not reprinted]. These
tables show how commonly a species differs significantly in average size on
different islands. The differences are often small, many individuals overlapping in measurements with those from other islands, but occasionally they
are very marked, with little or no overlap between extreme individuals. In
the large ground-finch *Geospiza magnirostris*, the individuals from the northern islands tend to be rather larger than those from the central islands. The
medium ground-finch *G. fortis* and the small *G. fuliginosa* show an opposite
tendency, the birds on the southern islands tending to be larger than those
on the northern islands, especially in wing-length. But these trends are by no
means completely regular, some island populations varying contrarily to the
main direction. Further, the differences are only in average size, and there is

Fig. 14. Variations in wing-length in *Geospiza fortis*. (Average wing-length in mm. for black males.)

marked overlap in the measurements of extreme individuals from different islands. The variations in the average length of the wing of *G. fortis* are shown in Fig. 14; for further details see Table XXIII.

The island variations found in the cactus ground-finch *G. scandens* are also erratic, since the small birds from James show no overlap in beak measurements with the large birds from Bindloe, the next island to the north; but on islands to the south occur populations which overlap widely in measurements with both the James and the Bindloe birds. The beak variations in *G. scandens* and the related species *G. conirostris* are shown in Fig. 15. Those of *G. conirostris* were discussed earlier.

The marked beak differences between the three subspecies of the sharp-beaked ground-finch *G. difficilis* were also considered in the last chapter, but further variations occur in this species, since each of the three subspecies inhabits two islands, and in each case there are small differences in average measurements between the two island populations (see Table XV and Table XXIII).

The vegetarian tree-finch *Camarhynchus crassirostris* and the small insectivorous tree-finch *C. parvulus* show little difference in average size on different islands, except for the unusually large form of *C. parvulus* on Chatham. On the other hand, both the large insectivorous tree-finch *C. psittacula* and the woodpecker-finch *C. pallidus* show marked island differences; in addition to the variations discussed earlier, both species are represented by a particularly

small form on Albemarle. Finally, in the warbler-finch *Certhidea* the birds of
the outlying Galapagos islands tend to be rather larger than those of the central islands as set out in Fig. 16. In this species the variations in plumage also
tend to a concentric distribution, as shown in Fig. 6 [not reprinted].

NATURE OF THE DIFFERENCES

In Darwin's finches some of the most marked beak differences between island forms of the same species are adaptive, as considered in the last chapter.
On the other hand, many of the smaller differences in beak or wing-length
are extremely difficult to relate to possible environmental factors. Some examples of this apparently pointless variation are set out in Table XV.

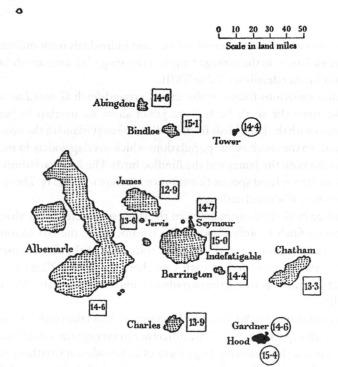

Fig. 15. Variations in beak in *Geospiza scandens* and *G. conirostris.* (Average length of culmen from nostril in mm.) (i) For *G. scandens*—islands stippled, figures in squares (ii) For
G. conirostris—islands black, figures in circles.

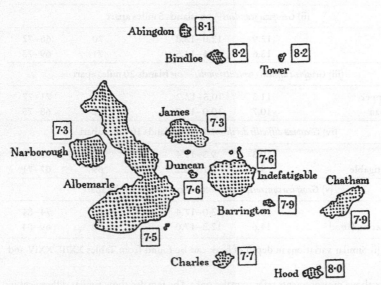

Fig. 16. Variations in beak in *Certhidea*. (Average length of culmen from nostril in mm.)

Table XV shows the existence of significant differences, both in average size and in extreme measurements, between populations of the same species living on islands only a few miles apart. In each case, so far as known, the finch occupies the same ecological niche on both islands, while the two islands provide similar physical and climatic conditions, similar habitats, and the same complement of other species of land birds. In such cases, more of which can be found in Table XXIII, it is almost impossible to believe that the differences between the island forms are adaptively related to possible differences in their environments. The size difference in *Geospiza conirostris* on islands only a mile apart is particularly striking, and recalls the yet more remarkable case of the white-eye *Zosterops rendovae* of the Solomons, in which well-defined island forms are separated by straits which are only 1, 2, 3 and 4 miles wide respectively (Mayr, 1942).

The rarity of regular trends of variation is another fact suggesting that in

TABLE XV. Beak and wing variations in forms on adjacent islands

Island	Culmen (from nostril) in mm.		Wing in mm. (black males)	
	Average	Limits	Average	Limits
(i) *Geospiza magnirostris*—on islands 5 miles apart.				
James	15.9	13.8–17.4	84	81–88
Jervis	15.3	13.9–17	83	78–90
(ii) *Geospiza scandens*—on islands 5 miles apart				
James	12.9	12.0–13.8	70	66–72
Jervis	13.6	12.4–14.5	71	69–73
(iii) *Geospiza difficilis septentrionalis*—on islands 20 miles apart				
Culpepper	11.3	10.5–12.2	73	71–77
Wenman	10.7	10.0–11.7	72	68–75
(iv) *Geospiza difficilis debilirostris*—on islands 10 miles apart				
James	10.3	9.5–11.4	72	65–76
Indefatigable	9.6	9.1–10.2	69	67–71
(v) *Geospiza conirostris conirostris*—on islands 1 mile apart				
Hood	15.4	13.0–17.4	80	74–84
Gardner nr. Hood	14.6	12.3–17.0	79	69–84

Notes: (i) Similar variations in depth of beak can be found from Tables XXIII, XXIV and XXVI.

(ii) The above measurements refer to males only. The females show similar differences in each case.

(iii) The statistical significance of the differences involved can be calculated from the number of specimens measured, and the standard deviations of the measurements given in Tables XXIX, XXX, and Supplementary Table C.

Darwin's finches many of the differences in beak and wing-length between island forms are without adaptive significance. Even where there is a suggestion of regularity, there are usually certain islands on which variation runs counter to the main trend. Examples of such irregularity can be found in every species.

It is extremely difficult to establish convincingly that a structural difference is non-adaptive. Also, it was a long time before I appreciated that some of the beak differences discussed in the last chapter were adaptive, so that there may well be other differences the adaptive significance of which has not yet been discovered. On the other hand, conditions are extremely similar on many of the Galapagos islands, and the differences in beak and wing-length shown by island forms so often seem haphazard and pointless, that it is probable that many of these differences are genuinely unrelated to possible environmental differences. The ways in which non-adaptive differences might arise between populations are considered in a later chapter.

...

* Chapter VIII: Size Differences between Species

The characters of the species of Geospiza, as well as of the following subgenera, run closely into each other in a most remarkable way.

Charles Darwin: *The Zoology of the Voyage of H.M.S. "Beagle"*

THE THREE GROUND-FINCHES

The large ground-finch *Geospiza magnirostris*, the medium *G. fortis*, and the small *G. fuliginosa* differ from each other solely in size and in relative size of beak. The probable connection of this difference with food has been discussed in Chapter VI, but the situation is so remarkable that it is here analysed in further detail.

The limits of measurement given in Fig. 17 and in Table XXIV show that on the northern Galapagos islands the three species are clearly separated from each other. But on some of the southern islands the separation is very narrow, and occasionally there is actual overlap in characters. Further, on some of the southern islands the medium *G. fortis* is highly variable, and the largest individuals come closer in measurements to small specimens of *G. magnirostris* than they do to small specimens of their own kind, while the latter come closer in measurements to large specimens of *G. fuliginosa* than they do to large individuals of their own kind. This situation is without parallel in birds. The three species, it must be remembered, differ not at all in plumage. It might therefore be wondered whether there are three distinct species, rather than a single highly variable form. But Fig. 17 shows that the measurements of these birds fall into three groups, each centred round a particular mean value, while individuals with intermediate measurements are very scarce.

OVERLAP IN MEASUREMENTS

Table XXIV [not reprinted] shows a further peculiarity, namely, that the gap between *G. magnirostris* and *G. fortis*, and the gap between *G. fortis* and *G. fuliginosa*, occur at rather different measurements on different islands. Thus black males with a wing-length of 64–66 mm. belong to *G. fortis* if collected on Abingdon or Bindloe, but to *G. fuliginosa* if from any other island. Similarly, males with a wing-length of 79–80 mm., a culmen from nostril of 14 mm. and a beak depth around 16.0–16.5 mm. belong to *G. magnirostris*

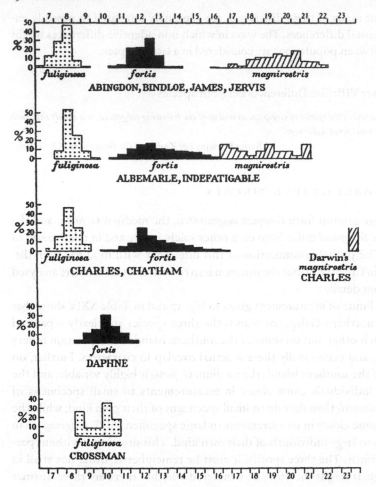

Fig. 17. Histograms of beak-depth in *Geospiza* species. Measurements in millimetres are placed horizontally, and the percentage of specimens of each size verticallv. Too few specimens are available for a histogram of Darwin's *magnirostris*. The Daphne and Crossman forms are discussed on pp. 84–86 [later in this chapter].

if collected on Abingdon, Bindloe or Jervis, but to *G. fortis* if collected on Charles.

As discussed in Chapter VI, the size differences between the species are probably correlated with their taking rather different foods. It is primarily on Charles and Chatham, where *G. magnirostris* is absent, that *G. fortis* attains a large size. However, *G. fortis* also overlaps in measurements with *G. magnirostris* on two islands where the latter species occurs, namely, Albemarle and Indefatigable—but though *G. magnirostris* is present there it is very scarce.

Thus of 511 specimens of the two species collected on Albemarle and Indefatigable, only 9 per cent belong to *G. magnirostris*, whereas of 455 specimens
from Abingdon, Bindloe, James and Jervis, 55 per cent belong to *G. magnirostris. G. fortis* does not normally attain a large size on any of the islands where
G. magnirostris is common.

Though now absent there, *G. magnirostris* probably occurred formerly
on Charles. But the specimens which Darwin collected are extremely large,
and are separated from the largest specimens of *G. fortis* on Charles by a gap
as wide as that between the two species on the northern Galapagos islands,
where both are smaller. This would suggest that the large individuals of
G. fortis on Charles did not compete for food with the unusually large Charles
form of *G. magnirostris*.

Except on Chatham, where there is a small overlap in their measurements, the gap between the medium *G. fortis* and the small *G. fuliginosa* is
more definite than that between *G. fortis* and *G. magnirostris*. But the position is complicated by the existence on the islet of Daphne of an intermediate
form, the smallest individuals of which are equal in size to large individuals of
G. fuliginosa and the largest to small specimens of *G. fortis*. A similar but
rather smaller intermediate form occurs on the Crossman islets. As shown in
Fig. 18, Daphne is not particularly near the Crossmans, which suggests that
these intermediate forms are of independent origin. Formerly, they were referred to a distinct species *G. harterti*, but Swarth (1931) considered that the
Daphne birds were unusually small specimens of *G. fortis* and the Crossman
birds unusually large specimens of *G. fuliginosa*. As the plumage of the two
species is identical, measurements provide the only clue to their specific identity. These are given in Fig. 17 and Table XVI.

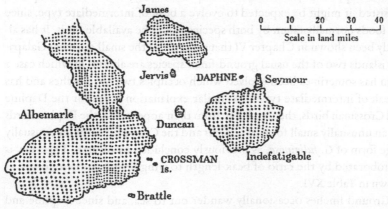

Fig. 18. Map showing Daphne and the Crossman Islets.

TABLE XVI. The peculiar forms of *Geospiza* on Daphne and Crossman

	Culmen from nostril in mm.		Wing in mm. (black males)		Ratio of culmen to wing	Coefficient of variability	
Form	Mean	Limits	Mean	Limits		Culmen	Wing
G. fortis (Indefatigable)	12.0	10.5–13.9	73	69–79	0.16	6.8	3.8
Daphne form	10.5	9.2–11.3	67	65–70	0.16	5.1	2.6
Crossman form	9.3	8.0–11.2	66	63–67	0.14	9.4	2.3
G. fuliginosa (Indefatigable)	8.4	7.5–9.3	64	61–66	0.13	4.5	2.2

Note: For number of specimens measured see Supplementary Table C.

Formerly I thought that the Daphne and Crossman forms might be of hybrid origin between *G. fortis* and *G. fuliginosa*. However, it would be highly remarkable if two species which occur together on many islands without interbreeding should, just in two places, give rise to a hybrid population. Further, it would be necessary to suppose that such interbreeding no longer takes place, since the Daphne and Crossman birds do not include any individuals in the upper range of size of *G. fortis* nor any in the lower range of size of *G. fuliginosa*. Moreover, the chief character in which these birds are intermediate is the beak, a structure which is particularly adaptable in Darwin's finches.

An alternative explanation is possible. Daphne is only half a mile in diameter and the Crossmans are rather smaller, so that they could support only very small populations of *G. fortis* or *G. fuliginosa*. Perhaps one of the two species became temporarily extinct there, or alternatively the islands may be too small to support populations of both species. If only one of the two species persisted, it might be expected to evolve a beak of intermediate type, since the foods normally taken by both species would be available for it. It has already been shown in Chapter VI that on some of the small outlying Galapagos islands two of the usual ground-finch species are absent, in which case a form has sometimes been evolved which occupies two food niches and has a beak of intermediate type. If a similar explanation holds for the Daphne and Crossman birds, then, to judge from their appearance, the Daphne birds are an unusually small form of *G. fortis* and the Crossman birds an unusually large form of *G. fuliginosa*, as previously concluded by Swarth. This view is corroborated by the ratio of beak-length to wing-length in these forms, as shown in Table XVI.

Ground-finches occasionally wander out to sea, and since Daphne and Crossman are close to large islands, it is probable that typical individuals of

G. fortis and *G. fuliginosa* occasionally wander there. Indeed, three typical specimens of *G. fuliginosa* have been taken on Daphne. Nevertheless, the peculiar
local forms persist, which presumably means that they are better adapted to
existence on small islets than are the typical forms of the species concerned.
A similar conclusion was reached regarding the persistence of forms such as
G. conirostris and *G. difficilis septentrionalis*, which combine two food niches on
the small remote Galapagos islands.

In addition to the cases of overlap in measurements so far considered,
there are four specimens from James and Bindloe which are so intermediate
between *G. magnirostris* and *G. fortis* that they cannot certainly be identified.
Their measurements are given in Table XXXII [not reprinted], but are omitted from Fig. 17 owing to doubt as to the species concerned. In the same table
are given the measurements of two specimens, from Hood and Chatham,
which are intermediate between *G. fortis* and *G. fuliginosa*; these are probably
freak specimens of the smaller species. Such intermediate specimens, which
are extremely rare, are discussed further in Chapter X.

DIFFERENCES IN PROPORTIONS

The three species *G. magnirostris*, *G. fortis* and *G. fuliginosa* differ from each
other not only in absolute size of beak and wing, but also in their proportions. The large *G. magnirostris* has proportionately the longest beak in relation to wing-length, and proportionately the deepest beak in relation to
beak-length. The small *G. fuliginosa* has proportionately the smallest beak in
relation to wing-length and proportionately the narrowest beak in relation
to beak-length.

Huxley (1927, 1942) has shown in a number of animals that the larger
individuals have proportionately larger external parts. This allometric relation holds in such diverse instances as antlers of deer and claws of crabs,
and suggests the possibility that the differences in proportions between the
three ground-finch species may be simply an indirect effect of their absolute
size. To test this, Tables XXVII and XXVIII [not reprinted] were prepared.
These rule out the above suggestion, showing that within each species the
larger individuals do not have relatively larger beaks, while those with longer
beaks do not have relatively deeper beaks. Instead, the proportions of wing-
length to beak-length, and of length to depth of beak, are approximately
the same for all the individuals of each species, whatever their absolute size.
Hence the differences between the three species do not depend simply
on general size factors with associated allometric relations affecting their
proportions.

Since the proportions of beak and wing are characteristic for each of the
three species, they provide an additional criterion for identification, but,

owing to the marked individual variation, only the average for a series of specimens is reliable. Because of their high variability in both proportions and absolute size, and owing to the narrowness of the gaps between their extreme measurements, and because their plumage is identical, these three ground-finch species show no completely certain criteria for identification. However, nearly every specimen can be safely identified by considering all its characters together, and knowing also the island where it was collected. Presumably the three species have markedly different hereditary constitutions, and so far as known they do not interbreed with each other, but they regularly give rise to individuals which are extremely similar to each other in all external characters, while very occasional individuals are so intermediate in appearance that they cannot safely be identified—a truly remarkable state of affairs. In no other birds are the differences between species so ill-defined.

The photographs in Plate IV [not reprinted] are of specimens in the California Academy of Sciences. They are all reproduced on the same scale, and show the marked difference in size between the largest and smallest specimen of *G. fortis*, and how closely the largest approaches to *G. magnirostris* and the smallest to *G. fuliginosa*. They also show the difference in the proportions of the beak in the three species.

CAMARHYNCHUS

The insectivorous tree-finches present a situation rather similar to that of the three ground-finches, though with certain differences. On most Galapagos islands there are two species, the large *Camarhynchus psittacula* and the small *C. parvulus*, which have similar plumage and differ solely in size and in proportionate size of beak. As in *Geospiza*, the larger species has proportionately the longer and deeper beak. On Charles there also occurs a third species *Camarhynchus pauper*, intermediate in size between the other two. The size variations in these three species are shown in Fig. 19 and Table XXV.

The small insectivorous tree-finch *C. parvulus* is of similar size on all the main Galapagos islands except Chatham. Its unusually large size on the latter is probably correlated with the absence there of *C. psittacula*. The largest individuals from Chatham overlap in all measurements with the Albemarle form of *C. psittacula*, but *C. parvulus* does not normally overlap in beak measurements with *C. psittacula* on any island where both occur together.

On the central Galapagos islands of James, Jervis and Indefatigable, the gap between the two species is clear and wide. On Albemarle it is still present, but it is much narrower because of the small size of Albemarle *C. psittacula*. Finally, on Charles the largest specimens of *C. parvulus* are only just separated in depth of beak from the smallest specimens of the medium-sized species *C. pauper*, and in length of wing and culmen from nostril they show a slight

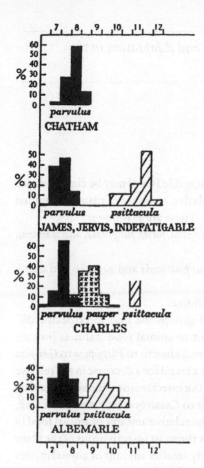

Fig. 19. Histograms of beak-depth in *Camarhynchus* species. Measurements in millimetres are placed horizontally, and the percentage of specimens of each size vertically. Too few specimens are available for a histogram of C. *psittacula* on Charles.

overlap. Similarly, the largest specimens of *C. pauper* overlap slightly in length of wing and culmen with the smallest specimens of *C. psittacula*, but they show no overlap in regard to depth of beak, though the gap is extremely narrow. Only sixteen adult specimens of *C. psittacula* have been collected on Charles, as compared with 142 of *C. pauper* and 124 of *C. parvulus*. Evidently the large *C. psittacula* is much less successful than the medium-sized *C. pauper* on this island.

In *Camarhynchus*, as in *Geospiza*, further overlap in measurements is provided by a few freak specimens which are intermediate in size between the small and the larger species. They cannot certainly be identified, but are probably abnormal specimens of *Camarhynchus parvulus*. Their measurements are given in Table XXXII.

Gloucester, MA: Peter Smith Publisher, 1968. Excerpts from chapters 6–8, pp. 55–90.

From *Morphological Differentiation and Adaptation in the Galápagos Finches* (1961)

✳ ROBERT I. BOWMAN

* Pages 70–71

SUMMARY OF FOOD ANALYSIS

The nine species of Geospizinae on Indefatigable Island may be classified into three main groups, on the basis of the relative amounts of plant and animal materials contained in the digestive tract. These are as follows:

　　1) Mainly plant eaters—*Geospiza magnirostris, fortis, fuliginosa,* and *scandens,* and *Platyspiza crassirostris*

　　2) Mainly animal eaters—*Camarhynchus psittacula* and *parvulus* and *Cactospiza pallida*

　　3) Entirely animal eaters—*Certhidea olivacea*

Within each of the afore-mentioned groups we find interspecific differences in the quantitative ratio of plant to animal food (table 11 [not reprinted]). Thus, in passing from *Geospiza magnirostris* to *Platyspiza* to *Geospiza scandens* to *G. fortis* to *G. fuliginosa* we note a trend for a decrease in the relative amount of plant food in the diet. Among the insectivorous finches, in passing from *Cactospiza* to *Camarhynchus psittacula* to *Camarhynchus parvulus*, we find, on the average, a trend for a decrease in the relative amount of animal food in the diet. It should be noted, however, that the most insectivorous of the plant eaters (*Geospiza fuliginosa*) takes a relatively smaller amount of animal matter than is consumed by the most herbivorous of the insect eaters (*Camarhynchus parvulus*).

. . .

　　Among the herbivorous finches, the species having the greatest diversity of diet, as indicated by the number of different recognizable plant items present in the sample, are *Geospiza fuliginosa* and *G. fortis* (cf. Lack, 1945:39); and the species with the least diversity of diet are *Geospiza scandens* and *G. magnirostris.* The diet of *Platyspiza* is only slightly less diverse than that of *Geospiza fortis.* Thus, within the genus *Geospiza* there is a trend in the direction of decreasing dietary diversity with increasing bill specialization.

　　Considering the number of different recognizable groups of insects present in the sample, the greatest diversity of diet among the insectivorous species is shown by *Camarhynchus parvulus*, and the least diversity by *Cactospiza pallida*. Thus, as in the case of *Geospiza*, we may recognize a trend in the direction of decreasing diversity with increasing bill specialization, namely, *Camarhynchus parvulus* to *C. psittacula* to *Certhidea* to *Cactospiza*.

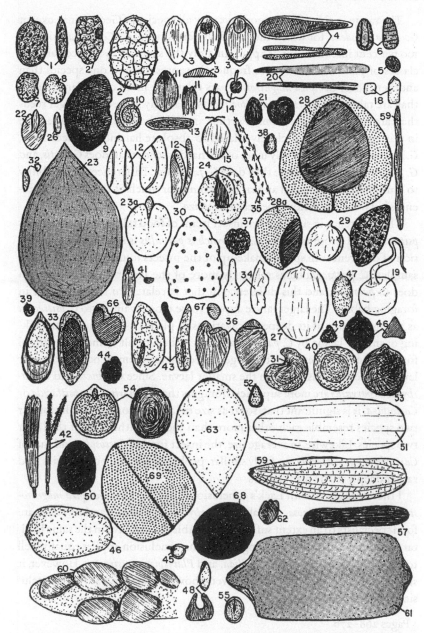

Fig. 3. Plant materials removed from the stomachs of eight species of Geospizinae collected on Indefatigable Island (× 5, except 23a and 28a, which are 2.5 ×). Most unidentified.

The range in size of plant food taken by all species of Geospizinae except *Geospiza scandens*, for which the sample is very small, and *Certhidea*, which is 100 per cent insectivorous, is approximately the same. However, certain size classes of plant items are more frequently represented in one species than another. Thus, *Geospiza magnirostris* eats relatively more of the larger seeds than are taken by *G. fortis*, and *G. fortis* eats relatively more of the larger seeds than are taken by *G. fuliginosa*. Further analysis shows that this difference in average size of seed is closely related to absolute hardness of seed. Thus, *G. magnirostris* eats relatively more of the harder seeds than does *G. fortis*, and *G. fortis* eats relatively more than does *G. fuliginosa*. Differences in the ability to crush hard seeds would appear to account for much of the dietary differences between the various species of *Geospiza*.

The difference in size between insect items of food taken by *Camarhynchus psittacula* and those taken by *Camarhynchus parvulus* is shown to be a function of differences in species make-up of the insects found in the respective samples. *C. psittacula* takes relatively more of the larger insect larvae than does *C. parvulus*, and this difference in diet is related to the difference in foraging habits. The larger larvae appear to be more available to *C. psittacula* as a consequence of its deeper excavations in plant tissue, whereas the taking of smaller larvae by *C. parvulus* appears to be a result of more superficial foraging on the vegetation; the larger insect larvae seem to occur at deeper levels in the plant tissues than the smaller larvae. Since *Certhidea* and *Camarhynchus parvulus* are more superficial foragers than *Camarhynchus psittacula* or *Cactospiza*, it is significant that, so far as size of larvae is concerned, the former two species resemble each other more than they do either one of the latter two species. Likewise, in average size, the insect food of *Cactospiza* and that of *Camarhynchus psittacula* are fairly similar, probably because both species feed rather deeply in the plant tissues.

I therefore agree with Lack's conclusion (1947:64) that *Geospiza magnirostris*, *fortis*, and *fuliginosa*, and also *Camarhynchus psittacula* and *parvulus*, are food specialists to an extent sufficient to enable them to live in the same habitat without effectively competing. This same conclusion applies equally well to *Certhidea olivacea*, *Cactospiza pallida*, and *Platyspiza crassirostris*. However, it should be noted that my reasons, given above, for arriving at these conclusions are different from his.

* Pages 268–276

Lack (1947:78–79) seems to be convinced that the differences between the island forms of the Geospizinae are not adaptively related to possible differences in their environments. His reasons are fourfold. First, as a rule, two neighboring island populations of the same species are "only a few miles apart." Second, "in each case, so far as is known, the finch concerned occupies the same ecological niche on both islands." Third, the islands provide "similar physical and climatic conditions, similar habitats, and the same complement of other species of land birds." Fourth, "there is a rarity of regular trends of variation . . . Even where there is a suggestion of regularity, there are usually certain islands on which variation runs counter to the main trend."

It cannot be denied that distance is an all-important factor in segregating two populations of the same species. Even though there are reports that geospizines—in many cases the birds were immatures—have been captured or seen many miles at sea, indicating their ability to fly over rather extensive areas of water (Gifford, 1919: 226, 230, and 237), this does not prove that the birds, once arrived on a nearby island, would survive in their new environment, for they must find an ecological niche comparable in most details to that which they occupied on the island of their origin. That the latter condition will in most instances be realized is rather unlikely, because of the marked botanical and physiographical differences between the various islands (see below). Habitat selection may be as effective a "barrier" to dispersal as an expanse of water, even though neighboring islands may be within easy flying range of the species (see Miller, 1942:32–33).

Closely related species of *Geospiza* live together in the arid coastal zone sharing the same habitat. The adaptations found in these sympatric species are primarily related to food getting. So far as I know, there are no studies on the feeding habits and general ecology of any species of Geospizinae presented in sufficient detail to justify the assumption that where a species occurs on two islands it occupies "the same ecological niche on both islands." Indeed, many of the published notes on feeding habits point to significant differences; observations of this nature for *Geospiza scandens* and *Camarhynchus psittacula* have been noted above, and there are other examples.

That the ecological conditions on many of the islands are very similar is not borne out by the evidence, most of which indicates that there are well-marked differences in the physical, climatic, and botanical conditions on the various islands of the Galápagos Archipelago. For example, comparing Jervis and James, two islands to which Lack's remarks (1947:78–79) refer, we see (fig. 1 of the present study [not reprinted]) that James Island is about twenty miles in length and Jervis about two miles in diameter, and that the former is more than twice as high as the latter (table 6 [not reprinted]). On James there are many very recent lava deposits, which are wanting on Jervis Island. Because of altitudinal differences, James Island receives more precipitation, and this is reflected in the greater elaboration of the vegetation zones. Concerning the diversity of the flora, James Island has 224 species of vascular plants, whereas Jervis Island has only 42 species (Stewart, 1911:237), and the two islands have only 29 species in common.

Beebe (1924:259) has correctly noted that "the most astonishing thing about the various islands of the Galápagos is their superficial similarity and their actual diversity." Because of the notable interisland floral differences and, therefore, differences in the kinds of food available, it is not surprising that a regular geographic trend or cline in bill variation could not be demonstrated by Lack (1947:79), but this does not mean that some form of nonfortuitous variation cannot be demonstrated. For example, in an earlier chapter (chap. viii) it was shown that certain interisland variations in plumage appear to be correlated with definite differences the environment. Such variation does not follow any neat system of north-south or east-west gradients; but rather, is intimately related to the particular conditions that prevail on the islands concerned, irrespective of their latitude, size, or isolation.

In an effort to account for the fact that a marked variability of the bill occurs in certain species of the Geospizinae but not in others, and to account for the interisland differences in variability, Lack (1947:90–95) compared population size with the degree of variability, on the rather unconvincing assumption, so far as it relates to Galápagos, that larger islands in general support larger populations of a given species than do the smaller islands. I have calculated the coefficients of variation for two bill dimensions from data given by Lack (1945:142–147), and these are presented in table 61 [not reprinted]. Using essentially these same data, Lack (1947:92) concludes that there is some support, although meager, for the idea that the more abundant forms of the Geospizinae are the more variable. It appears to me, however, that Lack's evidence, rather, denies such a fundamental relationship. Through an unfortunate error in calculation, Lack (1947:181) gives a value of 7.4 for the coefficient of variation of bill depth in *Geospiza conirostris* from Culpepper Island, one of the smallest islands of the Galapagos group (table 6). By basing our calculations on data given in his earlier work (1945:147) we find that the value should actually be 13.13! Indeed, this would appear to be the highest recorded value for a coefficient of variation in the Geospizinae, and hardly to be construed as support for the idea that larger populations are, in general, more variable than smaller populations. Such an exceptionally high value for the coefficient of variation in *Geospiza conirostris* (in the sense of Lack) on Culpepper Island, based on less than twenty-five specimens, would seem to suggest that Lack (1945:8–9) may have erred in "lumping" all the very thick-billed forms of *Geospiza* in one species (see Mayr et al., 1953:135). Formerly they were carried in two species, namely, *G. magnirostris* and *G. conirostris propinqua* (Swarth, 1931:149–150, 206). Obviously there is need for much new material from Culpepper Island if we are to arrive at any certain conclusion in this matter.

The conclusion reached by Lack (1945:135) that differences between island forms of the same species are "due primarily to the 'Sewall Wright effect,'" is a statement with which I cannot agree, for the following reasons. (1) There are interisland differences in feeding habits which are correlated with anatomical differences in the head region. (2) There are marked interisland differences in the food supply (flora). (3) Nearly every island has a different complement of geospizine species. (4) There are strong selective pressures on the populations on nearly all of the islands, in the form of predators. (5) Lastly, the size of most, if not all, of the geospizine populations, even at their lowest ebb is surely in excess of the few hundred maximum allowable for rapid fixation of genes.

VEGETATION DIFFERENCES AND NUMBER OF SPECIES OF THE GEOSPIZINAE

Previous workers have paid very little attention to the restraints imposed upon the variation, distribution, and relative abundance of the Geospizinae by the nature of their food plants. Because floristic differences between the various Galápagos islands have been the subject of much discussion by botanists (see Robinson, 1902; Stewart, 1911; and Kroeber, 1916), it is surprising that the importance of the botanical conditions in the evolution of the Geospizinae has not been recognized.

In the following discussion I shall attempt to show that the islands which are more diverse ecologically support a larger assemblage of geospizine species. The various Galápagos islands are so close to one another that irregularities in the distribution of the species of the Geospizinae are probably better attributed to ecological unsuitability of the islands than to failure of dispersal.

From table 6 it may be seen that, in general, the largest numbers of geospizine species are found on islands with the most diverse flora. On Albemarle, which has the largest number of species of vascular plants (329), there are 10 species of Geospizinae. James and Indefatigable islands, with 224 and 193 plant species respectively, each have 10 species of Geospizinae. It must be noted that Chatham and Charles, each with over 300 species of vascular plants, have only 7 and 9 species of Geospizinae, respectively; however, human settlements have long been established on both of these islands, and it is conceivable that some species of Geospizinae may have been exterminated. . . .

The islands of Abingdon, Duncan, Narborough, and Jervis, with totals of 119, 103, 80, and 42 species of vascular plants, respectively, each have 9 resident species of Geospizinae. The significant point here is that even though these islands are notably less diverse floristically than the islands previously mentioned, the number of species of Geospizinae present is not proportionately smaller. These four islands are each of sufficient height to support two or more vegetation zones. In other words, to a certain degree, ecologic diversity (i.e., variety of niches) is fully as important as floristic diversity in supporting a large geospizine fauna, if not more important. . . .

The evidence presented above seems to indicate rather clearly that the differences in the complements of finches on the various islands are functions of niche diversity. And when more detailed studies are made I think it will be found that in all species of the Geospizinae, interisland differences in variability of bill size and shape as well as differences in the numbers of geospizine species present are directly correlated with interisland differences in density, diversity, and availability of the food resources.

COMPETITION

But how is it that many of the immigrants have been differently modified, though only in small degree, in islands situated within sight of each other, having the same geological nature, the same height, climate, &c.? This long appeared to me a great difficulty: but it arises in chief part from the deeply-seated error of considering the physical conditions of a country as the most important; whereas it cannot be disputed that the nature of the other species with which each has to compete is at least as important, and generally a far more important element of success. (Darwin, 1875:355)

From these remarks it is apparent that Darwin considered the idea of "competition" necessary in order to account for the interisland differences in Galápagos organisms, and particularly the finches. Lack (1947) also attributes certain interisland differences in variability (pp. 63, 88, and 92), distribution (pp. 26–28), and pattern of adaptive radiation in the Geospizinae (pp. 113–114) to differences the nature of the "competing" species. Let us examine some of the examples he gives.

In discussing the variability of the bill in *Geospiza fortis* on small islands, he remarks (1947:92–93):

> It seems probable that the reduced variability . . . on Abingdon, Bindloe, James and Jervis is due primarily to the fact that . . . G. *magnirostris* is common there, and that this prevents G. *fortis* from attaining the large size which is possible for it on islands where G. *magnirostris* is absent or very scarce. . . . Thus on islands where . . . *Geospiza magnirostris* and . . . G. *fortis* are both common it might be expected that they would tend to have more restricted diets than on islands where one or the other is absent. The presence or absence of G. *fortis* may similarly affect the diet of the small G. *fuliginosa*.

It is implied in Lack's remarks that there is a "competition" between G. *fortis* and G. *magnirostris* for certain foods, and that the intensity of the competition is related to the size of the populations of the competing species. Although such a point of view is difficult to disprove, I believe that the conditions described may be interpreted in a different way, namely, that bill variation and population density are above all related to the availability and diversity of the food. Competition for food implies that there is an active or passive "struggle" on the part of one species to gain certain foods sought by another species, and that if more or less equal numbers of the two species are involved in the competition, then a kind of stability or equilibrium is reached and the food items under "dispute" are "shared."

One might question why on certain islands *G. magnirostris* is not present or is very scarce. A bird as specialized as this species is in its beak cannot live successfully in an area where the food for which it shows a singular adaptation (i.e., very hard seeds) is absent or not "superabundant" (to use the term of Lack). My study on the feeding habits of *G. magnirostris* and *G. fortis* on Indefatigable Island (presented in chapter iv) has shown that between these species there is a dietary overlap, just as there is between *G. fortis* and *G. fuliginosa*, and, indeed, between almost all the seed-eating species of Geospizinae. This overlap does not imply a "competition" for food so long as the supply of this food remains abundant and so long as it is not the only item in the diet of the species concerned....

The question arises, how much overlap in diet may be tolerated by two species living in the same environment? It is possible, theoretically, that two sympatric species might feed on "identical" foods, but with different frequencies of utilization of the different items. Should this occur, it would be the result of differences in the efficiency of the feeding mechanism and in the availability of each item of food. It is significant that during the most critical time of year (the end of the season), the species of *Geospiza* in the coastal zone feed on a greater variety of food than during the rainy season, when certain foods are very abundant (see table 13). However, it should also be mentioned that the number of kinds of food taken in common by *G. magnirostris* and *G. fortis* is somewhat higher during the dry season. This means that the intensity of utilization of any one plant food by two or more species is heightened somewhat at a time when one might suppose that the availability is reduced.

On the basis of my own field observations on Indefatigable Island during the extremely dry period of November–December, 1952, I had the definite impression that food was not so unavailable that closely related species of Geospizinae were thrown into "competition" with each other in search of it. Indeed, on numerous occasions I observed several individuals of at least four species (*Geospiza scandens*, *G. fortis*, *G. fuliginosa*, and *Camarhynchus parvulus*) feeding together in the very same shrub of *Scutia spicata* without any apparent interspecific strife. In other words, the birds were not making full utilization of the food resources available to them even at this seemingly "critical" time of year. The food supply was so plentiful ("superabundant"?) that it would seem to have been sufficient to support even a larger population of finches.

The biological advantages of maintaining the bird population considerably below the maximum level that the food resources of an area can support
at any given time, are obvious. Annual variations in the fruiting success of the
vegetation necessitates a utilization of the fruits by the birds (and other animals as well) at a level that guarantees a perpetuation of both the vegetation
and the animals dependent upon it.

Certain evidence suggests that there is a marked difference in the mechanical efficiency with which two sympatric species of Geospizinae feed on the
same kind of food. The study of the forces acting on the seed-crushing bill of
Geospiza during powerful adduction has shown that *magnirostris*, because of
the particular shape of its bill, is less likely to develop severe fracture-risks
along the culmen and gonys than, for example, is *G. fortis*. Also, the configuration of the skull proper, the position and size of the adductor muscles, and
the amount of bony reinforcement, all indicate that the feeding mechanism
in *G. magnirostris* is more efficiently designed to crush hard seeds than is the
feeding mechanism in *G. fortis*. That is to say, the amount of energy expended
in crushing a particular hard seed is relatively greater in a large-billed form of
G. fortis than in a small-billed form of *magnirostris*, and this energy difference
is disproportionately greater than the difference in absolute size of bill would
suggest. Therefore, one should not think of these two species as sharing the
same kind of food on an equal basis. "The problem of adaptation is merely
the problem of functional efficiency seen from a slightly different angle"
(Huxley, 1943:317). The anatomical differences between closely related species of *Geospiza* living in the same locality may be thought of as biological adjustments (adaptations) that prevent these species from competing with each
other. The mechanism by which these adjustments have evolved is unknown.
What seems to be clear, however, is that individuals of one species do not now
"compete" for food with certain other individuals of the same species (e.g.,
large-billed versus small-billed individuals of *G. fortis*), or with individuals
of another sympatric species, or at least not in any manner that has evolutionary significance today; and since there is no direct evidence that competition is occurring at the present time, I see no logical reason to assume that it
must have occurred in the past. If, as suggested by Lack (1947:162), closely
related species of the Geospizinae have differentiated in geographic isolation, then it may be reasoned that owing to botanical differences between
the islands the birds would have evolved different feeding habits and habitat preferences, with the consequence that when they were subsequently
brought together in the same area, the different "species" would select different niches in which to live. The same opinion is expressed by Andrewartha
and Birch (1954:463), who add that we are not obliged to suppose that dietary
preferences and associated structural differences were developed as a result
of "competition."

Major bill adaptations in species of the Geospizinae arose, presumably, in isolation and minor differences between island forms of the same species arose, presumably, on islands where the birds are now situated. Under the guidance of natural selection the numerous "adjustments" in bill size and shape have been "coordinated" with the particular array of available foods in such a way as to utilize each adjustment with the greatest mechanical efficiency possible. The differences in relative abundance of the various species on the various islands are probably due in large part to differences in the availability of the various food resources....

I conclude, therefore, from those cases which have been examined in some detail, that equal sharing of food between different species of finches living together in the same area is unknown, and that the inequality in food exploitation is simply due to the fact that the various species are differently disposed in structure and behavior. Where we witness two or more species feeding on the same kind of food we must assume that the items taken in common are in plentiful ("superabundant") supply. Those foods which are not "superabundant" appear not to be shared, or, at least, not equally shared, and thus their importance as objects of "competition" is minimized.

...

* Pages 295–296

Interisland differences in size and shape of the bill within one species are considered to be adaptively related to differences in the island environments. Such differences cannot be attributed to the "Sewall Wright effect," for several reasons . . . [as stated above].

In general, the islands of the Galápagos group that are more diverse ecologically support larger numbers of geospizine species.

TABLE 14. Plant species recorded in no more than one species of Geospizinae

Species	Number of different seeds	Plant species (Numbers refer to fig. 3)
Geospiza		
magnirostris	1	26
scandens	1	67
fortis	3	28, 32, 33
fuliginosa	4	15, 52, 55, 54
Camarhynchus		
parvulus	4	1, 9, 11, 22
psittacula	6	39, 44, 45, 46, 47, 48
Platyspiza		
crassirostris	8	56, 57, 59, 60, 61, 62, 63, 64

TABLE 15. Overlap of plant items in diet of eight species of Geospizinae on Indefatigable Island: combined data for all seasons and localities

Species	Number of different plant items in tract[a]	Geospiza				Platyspiza	Camarhynchus		Cactospiza
		magnirostris	fortis	fuliginosa	scandens		parvulus	psittacula	
Geospiza									
magnirostris	9	...	78% (7)[b]	33% (3)	11% (1)	22% (2)	44% (4)	44% (4)	11% (1)
fortis	26	27% (7)	...	46% (12)	8% (2)	23% (6)	35% (9)	31% (8)	12% (3)
fuliginosa	27	11% (3)	44% (12)	...	4% (1)	37% (8)	52% (14)	48% (13)	11% (3)
scandens	3	33% (1)	67% (2)	33% (1)	...	0% (0)	33% (1)	33% (1)	0% (0)
Platyspiza									
crassirostris	20	10% (2)	30% (6)	40% (8)	0% (0)	...	35% (7)	40% (8)	10% (2)
Camarhynchus									
parvulus	21	19% (4)	43% (9)	67% (14)	5% (1)	33% (7)	...	47% (10)	14% (3)
psittacula	22	18% (4)	36% (8)	59% (13)	5% (1)	36% (8)	45% (10)	...	14% (3)
Cactospiza									
pallida	3	33% (1)	100% (3)	100% (3)	0% (0)	67% (2)	100% (3)	100% (3)	...

[a]Data from table 13.

[b]Explanation: Of the 9 different items taken by G. magnirostris, 7, or 78 per cent. were also taken by G. fortis.

TABLE 19. Frequency of occurrence of plant items, by size in the diet of eight species of Geospizinae

Species	Number of occurrences	A		B		C		D		E		F		G		H[a]		I		Unidentified fragments
		No.	%	No.	%	No.	%	No.	%	No.	%	No.	%	No.	%	No.	%	No.	%	
Geospiza																				
magnirostris	70	1	1.4	4	5.7	...	...	2	2.8	26[b]	37.2	...	...	...	...	17	24.3	20	28.6	17
fortis	219	2	0.9	28	12.8	68[c]	31.2	9	4.2	14	6.5	...	...	...	...	72	32.9	25	11.5	28
fuliginosa	91	7	7.7	60	65.9	14	15.4	6	6.6	3	3.3	...	...	...	...	1	1.1	...	...	22
scandens	6	1	16.7	1	16.7	4	66.6	...	...	...	...	...	...	...	...	...	...	...	...	4
Platyspiza																				
crassirostris	100	1	1.0	51	51.0	23[d]	23.0	7	7.0	2	2.2	1	1.0	1	1.0	14	14.0	...	...	16
Camarhynchus																				
parvulus	91	13	14.3	58	63.7	18	19.8	1	1.1	...	...	...	...	...	...	1	1.1	...	...	37
psittacula	75	9	12.0	43	57.4	10	13.4	12[e]	16.0	...	...	...	...	...	...	1	1.4	...	...	17
Cactospiza																				
pallida	2	...	...	1	50.0	1	50.0	...	...	...	...	...	...	...	...	...	...	...	...	5

[a] 98 per cent of items in this seed class are seeds of *Scutia*. Records of occurrence of the soft outer covering of this fruit are not included in percentages.

[b] All items in this size class taken by *G. magnirostris* are seeds of *Zanthoxylum*.

[c] 76 per cent of items in this size class taken by *G. fortis* are seeds of no. 29 (fig. 3) and *Opuntia*.

[d] 70 per cent of items in this size class taken by *Platyspiza* are seeds of no. 29 (fig. 3).

[e] All items in this size class taken by *C. psittacula* are of nos. 24 and 40 (fig. 3).

Identification of insects from the digestive tracts of four species of Geospizinae from Indefatigable Island

Order	Occurrence[a]	Certhidea olivacea	Cactospiza pallida	Camarhynchus psittacula	Camarhynchus parvulus
Lepidoptera	A	Geometridae (larva) Phalaenidae (larva) Pyraustidae (larva) Arctiidae (larva)	Geometridae (larva) Phalaenidae (larva)	Geometridae (larva) Phalaenidae (larva) Pyraustidae (larva)	Geometridae (larva) Phalaenidae (larva) Pyraustidae (larva) Arctiidae (larva)
	B	Pterophoridae (larva)		Olethreutidae (larva)	Tischeriidae (pupa) Tortricidae (pupa) Tineidae (larva)
Coleoptera	A	Curculionidae (adult) *Lembodes*	Curculionidae (adult) *Pantomorus*	Curculionidae (larva & adult) *Pantomorus*	Curculionidae (adult) *Pantomorus* *Lembodes*
			Cerambycidae (larva & adult) *Estola*	Cerambycidae (larva & adult) *Estola*	Cerambycidae (adult) *Estola*
		Anobiidae (adult)		Anobiidae (adult)	
			Ostomidae (larva)	Ostomidae (larva)	
	B	Scolytidae (adult)		Anthribidae (adult) *Ormiscus*	Coccinellidae (adult)

Diptera	A	Otitidae (larva)	Otitidae (larva)		
	B			Stratiomyidae (larva) probably *Hermelia illucens*	
Hemiptera	A	Heteroptera (adult) Pentatomidae (eggs)	Pentatomidae (eggs)	Heteroptera (adult) Pentatomidae (eggs)	Heteroptera (adult) Pentatomidae (eggs)
	B	Aphididae (adult) Coccidae (adult) probably *Coccus* Diaspididae (adult) *Howardia biclavis* Pseudococcidae (adult) Dactylopiidae (adult) *Eriococcus*	Aphididae (adult)	Aphididae (adult)	Aphididae (adult)
Hymenoptera	A	Formicidae (adult)	Formicidae (adult)	Formicidae (adult)	Formicidae (adult)
Total number of groups		19	13	7	12

[a]A: groups common to two or more geospizines. B: groups found exclusively in geospizines indicated.

There is little, if any, significant "competition" for food today between individuals of the different species where they are sympatric. Anatomical differences between closely related species on any one island are best thought of as biological adjustments which have been evolved when the forms were in isolation. These adjustments could have prevented "competition" from occurring between the forms when subsequently they came together on the same island.

The following factors are considered to have been the most important in shaping the evolutionary pattern of the Galapagos finches: (1) interisland differences in physiography and flora; (2) the presence of avian and reptilian predators; and (3) the nature of the genetic constitution of the ancestral geospizine stock.

University of California Publications in Zoology 58 (1961): 1–302.

Ecological Character Displacement in Darwin's Finches (1985)

❋ DOLPH SCHLUTER, TREVOR PRICE, AND PETER R. GRANT

Abstract—Character displacement resulting from interspecific competition has been extremely difficult to demonstrate. The problem was addressed with a study of Darwin's ground finches (Geospiza). *Beak sizes of populations of* G. fortis *and* G. fuliginosa *in sympatry and allopatry were compared by a procedure that controls for any possible effects on morphology of variation among locations in food supply. The results provide strong evidence for character displacement. Measurement of natural selection in a population of* G. fortis *on an island (Daphne) lacking a resident population of* G. fuliginosa *shows how exploitation of* G. fuliginosa *foods affects the differential survival of* G. fortis *phenotypes.*

Ecological character displacement occurs when morphological differences between coexisting species are enhanced as a result of competitive interactions between them (1). Despite the wide range of conditions under which character displacement is predicted by coevolutionary models (2, 3) there are few good examples of its occurrence in nature (4, 5). Most evidence suggesting ecological displacement derives from a comparison of differences between species in locations of sympatry and allopatry. The main assumption used in inferring character displacement from such comparisons is that there are no other factors influencing morphological differences between locations—for example, differences in food supply. This assumption may often be false.

Darwin's finches provide one of the most familiar examples of apparent character displacement (6) (Fig. 1). *Geospiza fortis* and *G. fuliginosa* are sympatric on most Galápagos islands (for example, Santa Cruz), where they are very different in beak and body size. Their beak sizes are intermediate where they occur alone, on the islands of Daphne Major (*G. fortis*) and Los Hermanos (*G. fuliginosa*).

We reexamine this example of apparent character displacement by two approaches: (i) we compare the morphology of the species in sympatry and allopatry, controlling for any differences that might arise because of variation in food supply, and (ii) we use observations of selection pressures on one species in allopatry to show how the differential survival of phenotypes is affected by the use of foods exploited elsewhere by the second species. The results support the original interpretation of character displacement (6).

Detailed studies on the Galápagos show that seed supply strongly determines the distribution and survival of finch phenotypes (7–11). We can investigate the possibility that morphological differences in allopatry and sympatry result simply from variation in food supply (12) by quantitatively describing food characteristics of individual islands in terms relevant to the differential survival of finch phenotypes (13). The resulting "adaptive landscapes" (14) permit predictions of mean beak sizes for *G. fuliginosa* and *G. fortis* on different islands on the assumption that morphology is determined by food supply alone (15). Greater morphological differences in sympatry than expected from food supply would constitute evidence for character displacement.

The procedure for quantifying an island's food supply as an adaptive landscape (13) involves the estimation of expected density of a solitary finch population on an island as a function of mean beak depth. First, we identify those seeds that would be included in the diet of a finch population with a given mean beak depth by noting all edible seeds on a given island with size greater than a specific lower bound and with hardness softer than a specific upper bound. The size-hardness limits, determined from empirical data on 21 *Geospiza* populations, are both increasing functions of mean beak depth (13). Second, we compute the summed density of seeds on the island that fall between these two limits (13). Third, we estimate the number of individual finches which may be supported by that density of seeds, using an empirical relation between the density of finches in individual populations on islands and the available standing crop of the seeds they eat (13). These three steps are repeated for 0.05-mm increments of mean beak depth between 5.5 and 15 mm. The resulting curves describe expected population density on each island over the range of beak sizes spanned by actual populations of *G. fuliginosa* and *G. fortis* (16).

This procedure was used to calculate expected finch density as a function of beak size on eight Galápagos islands where both *G. fortis* and *G. fuliginosa* are present (Pinta, Marchena, San Salvador, Fernandina, Rábida, Plazas, Santa Fe, and Santa Cruz) as well as the single-species islands of Daphne and Los Hermanos. Almost all seed supply data were collected during visits in the dry season (June to January) when food supply limits finch density (8, 9). The exception is Plazas, where seed supply data are available for only the wet season (March). The essential features of expected density curves are not sensitive to seasonal or annual variation in food supply, as shown by analysis where repeated measurements of seed data are available (*17*).

Expected population density curves for Santa Cruz, Daphne, and Los Hermanos (Fig. 1) show two important features: first, there are distinct local maxima (peaks) in expected density and, second, a given island often has more than one maximum—that is, the distributions are polymodal.

Polymodality results from wide gaps in the frequency distribution of seed size and hardness classes on an island. A finch population with mean beak size directly under a given maximum has a higher expected density than a slightly smaller species because it is able to crack and consume one or more seed types that are too hard for a smaller finch to deal with. A finch population with a mean beak size to the right of this peak has lower expected density than a population at the peak because it includes no additional hard seeds in its diet, and it may include fewer smaller seeds. A population of larger finches has a lower expected density than a population of smaller finches when supported by the same biomass of seeds (*13*).

Expected density curves are important because, by the operation of natural selection, and in the absence of other interacting species, mean morphology in a population is expected to shift, with the consequence that population size is approximately maximized (*3*). There is indeed a general correspondence between mean beak sizes of finch species observed on an island and the positions predicted from peaks in expected density (*13*) (Fig. 1). Since peaks in expected density are determined solely by food, the associated beak sizes provide a standard against which actual beak sizes on different islands can be compared. Additional effects of interactions between particular species may be detected as discrepancies between observed and predicted beak sizes.

In Fig. 2, mean beak sizes of populations of *G. fuliginosa* and *G. fortis* are compared with the sizes predicted by density curves (*18*). The relatively small mean size of *G. fortis* observed on Daphne is close to the size predicted from the peak in the density curve on this island (Fig. 1). In contrast, sympatric populations of *G. fortis* are displaced from the positions predicted by the nearest peak on each island (Fig. 2) (*19*). On Los Hermanos, *G. fuliginosa* is large in relation to the size predicted from the density curve. But in this case a shift has occurred from one peak to the next: while *G. fuliginosa* populations on other islands are associated with comparable peaks in expected density at small beak size, the Los Hermanos population lies under a second peak that occurs at substantially larger mean beak size (Fig. 1) (*19*). The results support hypotheses of character displacement for both species.

A study of natural selection in the allopatric population of *G. fortis* on Daphne illuminates the mechanism of displacements. This population experienced a severe drought in 1977, when 85 percent of all individuals died

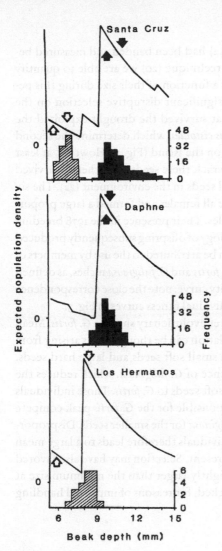

Fig. 1. Apparent character displacement in Darwin's finches, discovered by Lack (6). Beak depth histograms are for adult male *Geospiza fuliginosa* (dashed) and *G. fortis* (solid). Note greater difference in sympatry (Santa Cruz) than in allopatry (Daphne, Los Hermanos). Curves above histograms indicate expected population density of a ground finch species as a function of mean beak depth. Arrows indicate predicted (upward) and observed (downward) sizes for *G. fuliginosa* (pale) and *G. fortis* (dark).

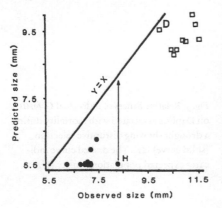

Fig. 2. Observed mean beak sizes for adult males in populations of *G. fuliginosa* (circle) and *G. fortis* (square) compared with the mean beak sizes predicted from peaks in expected population density. Allopatric populations are Los Hermanos (H) and Daphne (D). The arrowhead indicates the observed size of *G. fuliginosa* on Los Hermanos in relation to the second peak in expected density on this island (Fig. 1).

(*10*). A large sample of adults ($n = 642$) had been banded and measured before the drought. Using a regression technique (*20*) we are able to quantify the relative fitnesses of individuals as a function of their size during this period of mortality. The results show significant disruptive selection on the population (*21*). Most individuals that survived the drought exploited the large seeds, *Opuntia echios* and *Tribulus cistoides,* which determine the second peak in expected population density on the island (Fig. 1). However, at least six individuals that were not able to crack these seeds nevertheless survived by exploiting the few remaining small seeds in the environment (*22*). The six small individuals which survived were all females and formed a large proportion (20 percent) of all surviving females. Their presence in the 1978 breeding population affected the mean morphology of offspring subsequently produced (*23*). Thus, the disruptive selection can be attributed to the use by members of the Daphne population of both the *G. fortis* and *G. fuliginosa* niches, as defined by the two peaks in the expected density curve; note the close correspondence between the expected density and individual fitness curves in Fig. 3.

These observations help us interpret evolutionary shifts in *G. fortis.* Mean morphological size in a *G. fortis* population can be thought of as arising from a trade-off between feeding rates on small soft seeds and large hard seeds. The trade-off is affected by the presence of *G. fuliginosa,* which reduces the availability in the dry season of small soft seeds to *G. fortis.* Those individuals too small to crack the large seeds responsible for the *G. fortis* peak compete directly with the more efficient *G. fuliginosa* for the smaller seeds. Disproportionate mortality in these *G. fortis* individuals therefore leads to a large mean size on islands where *G. fuliginosa* is present. Selection may have also favored *G. fortis* individuals with beak size slightly larger than the minimum size at which the large hard seeds can be cracked, for reasons of improved handling efficiency (*9, 11, 24–26*).

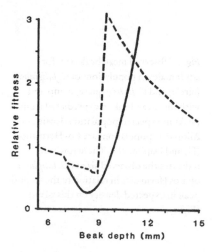

Fig. 3. Relative fitness of individual *G. fortis* on Daphne associated with mortality during a drought showing disruptive selection (solid curve) (*22*). The dashed curve indicates expected population density (Fig. 1).

Our observations on *G. fortis* in sympatry are not as detailed as those on Daphne but are consistent with the above explanation. Because of their greater size, *G. fortis* in sympatry are more efficient at consuming the large seeds responsible for peak position than are *G. fortis* on Daphne, and they less frequently abandon individual seeds without having successfully cracked them (*11, 25, 27*). Also their annual consumption of small seeds is lower than it is on Daphne (*7, 9, 25*).

The difference between mean beak size in the *G. fuliginosa* population on Los Hermanos and the beak size predicted by the second peak in expected density is very small (arrowhead in Fig. 2) (*19*). This situation is analogous to that on Daphne and is correlated with a similar absence of a smaller competitor for the small soft seeds. Thus, but for a small difference in the position of this second peak, the populations of *G. fuliginosa* on Los Hermanos and *G. fortis* on Daphne are morphologically and ecologically identical.

Size shifts in *G. fuliginosa* may also be explained by competition with *G. fortis*. Populations of *G. fuliginosa* in sympatry consume small soft seeds in the dry season (*9, 28*). Mean beak size in sympatry relative to the associated peak in expected density (Fig. 2) is thus probably largely a consequence of selection for efficient exploitation of these small seeds. In contrast, the allopatric population on Los Hermanos has a large beak size corresponding to a position determined by moderately large seeds (notably *Cenchrus*), consumed in sympatry by *G. fortis* (*9, 25*).

To summarize, previous work has shown that morphological patterns in Darwin's finches correspond reasonably well with those expected from the distribution of peaks in expected density curves (*13*). It also provided evidence of interspecific competitive effects. These effects could be the result of character displacement, an evolutionary process, or competitive exclusion, an ecological process. Our study carries the investigation a step further. First, we compare the sizes of particular species in sympatry and allopatry with the sizes predicted by peaks in the density curves. In agreement with an hypothesis of character displacement, *G. fortis* and *G. fuliginosa* were found to vary in their proximity to peaks according to whether they were sympatric or allopatric. Second, a study of marked *G. fortis* on an island lacking *G. fuliginosa* illustrates the connection between feeding and selection pressures associated with morphology. This approach to the study of character displacement should be useful, especially if the magnitude of a given displacement is small and therefore difficult to detect by the usual simple comparison of morphologies in sympatry and allopatry (*4, 29*). Indeed, applied to other systems the approach may show character displacement to be more common than is presently realized.

REFERENCES AND NOTES

1. W. L. Brown, Jr., and E. O. Wilson, *Syst. Zool.* 5, 49 (1956). Our use of the term character displacement subsumes both the origin and maintenance of morphological differences between species. The term thus applies whenever differences between species are greater in sympatry than allopatry as a result of interspecific competition. Initial populations may actually have diverged following sympatry or converged following allopatry. Although the results are the same, the latter process is sometimes considered separately as character release (4).

2. M. G. Bulmer, *Am. Nat.* 108, 45 (1974); M. Slatkin, *Ecology* 61, 163 (1980); M. L. Taper and T. J. Case, *ibid.* 65, (1984).

3. J. Roughgarden, *Theor. Popul. Biol.* 9, 388 (1976).

4. P. R. Grant, *Biol. J. Linn. Soc.* 4, 39 (1972); P. R. Grant, *Evol. Biol.* 8, 237 (1975).

5. T. W. Schoener, *Am. Nat.* 104, 155 (1970); R. B. Huey and E. R. Pianka,. *Am. Zool.* 14, 1127 (1974); T. Fenchel, *Oecologia (Berlin)* 20, 19 (1975); T. J. Case, *Fortschr. Zool.* 25, 235 (1979); A. E. Dunham, G. R. Smith, J. N. Taylor, *Evolution* 33, 887 (1979); W. Arthur, *Adv. Ecol. Res.* 12, 127 (1982); J. Fjeldså, *Ibis* 125, 463 (1983).

6. D. Lack, *Darwin's Finches* (Cambridge Univ. Press, Cambridge, England, 1983), pp. 82–86.

7. I. Abbott, L. K. Abbott, P. R. Grant, *Ecol. Monogr.* 47, 151 (1977).

8. J. N. M. Smith *et al.*, *Ecology* 59, 1137 (1978); P. R. Grant and B. R. Grant, *Oecologia (Berlin)* 46, 55 (1980); *Ecol. Monogr.* 50, 381 (1980); T. D. Price and P. R. Grant, *Evolution* 38, 483 (1984).

9. D. Schluter, *Ecology* 63, 1504 (1982).

10. P. T. Boag and P. R. Grant, *Science* 214, 82 (1981); T. D. Price, P. R. Grant, H. L. Gibbs, P. T. Boag, *Nature (London)* 309, 787 (1984).

11. P. R. Grant, B. R. Grant, J. N. M. Smith, I. J. Abbott, L. K. Abbott, *Proc. Natl. Acad. Sci. U.S.A.* 73, 257 (1976).

12. P. T. Boag and P. R. Grant, *Biol. J. Linn. Soc.*, in press.

13. D. Schluter and P. R. Grant, *Am. Nat.* 123, 175 (1984).

14. G. G. Simpson, *The Major Features of Evolution* (Columbia Univ. Press, New York, 1953); R. Lande, *Evolution* 30, 314 (1976).

15. R. I. Bowman, *Univ. Calif. Publ. Zool.* 58, 1 (1961).

16. Although empirical data from actual populations are used to generate expected population density curves, the shape of the curve for any island, given the seeds there, is independent of the number of species present and their mean beak sizes. Hence the problem of circularity [P. R. Grant and I. Abbott, *Evolution* 34, 332 (1980)] is avoided. To confirm this we repeated the procedure and generated curves without using the empirical data for *G. fuliginosa* and *G. fortis* on Santa Cruz, Daphne, and Los Hermanos, the islands compared in Fig. 1. The positions of the maximums in the expected density curves for all the islands, including these three islands, did not change.

17. Expected densities for given beak depths vary with season or year to some extent, but the shape of the expected density curve is generally invariant. For example, all six Daphne curves, derived from seed data collected in January and June of 1979 to 1981, were virtually identical, despite large seasonal and annual fluctuations in primary production and standing crop of seeds (8, 10). Wet season curves for Pinta, Marchena, Santa Cruz, and Los Hermanos were also similar to the corresponding dry season curves. In no case did the predicted optimum beak depth for either finch species depend on the choice of density curve.

18. Predicted beak size for each population is the beak size associated with the peak immediately to the left of the actual mean. The leftward criterion is necessary because a peak to the right of the observed mean beak size is determined by seeds which are not exploited by the population because of their hardness. With this criterion it was possible to associate 15 of the 16 sympatric populations of *G. fortis* and *G. fuliginosa* unambiguously with a particular peak (*13*) (Fig. 1). On Fernandina beak size in *G. fortis* did not correspond to any nearby peak, possibly because of insufficient seed sampling, and this island was excluded from further consideration. Predicted beak size for the two allopatric populations is the beak size associated with the peak to the left of the largest mean for the same species in sympatry.

19. The difference between observed and predicted beak size for *G. fortis* on Daphne (0.46 mm) is significantly less than the difference on other islands: 1.7 mm ± 0.11, standard error; Dixon's outlier test [R. R. Sokal and F. J. Rohlf, *Biometry* (Freeman, San Francisco, ed. 2, 1981), p. 413], one-tailed, $n = 9$, $r_{11} = 0.57$, $P < 0.05$. *Geospiza fuliginosa* on Los Hermanos is also close to the second peak on that island (0.20 mm), but it is significantly further away from the first peak (2.7 mm) than other *G. fuliginosa* populations (1.4 mm ± 0.08, $n = 12$, $r_{21} = 0.65$, $P < 0.01$). In addition, 54 of 121 male Daphne *G. fortis* were smaller than the predicted mean size on that island (Fig. 1). This compares with 12 of 454 for *G. fortis* on other islands combined [$\chi^2(1) = 110.2$, $P < 0.0001$].

20. R. Lande and S. J. Arnold, *Evolution 37*, 1210 (1983).

21. A regression analysis of relative fitness of individuals (w, being 0 for those that died and 6.6 for those that survived) on beak depth and the square of beak depth was carried out to obtain coefficients of selection (20). The regression equation $w = 21.49 - 5.13x + 0.31x^2$ ($n = 642$, $F = 27.7$, $P < 0.001$). The coefficient of directional selection was $\beta = 0.75 \pm 0.11$ standard error ($P < 0.001$) and of disruptive selection $\gamma = 0.62 \pm 0.21$ ($P < 0.01$). This selection event, and in particular the directional selection, is discussed in (*10*).

22. The smallest individuals on the island are unable to crack the hard seeds (*10, 11, 24*). Forty-six individuals were followed continuously for 30 to 90 minutes during a dry period in 1980. The proportion of time spent feeding on hard seeds was correlated with beak depth (Spearman's $r = 0.64$, $P < 0.001$). Some small individuals have never been seen to crack hard seeds, although some have occasionally attempted to do so and failed [(*24*); T. D. Price, in preparation].

23. P. T. Boag, *Evolution 37*, 877 (1983).

24. P. R. Grant, *Anim. Behav.* 29, 785 (1981).

25. D. Schluter, *Ecology 63*, 1106 (1982).

26. M. F. Willson, *Condor 73*, 415 (1971); I. Abbott, L. K. Abbott, P. R. Grant., *ibid. 77*, 332 (1975).

27. For example, Daphne *G. fortis* males took 34.0 seconds (1.3 seconds, standard error) to crack and consume *Opuntia* seeds ($n = 45$ observations) whereas the species on San Salvador took 16.0 seconds ± 2.0 ($n = 6$). Daphne *G. fortis* rejected on average 75 percent of *Tribulus cistoides* seeds (*24*) whereas *G. fortis* on Pinta, Marchena, and San Salvador were rarely or never observed to reject the large hard seeds determining the peak position on these islands (*Lantana*, ripe *Rhynchosia*, and *Opuntia*). On Daphne 42 percent of *G. fortis* that tried to crack *Bursera* stones were successful (*11*), whereas on Pinta 77 percent were successful (*25*).

28. D. Schluter and P. R. Grant, *Evolution 38*, 856 (1984).

29. T. J. Case and R. Sidell, *ibid. 37*, 832 (1983); P. R. Grant, in *Patterns of Evolution in Galapagos Organisms*, R. I. Bowman, M. Berson, A. Leviton, Eds. (AAAS, Pacific Division, San Francisco, 1983), pp. 187–199; P. R. Grant, and D. Schluter, in *Ecological Communities: Conceptual*

Issues and the Evidence, D. R. Strong, D. S. Simberloff, L. G. Abele, A. B. Thistle, Eds. (Princeton Univ. Press, Princeton, N.J., 1984), pp. 201–233.

30. We thank P. T. Boag for providing some data and P. T. Boag, B. R. Grant, R. Lande, T. W. Schoener, and J. N. M. Smith for comments and advice. Supported by NSF grants DEB 79–21119 and DEB 82–12515 to P.R.G.

Science 227 (1985): 1056–1059. Reprinted with permission from AAAS.

* 7 *
Darwin's Finches as a Case Study of Natural Selection

In 1898, when Bumpus first documented "the elimination of the unfit" in his sample of unfortunate frozen house sparrows in Providence, Rhode Island, he demonstrated that natural selection can occur episodically, with great intensity and with measurable consequence. Since then, the studies of natural selection on Darwin's finches, conducted primarily by Peter and Rosemary Grant and their colleagues, have become classic examples of how to document natural selection in progress. The biogeographic evidence for adaptive variation in the bill morphology of the finches, and the evidence of character displacement in particular, challenged researchers to measure the strength of natural selection on bill traits, and to identify causes and targets of natural selection.

How is evolution measured? A population geneticist measures evolution as a change in allele frequency. For example, one version, or allele, of a gene may give wrinkled seeds, another version smooth ones. An increase in the proportion (a.k.a. frequency) of the wrinkled allele as opposed to the smooth allele in a population qualifies as evolution. So does the appearance of a new allele through mutation. When a gene has a discernible effect on a phenotype, a change in the frequency of alleles for that gene will cause a change in the mean phenotype (value of a trait) of the population.

A quantitative geneticist measures evolution directly as a change in phenotype from one generation to the next. Traits with continuous variation typically do not show evidence of a simple genetic basis with Mendelian segregation ratios that can be studied by examining allele frequencies of particular genes, since such traits are frequently controlled by many genes. The quantitative-genetic approach derives from the very practical requirement of plant and animal breeders to be able to predict the traits of offspring in a breeding program when those traits do not show simple, discrete Mendelian segregation. Starting with a population of parents, and imposing a particular intensity of selection by allowing only a proportion of the population with particular desired trait values to breed, breeders need to predict the magnitude of change in the trait of interest they can expect in the following generation. Evolutionary biologists have appropriated the analytical tools of breeders to predict changes in the mean phenotypes of populations when natural selection, as opposed to artificial selection, is imposed.

Breeders use what is referred to as the "breeder's equation" to predict the response to selection. This equation describes the change in a phenotype from one generation to the next, R, as a function of the strength of selection, S, and the heritability[1] of the trait, h^2.

$$R = h^2 S$$

R is the difference between the mean phenotype of the parents (before selection) and the mean phenotype of the offspring generation, and it indicates the evolutionary response to selection. S is the difference between the mean phenotype of the parents before and after selection, and indicates the strength of selection. The heritability, h^2, can be thought of in two ways. First, it is the degree to which the parental phenotype predicts the offspring phenotype. That is, if one were to plot parental traits (the mean of the mother and father) on the x-axis and the traits of their corresponding offspring (the mean of all offspring) on the y-axis, and fit a line through those points, the heritability would be the slope of that line, provided the parental and offspring environments are decoupled. The heritability is also the proportion of total variation in the phenotype that is a genetically based variation. Variation in a trait can be caused not only by variation in the genes associated with that trait, but also by variation in the environment, since different individuals grow in slightly different environments. If all the variation in a trait were due to purely environmental influences, then selecting a subset of parents with a particular trait value would have no effect on the phenotype of the progeny (assuming parent and offspring environments are decoupled). The heritability would be zero, and there would be no change in the phenotype from one generation to the next, regardless of the strength of selection. In contrast, if all the variation in the trait were due to (additive) genetic variation, the heritability would be one, and the mean phenotype of the progeny would be the mean phenotype of their two parents.

It is important to note that heritability is not an intrinsic property of a trait; it is a property of a population. First, it is measured in terms of variance, a population characteristic by definition, and the variance is determined by allele frequencies. Thus if no genetic variance exists—if all members of the population have the same genotype—then heritability is zero. Second, it is measured as a proportion of total phenotypic variation, including variation caused by environmental variation. Therefore a given measure of heritability applies only to the environment in which it was measured. Third, differences

---

1. The notation of heritability is "h^2." Being based on a measure of variance, it is intrinsically a squared term; there is no such notation as "h."

among genotypes depend on the environment, making genetic variation dependent on environmental conditions. For example, in poor environments all genotypes may be small whereas in rich environments some genotypes may be larger than others. Thus heritability is specific to the population and the environment in which it was measured.

The breeder's equation makes explicit what the main ingredients of evolution by natural selection are. A change in phenotype from one generation to the next as a consequence of natural selection requires variation in a trait, inheritance and heritability of that trait, and an association between that trait and fitness.

When considering continuously varying phenotypes, such as the bill size and shape of Darwin's finches, the breeder's equation is helpful and has been widely applied within the field of evolutionary biology to predict changes in phenotypes over time. This is due largely to the efforts of Russell Lande. In 1983, Russell Lande and Stevan Arnold published a paper that developed an empirical method for measuring natural selection and for applying the breeder's equation to predict evolutionary responses to natural selection. Their contribution was twofold. First, they demonstrated that the strength of natural selection (S) can be measured as the slope of the regression line of fitness, frequently measured as number of offspring, against trait values. This is a very convenient extension of the breeding model in which only a select subset "survived" to breed, since it enables one to measure selection on continuous traits using continuous measures of fitness. These are metrics more suitable to analyzing populations in nature as opposed to controlled breeding programs.

Second, Lande and Arnold showed how to measure natural selection on characters that were correlated, and thereby quantify the degree to which a trait might directly influence fitness as opposed to the degree to which it might be selected simply because it is correlated with another trait that directly influences fitness. This distinction was a long-standing challenge that dates back to the 1890s when early biometricians, such as Francis Galton, Karl Pearson, and W. F. R. Weldon, first began documenting patterns of covariation among traits that influence survival in natural populations of organisms. For example, imagine a population of birds in which some individuals have red spots on their rumps and some have yellow spots, and many (but not all) birds with red rumps also happen to have red eyes, while those with yellow rumps happen to have yellow eyes. A field biologist who, having taken a fancy to red rumps and collects a sample of them, will also have collected a lot of red eyes. Red eyes were "indirectly selected" because they are correlated with red rumps. If the correlation is not perfect, one can discern that the sample criterion was red rumps as opposed to red eyes, because, presumably, some yellow-eyed red-rumps would have been collected as well. Multiple regression is the statistical method that quantifies the degree to which a trait value is "directly selected" as opposed to selected through its correlation with another selected trait. The multivariate version of the breeder's equation explained above is

$$\Delta \bar{z} = G\beta$$

$\Delta \bar{z}$ is the response to selection, and, like R, is the change in the phenotype from one generation to the next, but for *several* traits instead of a single trait (it is a vector of responses of each of the traits). G, similar to h^2, measures genetic variance but now also includes genetic covariances among traits (it is a matrix of genetic variances and covariances). β measures the strength of selection, like S, but it specifically measures selection acting *directly* on the trait, controlling for indirect selection acting through correlated traits (it is a vector of direct selection acting on each trait).[2] It is common for researchers to report both the magnitude of "direct" selection, β, and "total selection," S—the difference between them indicating the magnitude of "indirect selection" operating through correlated traits.

The method's efficacy opened a whole realm of analysis of natural selection that was formerly unavailable. Field biologists could now begin to quantify the strength of natural selection on particular traits and suites of traits. In that manner, they could begin to formulate and answer previously intractable questions such as: How strong is natural selection on morphological traits as opposed to life-history traits? How consistent is natural selection across the habitats that an organism encounters? How consistent is natural selection over the lifetime of an organism? How consistent is natural selection over generations? Is natural selection episodic? How strong is indirect selection compared to direct selection? Is natural selection stabilizing, favoring an intermediate phenotype, or directional, favoring ever increasing or decreasing phenotypes? Evolutionary biologists are still asking these questions.

The method, moreover, provides tools that help identify the actual targets of natural selection, that is, those traits that may directly influence fitness. As with any correlational analysis, the regression method cannot attribute causation to a trait, but it can significantly narrow down the suspects for further experimental manipulations. Importantly, the method reveals constraints on adaptive evolution. If two traits are positively correlated (say, red rumps and red eyes), and an increase in one trait is associated with higher fitness (say, red rumps), but a decrease in the other trait is associated with higher fitness (say, yellow eyes are advantageous), then there is a constraint. The correlation

2. β is actually $P^{-1}S$, where P is the phenotypic variance-covariance matrix. The multivariate breeder's equation can therefore also be written as $\Delta \bar{z} = GP^{-1}S$, which gives multivariate terms (G/P) analogous to h^2, which is (genetic variance)/(total phenotypic variance).

keeps most individuals from having the optimal phenotype for both traits. In this manner, selection on one trait can actually result in maladaptive evolution in other traits, at least in the short term. This is one big reason why life is not perfect.

READINGS

This chapter highlights the greatest success of Darwin's finches as a research system for evolutionary studies: the demonstration of natural selection in progress. The papers demonstrate the measurement of natural selection on particular bill traits and body size, the constraints imposed by correlations among traits, and responses to natural selection. Importantly, the studies demonstrate the episodic and inconsistent quality of natural selection, which varies with season, climate, age of the bird, and the presence of other species.

The first paper, by Peter Boag and Peter Grant, documents intense natural selection during a single episode. Predating the statistical techniques of Lande and Arnold, it shows an intuitive measure of natural selection as the difference in the mean phenotype of a population, with regard to a single trait,[3] before and after an episode of selection.[4]

The second paper, by the Grants, documents changes in the direction and intensity of selection on two finch species over a period of thirty years. It also presents long-term responses to natural selection as a function of natural selection and heritability, and as modified by immigration (in this case, the immigration of individuals from a closely related species that interbreeds with the native population). They compare predicted responses to selection, based on the breeder's equation, to those that they observe. These readings clarify how episodic natural selection can be, how variable in space and time, and how nondirectional it sometimes is over longer periods of time.

The final paper, also by the Grants, applies the measurement of natural selection on bill morphology specifically to the question of competitive character displacement. The contrast between this paper and the paper by Boag and Grant is an excellent illustration of how the widely applied Lande and Arnold technique changed studies of natural selection, calling attention to constraints on evolutionary responses to selection. This paper presents evidence of character displacement from the initial encounter, through the episodes of selection favoring morphological divergence between species, to the evolutionary response to this divergent natural selection. Remarkably, the

---

3. In this case, the trait is a composite trait in the form of a "principal component."
4. The paper also utilizes a measure, rarely used today, of the intensity of selection based on the estimated change in the mean fitness of the population caused by a change in the phenotype.

Grants document a change in the direction of natural selection on bill size in the small *Geospiza fortis*—a change that corresponds to the new presence and subsequent increase in abundance of a major competitor, *G. magnirostris*. This is a particularly striking study because the focal species evolved in the opposite direction of what had occurred in a previous drought in the absence of the other species, providing compelling evidence for the importance of competition in the evolution of bill morphology in this species.

Intense Natural Selection in a Population of Darwin's Finches (Geospizinae) in the Galápagos (1981)

✻ PETER T. BOAG AND PETER R. GRANT

Abstract—Survival of Darwin's finches through a drought on Daphne Major Island was nonrandom. Large birds, especially males with large beaks, survived best because they were able to crack the large and hard seeds that predominated in the drought. Selection intensities, calculated by O'Donald's method, are the highest yet recorded for a vertebrate population.

There are few well-documented examples of natural selection causing avian populations to track a changing environment phenotypically. This is partly because birds meet environmental challenges with remarkable behavioral and physiological flexibility (*1*), partly because birds have low reproductive rates and long generation times, and partly because it has been difficult for ecologists to quantify corresponding phenotypic and environmental changes in most field studies. In this report we demonstrate directional natural selection in a population of Darwin's finches and identify its main cause.

We studied Darwin's medium ground finch (*Geospiza fortis*) on the 40-ha islet of Daphne Major, the Galápagos, from July 1975 to June 1978. Each of more than 1500 birds was color-banded and measured for seven external morphological characters (*2*). Continuous records were kept of the banded birds and of rainfall. Each year during the breeding season (January to May) we banded nestlings and compiled nest histories. Three times a year (before, during, and after the breeding season) we collected the following data: (i) the number of seeds of each plant species in 50 randomly chosen 1.0-m² quadrats; (ii) a standardized visual census of finches over the entire island; and (iii) a minimum of 100 point records of feeding behavior, accumulated by noting food items eaten by banded birds encountered during nonsystematic searches (*2*).

During the early 1970's Daphne Major received regular rainfall, resulting in large finch populations and food supplies (*2*). From December through June in 1976 and 1978 we recorded rainfalls of 127 and 137 mm, respectively— sufficient for abundant production of plants, insects, and finches. However, in 1977 only 24 mm of rain fell on Daphne Major during the wet season (3, 4). *Geospiza fortis* did not breed at all in 1977 and suffered an 85 percent decline in population (Fig. 1A). The decline was correlated with a reduction in seed abundance ($r = .86, P < .01$) (Fig. 1B). Seeds form the staple diet of *G. fortis*, particularly in the dry season, when other plant matter and insects are scarce (2).

Between June 1976 and March 1978, the mortality, and possibly emigration (5), of *G. fortis* was nonrandom with respect to age, sex, and phenotype. Only one of 388 *G. fortis* nestlings banded in 1975 survived to 1978, and while the sex ratio was roughly equal in 1976, it had become skewed to six males to one female in 1978. Most significantly, the birds surviving into 1978 were considerably larger than those that disappeared (Fig. 1C). We use principal component 1 (6) as an index of overall body size because here, as in other avian studies (7), it explains a substantial portion (67 percent) of the phenotypic variance in the *G. fortis* population and has consistently high, positive correlations with the morphological variables it summarizes. The change is most obvious in the plot including all birds because it incorporates the changing sex ratio (most of the morphological characters are 4 percent larger in males than in females) and perhaps a small age effect, although all birds less than 12 weeks old were excluded from the analysis.

Small seeds declined in abundance faster than large ones, resulting in a sharp increase in the average size and hardness of available seeds (Fig. 1D). There was a corresponding change in feeding behavior. In May 1976 only 17 percent of feeding was on medium or large seeds [size-hardness index $\sqrt{DH} > 1.0$ (8)], while in May 1977 49 percent of feeding was on such seeds. During the present and related studies (2), large birds ate larger seeds than smaller birds, suggesting that small birds disappeared because they could not find enough food. For example, in a quantitative test of size-related feeding behavior, 198 birds that were only recorded eating seeds with a size-hardness index < 1.0 were significantly smaller than another 121 birds that routinely ate seeds with size-hardness indices ranging from 1.0 to 8.7 (8). In 1977, during the normally lush wet season, larger birds fed heavily on seeds extracted from the large, hard mericarps of *Tribulus cistoides* ($\sqrt{DH} = 8.68$), a food item ignored by almost all birds in earlier years (2). Many finches failed to molt that year, and their condition gradually deteriorated. Small birds fed heavily on *Chamaesyce* spp., the only producer of small seeds in 1977, and as a result their plumage often became matted with the latex of this euphorb. Several dead birds were found with completely bald heads from feeding on *Chamaesyce* and from digging in the soil for seeds. Such plumage loss may have led to increased energy loss during the cool nights of the dry season. The dependence of the finches on a declining seed supply ceased at the end of 1977, when *Opuntia* cactus began flowering and all birds fed heavily on its pollen and nectar (2).

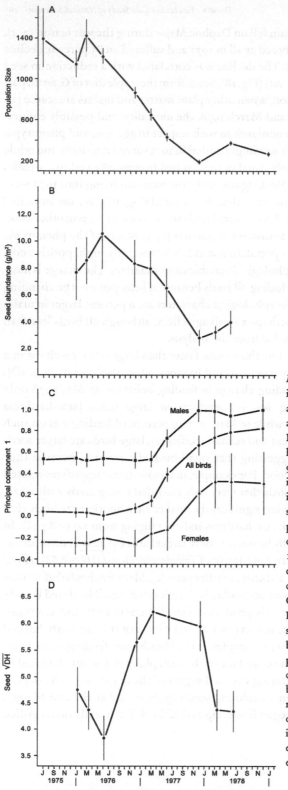

Fig. 1. Temporal changes in finch numbers, seed abundance, morphology, and average seed size on Daphne Major. (A) Population estimates (means ± 95 percent confidence limits) derived from a Lincoln index based on regular visual censuses of a marked population. (B) Estimates of seed abundance [means ± standard errors (S.E.)], excluding two seed species never eaten by any Galápagos finches. (C) Principal component 1 scores (means ± S.E.) for birds alive in each sample period, with coefficients calculated from the combined sample of all birds measured. (D) Estimates of the average $\sqrt{DH}$ index (means ± S.E.) of edible seeds available in each study period (8).

It is reasonable to infer natural selection from the greater survival of large birds because about 76 percent of the variation in the seven morphological measurements and in principal component 1 scores is heritable (*3, 9*). To calculate the intensity of selection we use O'Donald's method, $\Delta\bar{w}/\bar{w} = $ ($\bar{w}$ before selection $-$ $\bar{w}$ after selection) / $\bar{w}$ before selection $= V_w/\bar{w}^2$, where $\Delta\bar{w}/\bar{w}$ estimates the proportional increase in mean fitness of the population as a result of selection and V_w is the variance in fitness. O'Donald provides several functions relating fitness to phenotypic characters and gives formulas for calculating $\Delta\bar{w}/\bar{w}$ from the four moments of phenotypic distributions before and after selection.

Table 1 summarizes the phenotypic changes in the *G. fortis* population between June 1976 and January 1978. Changes in variance were small and none was statistically significant (*11*). Changes in means of most characters were significant and in the direction expected if larger birds survived best (*12*). A thorough examination of the data with both univariate and multivariate techniques suggests that the main differences between birds that survived and those that did not were in body size and bill dimensions, particularly bill depth (*8*). Table 1 includes standardized coefficients that show the relative contributions of each character to the discriminant functions separating survivors and nonsurvivors. Our analysis includes only adult finches measured before the 1976 dry season; the 1978 survivors are a subset of those 1976 individuals, and thus the 1978 range for any given variable falls entirely within the corresponding 1976 range.

Because selection acted primarily on character means, we assume a linear fitness function (*10*). The highest values for $\Delta\bar{w}/\bar{w}$ are observed in the discriminant functions and in variables weighted heavily by the functions (Table 1). Several of the selection intensities are considerably greater than any published to date. For example, O'Donald (*10*) reanalyzed H. Bumpus's data on a survival of house sparrows (*Passer domesticus*) during a particularly severe winter storm, and concluded that such values as the $\Delta\bar{w}/\bar{w} = .255$ he obtained for the change in discriminant score between the before-storm and after-storm sparrow samples indicated selection "more intense than any which has since been observed acting on particular quantitative characters" (*10*).

Table 1 and Fig. 1 show that females experienced stronger selective mortality than males, in agreement with the evidence that the sex ratio became skewed in favor of males. There is no question that the overall effect of selection in the two sexes was similar: larger individuals survived best. There is some evidence that slightly different aspects of "largeness" were favored in males over females (*13*). The results for the combined population illustrate how a large phenotypic shift can occur both as the result of changes in the frequency of discrete classes of individuals (males and females) and in the average measurements of individuals within those classes.

TABLE 1. Characteristics of finches surviving the 1977 drought. The sample measured before selection includes all mature *G. fortis* measured up to the end of May 1976. The sample measured after selection is the subset of the first sample still present on Daphne Major in March 1978. Standardized discriminant function coefficients (SDFC's) reflect the relative contribution of each univariate character to the discriminant function separating survivors and nonsurvivors. Principal component 1 is an index of body size and principal component 2 reflects bill pointedness: both are presented as standardized variables (7). Separate discriminant functions (11) were used to distinguish between survivors and nonsurvivors in the combined sample of birds emphasized here and in smaller samples of males only and females only, for which the discriminant and principal component scores alone are given here. The mean discriminant scores are unstandardized and, because different functions were used, scores of the three groups are not comparable. Values for $\Delta \bar{w}/\bar{w}$ give the proportionate increase in mean fitness as a result of selection, assuming a linear fitness function (10)

Variable	Sample size		Means		Variances		SDFC	$\Delta\bar{w}/\bar{w}$
	Before	After	Before	After	Before	After		
All birds								
Weight (g)	642	85	15.79	16.85	2.37	2.43	0.45	.49
Wing chord (mm)	642	85	67.71	69.22	5.89	5.01	0.35	.39
Tarsus length (mm)	642	85	18.76	19.11	0.57	0.49	0.13	.22
Bill length (mm)	641	85	10.68	11.07	0.55	0.57	0.14	.23
Bill depth (mm)	642	85	9.42	9.96	0.68	0.66	0.45	.44
Bill width (mm)	641	85	8.68	9.01	0.36	0.34	−0.56	.31
Bill length at depth of 4 mm	642	85	3.55	3.41	0.08	0.08	−0.35	.24
Principal component 1	640	85	0.00	0.73	1.06	0.95		.50
Principal component 2	640	85	0.01	−0.13	0.97	0.99		.02
Discriminant function	640	85	−11.55	−12.41	1.13	0.99		.66
Males only								
Principal component 1	198	56	0.54	0.98	0.92	0.76		.21
Discriminant function	198	56	7.35	6.87	0.75	0.64		.31
Females only								
Principal component 1	66	15	−0.21	0.32	0.97	0.96		.28
Discriminant function	66	15	−2.77	−3.71	1.57	1.81		.57

Our data provide a link between a specific environmental factor (size of available food) and phenotypic tracking of the environment. Others have consistently encountered difficulty in identifying the relation between complex and often rather small changes in body size and shape and general environmental parameters, such as temperature (14). Because of the high

correlations between the several characters we examined, it is difficult to specify the precise target of selection; univariate selection intensities and discriminant coefficients present in Table 1 and calculated in the similar analyses of
separate male and female groups (*13*) suggest that weight and bill dimensions
are most important. In addition to the relation between bill morphology and
changes in the food supply, it is likely that there were additional indirect selection pressures operating on, for example, body size for reasons associated
with energetics (*15*) and dominance behavior (*16*). Furthermore, it is likely
that a different set of selection pressures operates when food is abundant
and population size in increasing, thus giving rise to oscillating directional
selection (*2*).

Our results are consistent with the growing opinion among evolutionary ecologists that the trajectory of even well-buffered vertebrate species
is largely determined by occasional "bottle-necks" of intense selection during a small portion of their history (*17*). More specifically, given the many
small, isolated, relatively sedentary, and morphologically variable populations of Darwin's finches (*18, 19*) and the high spatial (*2, 19*) and
temporal (*4*) variability of the Galápagos, this type of event provided a
mechanism for rapid morphological evolution. Occasional strong selection
of heritable characters in variable environment may be one of the keys to explaining the apparently rapid adaptive radiation of the Geospizinae in the
Galapagos (*18, 20*).

REFERENCES AND NOTES

1. L. B. Slobodkin, in *Population Biology and Evolution*, R. C. Lewontin, Ed. (Syracuse Univ.
Press, Syracuse, N.Y., 1968), pp. 187–205; R. K. Selander and D. W. Kaufman, *Proc. Natl. Acad.
Sci. U.S.A.* 70, 1875 (1973).

2. P. R. Grant, J. N. M. Smith, B. R. Grant, I. J. Abbott, L. K. Abbott, *Oecologia (Berlin)*
19, 239 (1975); J. N. M. Smith, P. R. Grant, B. R. Grant, I. J. Abbott, L. K. Abbott, *Ecology* 59,
1137 (1978); P. R. Grant and B. R. Grant, *Oecologia (Berlin)* 45, 55 (1980); *Ecol. Monogr.* 50, 381
(1980).

3. P. T. Boag, thesis, McGill University, Montreal (1981); P. R. Grant, B. R. Grant, J. N. M.
Smith, I. J. Abbott, L. K. Abbott, *Proc. Natl. Acad. Sci. U.S.A.* 73, 257 (1976).

4. P. R. Grant and P. T. Boag, *Auk* 97, 227 (1980).

5. P. R. Grant, T. D. Price, H. Snell, *Not. Galápagos* 31, 22 (1980). This article reports the
only documented finch emigration from Daphne Major, involving two juvenile *G. scandens*.
There is indirect evidence that most of the missing *G. fortis* died on Daphne Major; the measurements of 38 *G. fortis* banded before June 1976 and found dead on Daphne in 1977 and
early 1978 are statistically indistinguishable from the other birds missing from the postselec-
tion population but not found. The individuals found dead also were significantly smaller
than the 1978 survivors, with both the entire sample and males and females separately showing patterns similar to those described for the entire set of missing birds (*12*).

6. Principal components were extracted from the covariance matrix of the seven log-

transformed variables, with all birds combined. The first component explained 67 percent of the total variance and had large correlations with the seven original variables; following the sequence used in Table 1, these component-character correlations were .88, .67, .60, .85, .94, .93, and −.49. Component 2 explained a further 16 percent of the variance and was strongly correlated with bill length at a depth of 4 mm ($r = .87$), followed by bill length ($r = .30$), with other characters showing low correlations of mixed signs. Correlations with length at a depth of 4 mm are reversed because this character is necessarily smaller in larger birds. Other analyses with these finches confirm that principal components 1 and 2 usually reflect overall body size and bill "pointedness," respectively, whether based on covariance or correlation matrices.

7. R. E. Blackith and R. A. Reyment, *Multivariate Morphometrics* (Academic Press, New York, 1971); A. R. Gibson, M. A. Gates, R. Zach, *Can. J. Zool.* 54, 1679 (1976); R. E. Ricklefs and J. Travis, *Auk* 97, 321 (1980); J. A. Wiens and J. T. Rotenberry, *Ecol. Monogr.* 50, 287 (1980).

8. The size-hardness index is the geometric mean of the depth (D) of a seed species in millimeters and its hardness (H) in newtons. In Fig. 1D, average $\sqrt{DH}$ values are obtained by weighting the mean number of seeds (of each species) per square meter by the average seed mass and $\sqrt{DH}$ for that species (2). The tests of hard- versus soft-seed feeders consisted of univariate *t*-tests on all seven morphological variables; in all cases birds feeding on large seeds were significantly larger ($P < .0001$) than those feeding only on small seeds. The seven-variable multivariate analysis of variance between the two groups was also highly significant [$F(7,311) = 7.94, P < .0001$]. The standardized coefficients of the discriminant function separating the two feeding groups weighted bill depth most heavily (.80), followed by wing chord (.57), with all other variables having coefficients under .25. This underlines the link between bill depth and feeding behavior, which persisted among the survivors at the end of 1977 (P. R. Grant, *Anim. Behav.* in press).

9. P. T. Boag and P. R. Grant, *Nature (London)* 274, 793 (1978).

10. P. O'Donald, *Theor. Popul. Biol.* 1, 219 (1970); *Evolution* 27, 398 (1973).

11. Separate linear discriminant functions were computed for males alone, for females alone, and for both combined with mature birds of uncertain sex, each maximizing the distance between the centroids of survivors and nonsurvivors in that group. Unstandardized discriminant scores were calculated for each bird by using the appropriate equation. The variances for the seven original variables, the first and second principal components, and the discriminant function of birds that survived from June 1976 to March 1978 were compared with those of birds that disappeared. The ten comparisons were made for males, for females, and for the combined group; in 21 of the 30 comparisons the selected group was less variable, but none of the 30 F-tests approached significance.

12. We computed *t*-tests for the 30 comparisons detailed in (*11*), again contrasting survivors and nonsurvivors to maintain sample independence. After the 1977 drought, males were significantly larger ($P < .01$) in all variables except wing chord, tarsus length, and principal component 2. Females were significantly larger ($P < .05$) in all variables except weight and tarsus length, with principal component 2 on the borderline ($P = .066$). The combined group was significantly larger ($P < .001$) in all variables except principal component 2.

13. The three largest standardized coefficients of the discriminant function for males alone were for bill depth (1.00), weight (.85), and bill width (.56), and the three largest $\Delta \bar{w}/\bar{w}$ values for male univariate characters were again for bill depth (.22), weight (.20), and bill width (.15). The corresponding results for the female group were different; the largest standardized discriminant function coefficients were for bill length (−1.21), bill length at a depth of 4 mm (.90), and bill depth. (.55), and the largest $\Delta \bar{w}/\bar{w}$ values were for bill length at a depth of 4 mm (.40), bill depth (.25), wing chord (.22), bill length (.20), and bill width (.20).

14. F. C. James, *Ecology 51*, 365 (1970); P. R. Grant, *Syst. Zool. 21*, 23 (1972); R. F. Johnston, D. M. Niles, S. A. Rohwer, *Evolution 26*, 20 (1972); P. E. Lowther, *ibid. 31*, 649 (1977); M. C. Baker and S. F. Fox, *Am. Nat. 112*, 675 (1978); D. M. Johnson *et al.*, *Auk 97*, 299 (1980).

15. S. C. Kendeigh, *Am. Nat. 106*, 79 (1972).

16. S. D. Fretwell, *Bird-Banding 40*, 1 (1969); D. H. Morse, *Am. Nat. 108*, 818 (1974); M. C. Baker and S. F. Fox, *Evolution 32*, 697 (1978).

17. J. A. Wiens, *Am. Sci. 65*, 590 (1977).

18. D. Lack, *Darwin's Finches* (Cambridge Univ. Press, Cambridge, 1947).

19. R. I. Bowman, *Univ. Calif. Berkeley Publ. Zool. 58*, 1 (1961); I. J. Abbott, L. K. Abbott, P. R. Grant, *Ecol. Monogr. 47*, 151 (1977).

20. K. Bailey, *Science 192*, 465 (1976).

21. This work was supported by the National Research Council of Canada and the Frank M. Chapman Fund of the American Museum of Natural History and was carried out with the permission of the Dirección General de Desarrollo Forestal, Quito, Ecuador. The Charles Darwin Research Station arranged logistics in the Galápagos. We thank B. R. Grant, E. Green, D. Nakashima, L. M. Ratcliffe, and R. Tompkins for assistance in the field and G. Bell, B. R. Grant, T. D. Price, L. M. Ratcliffe, and the reviewers for their advice on the manuscript. This is contribution 309 from the Charles Darwin Foundation.

Science 214 (1981): 82–84. Reprinted with permission from AAAS.

Unpredictable Evolution in a 30-Year Study of Darwin's Finches (2002)

❋ PETER R. GRANT AND B. ROSEMARY GRANT

Abstract—Evolution can be predicted in the short term from a knowledge of selection and inheritance. However, in the long term evolution is unpredictable because environments, which determine the directions and magnitudes of selection coefficients, fluctuate unpredictably. These two features of evolution, the predictable and unpredictable, are demonstrated in a study of two populations of Darwin's finches on the Galápagos island of Daphne Major. From 1972 to 2001, Geospiza fortis (medium ground finch) and Geospiza scandens (cactus finch) changed several times in body size and two beak traits. Natural selection occurred frequently in both species and varied from unidirectional to oscillating, episodic to gradual. Hybridization occurred repeatedly though rarely, resulting in elevated phenotypic variances in G. scandens and a change in beak shape. The phenotypic states of both species at the end of the 30-year study could not have been predicted at the beginning. Continuous, long-term studies are needed to detect and interpret rare but important events and nonuniform evolutionary change.

The value of long-term studies in ecology has become widely recognized among scientists and the media (*1–3*). Less widely appreciated is the similar value to be gained from long-term studies of evolution in nature. A classic study, spanning 49 years, was carried out by H. D. Ford and E. B. Ford (*4*) on phenotypic variation in a European butterfly, *Melitaea aurinia* (Marsh Fritillary). The most important discovery was made after about 40 years of monitoring that began with the collecting of specimens by amateur naturalists in 1881. In the early 1920s an outburst of phenotypic and presumed genetic variation occurred in association with a rapid increase in butterfly numbers from an extremely low density caused by parasitoids. Variation then declined to a lower and stable level, with phenotypes in the late 1920s being recognizably different from those in the same population at the beginning of the study. The inferred genetic reorganization helped to frame ideas about evolution in contemporary time, inspired other long-term studies of butterflies (5) and moths (6), and contributed to the development of at least one model of speciation (7).

Long-term studies of evolution involving annual or more frequent sampling have many potential benefits. These include documentation and understanding of slow and cryptic directional evolutionary change, perhaps in association with gradual global warming, reversals in the direction of evolution, rare events with strong effects such as genetic bottlenecks caused by population crashes, phenomena recurring at long intervals, and processes with high interannual variability such as erratic and intermittent gene flow. These benefits are beginning to be realized (8–15), but few studies have persisted long enough for us to be able to generalize about the temporal pattern and predictability of basic evolutionary processes in unconstrained natural populations.

Here, we report the results of a 30-year study of evolution of size and shape traits in two populations of Darwin's finches based on annual sampling and measurement. Distinctive features of the study are its length, continuity, entirely natural environmental setting, the availability of pedigree information to construct and interpret evolutionary change, and the macroevolutionary context of an adaptive radiation. The study reveals the irregular occurrence, frequency, and consequences of two evolutionary processes that are more often inferred than directly studied: natural selection and introgressive hybridization.

NATURAL SELECTION AND EVOLUTION

Populations of *Geospiza fortis* (medium ground finch) and *G. scandens* (cactus finch) have been studied on the Galápagos island of Daphne Major every year since 1973; adults that year were born (hatched) no later than 1972. Survival of marked and measured individuals has been recorded every year, and reproduction of most individuals has been recorded in most years (16). Six measured traits on adults whose growth has ceased have been reduced by principal components analyses to three interpretable synthetic traits: body size, beak size, and beak shape (17–20). The null expectation is that, subject to sampling error, means of these traits have remained constant across the period of study.

This expectation of no change is clearly not supported by the data (Fig. 1). Lack of independence of samples in successive years precludes year-by-year significance testing of the total samples. Nevertheless, comparisons across years show nonoverlapping 95% confidence estimates of the means at different times. Mean body size and beak shape were markedly different at the end of the period (2001) than at the beginning (1973) in both species (21). Between these two times mean body and beak size of *G. fortis* initially decreased, then increased sharply, and decreased again more slowly (Fig. 1, A

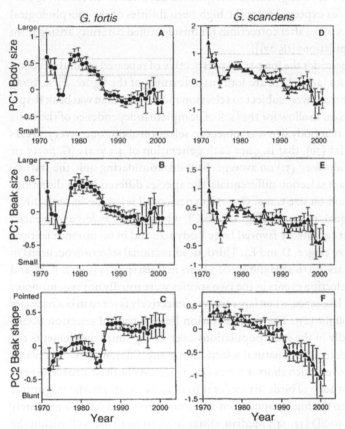

Fig. 1. Morphological trajectories of adult *Geospiza fortis* (*A* to *C*) and *G. scandens* (*D* to *F*). In the absence of change, mean trait values should have remained within the 95% confidence intervals (horizontal broken lines) of the estimates from the 1973 samples (body size: *G. fortis*, n = 115, *G. scandens*, n = 37; beak traits: *G. fortis*, n = 173, *G. scandens*, n = 62). Sample sizes varied from 45 (1997) to 976 (1991) for *G. fortis* and from 30 (1999) to 336 (1983) for *G. scandens*. The 1972 sample is composed of the adults (=1 year old) in 1973.

and B). Beak shape abruptly became more pointed in the mid-1980s and remained so for the next 15 years (Fig. 1C). *G. scandens*, a larger species, displayed more gradual and uniform trends toward smaller size and blunter beaks (Fig. 1, D to F), thereby converging toward *G. fortis* in morphology.

Apart from random sampling effects, annual changes in morphological means are caused by selective losses, as a result of mortality and emigration, and selective gains, as a result of breeding and immigration (22). Previous work has demonstrated directional natural selection on beak and body size traits associated with survival, in *G. fortis* at three times and in *G. scandens* once, when a scarcity of rain caused a change in the composition of the seed supply that forms their dry-season diets (23–25). Evolutionary responses of *G. fortis* to the two strongest selection episodes occurred in the following generations (26), as expected from the high heritabilities of the morphological traits [$h^2 = 0.5$ to 0.9 after corrections for misidentified paternity arising from extrapair copulations (18, 27)].

Figure 2 provides the long-term perspective of repeated natural selection in both species (28). There are four main features of the figure. First, body and beak size traits were subject to selection more often than was beak shape. Setting α at 0.01, to allow for the lack of complete independence of the traits (29), we find that body size was subject to selection about once every 3 years in both species (30), that is, once each generation of 4.5 years (*G. fortis*) or 5.5 years (*G. scandens*) (31) on average. Second, considering only the statistically significant selection differentials, the species differed in the directions of net selection on size traits. *G. fortis* experienced selection in both directions with equal frequency (Fig. 2, A to C), whereas *G. scandens* experienced selection that repeatedly favored large body size and in no instance favored small beak size (Fig. 2, D and E). Third, unidirectional selection occurred in successive years, up to a maximum of 3 years in both species (Fig. 2, A, D, and E). Fourth, selection events in the two species were usually not synchronous, except in the late 1970s, when large size was selectively favored in both species during a drought (23). The demonstration here of natural selection occurring repeatedly in the same populations over a long time complements the widespread detection of natural selection in many different species of plants and animals over much shorter times (32, 33, 34). As in these broad surveys, and in three studies of birds lasting for 11 to 18 years (15, 35, 36), the magnitude of selection on the finch populations was usually less than 0.15 SD and rarely more than 0.50 SD (33, 34). Median values (0.03 to 0.06) are well within the normal range (0.00 to 0.30) of other studies (34).

Evolution followed as a consequence of selection in both species because all traits are highly heritable (18, 20, 27). We compared the mean of a trait before selection with the mean of the same trait in the next generation by one-tailed *t* tests ($P < 0.05$) (26). Significant evolutionary events occurred in

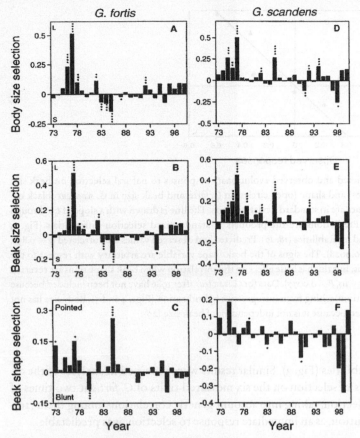

Fig. 2. (*A* to *F*) Standardized selection differentials, calculated for each sample surviving from year *x* to year *x* + 1. Positive values indicate selection for large size or pointed beaks; L, large; S, small. Significance levels are shown without correction for multiple testing or lack of independence (29). Males and females were combined with adults of unknown sex because separate selection analyses of males and females give similar results (23). Differentials are temporally autocorrelated to varying degrees; autocorrelation coeffcients are 0.416, 0.302, and 0.093 for *G. fortis* body size, beak size, and beak shape, respectively, and 0.171, 0.373, and 0.103 for the same traits in *G. scandens*. *$P < 0.05$; **$P < 0.01$; ***$P < 0.005$; ****$P < 0.001$.

G. fortis eight times (body size, four; beak size, three; beak shape, one) and in *G. scandens* seven times (body size, two; beak size, five). Evolution below the level of statistical detectability may have followed other instances of directional selection, may have been masked by annual variation in environmental effects on growth to final size (37), or may have been nullified by countervailing selection on correlated traits not included in the analyses (32). Magnitudes of evolution of the two independent beak traits (size and shape) are correlated with values predicted from the products of selection differentials

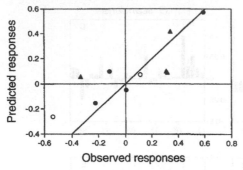

Fig. 3. Predicted and observed evolutionary responses to natural selection on beak size (black circles) and shape (open circles) in *G. fortis* and beak size in *G. scandens* (black triangles). Values are in standard deviation units. The line is drawn with a slope of 1.0 through the origin. Predictions are the products of standardized selection differentials (Fig. 2, $P < 0.01$) and heritabilities (*18, 27*). Predicted and observed values are correlated ($r = 0.832$, $n = 10$, $P = 0.0028$). The signs of the beak shape variable are arbitrary with respect to the beak size axis, but this has little effect on the correlation when beak shape signs are reversed ($r = 0.781$, $n = 10$, $P = 0.0077$). Data for *G. scandens* after 1986 have not been included because of complications arising from introgressive hybridization (Figs. 4 and 5). Body size has not been included because it is not independent of beak size (*29*).

and heritabilities (Fig. 3). Similar results were obtained in analyses of the direct effects of selection on the six measured traits of *G. fortis* at two times of intense selection, taking into account genetic correlations among them (*26*). Thus evolution, as an immediate response to selection, was predictable.

INTROGRESSIVE HYBRIDIZATION

Annual changes in morphology (Fig. 1) are largely but not entirely accounted for by selective losses. The greatest discrepancy is in the 1990s when the single occurrence of natural selection on beak shape in *G. scandens* at the beginning (at $P < 0.05$; Fig. 2F) does not account for the continuing change in mean beak shape over the decade (Fig. 1F). Therefore, we next consider selective gains as a result of nonrandom recruitment to the adult population.

There are four potential contributors to nonrandom additions: conspecific and heterospecific residents and immigrants. Breeding immigrants are not known in *G. scandens* and are extremely rare in *G. fortis* (*16*). Biased conspecific breeding has minor effects on morphological trajectories. Prior analyses indicate some evidence for sexual selection on morphological traits (*38*), yet little influence of morphological variation on lifetime fitness as measured by the production of offspring that survive to breed (*39*). However, hybridization does occur rarely between resident *G. fortis* and *G. scandens*, and

G. fortis also breeds with a rare immigrant species, *G. fuliginosa* (small ground finch) (40). In both cases there is generally little or no fitness loss (41, 42). After the dry period of the late 1970s, and beginning in the extraordinarily prolonged wet season of 1983 (El Niño year), successful breeding of F_1 hybrids and backcrosses was documented (40–43). Effects of introgression on morphological means and variances have not been tested before, but are to be expected in view of the large additive genetic variation underlying the size and shape traits of both *G. fortis* and *G. scandens* (18, 20, 27).

A specific prediction of the introgression hypothesis is an increase in variance and skewness in the morphological distributions, beginning in 1983 in *G. fortis* and 1987 in *G. scandens* (42). Increases are expected to be greater in *G. scandens* than in *G. fortis*, despite bidirectional gene exchange, because F_1 hybrids and first-generation backcrosses made a proportionately greater numerical contribution to the *G. scandens* samples (Fig. 4, A and B) (43). The predicted increases in beak shape variance and skewness are observed in *G. scandens* (Fig. 5, E and F) and are scarcely noticeable in *G. fortis* (Fig. 5, B and C). Beak shape variance in *G. scandens* doubled from 0.430 in 1973 (95% confidence intervals 0.311, 0.636; $n = 62$) to 1.026 (0.674, 1.762; $n = 35$) in 2001, whereas the variance in *G. fortis* beak shape remained stationary: 0.627 (0.493, 0.824; $n = 173$) in 1973, and 0.887 (0.744, 1.316; $n = 114$) in 2001. Beak size and body size variances (not shown) are not as well estimated and show no significant variation across the study period (95% confidence intervals broadly overlap). Skewnesses in beak size and body size distributions vary in parallel with beak shape skewness: negatively in *G. scandens*, projecting toward *G. fortis*; and positively in *G. fortis*, projecting toward *G. scandens*.

A second prediction is that variance and skewness will decrease if F_1 hybrids and first-generation backcrosses are deleted from the total samples. Deletions of these two classes of birds do indeed effectively eliminate the increases in skewness and variance in *G. scandens* beak shape, and reduce the degree to which the mean changed (Fig. 5, D to F), because on average F_1 hybrids and first-generation backcrosses are smaller in body size and beak size and have less pointed beaks than the parental *G. scandens* species (44). We conclude that introgressive hybridization caused a change in means and other moments of the frequency distributions of *G. scandens* measurements. Selection may have contributed as well but to a minor extent (Fig. 2F). Deletions had no obvious effect on the distributions of *G. fortis* traits (Fig. 5, A to C) (44, 45).

The proportionally greater gene flow from *G. fortis* to *G. scandens* than vice versa has an ecological explanation. Adult sex ratios of *G. scandens* became male biased after 1983 (Fig. 4C) as a result of heavy mortality of the socially subordinate females. High mortality was caused by the decline of their principal dry-season food, *Opuntia* cactus seeds and flowers; rampantly growing vines smothered the bushes (*16*). *G. fortis*, more dependent on small seeds of several other plant species, retained a sex ratio close to 1:1 (Fig. 4C). Thus, when breeding resumed in 1987 after 2 years of drought, competition among females for mates was greater in *G. fortis* than in *G. scandens*. All 23 *G. scandens* females paired with *G. scandens* males, but two of 115 *G. fortis* females paired interspecifically. All their F$_1$ offspring later bred with *G. scandens* (*43*) because choice of mates is largely determined by a sexual imprinting-like process on paternal song (*42*).

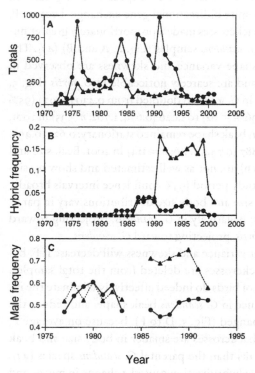

Year

Fig. 4. Total samples of measured birds (*A*), proportional contributions of F$_1$ hybrids and first-generation backcrosses to the totals (*B*), and minimal frequencies of adult males (*C*) before and after habitat and demographic changes were caused by the El Niño event of 1983; *G. fortis* (black circle) and *G. scandens* (black triangle). In (B), the decline in proportion of hybrids in the *G. scandens* samples after 1999 reflects the addition of birds of unknown parents. They have been classified as *G. scandens*, but some were probably unidentified hybrids and backcrosses. In (C), birds of unknown sex in female-like plumage (0 to 10%) have been added to the female samples, and therefore proportions of males are minima. Sexes of all birds were known in some years (e.g., 1989). Data for 1984 and 1985 are not shown because of the large proportion of birds of unknown sex (>15%) produced in 1983 and 1984, comprising both males and females.

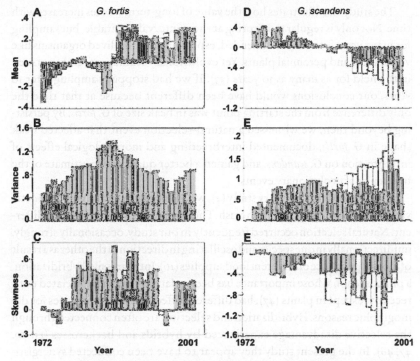

Fig. 5. Effects of hybrids and backcrosses on the mean, variance, and skewness of the beak shape frequency distributions of *G. fortis* (*A* to *C*) and *G. scandens* (*D* to *F*). Removal of F₁ hybrids and first-generation backcrosses (white bars) has a strong effect on *G. scandens* but a minor effect on *G. fortis* parameters (45). The proportion of known hybrids declined after 1991; not all hybrids and backcrosses could be identified with pedigree information, because some nestlings were not banded after 1991 (16), and therefore some may have been included as *G. fortis* and *G. scandens*. Negative skewness indicates a prolonged tail toward less pointed beaks. Additional removal of *G. fortis* × *G. fuliginosa* F₁ hybrids and backcrosses has scarcely detectable effects in these diagrams.

CONCLUSION

The long-term study of Darwin's finch populations illustrates evolutionary unpredictability on a scale of decades. Mean body size and beak shape of both species at the end of the study could not have been predicted at the beginning. Moreover, sampling at only the beginning and at the end would have missed beak size changes in *G. fortis* in the middle. The temporal pattern of change shows that reversals in the direction of selection do not necessarily return a population to its earlier phenotypic state. Evolution of a population is contingent upon environmental change, which may be highly irregular, as well as on its demography and genetic architecture (33, 46).

The study also illustrates how the value of long-term studies increases with time. Not only is regular monitoring at short intervals desirable, but sampling for many years is to be recommended, especially for long-lived organisms like vertebrates and perennial plants. Yet evolutionary studies are rarely pursued in the field for as many as 10 years (33). If we had stopped sampling after 10 years, our conclusions would have been different because at that time the only difference from the starting point was in beak size of *G. fortis*. By persisting beyond then, we witnessed a natural-selection event that affected beak shape in *G. fortis*, documented interbreeding and morphological effects of introgression on *G. scandens*, and gained a better quantitative estimate of the frequency of evolutionary events.

Unlike the Marsh Fritillary study (4), we did not witness a release of genetic variation following a population crash. The evolutionary dynamics were different. Natural selection occurred frequently in our study, occasionally strongly, unidirectionally in one species and oscillating in direction in the other as a result of their dependence on different food supplies (16). Introgressive hybridization, a phenomenon whose importance has been relatively underappreciated until recently, except in plants (47), had different effects on the two species for demographic reasons. Hybridization and selection are often connected through the selective disadvantage experienced by hybrids and backcrosses (13, 40, 47, 48). In the present study they appear to have been connected synergistically in the sense that interbreeding may have been facilitated in part by selection for more pointed, *G. scandens*–like, beaks in the *G. fortis* population in the mid-1980s. Choice of mates is partly determined by imprinting on parental beak morphology, as well as on paternal song (41). The principal causes of selection have been identified as changes in food supply (23–26) mediated in large part by droughts. The ultimate cause of repeated natural selection and introgressive hybridization may have been a change in the seasonal movement of water masses in the eastern subtropical and tropical Pacific (49, 50), triggering altered climatic patterns, including the intensification of El Niño and La Niña cycles.

Regardless of the precise chain of causality, field studies such as ours, in conjunction with multigenerational studies of microorganisms in the laboratory (51, 52) and experimental studies of selection in the field (53–55), provide an improved basis for extrapolating from microevolution to patterns of macroevolution; in the present case, from evolutionary dynamics of populations on the scale of decades to speciation and further adaptive radiation on the scale of hundreds of thousands of years (56). In conclusion, the long-term unpredictability of evolutionary change that arises from unpredictable ecological change, together with the need to strengthen generalizations about the frequency and importance of selection and hybridization, are reasons for encouraging additional, continuous, long-term studies of evolution in nature.

REFERENCES AND NOTES

1. G. Likens, Ed., *Long-term Studies in Ecology: Approaches and Alternatives* (Springer-Verlag, New York, 1989).

2. J. H. Brown, T. G. Whitham, C. K. M. Ernest, C. A. Gehring, *Science* 293, 643 (2001).

3. P. Kareiva, J. G. Kingsolver, R. B. Huey, Eds., *Biotic Interactions and Global Change* (Sinauer, Sunderland, MA, 1993).

4. H. D. Ford, E. B. Ford, *Trans. Entomol. Soc. London* 78, 345 (1930).

5. P. R. Ehrlich, L. G. Mason, *Evolution* 20, 165 (1965).

6. D. A. Jones, *Trends Ecol. Evol.* 4, 298 (1989).

7. H. L. Carson, in *Population Biology and Evolution*, R. C. Lewontin, Ed. (Syracuse Univ. Press, Syracuse, NY, 1968), pp. 123–137.

8. W. W. Anderson *et al.*, *Proc. Natl. Acad. Sci. U.S.A.* 88, 10367 (1991).

9. R. H. Cowie, J. S. Jones, *Biol. J. Linn. Soc.* 65, 233 (1998).

10. L. F. Keller, *Evolution* 52, 240 (1998).

11. M. E. N. Majerus, *Melanisms: Evolution in Action.* (Oxford Univ. Press, Oxford, UK, 1998).

12. M. E. Visser, A. J. van Noordwijk, J. M. Tinbergen, C. M. Lessells, *Proc. R. Soc. London B* 265, 1867 (1998).

13. S. E. Carney, K. A. Gardner, L. H. Rieseberg, *Evolution* 54, 462 (2000).

14. O. Halkka, L. Halkka, K. Roukka, *Biol. J. Linn. Soc.* 74, 571 (2001).

15. L. E. B. Kruuk, J. Merilä, B. C. Sheldon, *Am. Nat.* 158, 557 (2001).

16. P. R. Grant, B. R. Grant, in *Long-term Studies of Vertebrate Communities*, M. L. Cody, J. A. Smallwood, Eds. (Academic Press, San Diego, CA, 1996), pp. 343–390.

17. For each species, two principal-components analyses were performed on the correlation matrix with untransformed data from all birds [males, females, and birds of unknown sex of both species; see (*18*)]. Mass (weight), wing length, and tarsus length were used in the first analysis, and beak length, depth, and width (*19*) were used in the second analysis. For *G. fortis* (n = 3204), PC1 in analysis 1 is interpreted as a body size factor (69.2% variance explained), because factor loadings were high and of the same sign: 0.856 (mass), 0.839 (wing), and 0.800 (tarsus). PC1 in analysis 2 is interpreted as a beak size factor (85.8% variance explained); factor loadings are 0.890 (length), 0.951 (depth), and 0.937 (width). PC2 is a beak shape factor (10.1% additional variance explained) with factor loadings of different sign: 0.455 (length), −0.167 (depth), and −0.262 (width). Factor loadings and their interpretation were similar in analyses of *G. scandens* (n = 1037). Percentage variance was 67.5 for PC1 in the first analysis, 69.8 for PC1 in the second analysis, and 23.2 for PC2 in the second analysis. Separate analyses of males and females gave similar results.

18. L. F. Keller, P. R. Grant, B. R. Grant, K. Petren, *Heredity* 87, 325 (2001).

19. The six measured traits (*17*) have high repeatabilities (*20*). Nine specimens of *G. fortis* in the Charles Darwin Research Station museum, Galápagos, were measured by P.R.G. in 1975, 1976, and again in 2001 to check for possible changes in methods of linear measurement. No heterogeneity among years was found for any of the five traits [analyses of variance (ANOVAs), all $P > 0.67$], and no difference between pairs of years was found in any trait (Fisher's protected least significant difference post hoc tests, all $P > 0.38$).

20. P. R. Grant, B. R. Grant, *Evolution* 48, 297 (1994).

21. The first samples in 1973 (221 *G. fortis* and 72 *G. scandens*) were compared by ANOVAs with the last samples in 2001 (114 *G. fortis* and 35 *G. scandens*). The data were trimmed to 2.5 SD on either side of the mean by removing one to three individuals from the samples of each species. This corrected for skewness and unequal variances. Sex was included in

two-factor ANOVAs because males are generally larger than females. Mean body size was significantly smaller in *G. fortis* ($F_{1,132}$ = 7.773, P = 0.0061) and in *G. scandens* ($F_{1,50}$ = 11.272, P < 0.0001) in 2001 than in 1973. There was a significant effect of sex in each species (P < 0.002) but no sex-by-year interaction (P > 0.1). Mean beak size did not differ between years in either *G. fortis* ($F_{1,166}$ = 0.004, P = 0.9480) or *G. scandens* ($F_{1,50}$ = 3.108, P = 0.0840); sex effects were significant in both species (P < 0.007), but there were no sex-by-year interactions (P > 0.1). Beak shape differed between years in both species. For *G. scandens* there was a strong year effect ($F_{1,72}$ = 17.168, P < 0.0001), a weak sex effect ($F_{1,72}$ = 5.943, P = 0.0172), and no interaction. The *G. fortis* sexes do not differ in beak shape (P = 0.9715), and therefore a one-factor ANOVA was performed with adult males, females, and birds of unknown sex. It demonstrated a strong difference between years ($F_{1,287}$ = 30.246, P < 0.0001).

22. P. R. Grant, B. R. Grant, K. Petren, *Genetica 112–113*, 359 (2001).

23. P. T. Boag, P. R. Grant, *Science 214*, 82 (1981).

24. T. D. Price, P. R. Grant, P. T. Boag, H. L. Gibbs, *Nature 309*, 787 (1984).

25. H. L. Gibbs, P. R. Grant, *Nature 327*, 511 (1987).

26. P. R. Grant, B. R. Grant, *Evolution 49*, 241 (1995).

27. ———, in *Adaptive Genetic Variation in the Wild*, T. A. Mousseau, B. Sinervo, J. A. Endler, Eds. (Oxford Univ. Press, New York, 2000), pp. 3–40.

28. Directional selection differentials were calculated as the difference in trait means before and after selection, then standardized in each case by dividing the difference by the standard deviation of the sample before selection (*32*). After checking for normality, differences between survivors and nonsurvivors were tested by two-tailed *t* tests. Significant skewness (*57*) was eliminated by deleting outliers (more than 2.5 SD beyond the mean), with minor effects on probability values. Equality of variances was tested with *F* tests. Only beak traits displayed unequal variances. Correcting for skewness reduced the number of significant variance inequalities from three to two in *G. fortis* and from nine to four in *G. scandens*. Means in the remaining six cases were tested with a median test (*57*); five were nonsignificant (P > 0.1). Selection was diversifying in three of the five cases. Selection differentials measure the combined direct effects of selection on a trait and indirect effects of selection on correlated traits. Direct effects on the individual measured traits have been assessed by partial regression (selection gradient analysis) in some cases (*24–26*), yielding similar interpretations; body size, beak size, and beak shape are independently selected traits.

29. Body size and beak size are strongly correlated traits in *G. fortis* (r^2 = 0.464 to 0.650; median 0.540) and *G. scandens* (r^2 = 0.393 to 0.694; median 0.553) in all 30 years at P < 0.0001. Beak shape is occasionally and weakly correlated with body size in *G. fortis* (r^2 = 0.001 to 0.082; median 0.015) and *G. scandens* (r^2 = 0.001 to 0.179; median 0.026).

30. *G. scandens* body size was selected in 8 out of 22 years in which a minimum sample size of 75 was met, and beak size was selected in 7 of those years. Frequency of selection in *G. fortis* was 8 out of 25 years for body size, 5 out of 26 years for beak size, and 2 out of 26 years for beak shape. These are minimal frequencies; selection below the level of statistical detectability may have occurred in other years.

31. P. R. Grant, B. R. Grant, *Ecology 73*, 766 (1992).

32. J. A. Endler, *Natural Selection in the Wild* (Princeton Univ. Press, Princeton, NJ, 1986).

33. J. G. Kingsolver *et al.*, *Am. Nat. 157*, 245 (2001).

34. H. E. Hoekstra *et al.*, *Proc. Natl. Acad. Sci. U.S.A.* 98, 9157 (2001).

35. B. R. Grant, P. R. Grant, *Evolutionary Dynamics of a Natural Population: The Large Cactus Finch of the Galápagos* (Univ. of Chicago Press, Chicago, IL, 1989).

36. C. Barbraud, *J. Evol. Biol. 13*, 81 (2000).

37. J. Merilä, L. E. B. Kruuk, B. C. Sheldon, *Nature 412*, 76 (2001).

38. T. Price, *Evolution 38*, 327 (1984).

39. P. R. Grant, B. R. Grant, *Proc. R. Soc. London B 267*, 131 (2000).

40. ———, *Philos. Trans. R. Soc. London B 340*, 127 (1993).

41. ———, *Science 256*, 193 (1992).

42. ———, in *Endless Forms: Species and Speciation*, D. J. Howard, S. H. Berlocher, Eds. (Oxford Univ. Press, New York, 1998), pp. 404–422.

43. Eight F_1 hybrids (both sexes), all from families with a *G. scandens* father that sang a *G. scandens* song, backcrossed to *G. scandens* and produced 16 measured offspring. Six F_1 hybrids (both sexes), all from families with a *G. scandens* father that sang a *G. fortis* song, backcrossed to *G. fortis* and produced 26 measured offspring. Other F_1 hybrids and backcrosses were produced but not captured and measured.

44. The 1991 sample of *G. scandens* ($n = 112$), the last with complete pedigree information, included 20 hybrids (seven F_1 hybrids and 13 first-generation backcrosses). The hybrids were smaller in body size (ANOVA, $F_{1,109} = 38.522$, $P < 0.0001$) and beak size ($F_{1,109} = 34.583$, $P < 0.0001$) and had less pointed beaks ($F_{1,109} = 55.116$, $P < 0.0001$) than the rest. In the same year, one F_1 hybrid and nine first-generation backcrosses in the sample of 510 *G. fortis* were larger in body size than the rest ($F_{1,507} = 21.817$, $P < 0.0001$) and had more pointed beaks ($F_{1,507} = 8.323$, $P = 0.0041$) but did not differ in beak size ($F_{1,507} = 1.098$, $P = 0.2953$).

45. The fractional contribution to beak shape variance made by hybrids was on average 0.265 ± 0.140 SD for *G. scandens* (maximum 0.528; $n = 15$ years) and 0.024 ± 0.018 for *G. fortis* (maximum 0.064; $n = 17$ years).

46. H. Teotonio, M. R. Rose, *Evolution 55*, 653 (2001).

47. M. L. Arnold, *Natural Hybridization and Evolution* (Oxford Univ. Press, New York, 1997).

48. T. Veen *et al.*, *Nature 411*, 45 (2001).

49. T. P. Guilderson, D. P. Schrag, *Science 281*, 240 (1998).

50. R. H. Zhang, L. M. Rothstein, A. J. Busalacchi, *Nature 391*, 879 (1998).

51. R. E. Lenski, M. Travisiano, *Proc. Natl. Acad. Sci. U.S.A. 91*, 6608 (1994).

52. P. B. Rainey, M. Travisiano, *Nature 394*, 69 (1998).

53. D. N. Reznick, F. H. Shaw, F. H. Rodd, R. G. Shaw, *Science 275*, 1934 (1997).

54. B. Sinervo, E. Svensson, T. Comendant, *Nature 406*, 985 (2000).

55. J. R. Etterson, R. G. Shaw, *Science 294*, 151 (2001).

56. P. R. Grant, *Ecology and Evolution of Darwin's Finches* (Princeton Univ. Press, Princeton, NJ, 1999).

57. R. R. Sokal, F. J. Rohlf, *Biometry: The Principles and Practice of Statistics in Biological Research* (Freeman, New York, ed. 3, 1996).

58. We thank I. Abbott, L. Abbott, P. T. Boag, H. L. Gibbs, L. F. Keller, K. Petren, T. D. Price, J. N. M. Smith, and many field assistants acknowledged earlier (*31*); the Galápagos National Parks Service and Charles Darwin Research Station for permits and logistical support; M. Hau, K. Petren, D. L. Stern, and M. Wikelski for comments on the manuscript; and McGill University, Natural Sciences and Engineering Research Council (Canada, 1973 to 1978), and NSF (1978 to 2002) for the long-term financial support that has made this study possible.

Science 296 (2002): 707–711. Reprinted with permission from AAAS.

Evolution of Character Displacement in Darwin's Finches (2006)

PETER R. GRANT AND B. ROSEMARY GRANT

Abstract—Competitor species can have evolutionary effects on each other that result in ecological character displacement; that is, divergence in resource-exploiting traits such as jaws and beaks. Nevertheless, the process of character displacement occurring in nature, from the initial encounter of competitors to the evolutionary change in one or more of them, has not previously been investigated. Here we report that a Darwin's finch species (Geospiza fortis) on an undisturbed Galápagos island diverged in beak size from a competitor species (G. magnirostris) 22 years after the competitor's arrival, when they jointly and severely depleted the food supply. The observed evolutionary response to natural selection was the strongest recorded in 33 years of study, and close to the value predicted from the high heritability of beak size. These findings support the role of competition in models of community assembly, speciation, and adaptive radiations.

Character displacement (*1, 2*) is an evolutionary divergence in resource-exploiting traits such as jaws and beaks that is caused by interspecific competition (*3–5*). It has the potential to explain nonrandom patterns of co-occurrence and morphological differences between coexisting species (*6–10*). Supporting evidence has come from phylogenetic analyses (*11*) and from experimental studies of sticklebacks, in which the role of directional selection in character divergence has been demonstrated (*12*). The process of character displacement occurring in nature, from the initial encounter of competitors to the evolutionary change in one or more of them as a result of directional natural selection, has not previously been investigated.

The situation on the small Galápagos island of Daphne Major (0.34 km²) has been referred to as the classical case of character release (*1, 2, 13*), which is the converse of character displacement. Here, in the virtual absence of the small ground finch (*Geospiza fuliginosa*; weighing ~12 g) and released from competition, the medium ground finch (*G. fortis*; ~18 g) is unusually small in beak and body size. Lack (*14*) proposed that its small size reflects an evolutionary shift enabling *G. fortis* to take maximum advantage of small seeds made available by the absence of its competitor. Subsequent field studies demonstrated an association, previously only inferred, between beak sizes and seed diets (*13, 15*). In 1977, a drought on Daphne revealed that small seeds are preferred when they are abundant, but when they are scarce, finches turn increasingly to large and hard seeds that only the large-beaked members of the population can crack (*13, 15*). Most finches died that year, and mortality was heaviest among those with small beaks (*13, 16, 17*). Thus, a population's mean beak size is determined by the tradeoff in energetic rewards from feeding on small and large seeds, and the tradeoff is affected by variation in beak morphology and rates of seed depletion and replenishment (*7, 18, 19*). Competitors can modify the tradeoff (*7*).

The situation on Daphne changed in 1982 with the arrival of a new competitor species, setting up the potential for character displacement to occur. Between 1973 and 1982, a few individuals of the large ground finch (*G. magnirostris*; ~30 g) visited the island for short periods in the dry season but never bred (*15*). In late 1982, a breeding population was established by two females and three males at the beginning of an exceptionally strong El Niño event that brought abundant rain to the island (1359 mm) (*20–22*). *G. magnirostris* is a potential competitor as a result of diet overlap with *G. fortis* (Table 1), especially in the dry season when food supply is limiting (*9, 23*). The principal food of *G. magnirostris* is the seeds of *Tribulus cistoides*, contained within a hard mericarp and exposed when a finch cracks or tears away the woody outer covering (Fig. 1). Large-beaked members of the *G. fortis* population are capable of this maneuver—indeed, survival in the 1977 drought to a large extent depended on it (*13, 16*)—but on average they take three times longer than *G. magnirostris* to gain a seed reward (*13, 24*). The smallest *G. fortis* never attempt to crack them (*18, 24*). *G. magnirostris* compete with *G. fortis* by physically excluding them from *Tribulus* feeding sites and by reducing the density of *Tribulus* fruits to the point at which it is not profitable for *G. fortis* to feed on them, owing to handling inefficiencies in relation to search and metabolic costs (*7, 13, 18, 24*). By depleting the supply of *Tribulus* fruits, *G. magnirostris* was predicted to cause a selective shift in *G. fortis* in the direction of small beak size.

The predicted shift occurred in 2004 (Fig. 2). Initially, the population size of *G. magnirostris* was too small in relation to the food supply to have anything but a mild competitive effect on *G. fortis*. Their numbers gradually increased as a result of local production of recruits, augmented by additional immigrants (*22, 25*), and reached a maximum of 354 ± 47 (SE) in 2003 (Fig. 3). Little rain fell in 2003 (16 mm) and 2004 (25 mm), there was no breeding in either year, numbers of both species declined drastically, and from 2004 to 2005 *G. fortis* experienced strong directional selection against individuals with large beaks (*26*).

Selection differentials in *G. fortis* were uniformly negative for both males and females treated separately (Table 2). Average selection differentials in standard deviation units for the six measured traits that quantify bill size and shape and body size were 0.774 for males and 0.649 for females. Compared with values reported in other studies elsewhere (*27*), they are unusually large. The six traits are positively correlated to varying degrees. Selection gradient analysis helps to identify which particular traits were subject to selection independent of correlations among traits (*28*). However, bill depth and width are so strongly correlated in these samples ($r = 0.861$ for males, 0.946 for females) that their independent effects on survival cannot be distinguished.

TABLE 1. Proportions of seeds in the diets of three finch species. Small seeds are a composite group of 22 species, medium seeds are O[puntia] echios, and large seeds are T[ribulus] cistoides. N is the number of observations. There is strong heterogeneity in the G. fortis feeding data ($\chi^2_6 = 30.979, P < 0.0001$). The reduction in G. fortis feeding on Tribulus in 2004 makes a significant contribution ($\chi^2_1 = 3.912, P < 0.05$). Data were obtained by observations in the first 3 months of each year. In 1977 (only), when G. fortis experienced directional selection against small bill size, the proportion of large seeds in the diet rose to 0.304 (June) and 0.294 (December) (15)

Year	N	Small	Medium	Large
G. fortis				
1977	216	0.731	0.102	0.167
1985	205	0.805	0.000	0.195
1989	628	0.771	0.051	0.162
2004	97	0.804	0.113	0.082
G. magnirostris				
1985	27	0.185	0.000	0.815
1989	68	0.059	0.118	0.823
2004	110	0.045	0.264	0.691
G. scandens				
1977	115	0.852	0.148	0.000
1985	96	0.771	0.219	0.000
1989	145	0.234	0.697	0.000
2004	98	0.174	0.826	0.000

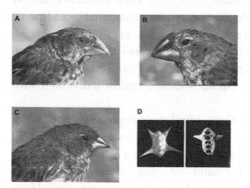

Fig. 1. Large-beaked *G. fortis* (*A*) and *G. magnirostris* (*B*) can crack or tear the woody tissues of *T. cistoides* mericarps (*D*), whereas small-beaked *G. fortis* (*C*) cannot. Five mericarps constitute a single fruit. In (*D*), the left-hand mericarp is intact. The right-hand mericarp, viewed from the other (mesial) side, has been exploited by a finch, exposing five locules from which seeds have been extracted. Mericarps are ~8 mm long and are shown at twice the magnification of the finches. [Photos are by the authors]

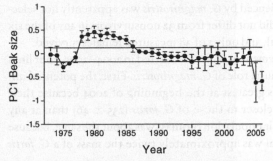

Fig. 2. Mean beak size PC1$_{bill}$ of adult *G. fortis* (sexes combined) in the years 1973 to 2005. Vertical lines show 95% confidence intervals for the estimates of the mean. Horizontal lines mark the 95% confidence limits on the estimate of the mean in 1973 to illustrate subsequent changes in the mean. Sample sizes vary from 29 (in 2005) to 950 (in 1987). Signs of the PC values are reversed so that mean size increases from the origin.

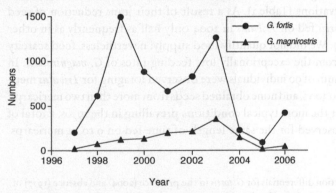

Fig. 3. Numbers of *G. fortis* and *G. magnirostris*. Breeding was extensive in 1997–1998 and 2002, and as a result finch numbers were elevated in the following years. There was no breeding in 2003 and 2004. Numbers before 1997 have been omitted because *G. magnirostris* were scarce (≤13 pairs) (25).

Selection gradient analysis without these two variables shows bill length to be the only significant entry into the gradient, for both males [partial regression coefficient (β) = -0.931 ± 0.334 SE, $P = 0.0079$; $R^2 = 0.190$] and females ($\beta = -0.814 \pm 0.295$, $P = 0.0130$; $R^2 = 0.455$). Inclusion of either bill depth or bill width made no difference to these results. Overall bill size rather than bill length is identified as the most important factor distinguishing survivors from nonsurvivors in each year, by the fact that PC1$_{bill}$ (bill size) was a selected trait in both sexes, whereas PC2$_{bill}$ (bill shape) was not selected in either. There was little effect on body size, unlike in the 1977 episode. In contrast to *G. fortis*, the heavy mortality experienced by *G. magnirostris* was apparently not selective: four surviving males did not differ from 32 nonsurvivors in any of the six measured traits (all $P > 0.1$), and only 1 of 38 measured females survived.

Thus, character displacement in *G. fortis* occurred in 2004–2005. Four lines of evidence support the causal role of *G. magnirostris*. First, the potential impact of *G. magnirostris* was greatest at the beginning of 2004 because their numbers (150 ± 19) were closer to those of *G. fortis* (235 ± 46) than at any other time (Fig. 3), and their population biomass was about the same, because a *G. magnirostris* individual was approximately twice the mass of a *G. fortis* individual.

Second, *G. magnirostris* are largely dependent on an important food resource, *Tribulus* seeds, which are not renewed during droughts. *G. magnirostris* deplete the *Tribulus* seed supply faster than do *G. fortis*. The seeds that are consumed by a *G. magnirostris* individual each day are sufficient for two *G. fortis* individuals if they feed on nothing else (*13*). Moreover, a much higher fraction of *G. magnirostris* than *G. fortis* feed on *Tribulus*, as inferred from feeding observations (Table 1). As a result of their joint reduction of seed biomass, *G. fortis* fed on *Tribulus* in 2004 only half as frequently as in other years (Table 1). We did not quantify food supply; nevertheless, food scarcity was evident from the exceptionally low feeding rates of *G. magnirostris*. In 2004, a minimum of 90 individuals were observed foraging for *Tribulus* mericarps for 200 to 300 s, and none obtained seeds from more than two mericarps; whereas under the more typical conditions prevailing in the 1970s, a total of eight birds observed for the same length of time fed on 9 to 22 mericarps,

TABLE 2. Selection differentials for *G. fortis* in the presence (2004) and absence (1977) of *G. magnirostris*. Statistical significance at $P < 0.05$, <0.01, <0.005, and <0.001 is indicated by *, **, ***, and ****, respectively

	2004		1977	
	Males	*Females*	*Males*	*Females*
Weight	−0.62*	−0.63	0.88****	0.84***
Wing length	−0.66*	−0.60	0.47***	0.71**
Tarsus length	−0.48	0.01	0.24	0.27
Beak length	−1.08****	−0.95*	0.75****	0.88***
Beak depth	−0.94***	−0.91*	0.80****	0.69*
Beak width	−0.87***	−0.81*	0.71****	0.62*
PC1 body	−0.67*	−0.52	0.69****	0.73**
PC1 beak	−1.02****	−0.92*	0.80****	0.74**
PC2 beak	−0.34	−0.26	0.23	0.29
Sample size	47	24	164	55
Proportion of survivors	0.34	0.54	0.45	0.42

with an average interval between successive mericarps of only 5.5 ± 0.5 s (SE) (24).

Third, numbers of *G. fortis* declined to a lower level (83) in 2005 than at any time since the study began in 1973, and numbers of *G. magnirostris* declined so strongly from the 2003 maximum that by 2005, only four females and nine males were left. The population was almost extinct, apparently as a result of exhaustion of the standing crop of large seeds and subsequent starvation. Of the 137 *G. magnirostris* that disappeared in 2004–2005, 13.0% were found dead, and so were 21.7% of 152 *G. fortis*. Consistent with the starvation hypothesis, the stomachs of all dead birds (23 *G. magnirostris* and 45 *G. fortis*, banded and not banded individuals combined) were empty.

The principal alternative food for both species is the seeds of *Opuntia* cactus, but production in 2004 was low, the fourth lowest since records were first kept systematically in 1982 (23). Not only were cactus seeds insufficient for the two granivore species to escape the dilemma of a diminishing supply of their preferred foods, they were insufficient for the cactus specialist *G. scandens* (~20 g), whose numbers, like those of *G. fortis*, fell lower (to 50) than in any of the preceding 32 years. The only escape was available to the smallest, most *G. fuliginosa*–like, members of the *G. fortis* population, which are known to feed like *G. fuliginosa* on small seeds with little individual energy reward (13, 18). We have no feeding observations to indicate that they survived as a result of feeding on the typical components of the *G. fuliginosa* diet: the very small seeds of *Sesuvium edmonstonei* and *Tiquilia fusca* (13, 15, 23). Nevertheless, it may be significant that two *G. fuliginosa* individuals were present on the island in 2004 and both survived to 2005.

The fourth line of evidence is the contrast between the directions of strong selection on the *G. fortis* population in the presence (2004) and near absence (1977) of *G. magnirostris*. In 1977, a year of only 24 mm of rain and no breeding, body size and beak size of both male and female *G. fortis* considered separately were subject to selection (Table 2). Average selection differentials were 0.642 for males and 0.668 for females, and they were uniformly positive. In the intervening years, 1978–2003, there was a weaker selection episode favoring *G. fortis* with small beaks when the food supply changed toward a predominance of small seeds and scarcity of large ones after the El Niño event of 1982–1983 (20, 21, 23). At that time, *G. magnirostris* were rare (22, 25); numbers varied from 2 to 24. The selection events of 1977 and 2004 stand out against a background of relative morphological stability (29) (Fig. 2). Immediately before 2004 there was no unusual rainfall to cause a change in the composition of the food supply and no other unusual environmental factor such as temperature extremes or an invasion of predators, yet with the same amount of rain as in 1977, and with the same community of plants in the environment, large finches survived at a high frequency in 1977 but survived at a low frequency in 2004. The conspicuous difference between these years was the number of *G. magnirostris*: 2 to 14 occasional visitors in 1977 (*15*) versus 150 ± 19 residents at the beginning of 2004.

Given the high heritability of beak size of *G. fortis* (*30, 31*), an evolutionary response is to be expected from strong directional selection against large size (*32*). This was observed. The mean beak size ($PC1_{bill}$) of the 2005 generation measured in 2006 was significantly smaller than that in the 2004 sample of the parental generation before selection ($t_{176} = 4.844$, $P < 0.0001$). The difference between generations is 0.70 SD, which is exceptionally large (*27, 29*). It may be compared with the range of values predicted from the breeders equation, namely the product of the average selection differential of the two sexes and the 95% confidence intervals of the heritability estimate. The observed value of 0.70 SD falls within the predicted range of 0.66 to 1.00 SD. Although a small component of the response is probably attributable to environmental factors [food supply and finch density (*30, 32*)], the major component is genetic. This is the strongest evolutionary change seen in the 33 years of the study.

The evolutionary changes that we observed are more complex than those envisaged by Lack. Nevertheless, they provide direct support for his emphasis on the ecological adjustments that competitor species make to each other, specifically in the final stages of speciation and more generally in adaptive radiations (*9–12, 14*). They also support models of ecological community assembly that incorporate evolutionary effects of interspecific competition, in contrast to null or neutral models (*6, 9*). Replicated experiments with suitable organisms are needed to demonstrate definitively the causal role of competition, not only as an ingredient of natural selection of resource-exploiting traits (*12*) but as a factor in their evolution (*33*). Our findings should prove useful in designing realistic experiments, by identifying ecological context (high densities at the start of an environmental stress) and by estimating the magnitude of natural selection.

REFERENCES AND NOTES

1. W. L. Brown Jr., E. O. Wilson, *Syst. Zool.* *5*, 49 (1956).
2. P. R. Grant, *Biol. J. Linn. Soc.* *4*, 39 (1972).
3. B. W. Robinson, D. S. Wilson, *Am. Nat.* *144*, 596 (1994).
4. D. C. Adams, F. J. Rohlf, *Proc. Natl. Acad. Sci. U.S.A.* *97*, 4106 (2000).
5. D. W. Pfennig, P. J. Murphy, *Ecology 84*, 1288 (2003).
6. T. W. Schoener, in *Ecological Communities: Conceptual Issues and the Evidence*, D. R. Strong, L.G. Abele, A.B. Thistle, Eds. (Princeton Univ. Press, Princeton, NJ, 1984), pp. 254–281.
7. D. Schluter, T. Price, P. R. Grant, *Science 227*, 1056 (1985).
8. J. B. Losos, *Proc. Natl. Acad. Sci. U.S.A.* *97*, 5693 (2000).
9. P. R. Grant, *Ecology and Evolution of Darwin's Finches* (Princeton Univ. Press, Princeton, NJ, 1999).

10. D. Schluter, *Am. Nat. 156*, S4 (2002).

11. J. B. Losos, *Evolution 44*, 588 (1990).

12. D. Schluter, *Science 266*, 798 (1994).

13. P. T. Boag, P. R. Grant, *Biol. J. Linn. Soc. 22*, 243 (1984).

14. D. Lack, *Darwin's Finches* (Cambridge Univ. Press, Cambridge, 1947).

15. P. T. Boag, P. R. Grant, *Ecol. Monogr. 54*, 463 (1984).

16. P. T. Boag, P. R. Grant, *Science 214*, 82 (1981).

17. T. D. Price *et al.*, *Nature 309*, 787 (1984).

18. T. Price, *Ecology 68*, 1015 (1987).

19. C. W. Benkman, *Ecol. Monogr. 57*, 251 (1987).

20. H. L. Gibbs, P. R. Grant, *J. Anim. Ecol. 56*, 797 (1987).

21. H. L. Gibbs, P. R. Grant, *Nature 327*, 511 (1987).

22. P. R. Grant, B. R. Grant, *Evolution 49*, 229 (1995).

23. P. R. Grant, B. R. Grant, in *Long-Term Studies of Vertebrate Communities*, M. L. Cody, J. A. Smallwood, Eds. (Academic Press, New York, 1996), pp. 343–390.

24. P. R. Grant, *Anim. Behav. 29*, 785 (1981).

25. P. R. Grant, B. R. Grant, K. Petren, *Genetica 112–113*, 359 (2001).

26. See methods in supporting material on *Science* Online.

27. J. G. Kingsolver *et al.*, *Am. Nat. 157*, 245 (2001).

28. R. Lande, S. Arnold, *Evolution 37*, 1210 (1983).

29. P. R. Grant, B. R. Grant, *Science 296*, 707 (2002).

30. P. T. Boag, *Evolution 37*, 877 (1983).

31. L. F. Keller *et al.*, *Heredity 87*, 325 (2001).

32. P. R. Grant, B. R. Grant, *Evolution 49*, 241 (1995).

33. P. R. Grant, *Science 266*, 802 (1994).

34. We thank K. T. Grant, L. F. Keller, K. Petren, and U. Reyer for help with recent fieldwork, and the Charles Darwin Research Station and Galápagos National Park Service for permission and support. The research was supported by grants from NSF.

SUPPORTING ONLINE MATERIAL

www.sciencemag.org/cgi/content/full/313/5784/224/DC1

Methods

References

Science 313 (2006): 224–226. Reprinted with permission from AAAS.

* 8 *

Sexual Selection in Darwin's Finches

Survival is one component of lifetime fitness. Another is mating success. Both influence the number of descendents that an individual produces. When an attribute influences the probability of mating or fertilizing gametes, it is subject to "sexual selection." Darwin conceived of sexual selection as an alternative to "natural selection," since it could explain the presence of attributes that seemed to increase the chances of mating even though they had adverse effects on survival. The classic example is the peacock tail, a thing of beauty that renders flight awkward, making its owner easy prey. Other examples are the colorful plumage of many birds, the decorative scales of fish or reptiles, and the chirping of birds, insects, and frogs, all of which make them more conspicuous to predators. Currently, sexual selection is usually considered a subset of natural selection, even though mating success can be in opposition to viability; sexual selection simply pertains to some of a larger set of components of fitness through which natural selection operates.

Mate choice does not always involve preference for dangerous or compromising traits, however. A major debate centers on whether or not females benefit from their choice of mates, either in terms of direct benefits for their own or their offspring's survival, or in terms of genetic benefits to their offspring, other than looking like their magnificent fathers. (In many birds and other organisms, females are believed to be choosier than males when it comes to selecting breeding partners, so at least in the ornithological literature, the focus has been on female mate choice.) It makes sense that females would choose mates that can better provide for them during incubation or gestation, or better provide for or defend their offspring. If a female bases her choice on direct indicators of these beneficial attributes (such as territory size) or indirect indicators (such as nuptial feeding), both she and her progeny are likely to be in better condition at the end of the breeding season. Moreover, if a female bases her choice on the vigor and health of the male, or reliable indicators of these (such as plumage brightness, the energy with which he courts, or even simply age), she and the offspring may be less likely to be infected by a sickly father, and the offspring might also inherit traits that, as in their father, increase their probability of survival. This would be a reasonable female.

In contrast, some females, like simple nitwits, may base their choice solely on something that strikes their fancy. If their preference for that special something is frequent enough, sons that inherit that attractive attribute from their father would be preferred, have higher mating success, and leave more offspring. Both the male attribute and the female preference for that attribute would become more common, since the attractive and increasingly more abundant males would have inherited the preference from their mothers, which they can pass to their daughters. As a consequence, those males would be preferred by even more of the females, and so on. This "runaway" process can theoretically result in the evolution of extreme and arbitrary, and even otherwise burdensome, male traits, as first shown by Ronald A. Fisher in the 1930s. Because the preferences are arbitrary, moreover, different isolated populations, sporting different male traits and female preferences, can race off in different directions and cause divergence and subsequent reproductive isolation. This nitwit scenario is appealing, not because it is more common or plausible than other theories, but because it is amusing to contemplate the idea that major evolutionary divergence can be initiated and exacerbated by nothing more than whim.

The two processes may also work together. Though originally male traits may have been reliable indicators of benefits to females or their offspring, they may lose their reliability (because of evolved cheaters), yet still become more common, provided enough of the females still prefer them.

Another explanation for the origin of female preferences is some sort of sensory bias in females, such that females prefer some traits because of a pre-existing inclination that evolved for some other reason. For example, females may have experienced natural selection to be able to hear a particular frequency of sound or see a particular wavelength of light, and then prefer males that create that sound or bear that color. This is another example of the evolution of correlated characters, whereby natural selection on sensory abilities results in correlated evolution of mate preference.

Sexual selection has a special status in evolutionary theory because it is an important mechanism not only of evolutionary divergence in morphology, but also of evolutionary divergence in mating preference. And divergence in mating preference can lead to positive assortative mating, whereby females tend to mate with males of their own species or population, and vice versa, contributing to the genetic separation of populations. Indeed, sexual selection begins to bridge the gap between morphological divergence and reproductive isolation, or speciation.

In Darwin's finches, the females seem to be somewhat reasonable. Females have been shown to prefer larger, blacker (i.e., older) males with larger territories and with more experience in breeding and raising chicks. These are all traits associated with better provisioning of mates and offspring and may also reflect a genetic quality that enables survival to old age.[1] But there is another very interesting component of female choice: song. Song in itself gives no tangible benefit to females—except, perhaps, pleasure. The readings explore some possible consequences of basing mate choice on such a trait.

READINGS

The earliest work on mate choice in Darwin's finches showed that the finches distinguish species based on morphology and body size, and that this morphology appears to be a cue for species recognition with regard to some aspects of mating behavior. The readings here, however, focus on song as a criterion of female mate choice. Song, like body size, may predict some aspect of male quality of direct or indirect benefit to mates or offspring. Song characteristics might also be subject to natural selection to increase the efficiency of song transmission in different sound environments. Such efficiency, for instance, could render the songsters better at defending territories or at attracting the attention of potential mates in a noisy environment while remaining inconspicuous to predators. Song apparently serves in species recognition too.

The first paper, by Robert Bowman, investigates sources of natural selection on song. This paper echoes his work on divergence in bill morphology, which argued that variation in bill shape and size reflects local variation in food items available to the finches. Here, Bowman argues that variation in birdsong also is the result of local variation in the vegetation structure, which alters sound transmission and thereby selects for different song attributes. Rather than divergence in song due to arbitrary sexual selection, Bowman sees convergence in song traits of birds inhabiting similar vegetative and therefore acoustic environments.

Bowman was also among the first to investigate how song is passed from parent to offspring in Darwin's finches. He discusses a very different mechanism of inheritance: namely, cultural transmission. Young birds learn the songs of their fathers—whether genetic or foster—so that misimprinting can cause lasting changes in birdsong. Culturally inherited traits may also evolve faster than purely genetically transmitted traits, simply because, as in the game whisper-down-the-lane, they are likely to have more transmission

---

1. Intriguingly, in *Geospiza fortis*, both natural selection via viability and sexual selection via mate choice favored larger males at one point. While it is theoretically possible that female choice of the better survivors may increase the future probability of the survival of her offspring, it has not been demonstrated in Darwin's finches that mate choice actually enhances offspring survival due to the inherited, preferred trait. Given the capricious and variable manner of natural selection on bill morphology and body size in some species, it is difficult to imagine that choosing mates based on these morphological traits during a given breeding season could accurately predict offspring survival in subsequent years.

errors. Cultural inheritance of song in Darwin's finches has received much attention subsequently, in particular by the Grants and their associates. Because these birds do use song for species recognition, Bowan argues, divergence in song can cause subsequent genetic divergence of populations of birds that sing and prefer slightly different songs.

The second paper, by Jeffrey Podos, shows a correlation between bill morphology (and body size) and birdsong across several species of Darwin's finches. Species with larger bills produce simpler songs with less modulation. The evolution of bill morphology, including its evolution through its effects on viability, therefore may be expected to result in altered songs, that is, altered signals of species recognition.

Whether natural selection on body size and bill traits causes correlated selection on birdsong has still not been demonstrated; the fact that some birds mistakenly sing the songs of other species suggests that the constraints of bill morphology are not absolute. To further complicate the matter, whether correlated evolution of birdsong would make members of other species progressively less acceptable as mates depends on whether song, despite its evolution, remains a major indicator of species identity and a criterion for mate choice. Female preference would have to keep pace with changes in conspecific male song. Learned preferences, for instance for paternal song structure, as the first paper demonstrates, may be flexible and track evolving signals closely. If so, the divergence of mating preferences in concert with divergence in male traits would be a plausible theory. In this scenario, females, like feathered Electras, would continue to prefer paternal songs, no matter what they are. In birds, this may be common; in other creatures without learned preferences, it is far less likely. As these inquiries remain unanswered, finch researchers continue to test if and how natural selection on bill traits and body size influences the evolution of birdsong, mate choice, and reproductive isolation.

From *The Evolution of Song in Darwin's Finches* (1983)

❊ ROBERT I. BOWMAN

* Pages 237–239

Abstract—*This study of the singing patterns of Darwin's finches was conducted over a 20-year period and has involved the spectrographic analysis of several thousand field recordings of the songs of all 14 species of the avian tribe Geospizini. Experiments were conducted on the progeny of captive finches to determine the relative importance of heredity and learning in the development of definitive vocalizations. Sound-pressure levels of the songs were measured in four species of free-living finches. Patterns of sound transmission were determined in various Galapagos environments and correlated with energy spectra of songs as well as with anatomical, ecological, and behavioral features of the birds.*

Three types of "vocal expression" (Darwin 1872) are associated with reproductive behavior. High-frequency "whistle" song is given during courtship and is structured so as to impede binaural localization of its source by nest predators, including short-eared owls and mockingbirds. "Basic" and "derived" songs are used in mate attraction and territorial defense, and appear in a variety of patterns associated with differing motivational states, transmission distances, and vegetation formations.

It is hypothesized that in the evolution of territorial vocal communication in Darwin's finches, natural selection has shaped the signal structure so as to minimize transmission loss of acoustical energy and thereby maximize the energetic response of the intended participants. Each Galapagos plant community transmits sound frequencies differently, i.e. exhibits a distinctive pattern of frequency-dependent attenuation. Regional differences in songs (dialects) reflect regional differences in sound transmission and ecology of the finches. Thus there is evidence that natural selection has shaped song form.

This study provides no evidence for "character displacement" in the songs of the finches. Song convergences (and parallelisms) are thought to be the result of sympatric or allopatric occupation of identical or very similar acoustical environments. Song divergences are the result of the occupation of different acoustical environments by different populations of the same or different species. Vocal convergences between Darwin's finches and their ecological counterparts in other regions of the world are apparent and indicate a correspondence between ecological niche and acoustical form. As in the feeding behaviors of the finches, there is a strong positive correlation in their vocal behaviors between structural and environmental diversity, i.e. the more variable the song, the broader the "acoustical aperture" of the transmission field, wherein differential frequency-dependent loss of sound energy is minimal at any given transmission distance.

Song-learning experiments, along with observations on heterospecific matings in the wild, suggest that specific adult mating preferences are conditioned early in life (age 10–40 days) through imprinting on the song of the attending parental male with whom a strong social bond has developed as a result of feeding. Should vocal misimprinting result from a heterospecific foster-father adoption, interspecific pairings may result in adulthood. Some experimental evidence indicates that female finches also imprint early in life on the songs of the attending adult male, and they vocalize this song as reproductive preparedness approaches. Thus heterosexual pairing may be the result of mutual recognition (matching) of conspecific song imprints. One wild male intergeneric hybrid ("Camarhynchus conjunctus") successfully backcrossed with a typical wild female Camarhynchus parvulus, producing three nestlings, thus proving that hybrids among Darwin's finches may be fertile and contribute to introgression of structural and vocal characteristics.

Inasmuch as vocal dialects in Darwin's finches appear to be local adaptations to prevailing acoustical environments, and since dialects are known to be culturally transmitted, they could, theoretically, be the initial cause and not the consequence of genic diversification. If local habitats are imprinted as are local song dialects, early in life, adult mobility could be reduced and genetic divergences affecting feeding and morphological adaptations enhanced.

The great ecological diversity of the Galapagos Islands, with their numerous insular worlds within a world, all differing in their florae and geospizine faunae, seems to have led to a rapid and varied proliferation of vocal (and feeding) adaptations, for which Darwin's finches are justly famous.

Ecological (acoustical) conditions that have made possible the remarkable radiation in Darwin's finches are not unique to insular settings such as the Galapagos Islands, but occur in differing degrees in all regions of the world where birds sing.

[INTRODUCTION]

Largely due to the classical writings of Darwin (1945) and Lack (1947), the 14 species of endemic songbirds of the Galapagos and Cocos islands have become famous as a textbook example of adaptive radiation in the class Aves (Sulloway 1982a, b). Although Darwin's finches are best known for their diversity of bill structure (Fig. 1 [not reprinted]) and associated feeding behavior (Bowman 1961), the present study has revealed an equally varied adaptive radiation in their song signals (Fig. 2). Like the feeding behaviors, finch vocal behaviors show clear correlations with the ecological conditions under which the birds live.

Following the methodology of ethologists (Eibl-Eibesfeldt and Kramer 1958), this study on the evolution of vocal communication in Darwin's finches was carried out in three stages. Initially, a large number of field recordings of songs were catalogued according to structure and behavioral context. Secondly, using aviary-bred, hand-reared individuals, the ontogeny and filial transmission of songs were studied under controlled acoustical environments. Thirdly, contributions to fitness of observed vocal characters were assessed from additional studies on sound attenuation in various Galapagos environments. Concurrently, a detailed investigation of the anatomy of the geospizine syrinx (Cutler 1970) made it possible to correlate certain details of syringeal structure with distinctive features of songs.

The results of these investigations are numerous and varied. New insights have been gained into the role of the acoustical environment in shaping vocal signal structure (Bowman 1979), the possible causes of convergence and parallelism in bird songs generally, the relative importance of nature and nurture in normal song development, and the role of song in mate selection.

For a number of reasons, Darwin's finches have been ideal subjects for behavioral studies. Genetic relationships among most members of the group are fairly well known (Polans, this volume; Yang and Patton 1981). All available evidence (Table 1 [not reprinted]) points to a monophyletic origin of

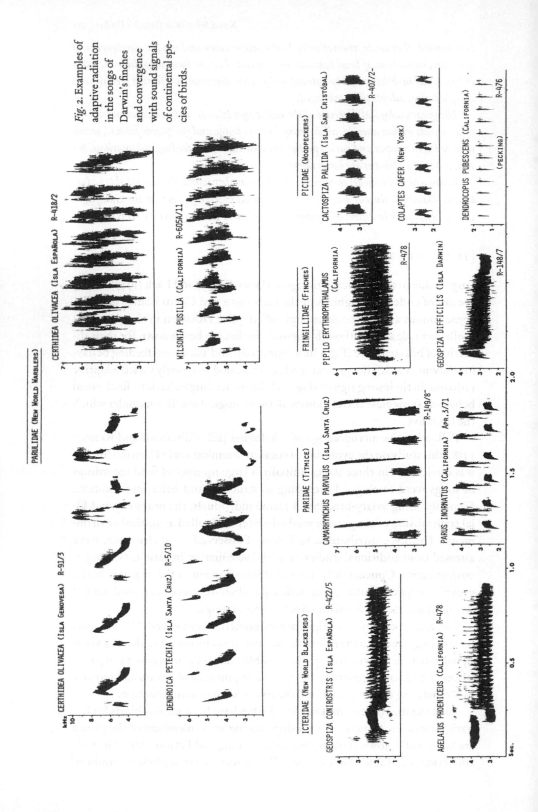

Fig. 2. Examples of adaptive radiation in the songs of Darwin's finches and convergence with sound signals of continental species of birds.

PARULIDAE (New World Warblers)

CERTHIDEA OLIVACEA (Isla Española) R-418/2

WILSONIA PUSILLA (California) R-605A/11

CERTHIDEA OLIVACEA (Isla Genovesa) R-91/3

DENDROICA PETECHIA (Isla Santa Cruz) R-5/10

PICIDAE (Woodpeckers)

CACTOSPIZA PALLIDA (Isla San Cristóbal) R-407/2-

COLAPTES CAFER (New York)

DENDROCOPUS PUBESCENS (California) R-476 (pecking)

FRINGILLIDAE (Finches)

PIPILO ERYTHROPHTHALAMUS (California) R-478

GEOSPIZA DIFFICILIS (Isla Darwin) R-148/7

PARIDAE (Titmice)

CAMARHYNCHUS PARVULUS (Isla Santa Cruz) R-149/8*

PARUS INORNATUS (California) Apr.3/71

ICTERIDAE (New World Blackbirds)

GEOSPIZA CONIROSTRIS (Isla Española) R-422/5

AGELAIUS PHOENICEUS (California) R-478

the finches, with major features of their adaptive radiation having arisen in their present oceanic island settings. The finches occur abundantly in most of the insular habitats, many of which are distinctive both floristically and vegetatively (Figs. 4–18 [not reprinted]). Because conditions on many of the islands are virtually pristine, correlations between the prevailing vegetation and local song dialects can be made directly, with little or no concern for the effects of past ecological disturbances by man or his domesticates. Sound pollution resulting from human activities is minimal. All finches sing at least two types of song, with a few singing three or more, all of which show some degree of geographic variation (Figs. 42–47 [not reprinted]). Collectively, the songs of Darwin's finches display structural convergences with those of ecological equivalents on continents (Figs. 2, 39–41 [Fig. 41 not reprinted]). Interisland differences in the composition of geospizine faunas (Fig. 3, Table 2 [not reprinted]) provide natural experiments in coexistence; the effects of these differences on song diversity can be readily evaluated. Except possibly *Certhidea olivacea* and *Cactospiza heliobates*, which have not been studied in captivity, all species adapt quickly to aviary conditions and thrive on simple diets consisting of millet seed, fresh greens, and occasional live insects. The birds are long-lived and may remain reproductively active for as long as 8 years in the wild (pers. comm., P. R. Grant) and over 15 years in captivity (Fig. 176 [not reprinted]). But surely one of the most important features that make the finches such a treasured resource is their extraordinary tameness toward humans (Darwin 1845), which facilitates the recording of their songs and of other aspects of their natural history. Such unbashful behavior should command our highest respect and our never-failing concern for its preservation.

* Pages 247–248

GENERAL DESCRIPTION OF GEOSPIZINE SONG

Subjective Descriptions

Before the widespread use of tape recorders and sound spectrographs in vocalization studies, definitive adult songs of Darwin's finches were variously described as follows: "Of primitive pattern, unmusical, and with no complex phrases" (Lack 1945); by no means do they "rank with those of even ordinary singing birds, and indeed anywhere else would scarcely pass for songs. One never hears from the Geospizas such songs as are uttered by the song-sparrow or house-finches" (Snodgrass and Heller 1904); their utterances, like those of all the finches, were monotonous, a series of double syllables, two notes apart (Beebe 1924). Whereas the sounds of most songbirds are generally narrow-band and tonal (Marler 1977), the songs of most of Darwin's finches are "drab," i.e. punctuated with repetitious "buzzy" wide-band unmusical syllables. Some populations are as variable in their vocalizations as they are in their bill dimensions, e.g. *Geospiza fortis* on Isla Santa Cruz (Fig. 31 [not reprinted]). As Harris (1974) has pointed out, intrapopulational variation in song may be as great as, if not greater than, interpopulational variation (cf. Figs. 31 and 32). Seemingly rampant variability in the songs of some species is accompanied by pronounced interspecific parallel resemblances within the geospizine tribe (Figs. 33–38 [Fig. 38 not reprinted]). At the same time, vocal divergences associated with various foraging niches would seem to have led to vocal convergences with ecologically equivalent species of songbirds on continents (Figs. 2, 39–41 [Fig. 41 not reprinted]). Despite the relatively colorless song character overall, some geospizine populations have given rise to what, in human terms, may be described as melodious rhythmical productions (Figs. 52G, H; 82A, B [not reprinted]).

Song Classification

Each geospizine song may be assigned to one of the following three categories: "whistle," "basic," or "derived" (see Fig. 42 [not reprinted]). Whereas basic and derived songs are used in territorial advertisement, whistle songs are not. Additional details on song structure and function are given in subsequent sections.

Whistle song (Fig. 42A). There are two forms of this song type: a long continuous hissing note (e.g. "hisssssssssssssssss") or a series of short hiss-like notes (e.g. "see-see-see-see-see-see"), both starting at a very high frequency (often inaudible to humans), descending gradually or rather precipitously, and ending at a lower (more audible) frequency.

Basic song (Fig. 42B). There are three forms of this song type: basic song proper, special basic song, and abbreviate basic song. The "basic song proper" (hereafter designated simply as "basic song") can be described syllabically by the city name "Chicago," with emphasis on the drawn-out middle, buzzy, "a" syllable (e.g. "chic-a.a.a.a.a.a.-go"). There is also a tremolant form of the basic song characteristic of *Geospiza magnirostris* on Isla Santa Cruz and other islands (e.g. "too-chee-oo-oo-oo-oo-oo-oo-oo-oo"). The "special basic song" consists of a drawn-out rasping syllable (e.g. "bizzzzzzzzzzzzzzzz"), as in the first syllable of the word "business." It is a rather high-pitched "growl." The "abbreviate basic song" is, in effect, a contracted form of the "basic song" in which the extended middle buzzy syllable has been much shortened (e.g. "ree-search, ree-search").

Derived song (Fig. 42C). There are essentially two forms of this song type: the polysyllabic form (e.g. "tee-you, tee-you"; "chee-tee, chee-tee") and the repetitive monosyllabic form (e.g. "churr-churr-churr-churr-churr-churr").

* Pages 265–266

FUNCTION AND ADAPTIVE MORPHOLOGY OF SONGS
OF DARWIN'S FINCHES

Adaptive Significance of Whistle Song

To broadcast a pair-bonding and nest-invitation signal beyond those individuals of immediate concern (i.e. the mate) is to provide an unnecessary and dangerous homing beacon for predators (Wilson 1975). Those features of sound that make spatial localization difficult to achieve for humans, and presumably also for Darwin's finches, have been described by Marler (1955, 1961), and are as follows: (1) Tonal purity combined with imperceptible fade-in and fade-out of a sustained stable frequency sound, thereby providing no abrupt discontinuities in the signal that would allow the differences in the time of arrival of the sound at the near and far ear to be compared. (Green [1976] is of the opinion that timing information is not present in vertebrate nervous systems at very high frequencies; cf. Busnel 1963). (2) High frequencies whose wavelengths are somewhat larger than the width of the head or the intertympanic membrane distance, thus making the signal somewhat too low-pitched for ideal detection of intensity differences, the latter best discerned when frequencies are smaller than the diameter of the head that serves to absorb or block the sound on its course to the far ear. The use of moderately high frequencies also makes ineffective the detection of phase differences between near and far ears—a process otherwise operative only at low frequencies. However, Knudsen (1980) claims that there are no behavioral data suggesting that birds do, in fact, exploit interaural phase differences for sound localization, which is most effective at low frequencies, according to Marler (1955, 1961). The use of high frequencies also promotes reflections from vegetation, resulting in a somewhat disorganized signal with high loss of sound pressure during transmission.

Based on these criteria, whistle songs of Darwin's finches appear to be ideally structured to hinder their auditory localization not only by the finches themselves, but also by sympatric passerine species, notably the mockingbirds. Whistles are largely sustained pure tones that have subtle intensity-graded beginnings and endings. They are high-frequencied and therefore with wavelengths that are not only small but also about 2 to 5 times greater than the interaural distances (Cocos finch excepted), which, presumably, makes them too high-pitched (i.e. wave-length too short) for effective detection of phase difference, yet too low-pitched (i.e. wavelengths too long) for effective detection of intensity difference (see Table 15 [not reprinted]). High frequencies, characteristic of the whistle songs, probably also promote rapid environmental attenuation of their small wavelengths with increasing distance from the sound source (see cumulative dB-loss isopleths in Fig. 29 [not reprinted]; cf. King et al. 1981). Even the *Asio* owl and the *Buteo* hawk possess interaural dimensions that probably forestall use of intensity differences at the two eardrums for signal localization. Skull diameter percentages (Table 15, Fig. 108 [not reprinted]) for these raptors do not substantiate this conclusion, but this parameter is probably not a true measure of the effective "frequency blocking" distance between left and right tympani.

* Pages 267–268

The critical function of whistle songs seems to be pair-bonding, and nest-invitation, and in order not to unintentionally facilitate the visual clues that could reveal the location of the singer or its mate at critical times (e.g. nest construction, incubation), vocalizations should provide some degree of "acoustical camouflage." Lack (1945) has correctly stated that nest-building in the Geospizini is closely linked with display, as occurs in some other birds that have no enemies at the nest. But are there really no sources of danger for the finches, no nest predators?

It is now known that a number of species occasionally focus their depredatory activities at geospizine nests, including the centipede (*Scolopendra*), colubrid snake (*Dromicus*), short-eared owl (*Asio*), and mockingbird (*Mimus* [*Nesomimus*]) (Gifford 1919; Bowman 1961; and personal observations). From direct and circumstantial evidence, the abundant and ubiquitous mockingbirds appear to be one of the most widespread predators on geospizine eggs and nestlings (Bowman and Carter 1971; Downhower 1978; Grant and Grant 1979, 1980), although the short-eared owl may be a serious threat to nestlings only on some islands (e.g. Genovesa; Grant and Grant 1980) and not on others (e.g. Wolf; Curio 1965). This owl is unreported from, or very scarce on, several islands (Swarth 1931). Mockingbirds have been observed by the writer to inspect the interior of newly constructed finch nests on islas Genovesa (*G. magnirostris*), Santa Cruz (*G. fortis*), and Española (*G. conirostris*) (cf. Grant and Grant 1980), and although no finch eggs were seen to be looted, the nest owners vigorously attacked the marauders, even riding on their backs for several seconds of flight! Curio (1969) remarks that mockingbirds "viciously" hunt after finches. . . .

On the basis of Marler's postulates (1955, 1961), Cutler's syringeal findings (1970), Konishi's auditory sensitivity studies on barn owls (1973), field observations of various workers, and lastly, inferences to be drawn from measurements of the interaural distances of the skulls in finches, mockingbirds, and owls (Fig. 108; Table 15), one may tentatively conclude the following: (1) *Asio* and *Buteo* do not hear the higher ranges of the whistle songs, or if they do, are incapable of locating their source binaurally. (2) Galapagos mockingbirds may or may not be capable of hearing the whistle songs, but if they do, like the finches which produce them, they probably would be incapable of locating their source binaurally. (3) Darwin's finches probably hear their own whistle songs within the limits of their territories, but most likely are incapable of spatial localization of their sources by auditory means alone. In other words, the public nature of sound communication has been circumvented and privacy achieved by the use of very high-frequency sound. Because of dispersion by vegetation, such small wavelengths are most vulnerable to amplitude reduction, and therefore would tend to reduce the probability of their reaching the ears of predators over moderate distances from the nest site. . . .

Adaptive Significance of Basic, Abbreviate Basic, and Derived Songs

* Pages 291–294

The Niche-Variation Hypothesis of Van Valen (1965)

This proposition states that the extent of morphological variation in a given population results in part from the size of that organism's ecological niche. Although the theory was formulated with structural characters in mind, such as bird bills and trophic appendages (cf. Grant et al. 1976; Grant and Price 1981), and the validity of the theory has been seriously questioned, most recently by Beever (1979), it may similarly be applied to the structure of signals (i.e. frequency bandwidth of advertising song) and with width of "sound niches" (i.e. "acoustical apertures"; see Hurlbert 1981, for a review of the "niche concept"). When formulated in this way, the niche-variation hypothesis might state that the variability of song bandwidths is directly correlated with the width of the available and "appropriate" acoustical aperture, i.e. "appropriate" with respect to such variables as modal frequency of song and territory size of the species. Thus, species populations using wide acoustical apertures available in the environment should show greater variability in frequency bandwidth of song than populations utilizing narrow apertures. In all cases the width of the acoustical aperture is a function of sound-transmission characteristics of the vegetation composing the species' niche.

Thus, according to the "adaptive variation hypothesis" of Van Valen (1965), there would be no single optimal song phenotype, but rather an array thereof whose individual population variants would have more or less equivalent fitness through differential use of appropriate segments of the environment (i.e. acoustical aperture). In the case of Darwin's finches, this would result in a highly polymorphic song population.

The validity of this concept may be tested in part by examining the correlation between diversity of vegetational and acoustical habitats of species and the widths of acoustical apertures and song bandwidths.

In the three examples presented in Table 29, there is a clear and positive relationship between these four variables, and, not surprisingly, even with bill variability (cf. Grant et al. 1976). Wherever habitat diversity is greatest and foraging niche broadest, variability of bill structure, breadth of song bandwidth, acoustical diversity of habitat, and width of acoustical aperture are also broadest. For example, if we compare *Geospiza conirostris* populations on islas Española and Genovesa, we find that on Española the bill is substantially more variable in length, the habitat (as measured by plant species diversity) is more diverse, the variety of vegetation patches utilized is greater, the breadth of the song bandwidth is greater, the acoustical diversity of the habitat is greater, and the breadth of the acoustical window is greater.

Songbirds of low-latitude oceanic islands tend to be smaller in body size and relatively more abundant in a given area than songbirds with similar ecologies on mainlands (Grant 1965, 1968). On the basis of population size alone, insular species might be expected to present a broader spectrum of vocal variation in a given area simply because of the increased need for individual recognition. Thus, among species of Darwin's finches that are obviously the most abundant on their respective islands (pers. obs.), e.g. *Geospiza fortis* (Isla Santa Cruz), *Geospiza conirostris* (Isla Española), and *Geospiza difficilis* (Isla Wolf), we should find some of the greatest individual variations in song, and this we do (cf. Figs. 31, 55–59, 61B, 62A, respectively [not reprinted]). These same populations exhibit the greatest extremes for their species in bill length (Bowman 1961, Table 61). Furthermore, the sympatric distribution of sibling species with their

TABLE 17a. Correlation of whistle and special basic songs of *Geospiza difficilis* with vegetation characteristics on five islands of the Galapagos archipelago

Island[1]	Whistle song separate from basic song	Typical basic song present	Region 5 of special basic song		Vegetation of nesting habitat	
			Whistle-like	"Normal"	Type	Density
Genovesa	yes	yes	special basic song absent		xeric	open
Darwin	yes	yes	special basic song absent		xeric to semi-xeric	semi-open
Wolf	whistle song absent	yes, but rarely	yes	no	xeric to semi-xeric	semi-open to dense
Santiago	whistle song absent	yes	yes	no	semi-xeric to mesic	dense to very dense
Pinta	whistle song absent	yes	no	yes	semi-xeric to mesic	dense to very dense

[1]Islands are arranged from top to bottom in order of decreasing xeric condition and increasing density of vegetation of nesting habitat.

TABLE 29. Correlation of bill variability with song bandwidth and habitat diversity and utilization in *Geospiza conirostris*, *Camarynchus parvulus*, and *Certhidea olivacea*

Species	Island	Bill length variability[1]		Habitat diversity[2]	Habitat utilization	Song bandwidth[3] (kHz)	Acoustical diversity[4]	Acoustical aperture[5]
		No.	CV					
Geospiza conirostris	Española	87	6.08	79	All areas of patchy arid zone	2.5	Medium	Wide
	Genovesa	43	5.38	22	Only in dense *Opuntia* patches[6]	1.0	Low	Narrow
Camarynchus parvulus (Isla Santa Cruz)	Coastal zone	—	—	Species-poor,[7] patchy distribution	All areas	4.0	High	Wide
	Scalesia forest zone	—	—	Species-rich, uniform distribution	All areas	1.5	Low	Narrow
Certhidea olivacea	Genovesa	41	2.5	22	Patchy areas of arid zone	4.0	High	Wide
	Santa Cruz	44	4.5	193	Lower transition through moist zones	3.25	Medium	Wide
	Española	44		79	All zones of arid zone	2.75	Medium	Wide

[1]Coefficients of variation from Bowman 1961, Table 61.

[2]The numbers of species of vascular plants are from Stewart (1911). Comparable totals based on Wiggins and Porter (1971) have not been compiled, but relative differences in numbers should not differ significantly.

[3]Data on *Geospiza conirostris* from Fig. 131, on *Camarynchus parvulus* from Fig. 143, on *Certhidea olivacea* from Fig. 139.

[4]Acoustical diversity is assumed to be high when the vegetation is patchy, strongly stratified, or distributed in altitudinal zones, and low when the vegetation is uniformly distributed, essentially unstratified, and contained within one vegetational zone.

[5]See sound-transmission isopleths, Figs. 133–134. The width of the acoustical aperture is determined on the basis of known territory size (Isla Genovesa) but estimated (for islas Española and Santa Cruz) on the basis of relative size of the bird (see Fig. 140 and Table 21).

[6]*Fide* Grant and Grant (1979).

[7]Reeder and Riechert (1975).

TABLE 30. Pragmatic analysis of some factors affecting the design of geospizine vocalizations

Feature	Physical effect on	Consequence on song
A. BODY SIZE	1. Diameter of syringeal vibratory membranes	Frequency of modal amplitude
Large	Large	Lower
Small	Small	Higher
	2. Territory and population size	Amplitude
Large	Large territory and small population size	Higher dB level
Small	Small territory and large population size	Lower dB Level
B. VEGETATION DENSITY	1. Visibility	Temporal and tonal characteristics
High	Poor	FM signal with high repetition rate of sharp transients; of relatively long duration; buzzy quality.
Low	Fair to good	Less note repetition and relatively more pure-tone whistle-like notes of variable form; relatively short song duration.
	2. Sound transmission	Frequency and modal amplitude
High	Selective attenuation at higher frequencies	Relatively less sound energy at higher frequencies and more concentrated about modal frequency.
Low	Attenuation less discriminatory of higher frequencies	Relatively more sound energy in higher frequencies; polymodal frequency distribution of energy may appear.
C. ATMOSPHERIC CONDITIONS	Frequency-dependent energy spectrum	Amount of modulation and modal frequency of signal
Wind and temperature		
1. Forest habitat (homogeneous)	Gradients less marked and slow to develop	Both AM and FM signal structures are effective.
2. Grassland and brush habitat (heterogeneous)	Wind and temperature gradients may be more marked, causing irregular transmission losses and degradation of features (via refraction); otherwise useful in binaural localization of signal source.	FM signal provides high repetition rate of sharp transients; AM avoided because of its greater susceptibility to distortion.

Feature	Physical effect on	Consequence on song
Relative humidity	Higher frequencies are attenuated less in moist air than in dry, but at any given humidity, lower frequencies are attenuated less than higher. With increasing distance from source, rate of sound attenuation for a given frequency increases more slowly in moist air than in dry.	Song energies are concentrated in lower frequencies.
D. TRANSMISSION DISTANCE	*Differential frequency-dependent attenuation*	*Signal band-width and frequency of modal amplitude*
Long-distance	Large	Energy concentrated in low frequency, narrow band-width, FM signal, or pure tone.
Short-distance	Small	Broad frequency spread of song in FM signal. Energy spectrum may be polymodal.
E. ELEVATION OF SINGING PERCH	*Broadcast area*	*Time of singing and modal frequency of song*
Low	Relatively small	For small species with relatively small territories, there is no particular advantage in singing in early a.m. over other times of day, or in emphasizing a particular frequency. Ditto for large species with relatively large territories. There may be no advantage in singing in early a.m. over other times of day, but of some advantage to emphasize lower frequencies.
High	Relatively large	For small and large species with relatively large territories, there is an advantage in singing in early morning over other times of day, and in using relatively low frequencies.

Editor's note: The last column, referencing figures not reprinted here, has been omitted.

sibling-like songs (e.g. Fig. 36 [not reprinted]) provides additional evolutionary "justification" for individuality in vocal signals. Possibly the widespread occurrence of basic and derived song patterns in their various forms (cf. Figs. 89–95, 99–103 [not reprinted]) can be partly attributed to this "need."

Finally, by having songs of various length and of slightly different modal frequency (the latter largely a consequence of body size differences), we may escape the problem of masking and psychological delay, as described by Brémond (1973), who states that a very "noisy" environment arising from a large number of conspecific and heterospecific individuals hinders communication to a degree that depends mainly on the continuous character of the noise. For species of Darwin's finches, there appears to be a margin of safety, derived from frequency and quality discontinuities within songs, sufficiently large to permit the birds to distinguish successfully between each other's vocal signals in all situations, despite human difficulties in doing likewise.

Song Convergence

Mention has already been made of the remarkable similarity in song structure between various groups of Darwin's finches on the Galapagos Islands and their ecological counterparts on continental areas (see Figs. 2, 39–41 [Fig. 41 not reprinted]). For example, "warbler-finches" (*Certhidea olivacea*) sing like New World warblers (*family Parulidae*); "blackbird-finches" (*Geospiza conirostris* and *G. scandens*) sing like New World blackbirds (family Icteridae); "tit-finches" (species of *Camarhynchus*) sing like New and Old

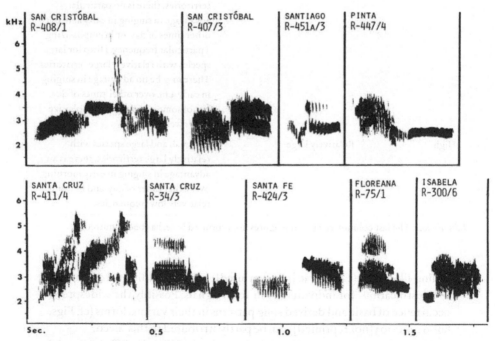

Fig. 32. Examples of population variation in the basic and derived songs of *Geospiza fortis.* Only the repetitive portion of each song is displayed.

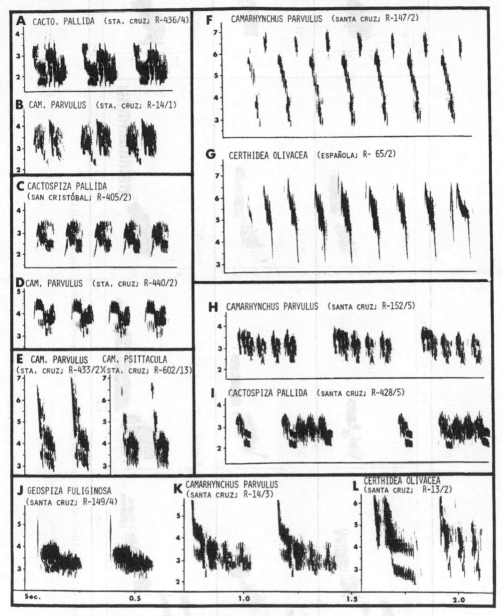

Fig. 33. Parallellism in the songs of Darwin's finches.

World titmice (family Paridae); "woodpecker-finches" (species of *Cactospiza*) sing like woodpeckers (family Picidae) and other trunk-foraging species (families Sittidae and Corvidae).

In song convergence, equivalent acoustical selective forces act in sympatry or allopatry upon species of differing genetical potential, so that analogous

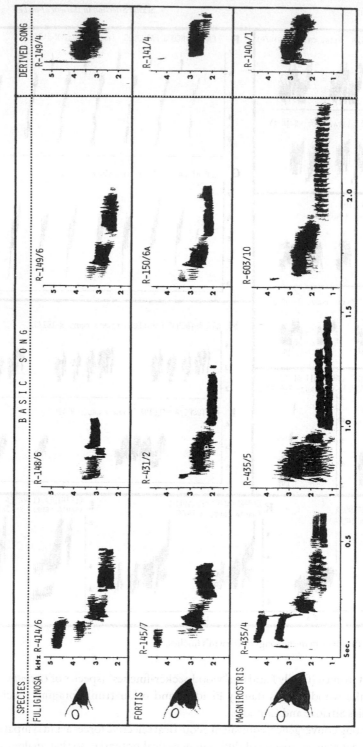

Fig. 36. Parallel structure in basic and derived songs of three sympatric songs of *Geospiza*, Academy Bay, Isla Santa Cruz.

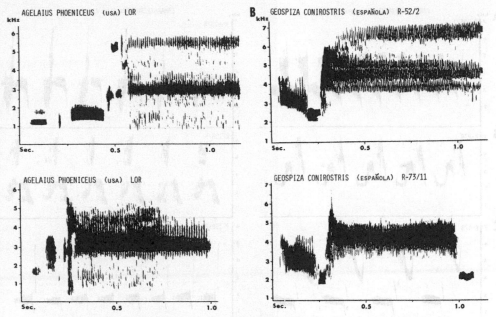

Fig. 39. Song convergence between *Agelaius phoeniceus* (USA) and *Geospiza conirostris* (Isla Española, Galapagos).

but not necessarily homologous song behaviors are produced. Convergence in bird songs between very different geographical areas containing genetically different bird faunas suggests that selection has reached optimal solutions in both areas, despite differences in histories, time scale, and genetic origins (Cody 1974b). . . . Thus it appears that the niches of avian species showing convergences in their advertising songs offer not only foraging niches with resources of similar kind and location, but probably also acoustical niches with similar sound-transmission characteristics. There is, of course, an obvious need to demonstrate acoustical similarity between Galapagos and continental niches, as well as between various continental niches where song convergences are known to occur (cf. Crowder 1980; Carr and James 1975; see also Figs. 2, 145, 147–150 [Figs. 149, 150 not reprinted]).

Origin and Function of Song Dialects in Darwin's Finches

Differentiation of populations is most likely to occur where environmental factors vary greatly in space, but are relatively stable in time (Levins 1962, 1973; Maynard Smith 1966). As shown in an earlier section of this report, more or less consistent structural differences in certain song populations (dialect) of Darwin's finches are closely correlated with regional differences in their

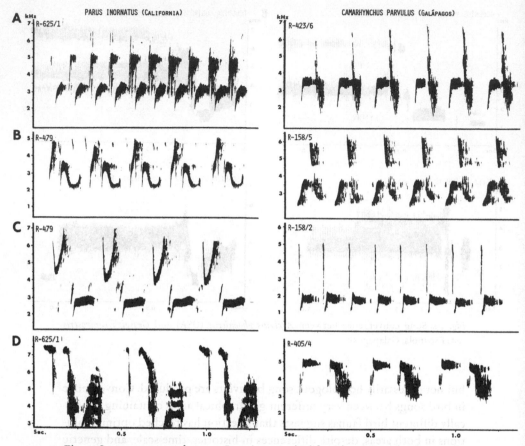

PARUS INORNATUS (CALIFORNIA)

CAMARHYNCHUS PARVULUS (GALÁPAGOS)

A R-625/1

R-423/6

B R-479

R-158/5

C R-479

R-158/2

D R-625/1

R-405/4

Fig. 40. Song convergence between *Parus inornatus* (California) and *Camarhynchus parvulus* (Galápagos).

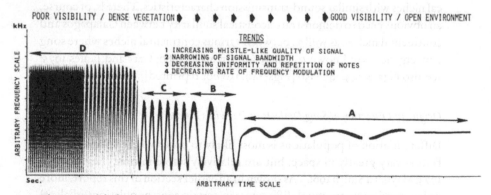

POOR VISIBILITY / DENSE VEGETATION ◆ ◆ ◆ ◆ ◆ ◆ ◆ ◆ GOOD VISIBILITY / OPEN ENVIRONMENT

TRENDS
1 INCREASING WHISTLE-LIKE QUALITY OF SIGNAL
2 NARROWING OF SIGNAL BANDWIDTH
3 DECREASING UNIFORMITY AND REPETITION OF NOTES
4 DECREASING RATE OF FREQUENCY MODULATION

D

C   B

A

Fig. 144. Model of song structures to show correlation between amount of frequency modulation and environmental conditions. Regions A–D should be compared with examples of bird song in Fig. 145.

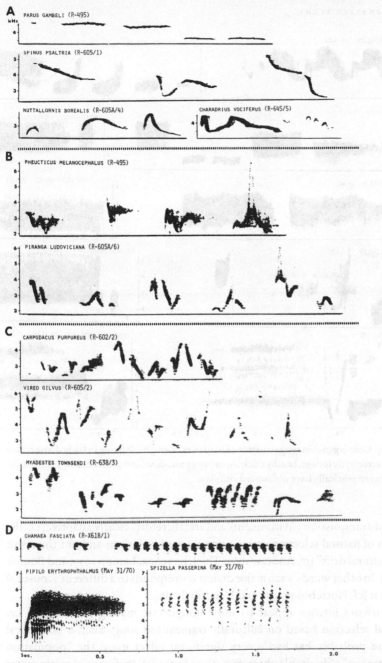

Fig. 145. Convergent vocalizations of North American birds that occupy similar acoustical environments. A. Whistle-like songs of species living in fairly open vegetation where visibility is good. B. Moderately undulating whistle-like songs of species living partly within or beneath the leaf canopy of mixed sclerophyll-broadleaf trees. C. Strongly undulating whistle-like songs of species living in moderately dense stands of broadleaf or coniferous trees. D. Staccato-like or buzzy songs of species living in dense vegetation where visibility is poor. Compare songs with model FM signals of same group letters (A–D) in Fig. 144. Narrow-band display.

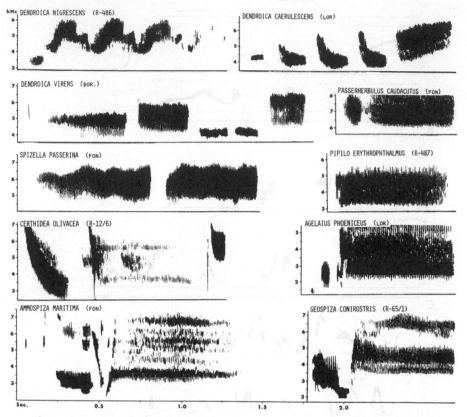

Fig. 147. Convergence in song structure of 10 species of songbirds living in fairly dense coarse grass, xerophytic foliage, brushy thickets, or boggy meadows and cattail swamps. Such habitats characteristically have obfuscated visibility.

sound-transmission environments and are, therefore, clearly adaptive, i.e. the result of natural selection acting upon learned variations and not the result of "cultural drift" (cf. Andrew 1962; Thielcke 1969a; Lemon 1975; and Bonner 1980). In other words, each major dialect corresponds to a different acoustical habitat (cf. Nottebohm 1975 and Handford 1981).

Darwin's finches appear to show nongenetic mating preferences, i.e. sexual selection based on culturally transmitted song. Such a behavioral system probably has had a very significant effect upon the "population mechanism" of heritable characters (Cushing 1941). Preferential mating rests on a complex basis involving the interaction of many factors. In the finches, sexual preference based on early song conditioning becomes fixed (imprinted) considerably in advance of sexual maturity, and thereafter remains immutable.

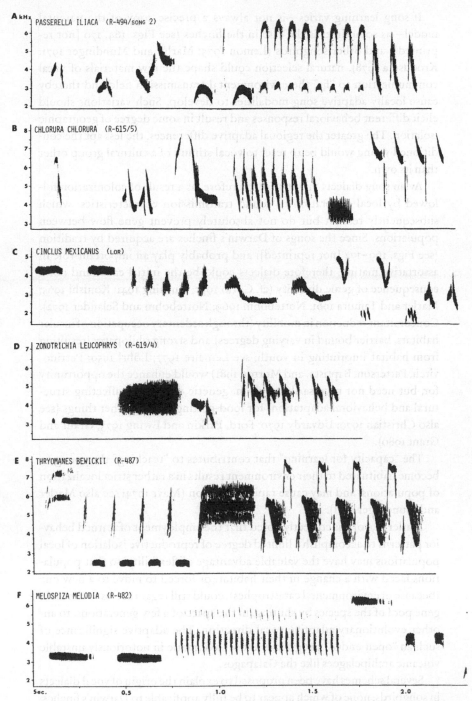

Fig. 148. Examples of the songs of six species of "edge" inhabiting birds.

If song learning varies—is not always a precise copy of the parental model—as sometimes is the case in the finches (see Figs. 164, 170 [not reprinted]) and other songbirds (Lemon 1975; Marler and Mundinger 1971; Kroodsma 1978), natural selection could shape the raw materials of vocal communication to "fit" the environmental transmission field and thereby cause locally adaptive song modalities to develop. Such variations should elicit different behavioral responses and result in some degree of geographic isolation. The greater the regional adaptive differences, the less apt the conditioned young would be to react to vocal stimuli of a cultural group other than its own.

Avian song dialects might arise, therefore, as a result of colonization followed by local adaptation to sound transmission characteristics, which subsequently restrict but do not absolutely prevent gene flow between populations. Since the songs of Darwin's finches are acquired by tradition (see Figs. 169–172 [not reprinted]) and probably play an important role in assortative mating, therefore dialects could be the initial cause and not a consequence of genic diversity (cf. Curio 1977; Cushing 1941; Konishi 1965; Marler and Tamura 1961; Nottebohm 1969; Nottebohm and Selander 1972). Concomitant reduction in mobility (through sedentary occupation of insular habitats, barrier bound in varying degrees) and strong philopatry (resulting from habitat imprinting in youth; see Lemaire 1977; Löhrl 1959; Petrinovitch, Patterson, Baptista, and Morton 1981) would enhance the opportunity for, but need not necessarily result in, genetic divergence affecting structural and behavioral adaptations for food-getting, among other things (see also Christian 1970; Udvardy 1970; Ford, Parkin and Ewing 1973; Grant and Grant 1980).

The "capacity for learning" that contributes to "teaching" the young to become habituated to their environment results in a rather strict localization of populations, and may cause rapid speciation (Mayr 1974; see also Marler and Mundinger 1971).

Marler (1970) has correctly noted that the employment of learned behavior patterns to accomplish a limited degree of reproductive isolation of local populations may have the valuable advantage of flexibility, so that populations faced with a change in their habitat, or forced to move to a new one (because of environmental catastrophes), could still regain access to the main gene pool of the species by shifting, in the space of a few generations, to another evolutionary adaptive vocal direction. The adaptive significance of such an "open-ended" program is almost axiomatic in notoriously unstable volcanic archipelagoes like the Galapagos.

Several schemes have been proposed to explain the origin of vocal dialects in songbirds, none of which appear to be fully applicable to Darwin's finches. For example, Lemon (1975) assumes that vocal dialects are ecologically nonadaptive or neutral vocal variants resulting from accidents of improper copying of male parental song during youth. Although "miscopying" does occur in geospizines, natural selection probably preserves only those variants that best fit local and transmission characteristics of the vegetation, and quickly rejects (through failure to obtain a conspecific mate) those that are acoustically nonadaptive. In other words, there is a disproportionate disappearance from the population of lineages less vocally adapted. Minor innovations are "tolerated" if they occur within the limits of the acoustical aperture and do not seriously alter species- and population-specific attributes of song.

From *Patterns of Evolution in Galápagos Organisms*, edited by R. I. Bowman, M. Berson, and A. E. Levinton (San Francisco: Pacific Division of the American Association for the Advancement of Science, 1983), 237–537. Excerpts as indicated by page numbers, plus tables 17a, 29, and 30 and figures 2, 32, 33, 36, 39, 40, 144, 145, 147, and 148.

Correlated Evolution of Morphology and Vocal Signal Structure in Darwin's Finches (2001)

✳ JEFFREY PODOS

Editor's note: Figure 2 has been modified for black-and-white reproduction.

Abstract—Speciation in many animal taxa is catalysed by the evolutionary diversification of mating signals.[1] According to classical theories of speciation, mating signals diversify, in part, as an incidental byproduct of adaptation by natural selection to divergent ecologies,[2,3] although empirical evidence in support of this hypothesis has been limited.[4–6] Here I show, in Darwin's finches of the Galápagos Islands, that diversification of beak morphology and body size has shaped patterns of vocal signal evolution, such that birds with large beaks and body sizes have evolved songs with comparatively low rates of syllable repetition and narrow frequency bandwidths. The converse is true for small birds. Patterns of correlated evolution among morphology and song are consistent with the hypothesis that beak morphology constrains vocal evolution, with different beak morphologies differentially limiting a bird's ability to modulate vocal tract configurations during song production. These data illustrate how morphological adaptation may drive signal evolution and reproductive isolation, and furthermore identify a possible cause for rapid speciation in Darwin's finches.

The avian vocal tract, comprised of the trachea, larynx and beak, acts as an acoustic resonance filter during sound production, attenuating harmonic overtones and emphasizing fundamental frequencies produced by the sound source.[7,8] The vocal tract thus enables birds to produce songs that have a high tonal purity or a "whistle-like" quality. While singing, songbirds (oscine Passeriformes) actively modify the configuration of their vocal tracts, in a manner closely coordinated with the sound source, so as to maintain the vocal tract's filtering function over a wide range of source frequencies.[7] Studies have shown that vocal tract reconfigurations are achieved largely through rapid changes in beak gape, with increases in beak gape accompanying increases in source frequencies, and vice versa.[9,10] Beak movements are normally very rapid and precise, and when birds' beaks are either temporarily immobilized or hampered, the tonal purity of songs becomes compromised.[11] These studies together suggest that limits on the dynamics of vocal tract movements during song production may shape patterns of vocal signal evolution.[12,13] For example, constraints on vocal tract dynamics in emberizid songbirds probably cause trade-offs between temporal and frequency-based song features,[13] and may set limits on the development and evolution of syllable repetition rate.[14]

Vocal performance capacities are predicted to vary as a function of vocal tract morphology, and particularly beak morphology.[12,13] Songbirds with comparatively large and strong beaks, such as those adapted for crushing hard seeds, should face relatively severe performance constraints on vocal tract dynamics. This is because of an intrinsic trade-off in jaw biomechanics between maximal force and velocity; as jaws become adapted for strength, they will be less able to perform the rapid movements required for the production of certain types of songs. By contrast, songbirds that have evolved smaller beaks, such as those adapted to probe for insects, should suffer less severe constraints on vocal performance. The objective of this study was to assess, in Darwin's finches, the extent to which adaptive diversification of the beak has shaped the evolution of song features related to vocal performance capacities. Darwin's finches are a particularly appropriate songbird group for this study not only because they express extensive diversity in beak morphology[15] and song[16] (Fig. 1), but also because beak morphology in this group has been shown to adapt precisely and rapidly, through natural selection, to ecological conditions such as food type availability and interspecific competition.[17-19]

I conducted correlation analyses, at intra- and interspecific scales, between morphological features and a composite measure of the temporal and frequency structure of songs, the "vocal deviation" (Fig. 2). High vocal deviations are indicative of poor vocal performance, and vice versa (see Methods). Intraspecific analyses were restricted to the medium ground finch, *Geospiza fortis*, for which the largest sample was obtained. *G. fortis* on Santa Cruz Island exhibits unusually high variation in morphological features relative to other Darwin's finches,[19] and produces a wide diversity of song types.[16] The vocal deviation correlated positively and significantly with all beak measurements and with body mass (Fig. 3; $P < 0.005$ for each correlation), but not with tarsus length ($P = 0.573$) or wing chord length ($P = 0.209$). For interspecific analyses, I calculated correlations between average values of morphological

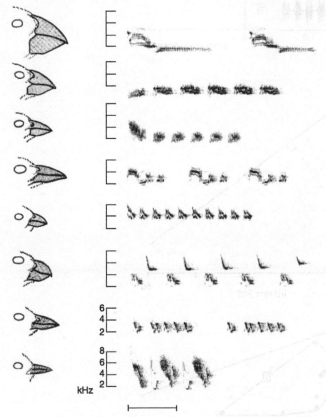

Fig. 1. Beak morphology (sketches reprinted[15]) and representative sound spectrograms of songs from eight Darwin's finch species on Santa Cruz Island (from top to bottom: *G. magnirostris*, *G. fortis*, *G. fuliginosa*, *G. scandens*, *C. parvulus*, *C. psittacula*, *C. pallida*, *C. olivacea*). Interspecific variation is apparent in both morphology and song structure. Comparability of the songs of different species is supported by the young age of the clade,[19] and the striking uniformity among species in the structure of the syrinx and associated musculature.[22] See ref. 16 for a discussion of homology among Darwin's finch songs. Spectrogram frequency resolution, 98 Hz; scale bar, 0.5 s.

variables and minimal values of vocal deviations (Fig. 4a). These correlations were not tested for statistical significance, however, because of the likelihood of statistical non-independence among species samples.[20] Correlations were thus re-evaluated in a phylogenetic context (Fig. 4b), using independent contrasts analyses.[20,21] Vocal deviation contrasts correlated positively with contrasts of all beak measurements, wing chord length, and body mass (Fig. 4c; $P < 0.051$ for all variables and all phylogenies), but did not correlate with tarsus length contrasts ($P = 0.159–0.207$).

The intra- and interspecific analyses are consistent in illustrating that vocal deviations have evolved in correlated fashion with beak morphology, in the

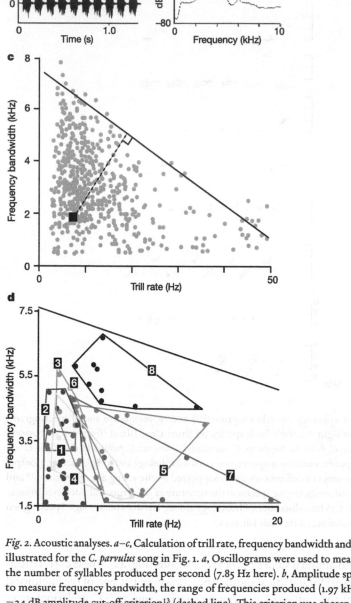

Fig. 2. Acoustic analyses. *a–c*, Calculation of trill rate, frequency bandwidth and vocal deviation, illustrated for the *C. parvulus* song in Fig. 1. *a*, Oscillograms were used to measure trill rate, as the number of syllables produced per second (7.85 Hz here). *b*, Amplitude spectra were used to measure frequency bandwidth, the range of frequencies produced (1.97 kHz here), using a −24 dB amplitude cut-off criterion[13] (dashed line). This criterion was chosen *a priori* to maximize the proportion of signal energy analysed while excluding background noise. Earlier analyses revealed that lower dB cut-off values (for example, −27 and −30 dB) regularly include background noise in spectral analyses, while higher cut-off values (for example, −21 and −18 dB) unnecessarily exclude signal energy. Amplitude spectra were computed at 32 kilopoints and smoothed to a frequency resolution of 300 Hz. *c*, For each trill type, average frequency bandwidth was plotted as a function of average trill rate (for example, filled square). Vocal deviations (dashed line) were calculated relative to an upper-bound regression (solid line),[13] which was in turn calculated relative to the distribution of trill types across 34 emberizid songbird species (grey dots).[13] *d*, Plot of all trill types for all species analysed, with minimum area convex polygons shown. Overlap in raw data among most species is apparent.

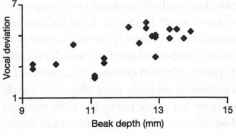

Fig. 3. Vocal deviation as a function of beak depth in *G. fortis*. Pearson product moment correlations: beak length, 0.763; beak depth, 0.733; beak width, 0.727; tarsus length, 0.481; wing chord length, 0.549; body mass, 0.792.

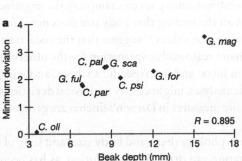

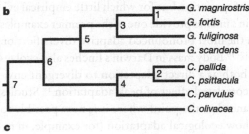

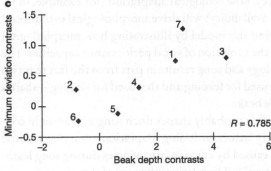

Fig. 4. Interspecific analyses. *a*, Minimum vocal deviation as a function of beak depth across eight species of Darwin's finch. *b*, One of four phylogenetic hypotheses used in independent contrasts analyses. Additional hypotheses differ in classifying *G. fortis* as sister taxon to *G. fuliginosa* and/or *C. psittacula* as sister taxon to *C. parvulus*. *c*, Minimum vocal deviation contrasts as a function of beak depth contrasts. Labels refer to phylogenetic nodes from Fig. 4b. Across the four phylogenies, ranges of Pearson product moment correlations were beak length, 0.757–0.809; beak depth, 0.751–0.784; beak width, 0.750–0.794; tarsus length, 0.535–0.588; wing chord length, 0.847–0.869; body mass, 0.835–0.847.

direction predicted by the vocal tract constraint hypothesis. Vocal deviations also correlated strongly and positively with body mass in both analyses, however, so the potential influence of body mass on vocal performance should be considered as well. Body mass varies positively and nearly isometrically with the size of the syrinx (the sound organ) in Darwin's finches,[22] and variation in syrinx size presumably influences the expression of fundamental frequency.[16,23] By contrast there are no clear predictions about how variation in body mass might shape the expression of vocal deviations. As an indirect test of the body mass hypothesis, I assessed the relationship between body mass and minimum vocal deviations across 30 emberizid species,[13] which express a range of body size variation comparable to Darwin's finches. Minimum vocal deviations were found to regress negatively on body mass (slope = -0.028, $r^2 = 0.038$), in the direction opposite to that predicted by the body mass hypothesis. Although the statistical significance of this regression could not be assessed (because of non-independence among species samples), the negative slope of the regression, along with the finding that body size does not influence emberizid upper-bound regression values,[13] suggest that the vocal tract constraint hypothesis offers a more reasonable explanation for the observed correlations. Body mass in both intra- and interspecific analyses, and wing chord length in the interspecific analyses, might covary with vocal deviations simply because beak and body size measures in Darwin's finches are generally highly correlated.[19]

The correlations between morphology (beak and body size) and song offer strong evidence that adaptation can drive signal evolution, as has been discussed in classical theories of speciation but for which little empirical evidence has been offered. Darwin's finches provide one of the premier examples of adaptation in nature, with the most pronounced adaptive diversification centering on beak evolution.[17–19] Body mass in Darwin's finches also evolves through adaptive processes, both as a direct adaptation to divergent environments and as an indirect correlated effect of beak adaptation.[19] Studies of finch adaptation have advanced to the point where it is now possible to predict, with high accuracy, how ecological adaptation (for example, in response to changing food availability) will drive morphological evolution.[24] The present findings extend this model by illustrating how morphological adaptation in turn drives the evolution of vocal performance capacities. The linkage between morphology and song results in part from the fact that two functional systems—that used for feeding and that used for singing—share a common morphology, the beak.

Morphological adaptation probably shapes finch song evolution in concert with other evolutionary factors, including adaptation to varying acoustic environments[16] and drift caused by copying inaccuracies during song learning (that is, cultural evolution[25]). The relative influence of these factors varies according to song feature, with, for example, copying inaccuracies exerting their greatest influence on the level of phonology[25] and with morphological constraints identified here exerting their greatest influence on higher-order timing and frequency features. A distinguishing feature of morphological influences on vocal evolution is that they appear to evolve in step with the primary locus of finch adaptation.

Because song presumably has a central role in reproductive isolation in

Darwin's finches,[19,26] the linkage between morphology and vocal performance capacities holds important implications for the dynamics of finch speciation. In island songbird clades, including the Darwin's finches, speciation is driven primarily, if not exclusively, by prezygotic isolating mechanisms, as evident from the regular occurrence of viable and fertile hybrids.[27] The effectiveness of prezygotic isolating mechanisms generally depends on the extent to which mating signals among incipient species are distinct, with more distinct signals increasing the probability of "correct" matings and thus enhancing probabilities of speciation.[6] My data suggest that magnitudes of ecological and morphological diversification among incipient Darwin's finch species will directly determine magnitudes of diversification in vocal features, and thus determine probabilities of speciation (to the extent that trill rate and frequency bandwidth are used by birds in mate recognition). Taking this hypothesis one step further, the high diversity of ecological opportunities for Darwin's finches on the Galápagos Islands may thus have promoted, through extensive morphological adaptation and correspondingly large evolutionary changes in vocal signal structure, conditions suitable for rapid speciation and the marked radiation that has defined the group.

METHODS

Sample

I collected morphological and acoustic data from nine species of Darwin's finch. Field work was conducted at coastal and upland sites on Santa Cruz Island during February and March 1999. Male birds, captured in mist nets, were banded with unique colour combinations, measured and released. Morphological measurements included: beak length, beak depth, beak width, tarsus length, wing chord length and body mass.[19] Songs of banded birds were recorded using Sennheiser K6/ME66 microphones and Sony TCD-5ProII tape recorders. The number of individuals banded and recorded were as follows: *Geospiza magnirostris*, $n = 4$; *Geospiza fortis*, $n = 18$; *Geospiza fuliginosa*, $n = 9$; *Geospiza scandens*, $n = 5$; *Platyspiza crassirostris*, $n = 2$; *Cactospiza pallida*, $n = 3$; *Camarhynchus psittacula*, $n = 2$; *Camarhynchus parvulus*, $n = 10$; *Certhidea olivacea*, $n = 7$. All song figures were generated and acoustic analyses conducted using SIGNAL version 3.1 sound analysis software (Engineering Design, Belmont, MA, 1999).

Acoustic Analyses

Eight species in the sample produce songs with trilled sequences (Fig. 1; I define a "trill" as a series of notes or note groups repeated in succession at a constant tempo). In emberizid songbirds, the family that includes the Darwin's finches, trilled sequences in particular are limited in their timing and frequency structure by constraints on vocal performance.[13] I thus focused on trilled sequences in Darwin's finch songs. For each recorded finch trill, I measured trill rate (rates of syllable repetition; Fig. 2a) and frequency bandwidth (ranges of frequencies produced; Fig. 2b). For each trill type, I then plotted average frequency bandwidth as a function of average trill rate (for example, Fig. 2c, filled square). Across the emberizids, trill rate by frequency bandwidth plots regularly yield triangular distributions at species, generic and family levels (Fig. 2c, grey dots illustrate the pattern at the family level[13]). These triangular distributions presumably occur because of motor trade-offs between rates and magnitudes of vocal tract reconfigurations during trill production.[13] Video analyses of vocal motor activity[9–11] provide evidence that these triangular distributions also reflect a gradient in vocal performance, with trill types that plot near the origin requiring low levels of vocal activity (for example, only minor vocal tract modulations) and with trill types closer to the upper edge of the distribution requiring higher levels of vocal activity (for example, more vigorous vocal tract modulations[13]).

The realized extreme for trill rate and frequency bandwidth production across the family can be characterized using an upper-bound regression (Fig 2c, solid line[13]). Minimum distances (Fig. 2c dashed line, referred to herein as "vocal deviations") between each Darwin's finch trill type and the family-wide upper-bound regression were used as a composite indicator of relative vocal performance, with higher deviation values reflecting lower levels of vocal activity, and vice versa. Use of the emberizid upper-bound regression for determining vocal deviations for Darwin's finches is supported by the fact that Darwin's finches are nested phylogenetically within the emberizids. For interspecific analyses, minimum vocal deviations provided an appropriate indicator of a species' vocal potential, because only a subset of vocal renditions are expected to challenge a species' vocal production capacities. Other possible summary values, including average vocal deviations, include submaximal events and thus less precisely reflect the vocal production capacities of a species.[13]

Phylogenetic Analyses

Independent contrast analyses used a punctuational or speciational assumption of evolutionary change, with all branch lengths set to unit length, as has been recommended for clades that have undergone adaptive radiations through the occupation of diverse niches.[5,28] Phylogenetic hypotheses were based on studies using molecular data and microsatellite DNA variation,[29,30] which have largely supported earlier hypotheses of branching relations among genera.[19] Within the constraints of available information four equally likely species-level phylogenies were identified (see Fig. 4b).

NOTES

1. West-Eberhard, M. J. Sexual selection, social competition, and speciation. *Quart. Rev. Biol. 58*, 155–183 (1983).

2. Dobzhansky, T. *Genetics and the Origin of Species* 3rd edn (Columbia Univ. Press, New York, 1951).

3. Mayr, E. *Animal Species and Evolution* (Harvard Univ. Press, Cambridge, MA, 1963).

4. Rice, W. R. & Hostert, E. E. Laboratory experiments on speciation: what have we learned in 40 years? *Evolution 47*, 1637–1653 (1993).

5. Schluter, D. & Nagel, L. Parallel speciation by natural selection. *Am. Nat. 146*, 292–301 (1995).

6. Rundle, H. D., Nagel, L., Boughman, J. W. & Schluter, D. Natural selection and parallel speciation in sympatric sticklebacks. *Science 287*, 306–308 (2000).

7. Nowicki, S. Vocal tract resonances in oscine bird sound production: evidence from birdsongs in a helium atmosphere. *Nature 325*, 53–55 (1987).

8. Fletcher, N. H. & Tarnopolsky, A. Acoustics of the avian vocal tract. *J. Acoust. Soc. Am. 105*, 35–49 (1999).

9. Westneat, M. W., Long, J. H. Jr., Hoese, W. & Nowicki, S. Kinematics of birdsong: functional correlation of cranial movements and acoustic features in sparrows. *J. Exp. Biol. 182*, 147–171 (1993).

10. Podos, J., Sherer, J., Peters, S. & Nowicki, S. Ontogeny of vocal tract movements during song production in the song sparrow. *Anim. Behav. 50*, 1287–1296 (1995).

11. Hoese, W. J., Podos, J., Boetticher, N. C. & Nowicki, S. Vocal tract function in birdsong production: experimental manipulation of beak movements. *J. Exp. Biol. 203*, 1845–1855 (2000).

12. Nowicki, S., Westneat, M. W. & Hoese, W. Birdsong: motor function and the evolution of communication. *Sem. Neurosci. 4*, 385–390 (1992).

13. Podos, J. A performance constraint on the evolution of trilled vocalizations in a songbird family (Passeriformes: Emberizidae). *Evolution 51*, 537–551 (1997).

14. Podos, J. Motor constraints on vocal development in a songbird. *Anim. Behav. 51*, 1061–1070 (1996).

15. Bowman, R. I. Morphological differentiation and adaptation in the Galápagos finches. *Univ. Calif. Publ.* Zool. *58*, 1–302 (1961).

16. Bowman, R. I. in *Patterns of Evolution in Galápagos Organisms* (eds Bowman, R. I., Berson, M. & Leviton, A. E.) 237–537 (American Association for the Advancement of Science, Pacific Division, San Francisco, 1983).

17. Schluter, D., Price, T. D. & Grant, P. R. Ecological character displacement in Darwin's finches. *Science 227*, 1056–1059 (1985).

18. Gibbs, H. L. & Grant, P. R. Oscillating selection in Darwin's finches. *Nature 327*, 511–513 (1987).

19. Grant, P. R. *Ecology and Evolution of Darwin's Finches* 2nd edn (Princeton Univ. Press, Princeton, 1999).

20. Felsenstein, J. Phylogenies and the comparative method. *Am. Nat. 125*, 1–25 (1985).

21. Martins, E. P. *COMPARE, version 4.2. Computer Programs for the Statistical Analysis of Comparative Data* (Univ. Oregon, Eugene, Oregon, 1999).

22. Cutler, B. *Anatomical Studies on the Syrinx of Darwin's Finches*. Thesis, San Francisco State Univ. (1970).

23. Ryan, M. J. & Brenowitz, E. A. The role of body size, phylogeny, and ambient noise in the evolution of bird song. *Am. Nat. 126*, 87–100 (1985).

24. Grant, P. R. & Grant, B. R. Predicting microevolutionary responses to directional selection on heritable variation. *Evolution 49*, 241–251 (1995).

25. Grant, P. R. & Grant, B. R. Cultural inheritance of song and its role in the evolution of Darwin's finches. *Evolution 50*, 2471–2487 (1996).

26. Ratcliffe, L. M. & Grant, P. R. Species recognition in Darwin's finches (*Geospiza*, Gould). III. Male responses to playback of different song types, dialects and heterospecific songs. *Anim. Behav. 33*, 290–307 (1985).

27. Grant, P. R. & Grant, B. R. Speciation and hybridization in island birds. *Phil. Trans. R. Soc. Lond. B 351*, 765–772 (1996).

28. Mooers, A. Ø., Vamosi, S. M. & Schluter, D. Using phylogenies to test macroevolutionary hypotheses of trait evolution in cranes (Gruinae). *Am. Nat. 154*, 249–259 (1999).

29. Petren, K., Grant, B. R. & Grant, P. R. A phylogeny of Darwin's finches based on microsatellite DNA length variation. *Proc. R. Soc. Lond. Ser. B. 266*, 321–330 (1999).

30. Sato, A. et al. Phylogeny of Darwin's finches as revealed by mtDNA sequences. *Proc. Natl Acad. Sci. USA 96*, 5101–5106 (1999).

ACKNOWLEDGMENTS

Field work was coordinated through the Charles Darwin Research Station and the Galápagos National Park Service. I thank M. Rossi-Santos, M. Moreano, H. Vargas, H. Snell and M. Hau for assistance in the field; S. Nowicki, L. Baptista, R. Bensted-Smith, R. Bowman, P. & R. Grant, A. Hendry, W. Hoese, S. Hopp, J. Jaenike, J. Lundberg, W. Maddison, L. McDade, C. Nufio, D. Papaj, S. Patek and R. Prum for discussion and feedback; and the National Science Foundation, the Univ. Arizona Foundation, the Univ. Arizona Office of the Vice President for Research and TAME airlines for financial support.

PART 3

The Origin and Maintenance of Species

Translating microevolutionary processes of adaptation into macroevolutionary patterns of speciation and phylogenesis is not a process of mere extrapolation. Divergent adaptation alone does not necessarily result in different species—evolutionarily independent lineages, genetically distinct from all other lineages. The formation of distinct species with independent evolutionary trajectories requires barriers to genetic exchange per se. What are these barriers, how do they evolve initially, and how are they maintained?

Perhaps the foremost question regarding the process of speciation, and the relationship between microevolution and macroevolution, is the role of natural selection. Evolution can proceed without natural selection; in principle, populations can diverge, even to the point of becoming incompatible with each other, through the process of mutation and random drift alone. The question of the nature of the involvement of natural selection in the process of speciation has taken the form of a controversy over the geography of speciation. Geography again. In order for alleles that contribute to incompatibilities between populations to be retained within a given population, some separation of the two populations is required. Otherwise, an allele associated with an incompatibility would be eliminated as soon as it arose, precisely because it is incompatible with some members of the population. The problem, then, is that the evolution of incompatibilities between species, effectively preventing genetic exchange, requires limited genetic exchange to begin with. How does the process get started? If the initial barrier to genetic exchange is geography, then natural selection need not have any role in that crucial first step.

Controversies that arise over the interplay of natural selection and geography dominate the literature on speciation. Resolution of these controversies would benefit from knowledge of the genetic basis of species differences and speciation itself—which genes cause incompatibilities of various sorts, and how they do so. Part 3 explores the process by which a single species becomes two, and how natural selection within a single species contributes to this process—or not.

Chapter 9 presents one approach to determining whether microevolutionary parameters of adaptation within species predict macroevolutionary outcomes. Using standard quantitative-genetic methods to predict changes in phenotypes within a population, this approach tests whether the genetic distance between species is proportional to the amount of selection required to cause the observed morphological divergence between them. In the case of Darwin's finches, it is.

While chapter 9 demonstrates a correspondence between morphological and genetic divergence, it does not address the question of how the species were able to diverge independently from one another—that is, how genetic exchange between them was prevented. Chapter 10 tackles this issue as it discusses the evolution of reproductive isolation between species of Darwin's finches.

Chapter 11 concerns hybridization, the magnitude of genetic exchange between species. The frequency of hybridization indicates the degree to which genetic exchange among incipient species still occurs. Hybridization is also a process that itself influences the dynamics of evolutionary diversification. One study in this chapter detects and quantifies the degree of hybridization; the other discusses some of the evolutionary consequences of hybridization.

Finally, chapter 12 addresses the molecular-genetic basis of differences between species, and how developmental processes contribute to these differences. More generally, it concerns how the genetic basis of variation influences the process of evolution.

Microevolution and Macroevolution: Does One Explain the Other?

How do the processes of adaptation and divergence within species contribute to the formation of different species? Can the latter be explained solely in terms of the former, or is the process of speciation qualitatively different from adaptive evolution within a species? Some taxonomists, who spend much of their time trying to find traits that distinguish one species from the other, have argued in the past that most differences between species are not adaptive but arbitrary—a bump on the hind leg here, a spot on the thorax there. If that is the case, then how important is natural selection to the process of speciation and to the accumulation of differences between species?

In theory, adaptive divergence is not necessary for speciation. Speciation occurs when genetic exchange between populations permanently ceases. When this happens, the two populations are said to be reproductively isolated. According to a simple genetic model, called the Bateson-Dobzhansky-Muller model, evolution within two isolated populations can produce genetic incompatibilities between the two populations when they subsequently interbreed, preventing genetic exchange between them. Consider one original population with the genotype *AA* at one locus and *BB* at another (genotype *AABB* hereafter). Assume, further, that this population somehow becomes split into two isolated populations, each with genotype *AABB*. In one population, a new allele *a* appears, which produces individuals *AaBB*. If the *a* allele goes to fixation (i.e., replaces the *A* allele and reaches a frequency of 100%), either through natural selection or genetic drift, that population will have individuals of genotype *aaBB*. Then imagine that the other population goes through a similar process, but at the other locus; in time, it will have the genotype *AAbb*. If the two populations subsequently interbreed, they will produce offspring of genotypes *AaBb*. Should the *a* allele and the *b* allele be incompatible, the hybrid will have reduced fitness, and if the hybrid cannot breed successfully, then alleles from one population cannot enter the other population through hybridization; the populations would be genetically isolated from each other. Such incompatibilities are an example of epistasis, in which the effect of an allele depends on what allele is present at other loci (*a* in combination with *B* is viable, but in combination with *b* it is incompatible);

that is, the two loci interact in a nonadditive manner. The key is that the *a* and *b* alleles never encountered each other before the two populations interbred[1] because the two populations were physically isolated from each other; neither of the isolated populations experienced a phase during which the low-fitness genotypes were present, so neither population experienced any maladaptive stage. It is only when the alleles come together through interbreeding of the two populations that offspring that carry both new alleles, *a* and *b*, are produced and consequently have low fitness, thereby contributing to reproductive isolation between the two populations.

In the above scenario, each of the new alleles could have been fixed by genetic drift alone. Thus natural selection within each population is not necessary for the evolution of reproductive isolation. If, of course, the new alleles were advantageous in each population, they would be fixed sooner, and genetic differences between the isolated populations would accrue more quickly. In this case, natural selection would speed the process of speciation. Note that natural selection does not have to differ between the two populations. That is, the two populations may inhabit the same ecological environment, but the *a* allele might pop up in the first population and the *b* allele in the second—or vice versa. In this case, the genetic difference between these populations is not the result of adaptive divergence, but rather of the random order in which the alleles appeared in each population. Alternatively, the two environments might differ, with natural selection favoring the *a* allele in the environment of the first population, and the *b* allele in the second. We do not really know whether incompatibilities are more likely to result from homogeneous or divergent natural selection, all else being equal, although some experimental evidence suggests that divergent selection might be more efficacious at fixing incompatibilities. In short, reproductive isolation can evolve with no natural selection at all within distinct populations, with consistent natural selection occurring in both populations, or with divergent natural selection in the two populations. The extent to which each scenario contributes to speciation remains a subject of investigation.

Since Lack's research, biologists have seen Darwin's finches as a prime example of how microevolutionary processes of divergent adaptation within a species result in speciation—the formation of two species. In his influential 1947 book Lack described the classic and purely neo-Darwinian gradualist process of speciation as a by-product of adaptation: the more adaptively divergent populations become, the less likely they are to be capable of genetic exchange. In this scenario, adaptive divergence increases the ecological isolation of the two incipient species—reducing the probability of encountering

---

1. If the two alleles did encounter each other before the populations interbred, it was not for long; if "b" arose after *a* became fixed within the first population, for instance, its carriers would be quickly eliminated.

the other incipient species—and also results in adaptively incompatible genotypes.

In later decades this view has competed with many creative alternatives, as evolutionary biologists argue that speciation is a process distinct from the gradual adaptive divergence of populations. One prominent alternative focused on the question of the gradual pace of speciation. Citing the fossil record, which often shows long periods without morphological change followed by short periods of rapid morphological change, Niles Eldridge and Stephen J. Gould proposed their theory of punctuated equilibrium. This theory originally described a process whereby morphological divergence could occur only during speciation, when the genetic ties between entities are being broken. Then (and only then) could divergence race away. Modified soon afterward, their theory became more a description of the punctuated pace of the morphological change they observed, a pace that contradicted the theory of gradual adaptive divergence. Other research suggested that genetic drift in newly colonized small populations might play some role in reproductive isolation, with the chance fixation of new combinations of alleles subsequently favoring different alleles throughout the genome to maintain genetic compatibility—a so-called genetic revolution—which could rapidly cause divergence and reproductive isolation. Biologists have thus come to consider adaptation not only to external ecological environments but also to internal genetic environments as crucial to the process of speciation.

The tenuous link between adaptive microevolution and macroevolution remains one of the greatest challenges in the field of evolutionary biology. As a consequence, recent researchers have shown considerable interest in exploring how microevolutionary processes of divergent adaptation account for observed differences between species, or conversely whether speciation is a distinct process in itself.

One approach is to compare patterns of morphological evolution within species to patterns of differences between species. If patterns of speciation—phylogenesis, or the splitting of one lineage into two independent lineages—follow patterns of adaptive phenotypic divergence between species, then this could be interpreted as consistent with the hypothesis that speciation rides along with phenotypic divergence. In such a case, explaining mechanisms of phenotypic evolution within a species could also provide insight into mechanisms of speciation. If, on the other hand, phenotypic divergence shows evolutionary patterns independent from phylogenetic divergence—with phenotypically similar taxa having arbitrary genetic similarity, for example—then speciation may occur independently of phenotypic divergence. Two questions thus emerge: Can the analytical tools that predict phenotypic change within a species, as conditioned by natural selection and genetic architecture, be applied to predict phenotypic differences among species? Can they be applied to predict genetic, or phylogenetic, relationships?

Dolph Schluter's paper, included here, shows that phenotypic divergence among taxa, specifically the total selection required to cause such divergence, reflects genetic differences and phylogenetic relationships among taxa. Schluter and others have continued to investigate the degree to which patterns of genetic variances and covariances observed within individual species constrain the evolution of differences between species. They have found evidence in other taxa, in fact, that macroevolutionary divergence proceeds along the "lines of least genetic resistance," that is, in directions for which there exists standing genetic variation and covariation. Steven Arnold and his colleagues (e.g., 2001) have even proposed that the quantitative-genetic study of microevolution and population divergence is the conceptual bridge to modeling macroevolution.

Crucial to this approach is the question of how constant the basic parameters of microevolutionary processes—natural selection, and genetic variances and covariances (the G-matrix presented in chapter 7)—remain over evolutionary time. How will these parameters change with alterations in both the external ecological environment and the genetic environment? In Schluter's study and those similar to it, genetic variances and covariances are estimated within a single taxon (frequently the basal taxon) for the purposes of predicting phenotypic changes over time. But what happens if those genetic parameters themselves evolve? Much thought and experimental work is currently devoted to testing how this fundamental component of evolutionary process—the G-matrix itself—evolves.[2] Changes in the G-matrix will influence evolutionary trajectories within species and trajectories of divergence between species, but our understanding of the evolutionary lability of G-matrices remains rudimentary and offers a rich topic for future investigation.

Phenotypic evolution within species may reflect phenotypic and genetic differences between species, but a vexing problem remains: phenotypic divergence is not reproductive isolation. In fact, the correspondence between phenotypic divergence and genetic divergence could reflect only the accumulation of phenotypic change *after* speciation has occurred rather than the mechanism of speciation. A major challenge in studies of speciation, therefore, is to distinguish processes that account for speciation itself as opposed to divergence after speciation has occurred. The genetic basis of species differences is not necessarily the same as the genetic basis of speciation. Chapter 10 tackles the problem of explaining the evolution of reproductive isolation per se.

2 An example of a change in the G-matrix, genetic variance and covariances among traits, is discussed in chapter 11. In this case, hybridization between two species altered genetic parameters in one of the species.

READINGS

Lack's description of the process of speciation in Darwin's finches opens this section on speciation. Chapter 14 of Lack's 1947 book presents many of his major concepts. He defines species as reproductively isolated populations, engages in the debate over the importance of geography, alludes to the pace and mechanism of speciation, and mentions processes of the reinforcement of reproductive isolation upon secondary contact. In particular, he takes a strong position on the geographic pattern of speciation, and thereby the implied mechanism of speciation.

Lack comes down firmly on the side of the allopatric speciation scenario, in which geographically isolated (allopatric) populations gradually diverge. He does consider scenarios in which species are not geographically isolated (currently referred to as "sympatric speciation" if the species inhabit the same geographic area, or "parapatric speciation" if they are geographically adjacent to one another), but in which they encounter each other infrequently because of differences in habits. He concludes, though, that ecological isolation alone is inadequate to prevent interbreeding. With his confidence in the geographic scenario of speciation, he infers its genetic trajectory—a gradual accumulation of reproductive isolation as a by-product of gradual adaptive divergence in allopatry. Incidentally, once species are reproductively isolated, the process of character displacement, championed by Lack, can cause further morphological divergence between species should they subsequently encounter each other.

Next, Dolph Schluter's study explores the question of whether speciation and phylogenesis accompanies gradual phenotypic divergence. His purpose was to test the importance of genetic constraints on patterns of morphological macroevolution (divergence between species), and to determine whether microevolutionary mechanisms of phenotypic change within a species predict differences between species as well. Applying the standard breeder's equation (introduced in chapter 7), which describes the change in phenotypes over time as a function of selection and heritability, he estimates the strength of selection required to cause a change in the phenotype of a magnitude equal to the phenotypic differences observed between species. This "selection distance" derives from his observed differences between species and his estimate of the genetic variances and covariances among traits. Significantly, the selection distance differs from purely morphological difference because it includes constraints of genetic variation and correlations among characters. Schluter then compares the selection distance among species to their genetic distance and finds a strong correlation between the two. Phylogenetic relationships, he concludes, are accurately predicted by estimates of natural selection. Therefore, the two main components of microevolution

within species—natural selection and genetic variation—do indeed predict
macroevolutionary patterns of divergence between species.

From *Darwin's Finches* (1947)

❋ DAVID LACK

* Chapter XIV: The Origin of Species

*No clear line has yet been drawn between species and sub-species ... or, again, between sub-species and
well-marked varieties, or between lesser varieties and individual differences. These differences blend
into each other in an insensible series; and a series impresses the mind with an actual passage.*

Charles Darwin: *The Origin of Species*, Ch. II

INCIPIENT SPECIES

The various specimens of Darwin's finches from any one island do not form a
continuously graded series from large to small, thick-billed to thin-billed, or
dark to pale. Instead they fall into distinct segregated groups, each group having a characteristic appearance, while the individuals of any one group do not
normally interbreed with the members of any other. A similar state of affairs
is found in the birds breeding in Britain, and for that matter in every other
region of the world. The segregated groups are, of course, the units termed
species, and their manner of origin has aroused discussion and controversy
ever since the theory of evolution came to be accepted.

The apparent fixity of species is most striking, and provides the basis for
systematic zoology. But with the full acceptance of the doctrine of evolution
there has arisen a tendency among general biologists, though not among taxonomists, to underestimate the definite nature of species, and a corresponding tendency to exaggerate the frequency of intermediate forms. Charles
Darwin and many after him are partly wrong when they assert that the determination of species is purely arbitrary. Provided the ornithologist keeps
within a limited district, he is usually in no doubt as to which birds should
be regarded as separate species, and the same holds in many other groups of
animals. Difficulty over intermediate forms arises mainly when the naturalist
compares related forms of birds or other animals from different districts, but
then the difficulty immediately becomes considerable, a fact which provides
the essential clue to the way in which new species originate.

Big evolutionary changes are normally achieved in a series of small steps,
so that it is to be expected that the gaps between species would come into
existence gradually, in which case some of the intermediate stages ought to be
visible. The closely related species of Darwin's finches differ from each other
in beak, in size of body, in the shade and amount of streaking of the female

plumage, and in the amount of black in the male plumage. It is in just these characters that island forms of the same species differ from each other and, as discussed in the last chapter, such island forms show every stage of divergence from differences that are barely perceptible to differences as marked as those which separate some of the species. Moreover, this is the only kind of incipient differentiation found among Darwin's finches. These facts strongly suggest that island forms are species in the making, and that new species have arisen when well-differentiated island forms have later met in the same region and kept distinct.

CAMARHYNCHUS ON CHARLES

An instance of such a manner of origin is provided by the two species of large insectivorous tree-finch, *Camarhynchus psittacula* and *C. pauper*, which occur together on Charles. These are undoubtedly two separate species, differing in size of beak, wing-length and shade of plumage. The differences between them are small, but constant and reliable, and each collected specimen can safely be allocated to one or the other type.

If *C. psittacula* (*sens. strict.*) did not occur on Charles, all the large insectivorous tree-finches could be included in one species *C. psittacula*, divided into four well-marked geographical races as follows: *psittacula* (*sens. strict.*) on the central islands of James, Indefatigable and Barrington, *habeli* to the north on Abingdon and Bindloe, *affinis* to the west on Albemarle and Narborough, and *pauper* to the south on Charles, as shown in Fig. 23. Of these four forms the Charles form *pauper* appears to be the most primitive, being much streaked and possessing the smallest and most finch-like beak. The Albemarle form *affinis* shows close resemblance to *pauper* both in plumage and beak, so links up with it. The central island form *psittacula* (*sens. strict.*) is less streaked, larger and with a more parrot-shaped beak, but links up with the Albemarle form through a population of intermediate type on the intervening island of Duncan. The northern form *habeli* shows most resemblance to *psittacula* (*sens. strict.*).

This simple situation is complicated by the fact that Charles is inhabited not only by the form *pauper* but also by the form *psittacula* (*sens. strict.*), the Charles individuals of the latter being indistinguishable from those found on the central islands. So far as known these two forms do not interbreed on Charles, and there are no specimens intermediate between them in appearance. The facts suggest that Charles has been colonized by the large insectivorous tree-finch on two separate occasions. Originally, the island was inhabited by the form *pauper*, or by a form which later turned into *pauper*, while more recently it has been invaded from the north by the form *psittacula* (*sens. strict.*). Formerly, *pauper* and *psittacula* (*sens. strict.*) were geographical races of the same species, but by the time that they met on Charles they

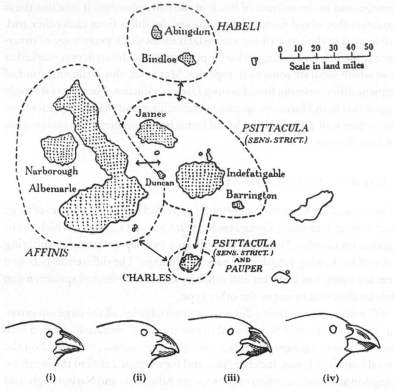

Fig. 23. The forms of *Camarhynchus psittacula* (*sens. lat.*). Showing the existence of two forms on Charles, the earlier form (*C. pauper*) being related to the Albemarle form (*affinis*), and the later form (*psittacula, sens. strict.*) coming from the central islands. The two have met, but do not interbreed. (i) *pauper* (Charles) (ii) *affinis* (Albemarle) (iii) *psittacula, sens. strict.* (James) (iv) *habeli* (Bindloe) Heads 2/3 natural size (after Swarth).

had become so different that they did not interbreed, and so they have become separate species. Similarly, if the Albemarle race *affinis* were now to colonize Abingdon, where the form *habeli* lives, the two are so distinctive that they might keep separate, in which case they also would have to be classified as separate species. But such an invasion has not yet occurred, so that it is more convenient to consider *affinis* and *habeli* as races of the same species....

PREVENTION OF INTERBREEDING

In birds generally, as in Darwin's finches, geographical forms show every stage of divergence, from differences which are barely perceptible to differences as marked as those which separate full species. From the evidence which has now accumulated, it is clear that the commonest method of species-formation in birds is through the meeting in the same region of two geographical forms which have become so different that they keep separate. The fundamental problem in the origin of species is not the origin of differences in appearance, since these arise at the level of the geographical race, but the origin of genetic segregation. The test of species-formation is whether, when two forms meet, they interbreed and merge, or whether they keep distinct.

As discussed in the last chapter, when two populations of the same form are isolated from each other, differences gradually arise between them. Muller (1940) considers further that such differences inevitably lead to some degree of sterility between the individuals of the two populations. If this view is correct, there should sometimes be partial sterility between geographical races of the same species, and this has now been established in the case of a number of insects, as summarized by Mayr (1942). Similar evidence is not yet available in birds, as they are rather unsuitable for quantitative breeding experiments.

If the members of two well-differentiated races meet later in the same region, and if they are partially intersterile, or if their hybrid offspring are at a disadvantage, then those individuals which breed with members of their own kind tend to leave more offspring than those which interbreed with individuals of the other race. Hence even if genetical segregation between two races is not complete when they first meet, natural selection will tend to deepen the gap between them. Indeed, it has even been claimed that, owing to the disadvantages possessed by hybrids, selection will initiate intersterility between forms which meet in this way. However, the latter view is not certain.

The above considerations show that any factors which prevent the interbreeding of forms have survival value, hence the frequency with which specific recognition marks have been evolved in birds, as discussed in Chapter V. Darwin's finches are unusual in that the beak is used as a recognition mark, but as this is the most prominent racial and specific difference, it is not surprising that the birds should have evolved behaviour responses relating to it.

Barriers to interbreeding might also be provided by differences in breeding season or habitat. But the various species of Darwin's finches breed at the same season and most of them are not separated in habitat, so that these factors have little or no importance. The latter conclusion applies to birds generally. Most related species found in the same region breed at the same season, and though they commonly occupy different habitats, the degree of isolation thus provided is usually quite inadequate to ensure genetic isolation. In other birds, as in Darwin's finches, the primary factors which prevent attempts at interbreeding are psychological ones correlated with breeding behaviour.

OTHER METHODS OF SPECIES-FORMATION

Darwin's statement quoted at the head of this chapter suggests that in animals every gradation exists between mere varieties and full species. Later knowledge has shown that, at least in birds, this statement is somewhat misleading. There is only one kind of variety, if such it should be called, which grades insensibly with the full species, namely the geographical race. In birds there is no other type of variety which can reasonably be termed a subspecies, indeed, the terms subspecies and geographical race have become synonymous. This strongly suggests that in birds the only regular method of species-formation is via races differentiated in geographical isolation. However, it has sometimes been claimed that, to produce the variety of species found in Darwin's finches, some quite peculiar method of evolution must have been involved. Even Rensch (1933), who was the first to advocate species-formation from geographical races as a widespread principle, was greatly puzzled by Darwin's finches, and considered that for them some different process must have operated.

The frequency with which closely related bird species occupy different habitats suggests that an alternative method of species formation is by ecological, instead of geographical, isolation. Since a bird tends to breed in the same type of habitat as that in which it was raised, it might have been expected that, where a species breeds in a variety of habitats, it would tend to become subdivided into populations each with a rather different habitat preference, and that with time this might result in the formation of new species. This is a plausible view and has been put forward by a number of writers, formerly including myself (1933). But there are two insuperable objections. First, the degree of isolation provided by differences in habitat is not usually at all complete, and the bird species which occupy separate habitats usually have numerous border zones where they come in contact with other species. To produce well-differentiated forms, complete isolation seems essential. Secondly, no cases are known in birds of incipient species in process of differentiation in adjoining habitats. All subspecies are isolated from each other geographically and, though they occasionally differ in habitat as well, geographical isolation is the essential factor. These points are treated in further detail by Mayr (1942). Finally, the frequent existence of habitat differences between closely related bird species has a quite different explanation, consideration of which is postponed to the next chapter.

Another alternative has been suggested by Lowe (1930, 1936), who supposes that Darwin's finches represent the varied products of interbreeding between a small number of original forms, as has happened in certain "species-swarms" in plants. The reasons for rejecting this view have been discussed in Chapter X. There is no evidence that hybridization has been of importance in species-formation in any group of birds.

Streseman (1936) is the only previous writer to suggest that species-formation has followed the same course in Darwin's finches as in other birds, i.e. that forms differentiated in geographical isolation have later met and kept distinct. With this conclusion, I fully agree. There is only one apparent case, in the large insectivorous tree-finches *Camarhynchus psittacula* and *C. pauper* on Charles. But such cases will rarely be apparent, since once a form has become firmly established in the range of another it will tend to spread rapidly right through that range, so that its place and means of origin quickly become obscured. The only type of incipient differentiation found in Darwin's finches is that shown by geographical races, and there is nothing to suggest that geographical isolation is not the essential preliminary to species-formation in this group. The existence of an unusually large number of similar species may be attributed first to the great length of time for which the finches have been in the Galapagos, secondly to the paucity of other land birds, and thirdly to the unusually favourable conditions provided by a group of oceanic islands, both for differentiation in temporary geographical isolation, and also for the subsequent meeting of forms after differentiation.

The primary importance of the geographical factor is strongly corroborated by the situation on Cocos Island. Here there occurs one, and only one, species of Darwin's finch, *Pinaroloxias inornata*. That it has been on Cocos a long time is suggested by the extent to which it differs from all the other species of Darwin's finches. Yet despite the length of time for which it has been there, despite the variety of foods and habitats which Cocos provides, and despite the almost complete absence of both food competitors and enemies, there is still only one species of Darwin's finch on Cocos. But Cocos is a single island, not an archipelago, and so provides no opportunity for the differentiation of forms in geographical isolation. . . .

Gloucester, MA: Peter Smith Publisher, 1968. Excerpt from chapter 14, pp. 125–133.

Morphological and Phylogenetic Relations among the Darwin's Finches (1984)

❋ DOLPH SCHLUTER

Despite the ecological and evolutionary significance of the Darwin's finches (Lack, 1947; Bowman, 1961; Abbott et al., 1977; Boag and Grant, 1981; Schluter and Grant, 1984), their phylogenetic relationships are incompletely understood. Yang and Patton's (1981) electrophoretic study on 11 of the 13 Galápagos species supported Lack's (1947) division of the finches into three major groups (Fig. 1): ground finches (*Geospiza*), tree finches (*Camarhynchus* and *Platyspiza*), and the warbler finch (*Certhidea olivacea*). Biochemical differ-

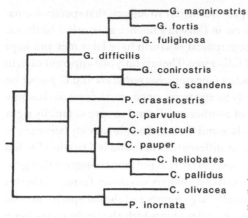

Fig. 1. Lack's (1947) phylogenetic tree, with branch lengths approximately as he drew them.

ences within these groups were slight, however, leaving details of species relationships largely unresolved. In this report I use a new method to estimate evolutionary relationships among the finches based on morphological data. The analysis includes all 13 Galápagos species and *Pinaroloxias inornata* from Cocos Island, putatively a warbler finch (Lack, 1947).

Ecological studies have indicated that morphology in the Darwin's finches has been strongly influenced by natural selection (Boag and Grant, 1981; Price and Grant, 1984; Schluter and Grant, 1984). Under these conditions morphological differences between species might be expected to reflect differences in environmental history rather than evolutionary relationships (Barrowclough, 1983). Indeed, poor concordance between estimates of phylogeny based on morphological data and estimates based on biochemical data (ostensibly neutral) has been noted in other groups (e.g., Zink, 1982) as well as in the Darwin's finches (Barrowclough, 1983). However, most standard measures of morphological distance (e.g., Cherry et al., 1982) lack explicit evolutionary justification, and this may partly explain poor concordance when it is observed. Here I employ a measure of morphological distance based on a theoretical study of multivariate evolution (Lande, 1979). The resulting evolutionary relationships inferred agree reasonably well with both the biochemical (Yang and Patton, 1981) and the traditional estimates (Lack, 1947), where comparison is possible.

MORPHOLOGICAL DISTANCE

Determining an appropriate measure of evolutionary distance between two forms hinges on finding a solution to two outstanding problems in phylogenetic inference: how to weight characters by their susceptibility to change, and how to account for correlations between characters (Felsenstein, 1982).

Recent models for the evolution of mean phenotype in populations (Lande, 1979) provide ways to deal with both these factors simultaneously, at least for morphological data. Under multivariate selection:

$$\beta = G^{-1}[z(t) - z(0)]. \tag{1}$$

$z(0)$ and $z(t)$ are vectors of character means in the population at times 0 and t. G is a matrix of genetic variances and covariances among characters, and it is assumed to remain constant through time. β is the net selection gradient, a vector whose elements β_i are the net forces of directional selection which have acted on each character i in order to produce the observed overall change in z. The magnitudes of individual β_i depend on the genetic variance in character i: a given change in the mean of a character requires more selection when genetic variance is low, than when it is high. Secondly, each β_i depends on the magnitude and direction of the indirect effects on z_i resulting from selection on genetically correlated characters. For example, a given increase in size in two characters z_1 and z_2 requires more selection if they are negatively correlated, than if they are uncorrelated or positively correlated (Lande, 1979). The length of the vector can be measured by the Euclidean distance,

$$B = [\Sigma \beta_i^2]^{1/2}, \tag{2}$$

which represents the total net force of directional selection that has acted to shift mean morphology from $z(0)$ to $z(t)$.

The above result can be applied to any modern group of related species, for example the Darwin's finches. For the pair of species a and b, mean morphologies $z(a)$ and $z(b)$ are substituted in place of $z(0)$ and $z(t)$ in (1) (examples are given in Price et al., 1984). In this situation B is a measure of "selection distance" between the two species: it estimates the total net force of directional selection which would be required to produce their observed morphological differences. In many situations B will therefore be an appropriate measure of morphological distance between them.

Because it is explicitly a measure of evolutionary distance, B may be used to estimate phylogenetic relations among the species. Such an application requires one further assumption: B will be directly proportional to t, the time interval involved in evolutionary change only if the total net force of directional selection averaged over generations (i.e., B/t) is approximately constant. Clearly there is no a priori reason to accept this assumption. However, the assumption is certainly worth stating as a null hypothesis, one which could be tested through comparison with independent estimates of phylogeny (e.g., Barrowclough, 1983).

Note that if we transform means for the original morphological characters z to yield $z' = G^{-1}z$, then for two species a and b,

$$\beta = z'(a) - z'(b)$$

and

$$B = [\Sigma \, (z_i{'}(a) - z_i{'}(b))^2]^{1/2}. \tag{3}$$

Hence, B is simply the Euclidean distance applied to the transformed mean vectors z', and it is given in units^{-1} of the original variables. A related distance is B^*, computed as in (3) using the standardized uncorrelated characters $z^* = G^{-1/2}z$. In this case B^* is the "generalized genetic distance" between two species (Lande, 1979), and it is dimensionless.

Whether one should use B or B^* as the measure of selection distance will depend on whether the selective environment is best described in terms of the units of the original variables (B), or in terms of population standard deviations (B^*). The latter distance is appropriate under the special case of artificial selection, or truncation selection (Lande, 1979), but B may be appropriate when dealing with natural environments. All analyses of the Darwin's finches were performed using both distance measures, and they gave very similar results. For this reason I present only results based on the unstandardized measure (B) in this report.

Note finally that the methods for computing selection distance account for only those characters included in the analysis. Hence there can be no certainty that inclusion of another character would not alter the estimates of individual and combined selection intensities (Lande and Arnold, 1983). The problem presented by unmeasured characters is not unique to the present distance measure; as always, a partial resolution is better than none (Lande and Arnold, 1983).

METHODS

Measurements of eight characters (Abbott et al., 1977) are available for all 14 species of Darwin's finches. These eight characters summarize most of the morphological differences between the species (Price et al., 1984). They are wing length (WNG), tarsus length (TRS), upper mandible length (UML), width (UMW) and depth (UMD), and lower mandible length (LML), width (LMW) and depth (LMD), all in mm. Measurements were made from museum specimens, and most populations of each species were included, 88 in all (Grant et al., unpubl.). Ground finches (*Geospiza*) were measured by I. J. Abbott and P. R. Grant; at least ten, and up to 278 individuals were measured per sex, per population available. Tree and warbler finches (including *Pinaroloxias*) were measured by P. R. Grant and R. L. Curry; ten adult individuals of each sex were measured per population. Identical methods were used throughout. All measurements were converted to the log(e) scale. Males are generally larger than females (Price, 1984; Grant et al., unpubl.), and I used only males when computing population means. Means for species are unweighted averages of (male) population means. Phenotypic variances and co-variances within species were computed by pooling individuals across sexes and populations, scaling for differences in subgroup means.

The distance measure, *B*, assumes that genetic variances and covariances remained approximately constant during divergence (Lande, 1979). Pheno-typic covariance matrices for different species were compared, by investi-gating the factor loadings of the different characters on the first two prin-cipal components (PC's). In general, character loadings were similar within the tree finches and ground finches, but differed slightly between the two groups. In all cases variables loaded positively onto PC1, but the highest load-ings on both PC1 and PC2 were for LMD in the tree finches, and for UMD in the ground finches. Relative to other characters UMD in the ground finches generally had a negative loading on PC2 and LMD had a positive loading. The signs were usually reversed in the tree finches. For this reason, a different covariance matrix was used to measure divergence within each of the two finch groups.

I used phenotypic variances and covariances in *G. fortis* and *C. psittacula* as representative of their respective groups, rather than a pooled sample of species. Each of these two species is intermediate in size in all measured char-acters with respect to other members of their group. Choice of these two spe-cies avoids high relative measurement error associated with the smaller spe-cies (Rohlf et al., 1983). Also, both species have a large number of individuals on which to base estimates of variance and covariance. *Geospiza fortis* is further appropriate in that some genetic parameters are known (Boag, 1983).

In *G. fortis*, I used two different estimates of *G*, the matrix of genetic vari-ances and covariances. In the first, genetic values were estimated from their phenotypic values. Genetic values for five characters were computed directly from data in Boag (1983): WNG, TRS, UML, LMW, and beak depth (UMD + LMD). Phenotypic variances and covariances recorded in the present study were correlated with the genetic variances ($r = .98, N = 5$) and covariances ($r = .99, N = 10$) in Boag (1983). Genetic parameters for the remaining charac-ters could thus be estimated using least squares regression.

For the second estimate of *G* in *G. fortis* I simply used the phenotypic val-ues. Both estimates gave very similar results in the evolutionary analyses, due to the similarity between the genetic and phenotypic matrices (Boag, 1983). I present results from only the first estimate. Because no genetic parameters are known for *C. psittacula*, and because the genetic and phenotypic parameters gave similar results in the ground finches, I used only the phenotypic values in the tree finches.

Phylogenetic relationships were inferred by computing minimum-length Wagner trees (Kluge and Farris, 1969) in PHYSYS (Farris and Mickevich, unpubl.) directly from the transformed mean characters for species (z'). I used this procedure instead of an alternative one suggested by Felsenstein (1981) primarily because his model of random divergence is inappropriate in this case. Divergence in the Galápagos finches probably occurred instead through sequential, stepwise adaptation to unoccupied feeding niches (i.e., stationary optima) along previously existing resource axes (Lack, 1947; Schluter and Grant, 1984; Price et al., 1984). Hence, though the statistical validity of parsimony remains questionable, this criterion seems intuitively the more appropriate for the adaptive model.

Unfortunately there are no procedures for computing minimum-length Wagner trees that use Euclidean distances. The phylogenetic trees presented here are based on Manhattan distances between species. However, Distance-Wagner procedures (Farris, 1972) using either distance metric gave essentially identical results.

The warbler finch, *Certhidea olivacea*, was used as an outgroup in both the ground finch and the tree finch analyses: Lack (1947) suggested that the warbler finches, *C. olivacea* and *P. inornata*, branched off prior to the ground finch-tree finch division (Fig. 1), and this is supported by Yang and Patton's (1981) results. *Pinaroloxias inornata* was also included in both analyses. Results were the same when it was excluded, or when it was used as an outgroup instead of *C. olivacea*.

RESULTS

Morphological Differences

Principal components analysis (Pimentel, 1979; Ricklefs and Travis, 1980) was used to represent morphological relationships among populations and species. The advantage of this procedure is that morphological relationships can be represented without distortion using a small number of dimensions. In the first analysis, populations of all 14 species were combined, in order to investigate morphological relationships between the ground, tree, and warbler finches. Components were determined on both the original means for populations (z) and on means transformed using genetic parameters derived from the ground finches (z'). Relationships among the three groups were similar when data were transformed using the corresponding matrix for the tree finches. Additional analyses were performed on the ground finches and tree finches separately, to more closely investigate patterns of divergence within the two main groups. Factor loadings for the ground finches and tree finches are listed in Tables 1 and 2, respectively. Mean species positions along the first two components (PC's) are shown in Figure 2 for the combined analysis. Population positions within the tree finches and ground finches are shown in Figures 3 and 4.

TABLE 1. Factor loadings on the first three principal components derived from the original (z) and transformed (z′) population means, for the ground finches

Variable	Original			Transformed		
	I	II	III	I	II	III
WNG	0.14	−0.08	0.36	−0.23	0.46	−0.74
UML	0.34	−0.60	−0.17	−0.51	−0.12	0.12
LML	0.32	−0.60	−0.17	−0.54	0.14	0.53
UMD	0.44	0.16	0.35	0.05	0.17	0.05
LMD	0.49	0.40	−0.57	0.19	0.18	0.06
UMW	0.35	0.11	0.35	0.00	−0.76	−0.22
LMW	0.43	0.25	0.01	0.60	0.16	0.28
TRS	0.12	−0.10	0.50	0.06	−0.30	−0.16
% variance	90.8	98.9	99.7	56.3	78.0	92.8

TABLE 2. Factor loadings on the first three principal components derived from the original (z) and transformed (z′) population means, for the tree finches

Variable	Original			Transformed		
	I	II	III	I	II	III
WNG	.25	.00	.37	.50	.38	.32
UML	.28	.62	.14	.19	.35	.35
LML	.19	.69	−.10	−.50	.75	−.22
UMD	.36	−.05	−.73	−.42	−.14	.58
LMD	.55	−.32	.07	.20	−.24	.19
UMW	.38	−.08	−.16	.18	.06	.00
LMW	.43	−.16	.12	.16	−.11	−.60
TRS	.25	−.05	.50	.44	.27	−.08
%variance	79.6	95.5	99.4	72.5	93.9	97.4

In all analyses PC1 and PC2 together accounted for at least 78% of the total variance among means. Hence PC1 and PC2 summarize the major trends of variation among populations and species (Figs. 2–4). In the combined analysis, 97% of the variance among untransformed means is accounted for by these two components. Figure 2 shows that the tree finches and ground finches overlap broadly along these dimensions (see also Grant et al., unpubl.). However, the two components derived from transformed means account for 85% of the variance but succeed in separating the two main groups (Fig. 2).

Factor loadings on principal components derived from the original means identify traits along which populations are morphologically most differentiated. Factor loadings on components derived from the transformed means (z′) identify the traits subject to the greatest net force of selection during morphological differentiation. Because different traits within populations

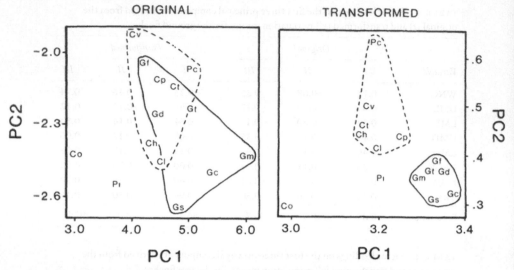

Fig. 2. Species means along the first two principal components (PC's) derived from original (*z*) and transformed (*z′*) population means. Lines delimit means for the ground finches (solid) and tree finches (dashed). Symbols are *Geospiza magnirostris* (Gm), *G. fortis* (Gt), *G. fuliginosa* (Gf), *G. difficilis* (Gd), *G. conirostris* (Gc), *G. scandens* (Gs), *Camarhynchus parvulus* (Cv), *C. pauper* (Cp), *C. psittacula* (Ct), *C. heliobates* (Ch), *C. pallidus* (Cl), *Platyspiza crassirostris* (Pc), *Certhidea olivacea* (Co), and *Pinaroloxias inornata* (Pi).

are not equally genetically variable, and because traits are correlated, these factor loadings are unlikely to correspond to those for the components derived from the original variables. Differences between loadings are evident in the ground and tree finch analyses. In both groups all untransformed variables load positively onto PC1 (Tables 1, 2), separating species by "size" (Figs. 3, 4). PC2 represents "shape" variation among species, particularly beak length relative to other variables. In contrast, principal components derived from the transformed variables indicate that variation in shape predominates along both axes. In the ground finches, the analysis indicates significant shape variation within species, particularly in *G. difficilis* (Fig. 3).

Some pairwise differences between species in a group also vary in the two analyses. For example, *C. psittacula* is morphologically more similar to *P. crassirostris* than to *C. parvulus*, with respect to the original variables (Fig. 4). However, the net force of selection necessary to actually bridge the difference between *C. psittacula* and *P. crassirostris* would be much greater than between the other pair, as indicated by their relative differences with respect to the transformed variables. Similarly, more net selection separates *G. fuliginosa* from *G. scandens* than from *G. magnirostris* (Fig. 3). With respect to the untransformed variables, these comparisons indicate that shape is considerably

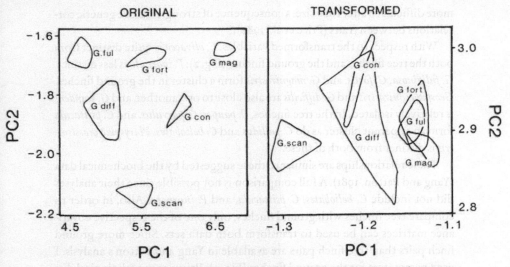

Fig. 3. Population means for the ground finches along the first two PC's derived from original and transformed data sets. Lines delimit means for separate species. Species abbreviations are for *G. magnirostris, G. fortis, G. fuliginosa, G. difficilis, G. conirostris,* and *G. scandens.*

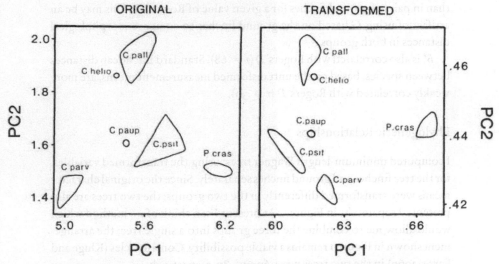

Fig. 4. Population means for the tree finches along the first two PC's derived from original and transformed data sets. Lines delimit means for species. Species abbreviations are for *C. parvulus, C. pauper, C. psittacula, C. heliobates, C. pallidus,* and *Platyspiza crassirostris.*

more difficult to alter than size, a consequence of strong, positive genetic correlations between traits (Price et al., 1984).

With respect to the transformed variables, *C. olivacea* is quite distinct from both the tree finches and the ground finches (Fig. 2); *P. inornata* is less distinct. *G. fuliginosa*, *G. fortis*, and *G. magnirostris* form a cluster in the ground finches. *Geospiza conirostris* and *G. difficilis* are also close to one another, and *G. scandens* is relatively isolated. In the tree finches, *C. pauper*, *C. parvulus*, and *C. psittacula* form an apparent cluster as do *C. pallidus* and *C. heliobates*. *Platyspiza crassirostris* is distinct from both of these.

These relationships are similar to those suggested by the biochemical data (Yang and Patton, 1981). A full comparison is not possible since their analyses did not include *C. heliobates*, *C. psittacula*, and *P. inornata*. Also, in order to compare tree finches with ground finches, only one of the respective covariance matrices can be used to transform both data sets. Since more ground finch pairs than tree finch pairs are available in Yang and Patton's analysis, I used parameters for the ground finches (Fig. 2). Pairwise morphological distances between species (B) are strongly correlated with the biochemical distances given in Yang and Patton (Rogers' D; $r = .74$; $N = 55$). This correlation probably underestimates the actual correspondence between morphological and biochemical differences: some of the residual variance can be attributed to a greater morphological distance between species in pairs of tree finches than in pairs of ground finches for a given value of Rogers' D. This may be an artifact of using G based on the ground finches to compare morphological distances in both groups.

B^* is also correlated with Rogers' D $(r = .68)$. Standard Euclidean distances between species, based on the untransformed measurement means, are more weakly correlated with Rogers' D $(r = .39)$.

Phylogenetic Relationships

I computed minimum-length Wagner trees, using the transformed variables, for the tree finches and ground finches separately. Since the original character means were transformed differently in the two groups, the two trees are also presented separately, in Figure 5. At present I can think of no method which would allow me to combine the three groups into a single tree; the arrangement shown in Figure 1 remains a viable possibility. Consistencies (Kluge and Farris, 1969) in the two trees were .67 and .80, respectively.

Overall, phylogenetic relationships within the tree finches and ground finches in Figure 5 are similar to those of Lack's tree (Fig. 1), which was based on morphology, plumage, and dietary considerations. They are also similar to relationships suggested by Yang and Patton's (1981) branching phenogram based on biochemical distances among 11 of these species. Similarities include

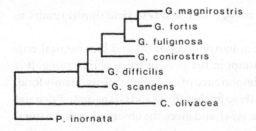

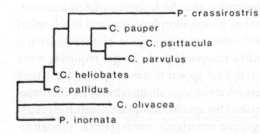

Fig. 5. Minimum-length Wagner trees for tree finches and ground finches, using *C. olivacea* as outgroup. Branch lengths for the two trees were scaled relative to the length of the *P. inornata* segment.

the inference that *C. parvulus*, *C. psittacula*, and *C. pauper* in the tree finches, and *G. fuliginosa*, *G. fortis* and *G. magnirostris* in the ground finches form monophyletic trios. *Camarhynchus pallidus* and *C. heliobates*, and *P. inornata* and *C. olivacea*, are grouped in both Lack's and my reconstructions. Figure 5 suggests that *C. pallidus* and *C. heliobates* are not monophyletic, as in Figure 1, but branch segments are short, and hence the local configuration uncertain.

Relationships among some species differ in Lack's and my reconstructions. For example Lack considered *G. conirostris* and *G. scandens* to be very closely related, but in terms of the transformed morphological variables *G. conirostris* is closer to *G. difficilis* than to *G. scandens*. Lack also suggested that *C. pallidus* and *C. heliobates* were highly modified forms (Fig. 1). However, morphologically they are closer to the hypothetical ancestral condition than are the other tree finches (Fig. 5).

DISCUSSION

In practice it is difficult to estimate G, the genetic covariance matrix. However, since phenotypic correlations tend systematically to underestimate genetic correlations (Cheverud, 1982; Cheverud et al., 1983; Boag, 1983; Grant, 1983), it will often be possible to estimate G from the phenotypic values. In other cases the observed phenotypic covariance matrix P can substitute for G when estimating selection distance. This assumes the phenotypic matrix to be a simpler scalar multiple of the genetic matrix. While unlikely, the resemblance will sometimes be great enough for P and G to yield similar results, as in the present study.

The correspondence between morphological (B) and biochemical estimates of evolutionary relationships in the present study is intriguing. It is especially surprising given the importance of ecological factors, mainly food, in determining morphology in these finches (Boag and Grant, 1981; Price and Grant, 1984; Schluter and Grant, 1984), and given the observation that natural selection can substantially affect mean morphology within one generation (Boag and Grant, 1981). Recall that the reliability of the present approach depends on the validity of the assumption that B/t is approximately constant, where t is the time since separation. Unless morphological and biochemical data do not provide independent estimates of phylogeny, this assumption is supported: in the Darwin's finches, morphological changes requiring more selection (e.g., changes in shape) indeed appear to have required more time than other changes involving less selection (e.g., changes in size). Differences in the amounts of selection required for specific morphological changes are determined entirely by G, the genetic covariance matrix. Hence, though selection has been responsible for morphological divergence in the Darwin's finches, the results are evidence that genetic parameters have exerted a powerful role in determining the direction and rate of change.

The results also suggest that an appropriate measure of morphological distance (i.e., B) may often allow reliable inference of phylogenetic relationships. Indeed, in some situations morphological data may have some advantages over biochemical data in estimating phylogeny. For example, rates of morphological divergence do not usually depend on effective population size, unlike short-term divergence rates in neutral characters (Lande, 1979; Felsenstein, 1981). Also, among closely related species (e.g., the Darwin's finches), estimates of relationships based on morphology will not be greatly influenced by small amounts of hybridization. Under neutrality, measured differences could reflect frequency of hybridization as much as actual phylogenetic distance.

Of course, good reasons remain why morphology will not always reflect phylogeny. For example, relationships may be obscured by morphological convergence, or by variable rates of morphological evolution. Morphological data may thus be most useful for phylogenetic purposes in providing complementary information to that derived from biochemical and other independent methods. However, even in situations where these other methods are thought to provide the true phylogenetic picture, it will be interesting to compare results with those of morphological analyses (Barrowclough, 1983). In this way it may be possible to clarify the importance of convergence and variable evolutionary rates in the history of morphological change. In such circumstances, because of its simple evolutionary interpretation, B may be a more appropriate measure of morphological distance than other more commonly used measures.

SUMMARY

A new measure of morphological distance is used to estimate morphological and phylogenetic relations among the Darwin's finches. The measure, B, is based on a model for multivariate evolution (Lande, 1979), and it estimates the total net force of directional selection acting on characters that is required to bridge the differences between any two species. This force depends on the amount of genetic variance in traits, and on genetic correlations between traits. "Selection distance" between species is shown to be correlated with biochemical distance, and the method produces a phylogenetic tree similar to the one originally suggested by Lack (1947). The results indicate that, in addition to natural selection, genetic parameters have strongly influenced the direction and rate of morphological divergence in the Darwin's finches.

ACKNOWLEDGMENTS

I thank P. Grant and I. Abbott for providing the morphological data, and J. Felsenstein, N. Heckman, R. Lande, and T. Price for valuable discussion and advice. D. Brooks, J. Farris, and R. O'Grady helped construct the phylogenetic trees. I. Abbott, G. Barrowclough, N. Barton, P. Grant, R. Lande, T. Price, and J. Smith read and commented on the manuscript. Computer funds were provided by the Institute of Animal Resource Ecology at UBC, and by the Division of Biological Sciences at The University of Michigan.

LITERATURE CITED

Abbott, I., L. K. Abbott, and P. R. Grant. 1977. Comparative ecology of Galápagos ground finches (*Geospiza* Gould): evaluation of the importance of floristic diversity and interspecific competition. Ecol. Monogr. 47:151–184.

Barrowclough, G. F. 1983. Biochemical studies of microevolutionary processes, p. 223–270. *In* Perspectives in Ornithology, Essays Presented for the Centennial of the American Ornithologists' Union. Cambridge Univ. Press, Cambridge.

Boag, P. T. 1983. The heritability of external morphology in the Darwin's finches (Geospizinae) of Daphne Major Island, Galápagos. Evolution 37:877–894.

Boag, P. T., and P. R. Grant. 1981. Intense natural selection in a population of Darwin's finches (Geospizinae) in the Galápagos. Science 214:82–85.

Bowman, R. I. 1961. Morphological differentiation and adaptation in the Galápagos finches. Univ. Calif. Publ. Zool. 58:1–302.

Cherry, L. M., S. M. Case, J. G. Kunkel, J. S. Wyles, and A. C. Wilson. 1982. Body shape metrics and organismal evolution. Evolution 36:914–933.

Cheverud, J. M. 1982. Phenotypic, genetic, and environmental morphological integration in the cranium. Evolution 36:499–516.

Cheverud, J. M., J. J. Rutledge, and W. R. Atchley. 1983. Quantitative genetics of development: genetic correlations among age-specific trait-values and the evolution of ontogeny. Evolution 37:895–905.

Farris, J. S. 1972. Estimating phylogenetic trees from distance matrices. Amer. Natur. 106:645–668.

Felsenstein, J. 1981. Evolutionary trees from gene frequencies and quantitative characters: finding maximum likelihood estimates. Evolution 35:1229–1242.

———. 1982. Numerical methods for inferring evolutionary trees. Quart. Rev. Biol. 57:379–404.

Grant, P. R. 1983. Inheritance of size and shape in a population of Darwin's finches, *Geospiza conirostris*. Proc. Roy. Soc. London B 214:219–236.

Kluga, A. G., and J. S. Farris. 1969. Quantitative phyletics and the evolution of anurans. Syst. Zool. 18:1–32.

Lack, D. 1947. Darwin's Finches. Cambridge Univ. Press, Cambridge.

Lande, R. 1979. Quantitative genetic analysis of multivariate evolution, applied to brain : body size allometry. Evolution 33:402–416.

Lande, R., and S. J. Arnold. 1983. The measure of selection on correlated characters. Evolution 37:1210–1226.

Pimentel, R. A. 1979. Morphometrics, the multivariate analysis of biological data. Kendal/Hunt, Dubuque.

Price, T. D. 1984. The evolution of sexual size dimorphism in Darwin's finches. Amer. Natur. 123:500–518.

Price, T. D., and P. R. Grant. 1984. Life history traits and natural selection for small body size in Darwin's finches. Evolution 38:483–494.

Price, T. D., P. R. Grant, and P. T. Boag. 1984. Genetic changes in the morphological differentiation of Darwin's ground finches, p. 49–66. *In* K. Wohrman and V. Loscheke (eds.), Population Biology and Evolution. Springer, N.Y.

Ricklefs, R. E., and J. Travis. 1980. A morphological approach to the study of avian community organization. Auk 97:321–338.

Rohlf, F. J., A. J. Gilmartin, and G. Hart. 1983. The Kluge-Kerfoot phenomenon—a statistical artifact. Evolution 37:180–202.

Schluter, D., and P. R. Grant. 1984. Determinants of morphological patterns in communities of Darwin's finches. Amer. Natur. 123:175–196.

Yang, S. Y., and J. L. Patton. 1981. Genetic variability and differentiation in the Galápagos finches. Auk 98:230–242.

Zink, R. M. 1982. Patterns of genic and morphological variation among sparrows in the genera *Zonotrichia*, *Melospiza*, *Junco*, and *Passerell*a. Auk 99:632–649.

Evolution 38 (1984): 921–930. Reprinted with permission of Wiley-Blackwell.

* 10 *
The Evolution of Reproductive Isolation

Reproductive isolation is the cessation of genetic exchange between two populations. It has been the ultimate diagnostic of species ever since the "biological species concept" became established with Ernst Mayr's 1942 publication. Since then, for proponents of the "biological species concept," the process of speciation has been equated with the evolution of reproductive isolation.

It is helpful to distinguish between two sorts of reproductive isolation. Prezygotic reproductive isolation prevents the formation of hybrid offspring by impeding mating or fertilization between two species (heterospecific mating or fertilization). Mate choice is an important component of prezygotic reproductive isolation in many organisms, including plants and fungi. Another effective mechanism of prezygotic reproductive isolation is ecological isolation, which occurs when different incipient species cannot inhabit the same location or do not encounter each other in the same location because of divergent adaptations. But perhaps the most obvious mechanism of prezygotic reproductive isolation is simple spatial isolation due to some geographic barrier.

Postzygotic reproductive isolation, in contrast, is the production of inviable or infertile hybrid offspring. Even if two populations were to come into contact, mate, and fertilize gametes, postzygotic reproductive isolation would prevent genetic exchange between them. Natural selection cannot favor postzygotic reproductive isolation per se, as it is never advantageous to produce offspring with zero fitness.[1] While intrinsic genetic incompatibilities have been shown to contribute to low hybrid fitness regardless of the ecological environment, hybrids can also have reduced fitness due to the disruption or dilution of adapted genotypes; intermediate phenotypes of hybrids may be less adapted than either parent to their particular ecological environment, or novel combinations of alleles or traits may produce extreme phenotypes that are maladaptive in particular environments.

Regarding the evolution of reproductive isolation, two questions are es-

1. It is not advantageous to produce offspring with zero fitness, unless through some unusual practice of parental provisioning and sibling cannibalism, or other oddity. There are always exceptions to any rule.

pecially prominent. First, what is the contribution of natural selection to the evolution of reproductive isolation? The previous chapter discussed the contribution of natural selection to the accumulation of postzygotic reproductive isolation via Bateson-Dobzhansky-Muller incompatibilities. Equally engaging is the question of what contribution natural selection makes, if any, to the evolution of prezygotic isolation. This issue is especially controversial, since purely geographic barriers alone can conceivably prevent members of different populations from encountering each other. The second question, then, is, how important is geographic isolation for the evolution of reproductive isolation?

Considering first the contribution of natural selection to prezygotic reproductive isolation, it is clear that natural selection is in no way necessary for purely geographic isolation; dispersal alone or some change in the physical landscape itself is adequate to cause geographic isolation. Natural selection does, however, contribute to ecological isolation, with divergent adaptation to different environments frequently resulting in spatial isolation.

Moreover, natural and/or sexual selection may or may not contribute to divergence in mate choice. First, mate choice may evolve in divergent directions because of genetic or cultural drift occurring independently in the isolated populations, with natural selection playing no role. Second, mate choice that evolves through differences in arbitrary, nonadaptive preferences would not require divergent ecological adaptation to diverge (see chapter 8). In contrast, if characters associated with reproduction are subject to divergent natural selection, then mating patterns can also diverge. For example, if early reproduction is favored in one location because of food availability, but later reproduction is favored in another for the same reason, then by adapting to food availability, reproductive timing also diverges, possibly leading to separate early- and late-reproducing populations. Similarly, mate choice itself might be under selection within an isolated population, and this might influence the probability of mating with members of another population upon secondary contact. Finally, mate discrimination against members of a different population might conceivably be under direct selection when two populations encounter each other. This would occur only if hybrids had low fitness, since otherwise hybridization would have no adverse effect (assuming the mate pays no direct cost). This last process is called reinforcement, since postzygotic reproductive isolation (low hybrid fitness) is reinforced by additional prezygotic reproductive isolation. In short, divergence in mating preference can result from drift alone, runaway sexual selection, indirect selection against mating with different species, or direct selection against hybridization.

Regarding the role of geographic isolation in speciation, Jerry Coyne and Allen Orr stated in 2004 that the biogeography of speciation was the most hotly contested question about the process. The purely allopatric model of speciation has been challenged by models of parapatric speciation (adjacent populations with some genetic exchange) and sympatric speciation (genetic exchange within populations at the same location). The process of reinforcement is also pertinent to this debate, since prezygotic reproductive isolation evolves in sympatry as a direct outcome of hybridization and selection against hybrids. In Darwin's finches, the debate over the geography of speciation received some attention in the 1970s and 1980s. According to models of sympatric speciation, disruptive natural selection, favoring two distinct phenotypes, can lead to bimodality in a population. The most important requirement for speciation within this context is that the trait under disruptive selection is linked to a trait governing mate choice; positive assortative mating (similar phenotypes mating with each other) must occur somehow. Interestingly, in 1979 the Grants noticed a pattern of polymorphism in bill morphology and song type in *Geospiza conirostris*, coupled with mating based on song type. They hypothesized that sympatric speciation was in process. In 1983, however, they reported that the pattern of subdivision into morphological and song types had dissolved. While the empirical patterns did not persist in the long term, it may well be that associations, or linkage disequilibrium, can accrue in sympatry between adaptively important morphological characters and characters associated with assortative mating. The question is whether it can persist long enough to effect speciation.

To understand the role of natural selection in speciation, both prezygotic and postzygotic reproductive isolation need to be characterized. How important is drift (both genetic and cultural) for the evolution of pre- and postzygotic reproductive isolation? Is postzygotic reproductive isolation caused by intrinsic genetic incompatibilities that result in developmental anomalies, or is it due to the disruption or dilution of adaptive genotypes, causing deficiencies in adaptation to particular local ecological conditions? Does each evolve in allopatry first, or does prezygotic reproductive isolation evolve in sympatry through the process of reinforcement? The papers here investigate premating and postzygotic reproductive isolation in Darwin's finches and how they contribute to the evolutionary radiation in the Galápagos.

READINGS

The first two papers investigate prezygotic and postzygotic reproductive isolation in Darwin's finches. The paper by Sarah Huber and colleagues documents in a single population (i.e. sympatric birds) the presence of positive assortative mating based on bill size, which is known to be subject to strong natural selection for feeding efficiency. They also show that genetic divergence exists between groups that have different bill morphology. This pattern is similar to those documented in the earlier studies of the Grants mentioned above, but unlike those, this association was consistent over three years of study. Thus, the authors argue, divergence in bill size is accompanied by divergence in mating patterns, which can lead to further genetic divergence. In this scenario, natural selection contributes discernibly to the evolution of prezygotic reproductive isolation and genetic divergence.

The second paper, by the Grants, shows that postzygotic reproductive isolation in Darwin's finches is negligible. First, intrinsic genetic incompatibilities between hybridizing species are not apparent. Second, the intermediate phenotypes of hybrids are not always disadvantageous and may even be advantageous under some conditions. The intermediate phenotypes are associated with intermediate diets and feeding behaviors, both in terms of preferred food items and diet breadth. In some ecological circumstances, such intermediate morphology increases survival. The paper rules out, therefore, intrinsic genetic incompatibilities and provides evidence against the disruption of phenotypes that are adapted to local environmental conditions.

For Darwin's finches, then, reproductive isolation is apparently primarily premating isolation. If mate recognition is based on body and bill morphology, then divergent natural selection on morphology may cause corresponding divergence in mate preference. However, when mate recognition occurs primarily through song, as suggested by Bowman (chapter 8) and other work by the Grants, natural selection and adaptation to different local environments may be relatively insignificant for the evolution of reproductive isolation. That is, adaptation may not contribute much to the actual process of speciation. This would be a surprise.

Nonetheless, the coexistence of two competing species requires some divergence whenever resources are limiting. While song alone might lead to reproductive isolation, it would not lead to the stable coexistence of both species if they were to co-occur. For this to happen, both functionally significant phenotypic divergence and reproductive divergence must occur. Thus, even if mate choice evolves by fancy alone, divergent natural selection plays a role in the emergence of two viable species from one, since it enables their coexistence. In the third selection, Lack emphasizes that the persistence of diverged species is as important for explaining the dynamics of speciation as is the creation of those differences.

Reproductive Isolation of Sympatric Morphs in a Population of Darwin's Finches (2007)

✳ SARAH K. HUBER, LUIS FERNANDO DE LEÓN, ANDREW P. HENDRY, ELDREDGE BERMINGHAM, AND JEFFREY PODOS

Abstract—Recent research on speciation has identified a central role for ecological divergence, which can initiate speciation when (i) subsets of a species or population evolve to specialize on different ecological resources and (ii) the resulting phenotypic modes

become reproductively isolated. Empirical evidence for these two processes working
in conjunction, particularly during the early stages of divergence, has been limited.
We recently described a population of the medium ground finch, Geospiza fortis,
that features large and small beak morphs with relatively few intermediates. As in
other Darwin's finches of the Galápagos Islands, these morphs presumably diverged in
response to variation in local food availability and inter- or intraspecific competition.
We here demonstrate that the two morphs show strong positive assortative pairing, a
pattern that holds over three breeding seasons and during both dry and wet conditions.
We also document restrictions on gene flow between the morphs, as revealed by genetic
variation at 10 microsatellite loci. Our results provide strong support for the central
role of ecology during the early stages of adaptive radiation.

1. INTRODUCTION

Bimodal populations, although rare, provide outstanding opportunities
to study the early stages of adaptive diversification (Smith 1993; Smith &
Skulason 1996; Orr & Smith 1998; Gislason *et al.* 1999; Rundle & Nosil 2005).
We have recently described a bimodal population of the medium ground
finch (*Geospiza fortis*) at El Garrapatero on Santa Cruz Island, Galápagos,
Ecuador (figure 1). This population features birds that fall mainly into large
and small beak size morphs, with relatively few intermediates, a pattern that
has been confirmed statistically (Hendry et al. 2006; Huber & Podos 2006). If
other *G. fortis* populations are any guide (Price 1987; Grant 1999; Keller et al.
2001; Grant & Grant 2006), this variation has a strong additive genetic basis
and reflects selection imposed by variation in the size and hardness of seeds.
The bimodality has almost certainly arisen owing to specialization by the two
morphs on different food types, perhaps coupled with intra- or interspecific
competition, and reflecting processes thought to have driven the adaptive
radiation as a whole (Lack 1947; Boag & Grant 1981; Schluter 2000; Grant &
Grant 2002; Herrel et al. 2005). Moreover, the structure of vocal mating signals (songs) of males at El Garrapatero differs between the morphs in acoustic
parameters that correspond to differences in beak size and vocal performance
(Huber & Podos 2006). The presence of ecologically driven bimodality
in beak size, coupled with divergence in mating signals, suggests that this
population might be in an early stage of speciation, a possibility that we investigate here.

We examined three factors that may influence incipient ecological speciation in El Garrapatero *G. fortis*: the strength of assortative pairing; the
persistence of assortative pairing over time and across variable ecological
conditions; and levels of gene flow between the morphs. Our study focused
on breeding pairs of *G. fortis* during 2004–2006. Climatic conditions varied
widely during these years, which allowed us to test for the strength and stability of assortative pairing under variable ecological conditions. Virtually no

rain fell in 2004 and in the first two months of 2005, making this the most extreme drought in the 40-year period of record (Grant & Grant 2006). Some breeding occurred during this period but at a low rate. Heavy rains fell in March 2005, and the number of breeding pairs increased considerably. More typical rainfall prevailed in 2006.

2. MATERIAL AND METHODS

We studied pairs of *G. fortis* during the breeding season in January–April 2004, January–May 2005 and January–March 2006 at El Garrapatero, Santa Cruz Island, Galápagos, Ecuador (GPS coordinates: 00°40′20″–41′20″ S; 90°13′10″–14′40″ W). Birds were captured in mist nets and banded with unique combinations of one metal and three colour bands. We took the following measurements on each bird (Grant et al. 1985): beak length; beak depth; and beak width. We then collected a small volume of blood from the ulnar vein of each bird, using a 27-gauge needle and filter paper treated with EDTA.

Focal observations of individuals were used to determine pairing status. Repeat observations of pairs were made every 3–4 days throughout the breeding season or until nestlings fledged. The occurrence of two or more of the following behaviours was used to identify mated pairs: nest building by both the male and female; copulation; mate guarding; feeding of the female by the male during incubation or courtship; feeding of nestlings; back-and-forth calling between the male and the female. This study was restricted to pairs that bred (i.e. females that laid eggs). Nests of the two morphs are fully interspersed at our study site (S. K. Huber 2005, unpublished data), and patterns of assortative pairing could thus be attributed to assortative mate choice rather than spatial segregation.

To test for assortative pairing, we first calculated a composite measure of beak size, using a principal components analysis that included beak length, depth and width (as in Grant 1999). Across all birds at El Garrapatero banded between 2004 and 2006, PC1 explained 88.3% of the variation in beak measurements (eigenvalue = 2.65). Assortative pairing was then tested by plotting PC1 for males against PC1 of the females with which they were paired. Nonparametric Spearman's rank correlations were used to determine the degree of assortative pairing. These correlations were calculated based on pairs formed under dry conditions (2004 to early 2005, $n = 21$ pairs), under very wet conditions (late 2005, $n = 33$ pairs) and under moderately wet conditions (2006, $n = 26$ pairs). The 2004 to early 2005 dataset did not contain any duplicate individuals. Some individuals were included in more than one of the three datasets. However, no individuals paired with multiple mates within a given year, and all birds that bred in multiple years changed mates from one year to the next.

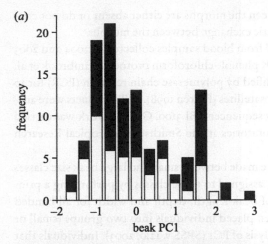

Fig. 1. (*a*) The bimodal distribution of beak sizes for *G. fortis* at El Garrapatero in 2004 (white bars, females; black bars, males). Bimodality has been inferred for this population by statistical comparison of fits with unimodal and bimodal distributions (Hendry *et al.* 2006). (*b*) Representative small morph (left) and large morph (right) birds; both are mature males caught at the same time in the same mist net. (*c*) A large ground finch (*G. magnirostris*) and the same large morph *G. fortis* shown in (*b*). (*d*) A small ground finch (*G. fuliginosa*). Scale bar = 5 mm. Photo credits Andrew Hendry.

The consequences of assortative pairing depend largely on the extent of extra-pair fertilizations (EPFs) and whether EPFs occur within or between the morphs. EPFs have been documented in *G. fortis* of another, small island population (Keller et al. 2001). Even low rates of EPFs would presumably eliminate any genetic differences between the morphs that accrue through assortative mating. To assess levels of intermorph gene flow, we divided the birds into small and large beak size classes, between which we examined patterns of genetic variation across 10 microsatellite loci. If genetic differences are present, then EPFs between the morphs are either absent or do not contribute substantially to genetic exchange between the morphs.

Total DNA was extracted from blood samples collected in 2004 and 2005 using a modified proteinase K phenol–chloroform protocol (Sambrook et al. 1989). Fragments were amplified by polymerase chain reaction (PCR) for 10 unlinked dinucleotide microsatellites (Petren 1998). PCR products were analysed using a multi-capillary sequencer ABI 3100. Genetic work was carried out at Naos Molecular Laboratories at the Smithsonian Tropical Research Institute in Panama.

Genetic comparisons were made between small and large beak size classes within each year. Birds were assigned to these classes by performing a principal components analysis of beak length, depth and width for all banded birds in a given year. We then placed individuals into two groups (small or large) based on a cluster analysis of PC1 (SPSS v. 12.0, 2003). Individuals that were within 0.5 s.d. of the small/large "cutoff" were considered intermediate and removed from the analysis ($n = 47$). These intermediate birds were encompassed by both the upper tail of the small beak morph distribution and the lower tail of the large beak morph distribution, and thus could not be assigned reliably to either class. If gene flow is not influenced by beak size (the null expectation), then removing birds with intermediate beak sizes should have no effect on our results.

Measures of genetic variation, including allelic diversity, frequency and heterozygosity, were calculated using GENEPOP v. 3.4 (Raymond & Rousset 1995), FSTAT v. 2.9.3.2 and ARLEQUIN v. 3.0 (table 1). We tested for Hardy-Weinberg equilibrium using the Markov chain method as implemented in

TABLE 1. Overall genetic diversity for large and small morph *G. fortis* sampled at El Garrapatero (small morph, $n = 197$; large morph, $n = 59$). (Shown are the number of birds analysed (N), number of alleles (N_A), observed heterozygosities (H_O), expected heterozygosities (H_E), F_{IS} and p values from a Hardy-Weinberg test for heterozygote deficits across all birds. Italicized p values indicate those that remained significant after a sequential Bonferroni correction.)

Locus	N	N_A	H_O	H_E	F_{IS}	p
Gf03	254	14	0.818	0.854	0.041	0.017
Gf04	254	5	0.472	0.474	−0.002	0.502
Gf05	254	10	0.677	0.681	0.006	0.255
Gf07	245	19	0.861	0.870	0.010	0.007
Gf08	253	24	0.885	0.925	0.043	*0.004*
Gf09	256	15	0.578	0.601	0.039	0.266
Gf11	243	29	0.844	0.937	0.100	*<0.001*
Gf12	252	16	0.885	0.901	0.017	0.204
Gf13	255	13	0.878	0.870	−0.010	0.751
Gf16	255	11	0.808	0.789	−0.025	0.673

TABLE 2. Pairs of loci that showed significant ($p < 0.05$) linkage disequilibrium across all loci. (Italicized p values indicate those that remain significant after sequential Bonferroni corrections.)

Loci	α^2	d.f.	p
Gf05 & Gf11	infinity	2	*<0.0001*
Gf04 & Gf13	19.44	2	*<0.0001*
Gf03 & Gf05	8.24	2	0.016
Gf05 & Gf08	7.04	2	0.030
Gf04 & Gf08	6.95	2	0.031
Gf08 & Gf11	6.67	2	0.036
Gf07 & Gf08	6.37	2	0.041

TABLE 3. Genetic differences between small and large beak size classes. (Genic differentiation p values were obtained using GENEPOP. Italicized p values indicate those that remained significant after a sequential Bonferroni correction.)

Locus	F_{ST}	R_{ST}	Genic differentiation (p)
Gf03	0.026	0.032	*0.005*
Gf04	0.030	0.016	*0.006*
Gf05	0.012	0.038	*<0.001*
Gf07	0.038	0.001	*<0.0001*
Gf08	0.015	0.051	*0.0001*
Gf09	0.005	−0.000	*0.009*
Gf11	0.017	0.050	*<0.0001*
Gf12	0.007	0.041	0.064
Gf13	0.006	0.049	*0.006*
Gf16	0.012	0.012	*0.006*
overall	0.017	0.040	*<0.0001*

GENEPOP v. 3.4 (dememorization = 10 000, batches = 100, iterations per batch = 5000; table 1). We tested for linkage disequilibrium between pairs of loci using GENEPOP v. 3.4 (table 2).

Genetic differences between beak size classes were analysed in several ways. First, we tested for significant differences in allele frequencies by using the "genic differentiation" option in GENEPOP v. 3.4. Second, we estimated genetic distances using F-statistics (F_{ST}) and R-statistics (R_{ST}; Weir & Cockerham 1984; Slatkin 1995). These analyses used 10 000 randomizations. For F_{ST} across all loci (table 3; $F_{ST} = 0.017$), the 95% confidence interval was computed to be 0.011–0.024 using FSTAT.

Multilocus genotypes were also used to assess population structuring. First, we used factorial correspondence analysis in GENETIX v. 4.05 (Belkhir *et al.* 2004) to determine the similarity of allelic states between the morphological classes. We used a t-test for differences between beak morphs in scores for factors 1 and 2. Second, we used the Bayesian approach implemented in STRUCTURE v. 2.1 (Pritchard *et al.* 2000). We ran five simulations for each putative number of clusters ($K = 1–5$). In each case, we used the admixture model with burn-in of 100 000 and Monte Carlo Markov chain iteration value of 500 000. The most probable number of clusters was always $K = 1$, but visual inspection suggested some differences between the clusters. We therefore tested for significant differences in cluster placement when $K = 2$. Specifically, the results of a STRUCTURE analysis give a probability between one and zero that an individual belongs to cluster 1 or cluster 2 (the sum of probabilities for both clusters is one). Probabilities were arcsine square root transformed, and a *t*-test was used to compare values between the two putative clusters for each morphs in all five iterations. This analysis revealed whether morphs were being randomly placed into the two clusters.

3. RESULTS

We found strong positive assortative pairing by beak size (figure 2a, b) for dry conditions in 2004 and early 2005 ($r = 0.742$, $p = 0.001$; figure 2c), very wet

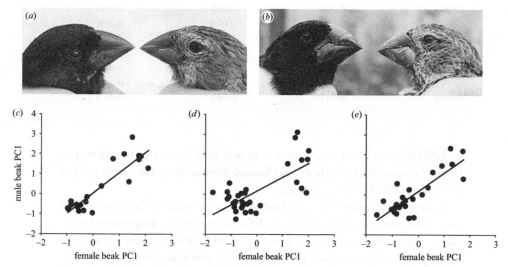

Fig. 2. Assortative pairing by beak size in pairs of *G. fortis* at El Garrapatero. (*a*) A breeding pair of small morph individuals (photo credit Eric Hilton) and (*b*) a breeding pair of large morph individuals (photo credit Sarah Huber) photos not to scale. (*c*) Assortative pairing under dry conditions (2004, early 2005). (*d*) The pattern under very wet conditions (late 2005) and (*e*) the pattern under moderately wet conditions (2006). Male and female "beak PC1" values are scores along the first principal component based on beak length, depth and width.

conditions in late 2005 ($r = 0.390$, $p = 0.025$; figure 2d) and moderately wet conditions in 2006 ($r = 0.705$, $p < 0.001$; figure 2e).

Multiple lines of evidence, consistent across loci, indicate restricted gene flow between the large and small morphs. First, we found several signatures of population admixture (Hardy–Weinberg deficits and linkage disequilibrium) when pooling all of the birds (tables 1 and 2). Second, allele frequencies differed significantly between beak size classes at nine of the ten loci (table 3). Third, genetic divergence measures between the large and small beak size classes were non-trivial ($F_{ST} = 0.017$; $R_{ST} = 0.040$) and differed significantly from zero (table 3).

Some population structure was also evident based on multilocus genotypes. In particular, GENETIX revealed significant differences between the large and small beak size classes on each of the first two factors (factor 1: $t = -2.30$, $p = 0.02$; factor 2: $t = 2.56$, $p = 0.01$). STRUCTURE found few differences between the morphs (with a single cluster always being most likely; mean $\ln(p) = -10318.92$), but this is expected when groups are only moderately differentiated (Pritchard *et al*. 2000). Yet, the assignment of individuals to clusters when $K = 2$ was not random with respect to beak size for large morph individuals in two of the five iterations (iteration 1: $t = -3.24$, $p = 0.001$; iteration 2: $t = 2.47$, $p = 0.014$) and for small morph individuals in one of the five iterations ($t = 5.65$, $p < 0.001$).

4. DISCUSSION

Our genetic data reveal that the large and small beak morphs at El Garrapatero represent two partially distinct gene pools. Our behavioural data suggest that this genetic divergence can be attributed, at least in part, to females' choice of males with similar beak sizes. With this evidence, we are in a position to consider factors that might promote and maintain the observed bimodality.

One possible factor promoting bimodality is disruptive selection in sympatry (Rueffler *et al*. 2006), in this case against birds with intermediate beak sizes. Indeed, we have found that intermediate birds survive at lower rates between years in comparison with large and small morphs (A. P. Hendry & J. Podos 2006, unpublished data). Such disruptive selection could, in principle, lead to a purely sympatric origin of reproductive isolation (Higashi *et al*. 1999; Kondrashov & Kondrashov 1999; Ryan *et al*. 2007). This process is especially probable under two conditions. The first condition is that traits under divergent selection (here beak size) are the same as, or are genetically or phenotypically linked to, traits that influence mate choice (here beak size, a visual cue and song, a vocal mating signal; see Ratcliffe & Grant 1983; Podos 2001). The second condition is that mating is assortative with respect to those traits (Grant *et al*. 2000), as shown here.

Another factor that may promote bimodality is initial divergence during a period of allopatry. The El Garrapatero *G. fortis* morphs may have originated at different places on the same island, or on different islands, under distinct ecological conditions and divergent selection regimes. Following secondary contact, these differences could have led to assortative mating and reduced gene flow through the sympatric processes described above. Indeed, this scenario of initial allopatric divergence followed by further sympatric divergence mirrors a widely accepted model of speciation in many taxa, including Darwin's finches (Grant 1999; Schluter 2000).

Yet another factor potentially influencing bimodality is introgression with other Darwin's finch species. At our study site, *G. fortis* is sympatric with a smaller ground finch species, *Geospiza fuliginosa*, and a larger ground finch species, *Geospiza magnirostris*. Perhaps large *G. fortis* historically hybridized with *G. magnirostris* or small *G. fortis* hybridized with *G. fuliginosa*. We have identified at least one instance of hybridization in our population, in which a large morph *G. fortis* female mated with a *G. magnirostris* male. This sort of interspecies mating could increase phenotypic and genetic variation in *G. fortis*, which might then facilitate the emergence of bimodality (Seehausen 2004). Indeed, the Galápagos ground finches may be a promising system for determining how hybridization facilitates speciation (Mallet 2007) rather than just hampering it (Grant & Grant 2002).

In conjunction with previous work on Darwin's finches, our results support the role of ecologically mediated phenotypic divergence as an important driving force in the early stages of adaptive radiation. Divergence is initiated when variation in food types, food availability or competition imposes divergent selection (in allopatry) or disruptive selection (in sympatry) on beak morphology (Boag & Grant 1981; Grant 1999). Resulting adaptive divergence then imposes secondary consequences on the evolution of mating signals (Ratcliffe & Grant 1983; Podos 2001; Podos & Nowicki 2004; Podos *et al.* 2004; Huber & Podos 2006). This divergence in mating signals may then cause assortative mating and thus help maintain reproductive isolation in sympatry. Beak morphology in Darwin's finches may therefore be regarded as one of the elusive "magic traits" of speciation (Gavrilets 2004), given that it is both a target of divergent selection and a component in mating signals that drive assortative mating (Grant & Grant 1997; Podos & Hendry 2006). We have identified a population that seems to be in the early stages of this process.

It is uncertain whether or not the two morphs at El Garrapatero will ultimately diverge to the level of well-defined species. For instance, immigration from a nearby unimodal population (Hendry *et al.* 2006), which might itself have considerable gene flow between birds with large and small beaks, could hamper further divergence between the morphs at El Garrapatero. Additionally, environmental conditions may eventually change to the extent that "hybrid" offspring no longer have reduced fitness, as has been the case for established species of Darwin's finches (Grant & Grant 1996). Regardless, our data support the hypothesis that the early stages of assortative mating and reproductive isolation are driven by ecological divergence.

The collection of data in this study was done in concordance with Animal use Protocols approved by the University of Massachusetts Amherst.

We thank the Galápagos National Park and the Charles Darwin Research Station for providing support to this research. Anthony Herrel, Ana Gabela, Eric Hilton, Haldre Rogers and Steve Johnson provided assistance in the field. Funding was provided by National Science Foundation grant IBN 0347291 to J.P., and National Science Foundation grant DDIG 0508730 to J.P. and S.K.H. Additional funding for S.K.H. was provided by an American Ornithologists' Union Student Research grant, an Explorer's Club grant, a Sigma Xi Grant-in-Aid of Research, an Animal Behaviour Society Student Research grant and a University of Massachusetts Woods Hole Fellowship. Support for L.F.D. was provided by Instituto para la Formación y Aprovechamiento de Recursos Humanos and Secretaria Nacional de Ciencia, Tecnología e Innovación.

REFERENCES

Belkhir, K., Chikhi, L., Raufaste, N. & Bonhomme, F. 2004 *GENETIX logiciel sous Windows TM pour la génétique des populations*, 4.05. Montpellier, France: Laboratoire Génome, Populations, Interactions CNRS UMR 5000, Université de Montpellier II.

Boag, P. T. & Grant, P. R. 1981 Intense natural selection in a population of Darwin's finches (Geospizinae) in the Galápagos. *Science 214*, 82–85. (doi:10.1126/science.214.4516.82)

Gavrilets, S. 2004 *Fitness landscapes and the origin of species.* Princeton, NJ: Princeton University Press.

Gislason, D., Ferguson, M. M., Skulason, S. & Snorrason, S. S. 1999 Rapid and coupled phenotypic and genetic divergence in Icelandic Arctic char (*Salvelinus alpinus*). *Can. J. Fish. Aquat. Sci. 56*, 2229–2234. (doi:10.1139/ cjfas-56–12–2229)

Grant, P. R. 1999 *Ecology and evolution of Darwin's finches.* Princeton, NJ: Princeton University Press.

Grant, B. R. & Grant, P. R. 1996 High survival of Darwin's finch hybrids: effects of beak morphology and diets. *Ecology 77*, 500–509. (doi:10.2307/2265625)

Grant, P. R. & Grant, B. R. 1997 Mating patterns of Darwin's finch hybrids determined by song and morphology. *Biol. J. Linn. Soc. 60*, 317–343. (doi:10.1006/bijl.1996.0103)

Grant, P. R. & Grant, B. R. 2002 Unpredictable evolution in a 30-year study of Darwin's finches. *Science 296*, 707–711. (doi:10.1126/science.1070315)

Grant, P. R. & Grant, B. R. 2006 Evolution of character displacement in Darwin's finches. *Science 313*, 224–226. (doi:10.1126/science.1128374)

Grant, P. R., Abbot, I., Schluter, D., Curry, R. L. & Abbott, L. K. 1985 Variation in the size and shape of Darwin's finches. *Biol. J. Linn. Soc. 25*, 1–39.

Grant, P. R., Grant, B. R. & Petren, K. 2000 The allopatric phase of speciation: the sharp-beaked ground finch (*Geospiza difficilis*) on the Galápagos islands. *Biol. J. Linn. Soc. 69*, 287–317. (doi:10.1006/bij1.1999.0382)

Hendry, A. P., Grant, P. R., Grant, B. R., Ford, H. A., Brewer, M. J. & Podos, J. 2006 Possible human impacts on adaptive radiation: beak size bimodality in Darwin's finches. *Proc. R. Soc. B 273*, 1887–1894. (doi:10.1098/rspb.2006.3534)

Herrel, A., Podos, J., Huber, S. K. & Hendry, A. P. 2005 Bite performance and morphology in a population of Darwin's finches: implications for the evolution of beak shape. *Funct. Ecol. 19*, 43–48. (doi:10.1111/j.0269–8463.2005.00923.x)

Higashi, M., Takimoto, G. & Yamamura, N. 1999 Sympatric speciation by sexual selection. *Nature 402*, 523–526. (doi:10.1038/990087)

Huber, S. K. & Podos, J. 2006 Beak morphology and song features covary in a population of Darwin's finches (*Geospiza fortis*). *Biol. J. Linn. Soc. 88*, 489–498. (doi:10.1111/j.1095–8312.2006.00638.x)

Keller, L. F., Grant, P. R., Grant, B. R. & Petren, K. 2001 Heritability of morphological traits in Darwin's finches: misidentified paternity and maternal effects. *Heredity 87*, 325–336. (doi:10.1046/j.1365–2540.2001.00900.x)

Kondrashov, A. S. & Kondrashov, F. A. 1999 Interactions among quantitative traits in the course of sympatric speciation. *Nature 400*, 351–354. (doi:10.1038/22514)

Lack, D. 1947 *Darwin's finches*. Cambridge, UK: Cambridge University Press.

Mallet, J. 2007 Hybrid speciation. *Nature 446*, 279–283. (doi:10.1038/nature05706)

Orr, M. R. & Smith, T. B. 1998 Ecology and speciation. *Trends Ecol. Evol. 13*, 502–506. (doi:10.1016/S0169–5347(98)01511–0)

Petren, K. 1998 Microsatellite primers from *Geospiza fortis* and cross-species amplification in Darwin's finches. *Mol. Ecol. 7*, 1782–1784.

Podos, J. 2001 Correlated evolution of morphology and vocal signal structure in Darwin's finches. *Nature 409*, 185–188. (doi:10.1038/35051570)

Podos, J. & Hendry, A. P. 2006 The biomechanics of ecological speciation. In *Ecology and biomechanics: a mechanical approach to the ecology of animals and plants* (eds A. Herrel, T. Speck & N. Rowe), pp. 301–321. Boca Raton, FL: CRC Press.

Podos, J. & Nowicki, S. 2004 Beaks, adaptation, and vocal evolution in Darwin's finches. *Bioscience 54*, 501–510. (doi:10.1641/0006–3568(2004)054[0501:BAAVEI]2.0.CO;2)

Podos, J., Southall, J. A. & Rossi-Santos, M. 2004 Vocal mechanics in Darwin's finches: correlation of beak gape and song frequency. *J. Exp. Biol. 207*, 607–619. (doi:10.1242/jeb.00770)

Price, T. 1987 Diet variation in a population of Darwin's finches. *Ecology 68*, 1015–1028. (doi:10.2307/1938373)

Pritchard, J. K., Stephens, M. & Donnelly, P. 2000 Inference of population structure using multilocus genotype data. *Genetics 155*, 945–959.

Ratcliffe, L. M. & Grant, P. R. 1983 Species recognition in Darwin's finches (*Geospiza*, Gould). I. Discrimination by morphological cues. *Anim. Behav. 31*, 1139–1153. (doi:10.1016/S0003–3472(83)80021–9)

Raymond, M. & Rousset, F. 1995 GENEPOP version 1.2: population genetics software for exact tests and ecumenicism. *J. Hered. 86*, 248–249.

Rueffler, C., Van Dooren, T. J. M., Leimar, O. & Abrams, P. A. 2006 Disruptive selection and then what? *Trends Ecol. Evol. 21*, 238–245. (doi:10.1016/j.tree.2006.03.003)

Rundle, H. D. & Nosil, P. 2005 Ecological speciation. *Ecol. Lett. 8*, 336–352. (doi:10.1111/j.1461–0248.2004.00715.x)

Ryan, P. G., Bloomer, P., Moloney, C. L., Grant, T. J. & Delport, W. 2007 Ecological specia-

tion in South Atlantic island finches. *Science 315*, 1420–1423. (doi:10.1126/science
.1138829)

Sambrook, J., Fritsch, E. F. & Maniatis, T. 1989 *Molecular cloning: a laboratory manual*, 2nd
edn. New York, NY: Cold Spring Harbor Laboratory Press.

Schluter, D. 2000 *The ecology of adaptive radiation*. Oxford, UK: Oxford University Press.

Seehausen, O. 2004 Hybridization and adaptive radiation. *Trends Ecol. Evol. 19*, 198–207.
(doi:10.1016/j.tree.2004.01.003)

Slatkin, M. 1995 A measure of population subdivision based on microsatellite allele frequencies. *Genetics 139*, 457–462.

Smith, T. B. 1993 Disruptive selection and the genetic basis of bill size polymorphism in
the African finch *Pyrenestes*. *Nature 363*, 618–620. (doi:10.1038/363618a0)

Smith, T. B. & Skulason, S. 1996 Evolutionary significance of resource polymorphisms in
fishes, amphibians, and birds. *Annu. Rev. Ecol. Syst. 27*, 111–133. (doi:10.1146/annurev.
ecolsys.27.1.111)

Weir, B. S. & Cockerham, C. C. 1984 Estimating *F*-statistics for the analysis of population
structure. *Evolution 38*, 1358–1370. (doi:10.2307/2408641)

Proceedings of the Royal Society B 274 (2007): 1709–1714. Reprinted with permission from the
Royal Society.

High Survival of Darwin's Finch Hybrids: Effects of Beak Morphology and Diets (1996)

❋ B. ROSEMARY GRANT AND PETER R. GRANT

Abstract—Three species of Darwin's finches (Geospiza fortis, *G.* scandens, *and*
G. fuliginosa) *hybridize rarely on the small Galápagos island of Daphne Major. Following the exceptionally severe El Niño event of 1982–1983, hybrids survived as well
as, and in some cases better than, the parental species during dry seasons of potential
food limitation. They also backcrossed to two of the parental species. This study was
undertaken to compare the diets of hybrids with the diets of the parental species in
order to assess possible reasons for the high hybrid survival. Diets of F₁ hybrids and first
generation backcrosses to* G. fortis *were intermediate between the diets of the respective parental species. Distinctiveness of the hybrid diets was most pronounced where
the diets of the parental species differed most. A strong determinant was beak morphology; hybrids inherit beak traits from both parents, and, on average, have intermediate
beak sizes. Among the combined groups of species and hybrids, and among the hybrids
alone, dietary characteristics covaried with beak morphology. Hybrids that differ most
from* G. fortis *in beak morphology, the* G. fortis × G. scandens *F₁ hybrids, experience a feeding efficiency advantage when feeding on* Opuntia echios *seeds, commonly
consumed in the dry season.*

*These findings are used to interpret the higher survival of hybrids after 1983
than beforehand. The El Niño event that year led to an enduring (10-yr) change in
the habitat and plant composition of the island. A decrease in absolute and relative
abundance of large and hard seeds apparently caused relatively high mortality among*

G. scandens *and the largest* G. fortis *individuals. Hybrids were favored by an abundance of small seeds. The high survival of* G. fortis × G. scandens F_1 *hybrids may have been due, additionally, to a broad diet and to efficient exploitation of Opuntia seeds. The study demonstrates long-term ecological and evolutionary consequences of large-scale fluctuations in climate, and the role of ecological (food) factors in determining hybrid fitness.*

INTRODUCTION

Hybridization can have significant ecological and evolutionary consequences for the species that interbreed and for others that interact with them (Harrison 1990, 1993, Arnold 1992, Floate et al. 1993). In plants and animals, there are many examples in which a few members of a population interbreed with individuals of a closely related species, and gene flow occurs through backcrossing to one or both of the parental species. Occasionally, new species are formed by hybridization, but more often the genetic and ecological characteristics of the interbreeding species are transformed to varying degrees. Some of these examples occur naturally, others occur as a consequence of human alteration of the environment that brings into contact closely related species that were previously separated (Anderson and Stebbins 1954, Harrison 1990, Arnold 1992).

The main framework for studying hybridization is a theory of gene flow across environmental gradients (Endler 1977, Barton and Hewitt 1985, Harrison 1993). This has been developed and used to account for many features of zones of hybridization that separate broad distributions of the hybridizing species, especially terrestrial species. These features include the width of the zone and the concordant or discordant clinal variation of morphological, biochemical, and life history traits across the zone.

The fitness of hybrids is usually lower than the fitness of the parental species (Hewitt 1989). When hybrid fitness is not detectably different from parental species' fitnesses, or even higher, hybrids are usually restricted to narrow zones (Moore and Koenig 1986, Saino and Villa 1992). The high fitness may be due to the particular ecological conditions in the zone, comprising elements of the conditions experienced by each of the interbreeding species in their respective geographical ranges (Moore 1977, Harrison 1990). The usual reason postulated for why the zone does not increase in width is that the boundaries are fixed by spatial variation in the ecological variables governing fitness (Moore 1977, Barton and Hewitt 1989). Presumably, when the zones are static (e.g., Moore and Buchanan 1985), the ecological variables remain geographically constant. When there is temporal variation in those ecological variables, the zones or local boundaries move (Gill 1980, Arntzen and Wallis 1991). Identifying those variables has not been easy. Although temperature, features of the habitat, and food are obvious candidate variables (Bert and

Harrison 1988, Harrison and Rand 1989, Arntzen and Wallis 1991, Saino 1992, Saino and Villa 1992), surprisingly little quantitative work has been done on them in either fixed or moving hybrid zones.

Ecological shortcomings of the theory are evident in those situations where hybridization occurs locally but not uniformly across the region of contact of the interbreeding species (Sibley 1954, Dowling et al. 1989, Dowling and Hoeh 1991, Aspinwall et al. 1993); where the zone is a mosaic of patches of hybridization (Harrison and Rand 1989, Spence 1990, Arntzen and Wallis 1991, Howard et al. 1993); where the environment is dendritic, as is the case with rivers, rather than planar (Smith 1992); and in environments like lakes, where distributions of the species are coextensive and there is no zonation in the distribution of hybrids (Smith 1992, Aspinwall et al. 1993). In all these cases, the key to understanding the spatial patterning of hybridization and the relative fitness of hybrids may lie in the ecology and breeding behavior of the hybridizing species, although genetic heterogeneity is also clearly relevant (Howard et al. 1993, Scribner 1993).

To be able to predict the consequences of hybridization, we need to understand not only the genetic characteristics of the hybridizing species but also the ecological conditions under which hybrid offspring survive long enough to reproduce. This paper examines the second topic in populations of Darwin's finches *(Geospiza:* Geospizinae) on the undisturbed Galápagos island of Daphne Major. Like some fish species in lakes, finches hybridize without setting up a hybrid zone. Temporal variation in environmental conditions, rather than spatial variation, governs the outcome of hybridization.

Background

Daphne Major is a small island of 34 ha. The island has no human settlement or introduced plants or animals. It supports breeding populations of four species of Darwin's finches. These species differ from each other in body size, bill dimensions, and song. They resemble each other in plumage coloration, type of nest, and in behavioral characteristics associated with courtship (Grant 1986).

Survival and breeding of uniquely banded individuals of these species were studied intensively from 1976 to 1992 (Grant and Grant 1992a, 1995a). Low levels of hybridization occurred among three of the species: between *G. fortis* (Medium Ground Finch, body mass $\approx$ 17 g) and *G. fuliginosa* (Small Ground Finch, $\approx$14 g), which have similar allometries; and between *G. fortis* and *G. scandens* (Cactus Finch, $\approx$21 g), which have different allometries (Grant and Grant 1994). Interbreeding involved, on average, 1.8% of the breeding *G. fortis* population (harmonic mean of 198 breeding individuals), 0.8% of the *G. scandens* breeding population (harmonic mean of 80), and 73% of the very small *G. fuliginosa* population (harmonic mean of 3) (Grant 1993).

From 1976 to 1983, *G. fortis* × *G. fuliginosa* pairs produced a total of 32 fledglings. However, only two of these survived to 1983, and none bred before that year (Grant and Grant 1993). Similarly, the offspring of two *G. fortis* × *G. scandens* pairs born during this period failed to survive long enough to breed. Environmental conditions changed in 1983 as a result of an exceptionally large amount of rain associated with a severe El Niño event (Gibbs and Grant 1987a). During and after 1983, a total of 92 *G. fortis* × *G. fuliginosa* fledglings and 27 *G. fortis* × *G. scandens* fledglings were produced, of which 59 and 15, respectively, survived to breed. Between 1983 and 1992, there was extensive backcrossing of hybrids into the *G. fortis* population without loss of fitness, and in 1992, the first cases of backcrossing to *G. scandens* were recorded. Both sexes of F_1 hybrids and the backcrosses were viable and fertile, showing no indication of the Haldane effect of sex-related genetic incompatibility (Grant and Grant 1992b).

After 1983, survival and reproduction of the hybrids and the products of the backcrosses to *G. fortis* and *G. scandens* were as good as, or superior to, those of the parental species (Grant and Grant 1992b, Grant and Grant 1993). Both genetic and ecological factors might have contributed to the high success of hybrids at this time (Grant and Grant 1992b). An obvious ecological factor of potential importance in the high survival of hybrids is their diet. This is the focus of our paper.

Diet variation among species on this island is related to beak size and shape (Grant 1981, Boag and Grant 1984a, Price 1987), especially in the dry season when populations are most likely to be food limited (Smith et al. 1978, Grant 1986). Beak dimensions are heritable traits and, not surprisingly, hybrids have beak characteristics intermediate between those of the respective parental species (Grant and Grant 1994). Therefore, we investigated dry-season diets of hybrids and backcrosses in relation to diets of the parental species, paying attention to the association between diet and beak size and shape. Our expectation was that hybrids and backcrosses would have dietary characteristics intermediate between those of the hybridizing species.

METHODS

Methods have been described in detail elsewhere (Boag and Grant 1984a, b, Gibbs and Grant 1987a, Grant 1993). From 1976 to 1991, either we or our assistants spent the first 4–8 mo of each year on the island. Breeding normally occurs in the first few months of the year, in response to rainfall (Grant and Boag 1980). An attempt was made to find every nest and to identify every banded individual. We spent January–March on the island in 1992 and 1993, and January–February in 1994 and 1995. Rainfall was recorded daily in a rain gauge; in 1992, a permanent rain gauge was installed to record rainfall while we were not present.

Finch Morphology

Adults and immatures were captured in mist nets every year since 1975. Before release, they were measured and given a unique combination of one numbered metal band and three colored plastic (PVC) leg bands coded to correspond to the number on the metal band. They were weighed to the nearest 0.1 g with a Pesola spring balance, and five dimensions were measured in millimetres: length of wing and tarsus, and length, depth, and width of bill. Since 1976, nestlings were banded in every year when breeding occurred; in some drought years there was no breeding (Grant and Grant 1992a). They were measured again when captured in nets fully grown, i.e., after day 60 (Boag 1984). By 1991, all finches on the island except two had been banded (Grant and Grant 1992a).

Hybrids were identified from pedigrees based on observations of known adults feeding nestlings. Incorrect assignment of parents is likely to be a minor problem, at most, for two reasons. First, observations of extra-pair copulations are extremely rare in the week before and during the egg-laying period (Price 1984, Grant and Grant 1995b), when the female is presumed to be receptive (cf. Birkhead and Møller 1992), and egg-laying in the nests of other females has never been observed. Second, regressions of offspring measurements on the measurements of the apparent father are no different from those on the measurements of the apparent mother, and both sets are very high (Boag 1983). For the diagnosis of species and a full list of hybrids and backcrosses, see Grant (1993).

Diets

From 1976 to 1993, feeding observations of uniquely banded birds were made between 0700 and 1000 every 7–10 d in the months January–May, until rain fell and birds started to breed (see Boag and Grant 1984b, Price 1987, for details). Only the first identified food item consumed by each bird was recorded on any one day. Feeding data were collected on "feeding walks" throughout the island. The tameness of the birds and the simplicity of the habitat usually made food identification easy, and it is unlikely that the data were biased by overrepresentation of the most conspicuous feeding activities.

For statistical purposes, hybrids were compared with parental species over the period when hybrids were observed feeding. With three trivial exceptions, feeding observations of F_1 hybrids were not made before 1983, diets of first generation backcrosses were not known before 1987, and diets of second generation backcrosses were not known before 1991. Diets were compared in two ways: first, by using the first observation made on each bird; and second, by using all observations of birds seen at least five times. When a group was compared with more than one other, N times, the decision to accept or reject the null hypothesis was based on $P = 0.05/N$. All tests were two-tailed, except where indicated.

Feeding Efficiency

Timed observations of the handling of two common seed species (*Chamaesyce amplexicaulis*, a small and soft seed (fruit), and *Opuntia echios*, a medium seed) were made with stopwatches in 1994 and 1995. We repeatedly recorded the number of *C. amplexicaulis* fruits removed from a bush in 30 s by as many identified birds as possible, and averaged the counts per bird. Finches were also timed while feeding on seeds obtained from green or yellow (moist) *Opuntia* fruits that had fallen to the ground. They first removed the surrounding aril, then cracked the seed and extracted the kernel. Total handling time per seed was recorded as often as possible, and averages were calculated for different birds.

RESULTS

Diets of the Species

Seeds dominate dry season diets. There are <50 seed-producing plant species on the island, most of them uncommon or rare (Boag and Grant 1984b). Seeds consumed by the finches fall into three discrete categories of size (depth, D, in millimetres) and hardness (H, in newtons) combined (see Abbott et al. 1977 and Grant and Grant 1993 for details). The category of small and soft seeds ($DH^{1/2} < 4$) comprises 22 species; three species, *Bursera graveolens*, *Cenchrus platyacanthus*, and *Opuntia echios*, constitute the medium category ($DH^{1/2} = 9–14$); and the third category is represented by a single species, *Tribulus cistoides*, whose woody mericarp containing four to six seeds has an average depth-hardness value of 27. *Bursera* and *Cenchrus* seeds are consumed preferentially in the wet season (Abbott et al. 1977), with the result that few remain in the dry season. Finches consume, in addition to seeds, nectar and pollen from *Opuntia* flowers (Grant 1996), and arthropods obtained from the surface of various plants or excavated from rotting *Opuntia* pads and stems. These various food types are exploited in different ways that depend on beak morphology (Grant 1986). We have used data on adults only, i.e., excluding those birds observed feeding in the calendar year of hatch, to characterize diets and compare finch species and hybrids statistically.

The four species of finches have clearly different diets in the dry season. The three main hybridizing species are shown in Fig. 1. Most *G. fuliginosa* were observed feeding on small seeds, most *G. magnirostris* ($N = 108$, not shown) were seen feeding on the largest and hardest seeds (*Tribulus cistoides*), and most *G. scandens* were recorded feeding on seeds or flowers of *Opuntia* cactus. Like *G. fuliginosa*, the fourth species, *G. fortis*, feeds mainly on small seeds but has a more generalized diet. Overall, there is a strong statistical heterogeneity among the four species when diets are represented in six categories ($\chi^2_{15} = 1031.38$, $P < 0.0001$): small seeds, *Opuntia* seeds, *Opuntia* flowers, large seeds, arthropods, and all other items. Taken separately, the two most similar species, *G. fortis* and *G. fuliginosa*, differ significantly ($\chi^2_5 = 18.85$, $P = 0.0021$).

G. magnirostris is included in the comparison for completeness, even though it has been known to hybridize (with *G. fortis*) only once and without fledging any offspring (Grant 1993). It will not be considered further.

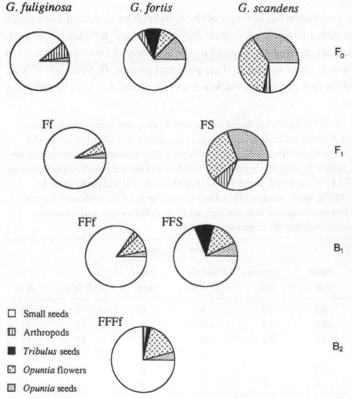

Fig. 1. Diets of Darwin's finch species and three generations of hybrids constructed from the first observation of food consumed by each identified bird in the period 1984–1994. A miscellaneous category, comprising mainly the soft pulp of *Opuntia* cactus pads and stems, has been omitted. The groups and sample sizes (in parentheses) are *Geospiza fortis* (1981), *G. fuliginosa* (44), and *G. scandens* (657); *G. fortis* × *G. fuliginosa* F_1 hybrids (Ff; 36) and *G. fortis* × *G. scandens* F_1 hybrids (FS; 13); first generation backcrosses (B_1) to *G. fortis* of Ff (FFf; 38) and FS (FFS; 32); and a second generation backcross (B_2) to *G. fortis* (FFFf; 47).

Diets of Hybrids

Diets of F_1 hybrids are intermediate, in their main components, between the diets of their respective parental species, according to compiled records of the first observations of each bird. Comparisons are illustrated with the full data set in Fig. 1, and analyzed statistically with a reduced data set obtained under strictly comparable conditions (same ages and years) given in Table 1.

G. fortis and G. fuliginosa differ significantly from each other, although not strongly ($\chi^2_1 = 13.56$, $P = 0.0187$). However, the F_1 hybrids do not differ significantly from either G. fortis ($\chi^2_5 = 7.91$, $P > 0.1$) or G. fuliginosa ($\chi^2_1 = 0.86$, $P > 0.3$); for the latter test, observations were grouped into two categories, small seeds vs. all else, because of the small numbers of observations in several cells (Table 1).

The first generation backcrosses of these hybrids to G. fortis do not differ in diet from either G. fortis ($\chi^2_5 = 2.16$, $P > 0.5$) or the F_1 hybrids ($\chi^2_1 = 0.01$, $P > 0.5$); diets are dominated by small seeds. Similarly, the second generation backcrosses to G. fortis resemble their parental groups, G. fortis ($\chi^2_5 = 3.18$, $P > 0.5$) and the first generation backcrosses ($\chi^2_1 = 0.01$, $P > 0.5$). Backcrosses

TABLE 1. Diets of hybrid finches, based on the first feeding observation made on each identified bird. Numbers of birds in each category are shown; no individual enters the table twice. The category Other is mainly *Opuntia* pulp (Fig. 1). Abbreviations refer to the species; to G. fortis × G. fuliginosa hybrids (Ff) and first and second generation backcrosses to G. fortis (FFf, FFFf); and to G. scandens × G. fortis hybrids (FS) and backcrosses to G. fortis (FFS, FFFS) and G. scandens (SSF). Data for species and F_1 hybrids are from 1984 onwards, for first generation backcrosses they are from 1988 onwards, and for second generation backcrosses they are from 1993 onwards

	Diet categories					
Species	Small seeds	Opuntia seeds	Opuntia flowers	Large seeds	Arthropods	Other
G. fuliginosa f	33	0	0	0	1	0
G. fortis F	907	59	130	98	65	75
G. scandens S	88	79	187	6	17	32
F_1 Hybrids						
Ff	30	1	1	0	0	2
FS	4	4	4	0	1	0
Backcross 1						
FFf	28	2	4	0	2	2
FFS	21	2	4	3	0	2
SSF	1	3	12	0	0	0
Backcross 2						
FFFf	34	2	8	1	0	2
FFFS	6	0	5	1	0	0

to *G. fuliginosa* have been produced rarely, and they have not been observed feeding.

Hybridization between *G. fortis* and *G. scandens* produces birds with distinctive diets (Fig. 1, Table 1). The F_1 hybrids differ from *G. fortis* ($\chi^2_5 = 29.36$, $P = 0.0001$) in eating small seeds relatively infrequently. They also differ from the *G. fortis* × *G. fuliginosa* F_1 hybrids ($\chi^2_4 = 21.36$, $P = 0.0003$). Their diets resemble those of *G. scandens* ($\chi^2_5 = 3.53$, $P > 0.1$) in their concentration on *Opuntia* cactus and in the absence of large seeds. Hybrids have backcrossed to both parental species. The first generation backcrosses to *G. fortis* have diets indistinguishable from those of *G. fortis* ($\chi^2_5 = 2.94$, $P > 0.5$) but differing from those of *G. scandens* ($\chi^2_5 = 60.08$, $P = 0.0001$). Reciprocally, backcrosses to *G. scandens* do not differ from *G. scandens* ($\chi^2_5 = 0.25$, $P > 0.5$), but do differ strongly from *G. fortis* ($\chi^2_5 = 63.77$, $P = 0.0001$).

Most of these results were confirmed by analyzing the total observations made of hybrids and first generation backcrosses, but restricting the data set to those individuals observed a minimum of five times. For these comparisons, it was important to minimize extraneous variation due to sample size effects, annual changes in feeding conditions, and, possibly, experience. This was done by matching each hybrid with a *G. fortis* or *G. scandens* observed the same number of times (5–29) and closest to it in band number (and age). The smaller sample sizes precluded analyses of second generation backcrosses. Only two differences from the first set of results were found. First, the *G. fortis* × *G. scandens* F_1 hybrids differed from the matched sample of *G. fortis* ($\chi^2_4 = 38.48$, $P = 0.0001$), as was found in the previous analysis, but they also differed strongly from *G. scandens* ($\chi^2_3 = 35.49$, $P = 0.0001$). Second, first generation backcrosses of these hybrids to *G. fortis* differed from *G. fortis* ($\chi^2_4 = 20.46$, $P = 0.004$) in being less restricted to small seeds. The frequency of consumption of large (*Tribulus*) seeds was notably high (10.5%).

In combination, analyses of the full and restricted samples of species and hybrids show that diets of F_1 hybrids are intermediate between diets of the parental species. Differences between hybrids and parental species are pronounced and statistically demonstrable when the parental species differ strongly (*G. fortis* and *G. scandens*).

Diet Breadths

Breadths of hybrid diets were compared with breadths of species diets by considering only those identified individuals observed feeding a minimum of five times. Two measures of breadth, or diversity (Magurran 1988), were used: C, the number of food categories in the diet (listed in Table 1) but excluding "other," and Simpson's index of diversity, $D = 1/\Sigma p_{ij}^2$ where p_{ij} refers to the proportion of the ith category in the diet of species j. Both measures tend to be positively correlated with number of observations in all species and hybrid groups, although the scatter of points is wide. To minimize variation due to sample size and annual variation in feeding conditions, each hybrid was matched with a *G. fortis* or *G. scandens* individual observed the same number of times and closest to it in band number, as was done in the analysis of diets using the restricted samples.

Mean diet breadth (C) was about one category (small seeds) for *G. fuliginosa*, two for *G. fortis*, and three for *G. scandens*. *G. fortis*, the middle species, differed significantly from both *G. fuliginosa* (paired $t_6 = 3.29$, $P = 0.0167$) and *G. scandens* (paired $t_{15} = 3.49$, $P = 0.0045$) at the extremes. The same results were obtained with interspecific comparisons of diversity (D), and at the same significance levels. Both tests were two-tailed. Results justify the use of one-tailed tests of directional hypotheses involving hybrids.

As expected, hybrids are observed to occupy intermediate positions on each of the two diet breadth axes (Fig. 2). *G. fuliginosa* $\times$ *G. fortis* F_1 hybrids exploit a smaller number of categories (C) than *G. fortis* (paired $t_{13} = 1.963$, $P = 0.0357$); *G. fortis* $\times$ *G. scandens* F_1 hybrids exploit a smaller number of categories than *G. scandens* (paired $t_9 = 2.71$, $P = 0.0119$); and the backcrosses of *G. fortis* $\times$ *G. scandens* hybrids to *G. fortis* exploit a larger number of categories

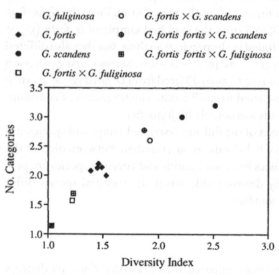

Fig. 2. Diet diversity of Darwin's finch species and hybrids. The mean number of diet categories recorded for each group of finches is plotted against Simpson's index of diversity, *D*, which weights the diet categories by their proportional use. *G. fortis* and *G. scandens* are represented by more than one sample, each matched with a different hybrid group. There is a positive correlation between the two diversity measures ($r = 0.966$, $N = 7$, $P < 0.005$) when the largest *G. fortis* and *G. scandens* samples are used.

than *G. fortis* (paired t_{12} = 2.635, P = 0.0109). The same results are obtained with diversity (*D*), except that with this measure the *G. fortis* × *G. scandens* F$_1$ hybrids have a more diverse diet than *G. fortis* (paired t_9 = 1.848, P = 0.0488). All tests with hybrids were one-tailed.

Diets in Relation to Beak Morphology

Intermediate diets of hybrids are likely to be determined in part by intermediate beak morphology, just as diets of parental species are determined by their beak morphologies (Grant 1981, Boag and Grant 1984a, Price 1987). Fig. 3 (upper) shows the positions of the three hybridizing species, their F$_1$ hybrids, and two groups of first generation backcrosses on two beak size axes. The lower portion of this figure shows the same groups arrayed on two principal niche axes. Here, the proportion of individuals in each group observed feeding at least once on *Opuntia* flowers and *Tribulus* seeds has been used to characterize the importance of these two very different food items in the diet of each group. This characterization was chosen to maximally include relatively infrequent items in the recorded diet of individuals.

There is a strong overall correspondence between beak size and shape and diet, as revealed by a high canonical correlation (*R* = 0.972) between the two sets of data; all canonical roots combined are highly significant (χ^2_4 = 13.907, P = 0.0076).

The species are well separated on the *Opuntia* flower axis, and F$_1$ hybrids and backcrosses occupy intermediate positions. A less pronounced separation is seen on the seed axis. Positions on the individual niche axes can be predicted from beak dimensions (Table 2). Multiple regression analyses were performed with arcsine-transformed proportions of birds observed feeding on *Opuntia* flowers or *Tribulus* seeds, taken from the restricted data set and regressed on mean beak dimensions. Proportions of birds in the different groups (*N* = 7) observed feeding on nectar and/or pollen from *Opuntia* flowers are well predicted (R^2 = 0.994). The slope coefficient (β) for bill length has a positive sign and is the only one significantly different from zero (Table 2). Position on the seed axis is also predicted by beak dimensions (R^2 = 0.965). *Tribulus* proportions increase with an increase in bill width and, independently, a decrease in bill depth. There is no association with bill length. Bill width has a larger standardized partial regression coefficient (6.031) than bill depth (−5.239), and therefore is the better predictor of feeding proportions. The opposite signs of the coefficients can be interpreted as reflecting a shape factor.

The association between diet and bill dimensions is exhibited by the sample of hybrids alone (*N* = 51), as demonstrated by two MANOVA. Hybrids observed feeding on *Opuntia* flowers differed from those not observed doing so

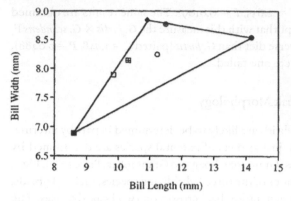

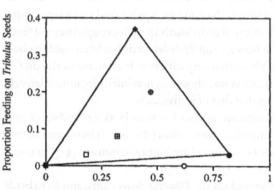

Fig. 3. The association between beak morphology (upper) and diet (lower) among species and hybrids of Darwin's finches. Points on the *x*-axis represent the proportion, from the total data set, of birds in each species and hybrid group observed feeding at least once on *Opuntia* flowers. Points on the *y*-axis represent proportions of the seed-eating subset observed feeding on *Tribulus* seeds, which are the largest and hardest seeds. Hybrids occupy similar relative positions on the morphological and niche axes. Symbols as in Fig. 2.

TABLE 2. Results of Multiple Linear Regression of proportions of birds observed feeding on (1) *Opuntia* flowers for pollen and nectar and (2) *Tribulus* seeds, on three beak dimensions. Proportions of birds in each of the three species, two F_1 hybrid groups and two first generation backcross groups were arcsine-transformed prior to analysis. β values with standard errors (1 SE) are partial regression coefficients

	β	SE	t	P
1. *Opuntia* flowers ($F_{1,5}$ = 153.603, P = 0.0009, R^2 = 0.994)				
Bill length	0.328	0.019	17.256	0.0004
Bill depth	0.213	0.186	1.142	0.3362
Bill width	−0.430	0.251	1.715	0.1848
2. *Tribulus* seeds ($F_{1,5}$ = 27.645, P = 0.011, R^2 = 0.965)				
Bill length	−0.022	0.011	2.045	0.1333
Bill depth	−0.717	0.104	6.911	0.0062
Bill width	1.081	0.140	7.744	0.0045

(Wilks' lambda = 0.783; F = 4.338; df =3, 47; P = 0.0089). Hybrids observed feeding on *Tribulus* seeds differed from those that were not (Wilks' lambda = 0.678; F = 7.347; df = 3, 47; P = 0.0004). Student-Newman-Keuls post hoc tests revealed that, in the first case, the difference resided solely in bill length (P < 0.01; for the other two bill dimensions, P > 0.2): the *Opuntia* exploiters had longer beaks on average. In the second case, each of the three bill dimensions was larger (P < 0.01) in the group observed eating *Tribulus* seeds.

Feeding Efficiency

In addition to differing in the composition of their diets, species and their hybrids may differ in the efficiency with which they exploit the same food type. We examined this possibility by quantifying feeding skills on two common food types, *Chamaesyce amplexicaulis* fruits ($DH^{1/2}$ < 4) and *Opuntia echios* seeds ($DH^{1/2} \approx$ 10).

We recorded the number of *Chamaesyce* fruits removed from a bush in 30 s (maximum 18, mean 7.4) for a minimum of four times per bird, and averaged the counts. Four F_1 hybrids between *G. fortis* and *G. fuliginosa* and seven backcrosses to *G. fortis* were compared with 30 *G. fortis*. There was significant heterogeneity among the groups ($F_{2,38}$ = 3.886, P = 0.0291). The F_1 hybrids (10.8 ± 1.77 fruits/30 s, x̄ ± 1 SE) and the backcrosses (11.1 ± 0.74) took two more fruits on average than *G. fortis* (8.9 ± 0.35). Student-Newman-Keuls post hoc tests showed no significant differences between any two groups. However, the combined sample of F_1 hybrids and backcrosses differed significantly from *G. fortis* ($F_{1,39}$ = 7.903, P = 0.0077). Two *G. fuliginosa*, not included in the analyses because they were recorded feeding only 11 times in total, fed at a high average rate (10.9 ± 2.10), as did the hybrids.

Four *G. scandens* × *G. fortis* F_1 hybrids were timed while feeding on *Opuntia* seeds. The average total handling time per seed of these four birds (N = 39 observations) was compared with the handling times of six *G. scandens* (N = 61 observations) and five *G. fortis* (N = 32 observations). Hybrid handling times (26.60 ± 2.36 s/seed, x̄ ± 1 SE) were intermediate between those of *G. scandens* (16.57 ± 1.10 s) and *G. fortis* (38.08 ± 2.08 s). Despite the small sample sizes, variation among the three groups was statistically significant ($F_{2,12}$ = 39.97, P = 0.0001). SNK post hoc tests revealed pairwise differences (P < 0.01) between the hybrids and each of the parental species.

Feeding Efficiency in Relation to Beak Morphology

We tested, with multiple regression analysis, for an association between feeding efficiency on each of the two types of seeds and beak morphology. All variables were ln-transformed prior to analysis. For the combined group of *G. fortis*, F_1 hybrids, and backcrosses timed while feeding on *Chamaesyce* fruits, feeding efficiency was not related to beak morphology ($F_{3,23} = 0.425$, $P = 0.7368$), and none of the partial regression coefficients was significant ($P > 0.1$ in all cases). A similar analysis performed with the data on *Opuntia* seed handling demonstrated a significant relationship with morphology ($F_{3,10} = 4.725$, $P = 0.0265$). The partial regression coefficient for beak length was statistically significant ($\beta = -2.13 \pm 0.63$, $t = 3.388$, $P = 0.0069$), whereas the coefficients for beak depth ($P = 0.3491$) and beak width ($P = 0.8421$) were not. Thus, birds with the longest beaks, principally *G. scandens*, took the shortest time to crack, open, and consume a seed; variation in the other beak dimensions was unrelated to performance at these tasks.

DISCUSSION

The study has three main findings. First, hybrids are intermediate between their respective parental species in dietary characteristics. They are also intermediate in beak characteristics (Grant and Grant 1994). Second, dietary intermediacy is clearest where beak differences between the parental species are most pronounced. Third, there is evidence that the basis of dietary differences among one set of hybrids and parental species is beak-related differences in feeding efficiency.

We first consider the implications of these results for an understanding of high survival of finch hybrids, and then discuss their more general implications for hybrid zones.

The Ecology of Hybrid Fitness

Before the 1982–1983 El Niño event, the seed supply available to finches on Daphne was dominated in biomass by the single large and hard type, *Tribulus cistoides*. After the event, small seeds predominated. The survival of finch hybrids was much higher after than before the event (Grant and Grant 1993). The determination of diets of the hybrids, their feeding efficiencies, and the relationship of both to beak morphology helps us to interpret the change in hybrid survival.

G. fortis × *G. fuliginosa* hybrids feed predominantly on small seeds. Very few individuals have been observed eating the larger and harder *Opuntia* seeds, and only one has been recorded feeding on the even larger and harder *Tribulus* seeds. They are able to feed at least as efficiently as *G. fortis* on small seeds, perhaps more efficiently. Our data, while pointing in that direction, are insufficient for drawing firm conclusions (timed observations of captive birds under controlled conditions would be preferable, as they would eliminate extraneous variation; cf. Benkman 1993). Therefore, higher survival of the hybrids after 1983 than before can be explained by a shift in the seed supply towards their preferred food.

This explanation is supported by the documentation of directional natural selection on beak dimensions of *G. fortis* before and after the El Niño event, at two times of heavy mortality. Birds with large beaks were selectively favored during the 1976–1977 drought, as the supply of small and soft seeds declined absolutely and in relation to the supply of large and hard seeds (Boag and Grant 1981, Price et al. 1984). Birds with small beaks were selectively favored in the 1984–1985 drought, during which the supply of large seeds was minimal (Gibbs and Grant 1987b, Grant and Grant 1993).

Hybrids were not included in those analyses, and the sample sizes of measured hybrids were too small for comparable analyses of selection in the group of hybrids alone. Nevertheless, their small average beak sizes compared with *G. fortis* are consistent with their low survival when large beak size was selectively favored in the *G. fortis* population, and with their high survival when small beak size was selectively favored.

The high survival of *G. fortis* × *G. scandens* hybrids after 1983 can be partly accounted for by the same factor, an abundance of small seeds. They survived better than *G. scandens* at this time, probably because they were less dependent upon a food supply, *Opuntia echios* seeds, that declined after 1983 (Grant and Grant 1993). Higher survival of these hybrids than of *G. fortis* can be explained by two reasons: hybrids are more efficient when feeding on *Opuntia* seeds, and they have a more diverse diet.

Another possible explanation for their high survival remains untested: that they experienced a hybrid vigor (heterosis) owing to their generally heterozygous condition. Other factors that can influence the relative fitness of hybrids are predators (Wake et al. 1989) and parasites (Sage et al. 1986, Moulia et al. 1991, 1995, LeBrun et al. 1992), although we have no evidence that either do so in our study system (Grant 1986).

General Intermediacy of Hybrid Diets

A small number of studies of other organisms have shown that, as in the finches, hybrid diets are generally intermediate between the diets of the parental species, and sometimes broader. Most studies have been conducted in the laboratory. Weider (1993) showed that hybrid *Daphnia* exhibited broad and generally intermediate diets. Schluter (1993) found that hybrid sticklebacks took similar foods and were somewhat less efficient than either of the hybridizing species in their preferred habitats, or were intermediate in feeding efficiency (Schluter 1995). Some degree of food and feeding differentiation was reported by Vrijenhoek (1978, 1984), both among hybrid asexual clones of *Poeciliopsis* fish and between them and the sexual parental species.

Using stomach contents, Clarkson and Minckley (1988) inferred diets of morphologically intermediate and, hence, presumed hybrids between a carnivorous and a herbivorous species of catastomid fish. They found that hybrids used a broad food base, with feeding habits that collectively spanned those of each parent. An unusual feature of their study was the tendency for hybrid diets to be more similar to those of one parental species than to the other, apparently in association with feeding morphology. Hybridization of fish does not always produce additive and intermediate effects on morphological traits (Smith 1992).

The studies described were conducted in the laboratory because it is difficult to observe the feeding of aquatic organisms in nature. Birds are more visible, yet we have been able to find only one report of a feeding study of hybrids in the wild. Saino (1992) found two crow species in Europe to be partially segregated in a narrow zone of hybridization. Each species occurred most frequently in the habitat it occupied in allopatry. Hybrids between them showed no such segregation. In one habitat, hybrids took two classes of prey in equal frequency, whereas each of the species specialized on one (different) class (see also Rolando and Laiolo 1994).

Temporal and Spatial Variation in Hybrid Fitness

Limits to hybrid zones are set by ecological and genetic factors (Harrison 1990). The various studies reviewed here are relevant to the ecological factors. Their chief implication is that, outside the region of hybridization, the hybrids, with broad and intermediate diets, would be at a competitive disadvantage with the more specialized parental species (see especially Schluter 1993, 1995). The situation on small islands like Daphne Major is not directly comparable, being more akin to hybridizing populations of fish (Vrijenhoek 1989, Smith 1992, Aspinwall et al. 1993) and cladoceran crustacea (Weider 1993, Schwenk and Spaak 1995) in lakes, in that species and their hybrids are intermixed and well-distributed in their environment rather than spatially separated into zones (although these can occur in lake fish; Schluter 1993). In these cases, spatially varying ecological factors do not set the limit to the frequency of hybrids. Our study of Darwin's finches suggests that temporal variation can do for these situations what spatial variation does for hybrid zones: temporal variation in ecological factors causes fluctuation in the frequency of hybrids.

Although we have contrasted the spatial variation of hybrid zones with the temporal variation in lakes and small islands, we do not wish to imply that these are mutually exclusive. Ecological conditions are heterogeneous in both space and time to varying degrees in different systems (e.g., see Bell 1992, Levin 1992), and it is their interplay that governs the biodiversity we observe, including the prevalence, distribution, fitness, and fate of hybrids.

ACKNOWLEDGMENTS

We thank the more than 20 people who have assisted us in the field. The study was conducted with permission and support from the Dirección General de Desarrollo Forestal, Quito, the Galápagos National Park Service, and the Charles Darwin Research Station. It has been funded by grants from NSERC (Canada) and NSF, most recently DEB-9306753. D. Schluter gave help with an earlier version of the manuscript.

LITERATURE CITED

Abbott, I., L. K. Abbott, and P. R. Grant. 1977. Comparative ecology of Galápagos Ground Finches (*Geospiza* Gould): evaluation of the importance of floristic diversity and inter-specific competition. Ecological Monographs 47:151–184.

Anderson, E., and G. L. Stebbins, Jr. 1954. Hybridization as an evolutionary stimulus. Evolution 8:378–388.

Arnold, M. L. 1992. Natural hybridization as an evolutionary process. Annual Review of Ecology and Systematics 23:237–261.

Arntzen, J. W., and G. P. Wallis. 1991. Restricted gene flow in a moving hybrid zone of the newts *Triturus cristatus* and *T. marmoratus* in western France. Evolution 45:805–826.

Aspinwall, N., J. D. McPhail, and A. Larson. 1993. A longterm study of hybridization between the peamouth, *Mylocheilus caurinus*, and the redside shiner, *Richardsonius balteatus*, at Stave Lake, British Columbia. Canadian Journal of Zoology 71:550–560.

Barton, N. H., and G. M. Hewitt. 1985. Analysis of hybrid zones. Annual Review of Ecology and Systematics 16:113–148.

———. 1989. Adaptation, speciation and hybrid zones. Nature 341:497–503.

Bell, G. 1992. Five properties of environments. Pages 33–56 *in* P. R. Grant and H. S. Horn, editors. Molds, molecules, and metazoa. Growing points in evolutionary biology. Princeton University Press, Princeton, New Jersey, USA.

Benkman, C. W. 1993. Adaptation to single resources and the evolution of Crossbill (*Loxia*) diversity. Ecological Monographs 63:305–325.

Bert, T. H., and R. G. Harrison. 1988. Hybridization in western Atlantic stone crabs (genus *Menippe*): evolutionary history and ecological context influence species interactions. Evolution 42:528–544.

Birkhead, T. R., and A. P. Møller. 1992. Sperm competition in birds. Evolutionary causes and consequences. Academic Press, New York, New York, USA.

Boag, P. T. 1983. The heritability of external morphology in Darwin's Ground Finches (*Geospiza*) on Isla Daphne Major, Galápagos. Evolution 37:877–894.

———. 1984. Growth and allometry of external morphology in Darwin's finches (*Geospiza*) on Isla Daphne Major, Galápagos. Journal of Zoology, London 204:413–441.

Boag, P. T., and P. R. Grant. 1981. Intense natural selection in a population of Darwin's Finches (Geospizinae) in the Galápagos. Science 214:82–85.

Boag, P. T., and P. R. Grant. 1984a. The classical case of character release: Darwin's Finches (*Geospiza*) on Isla Daphne Major, Galápagos. Biological Journal of the Linnean Society 22:243–287.

Boag, P. T., and P. R. Grant. 1984b. Darwin's Finches (*Geospiza*) on Isla Daphne Major, Galápagos: breeding and feeding ecology in a climatically variable environment. Ecological Monographs 54:463–489.

Clarkson, R. W., and W. L. Minckley. 1988. Morphology and foods of Arizona catasto-
    mid fishes: *Catastomus insignis*, *Pantosteus clarki*, and their putative hybrids. Copeia
    1988:422–433.
Dowling, T. E., and W. R. Hoeh. 1991. The extent of introgression outside the contact zone
    between *Notropis cornutus* and *Notropis chrysocephalus* (Teleostei: Cyprinidae). Evolution
    45:944–956.
Dowling, T. E., G. R. Smith, and W. M. Brown. 1989. Reproductive isolation and introgres-
    sion between *Notropis cornutus* and *Notropis chrysocephalus* (Family Cyprinidae): com-
    parison of morphology, allozymes, and mitochondrial DNA. Evolution 43:620–634.
Endler, J. 1977. Geographic variation, speciation, and clines. Princeton University Press,
    Princeton, New Jersey, USA.
Floate, K. D., M. J. C. Kearsley, and T. G. Whitham. 1993. Elevated herbivory in plant
    hybrid zones: *Chrysomela confluens*, *Populus*, and physiological sinks. Ecology
    74:2056–2065.
Gibbs, H. L., and P. R. Grant. 1987a. Ecological consequences of an exceptionally strong
    El Niño event on Darwin's Finches. Ecology 68:1735–1746.
———. 1987b. Oscillating selection on Darwin's Finches. Nature 327:511–513.
Gill, F. B. 1980. Historical aspects of hybridization between Blue-winged and Golden-
    winged Warblers. Auk 97:1–18.
Grant, B. R. 1996. Pollen digestion by Darwin's finches and its importance for early breed-
    ing. Ecology 77:489–499.
Grant, B. R., and P. R. Grant. 1993. Evolution of Darwin's Finches caused by a rare climatic
    event. Proceedings of the Royal Society of London B 251:111–117.
Grant, P. R. 1981. The feeding of Darwin's Finches on *Tribulus cistoides* (L.) seeds. Animal
    Behaviour 29:785–793.
———. 1986. Ecology and evolution of Darwin's Finches. Princeton University Press,
    Princeton, New Jersey, USA.
———. 1993. Hybridization of Darwin's Finches on Isla Daphne Major, Galápagos. Philo-
    sophical Transactions of the Royal Society of London B 340:127–139.
Grant, P. R., and P. T. Boag. 1980. Rainfall on the Galápagos and the demography of
    Darwin's Finches. Auk 97:227–244.
Grant, P. R., and B. R. Grant. 1992a. Demography and the genetically effective sizes of two
    populations of Darwin's Finches. Ecology 73:766–784.
———. 1992b. Hybridization of bird species. Science 256:193–197.
———. 1994. Phenotypic and genetic effects of hybridization in Darwin's Finches. Evolu-
    tion 48:297–316.
———. 1995a. The founding of a new population of Darwin's Finches. Evolution 49:
    229–240.
———. 1995b. Predicting microevolutionary responses to directional selection on heritable
    variation. Evolution 49:241–251.
Harrison, R. G. 1990. Hybrid zones: windows on evolutionary processes. Pages 69–128 *in*
    D. J. Futuyma and J. Antonovics, editors. Oxford surveys in evolutionary biology.
    Volume 7. Oxford University Press, Oxford, UK.
———. 1993. Hybrids and hybrid zones: historical perspective. Pages 3–12 *in* R. G. Har-
    rison, editor. Hybrid zones and the evolutionary process. Oxford University Press,
    Oxford, UK.
Harrison, R. G., and D. M. Rand. 1989. Mosaic hybrid zones and the nature of species
    boundaries. Pages 111–133 *in* D. Otte and J. A. Endler, editors. Speciation and its conse-
    quences. Sinauer, Sunderland, Massachusetts, USA.

Hewitt, G. M. 1989. The subdivision of species by hybrid zones. Pages 85–110 *in* D. Otte and J. Endler, editors. Speciation and its consequences. Sinauer, Sunderland, Massachusetts, USA.

Howard, D. J., G. L. Waring, C. A. Tibbets, and P. G. Gregory. 1993. Survival of hybrids in a mosaic hybrid zone. Evolution 47:789–800.

LeBrun, N., F. Renaud, P. Berrebi, and A. Lambert. 1992. Hybrid zones and host-parasite relationships: effect on the evolution of parasite specificity. Evolution 46:56–61.

Levin, S. A. 1992. The problem of pattern and scale in ecology. Ecology 73:1943–1983.

Magurran, A. E. 1988. Ecological diversity and its measurement. Princeton University Press, Princeton, New Jersey, USA.

Moore, W. S. 1977. An evaluation of narrow hybrid zones in vertebrates. Quarterly Review of Biology 52:263–277.

Moore, W. S., and D. B. Buchanan. 1985. Stability of the Northern Flicker hybrid zone in historical times: implications for adaptive speciation theory. Evolution 39:135–151.

Moore, W. S., and W. D. Koenig. 1986. Comparative reproductive success of Yellow-shafted, Red-shafted, and hybrid Flickers across a hybrid zone. Auk 103:42–51.

Moulia C., J. P. Aussel, F. Bonhomme, P. Boursot, J. T. Nielsen, and F. Renaud. 1991. Wormy mice in a hybrid zone: a genetic control of susceptibility to parasite infection. Journal of Evolutionary Biology 4:679–687.

Moulia, C., N. LeBrun, C. Loubes, R. Marin, and F. Renaud. 1995. Hybrid vigour against parasites in interspecific crosses between two mice species. Heredity 74:48–52.

Price, T. D. 1984. Sexual selection on body size, territory and plumage variables in a population of Darwin's Finches. Evolution 38:327–341.

———. 1987. Diet variation in a population of Darwin's Finches. Ecology 68:1015–1028.

Price, T. D., P. R. Grant, H. L. Gibbs, and P. T. Boag. 1984. Recurrent patterns of natural selection in a population of Darwin's finches. Nature 309:787–789.

Rolando, A., and P. Laiolo. 1994. Habitat selection by Hooded and Carrion Crows in the alpine hybrid zone. Ardea 82:193–199.

Sage, R. D., D. Heyneman, K.-C. Lim, and A. C. Wilson. 1986. Wormy mice in a hybrid zone. Nature 324:60–62.

Saino, N. 1992. Selection of foraging habitat and flocking by crow *Corvus corone* phenotypes in a hybrid zone. Ornis Scandinavica 23:111–120.

Saino, N., and S. Villa. 1992. Pair composition and reproductive success across a hybrid zone of Carrion Crows and Hooded Crows. Auk 109:543–555.

Schluter, D. 1993. Adaptive radiation in sticklebacks: size, shape, and habitat use efficiency. Ecology 74:699–709.

———. 1995. Adaptive radiation in sticklebacks: trade-offs in feeding performance and growth. Ecology 76:82–90.

Schwenk, K., and P. Spaak. 1995. Evolutionary and ecological consequences of interspecific hybridization in cladocerans. Experientia 51:465–481.

Scribner, K. T. 1993. Hybrid zone dynamics are influenced by genotype-specific variation in life-history traits: experimental evidence from hybridizing *Gambusia* species. Evolution 47:632–646.

Sibley, C. G. 1954. Hybridization in the red-eyed towhees of Mexico. Evolution 8:252–290.

Smith, G. R. 1992. Introgression in fishes: significance for paleontology, cladistics, and evolutionary rates. Systematic Biology 41:41–57.

Smith, J. N. M., P. R. Grant, B. R. Grant, I. Abbott, and L. K. Abbott. 1978. Seasonal variation in feeding habits of Darwin's Ground Finches. Ecology 59:1137–1150.

Spence, J. R. 1990. Introgressive hybridization in heteroptera: the example of *Limnoporus*

Ståhl (Gerridae) species in western Canada. Canadian Journal of Zoology 68:1770–1782.

Vrijenhoek, R. C. 1978. Coexistence of clones in a heterogeneous environment. Science 199:549–552.

———. 1984. Ecological differentiation among clones in a heterogeneous environment: the frozen niche model. Pages 217–231 *in* K. Wöhrmann and V. Loeschcke, editors. Population biology and evolution. Springer-Verlag, New York, New York, USA.

———. 1989. Genotypic diversity and coexistence among sexual and clonal lineages of *Poeciliopsis*. Pages 386–400 *in* D. Otte and J. Endler, editors. Speciation and its consequences. Sinauer, Sunderland, Massachusetts, USA.

Wake, D. B., K. P. Yanev, and M. M. Frelow. 1989. Sympatry and hybridization in a "ring species": the plethodontid salamander *Ensatina eschscholtzii*. Pages 134–157 *in* D. Otte and J. Endler, editors. Speciation and its consequences. Sinauer, Sunderland, Massachusetts, USA.

Weider, L. J. 1993. Niche breadth and life history variation in a hybrid *Daphnia* complex. Ecology 74:935–943.

Ecology 77 (1996): 500–509. Reprinted with permission from the Ecological Society of America.

From *Darwin's Finches* (1947)

✳ DAVID LACK

* Chapter XV: The Persistence of Species

If a variety were to flourish . . . it might come to supplant and exterminate the parent species; or both might co-exist, and both rank as independent species.

Charles Darwin: *The Origin of Species*, Ch. II

WHEN SPECIES MEET

Closely related species of animals often differ from each other only in small and apparently trivial ways. After a careful survey of the evidence at that time available, Robson and Richards (1936) concluded that the differences between such species are not usually adaptive, and that adaptive differences tend to appear only at the level of divergence represented by the genus or subgenus. This view has been widely accepted, and, if true, constitutes one of the most puzzling features of the species problem. But further consideration has led me to realize that the absence of adaptive differences is only apparent, and that in fact closely related species differ from each other in ways which play an extremely important part in determining their survival.

When two forms, originally geographical races of the same species, meet later in the same region and keep distinct, thus forming separate species, there is raised not only an important genetical problem, the prevention of interbreeding, but an even more important ecological problem, since the two forms will tend to compete against each other. The chance is negligible that, after their differentiation in isolation, both forms should be equally efficient in every respect, so that the following possibilities exist.

First, one of the two forms may be so much better adapted than the other that it spreads rapidly right through the range of the other and exterminates it. This seems the most likely possibility, but will rarely be observed as it leaves no trace.

Secondly, one form may be better adapted than the other in the region where they meet but, after it has eliminated the less successful form from part of its range, it may come to a region where the environment is better suited to the other form. In this case the two species will come to occupy separate but adjoining geographical regions. Since environmental factors tend to change gradually, there may be a region where both forms are about equally well adapted, and here their ranges will overlap.

Thirdly, one form may prove better adapted to one section of the original habitat, and the other to the rest. In this case each will tend to spread through the geographical range of the other, each eliminating the other from part of its original habitat, so that they come to occupy separate but adjacent habitats in the same regions.

Fourthly, one form may prove better adapted for obtaining certain foods, the other for taking other foods. In this case, if their numbers are limited primarily by food supply, the two species may be able to co-exist in the same habitat, dividing the available foods. The evidence discussed in Chapter VI suggests that in such cases the foods taken by the two species need be only partly and not wholly different, and that a difference in food habits is commonly associated with a marked difference in size, including size of beak.

Particularly in migratory birds, the above possibilities may be modified by the seasonal factor. Two species need not be isolated from each other in the same way at all seasons; for instance, they might occupy different habitats in the same region when breeding, but a similar habitat in separate regions in winter. Further, if mutual competition would seriously affect their numbers at only one time of year, for instance, in the breeding season, they might be able to mix freely at other seasons without effectively competing. Another but less likely possibility is that two species might inhabit the same place, but breed at different times of year. It is also possible for two species to be isolated in different ways in different parts of their range, in one part separated in habitat and elsewhere geographically.

Dr G. C. Varley has pointed out to me that the above possibilities apply primarily to animals which are limited by food supply. In the case of species whose numbers are controlled by parasites or predators, the population density may be greatly below the limit set by food, so that two species could live in the same habitat and eat the same foods without effectively competing. This type of situation seems much more likely to be important in animals such as insects than in birds.

ECOLOGICAL ISOLATION IN DARWIN'S FINCHES

Darwin's finches provide considerable support for the correctness of the above views, since all the closely related species appear to be isolated from each other in one way or another. This is shown in the following summary, which is based on the data given previously in Chapters II, III and VI. The sharp-beaked ground-finch *Geospiza difficilis* is of particular interest, as on the central islands it differs from the small ground-finch *G. fuliginosa* in habitat, while on the northern islands it is separated from it geographically.

Geographical Separation: 2 Cases

(i) The cactus ground-finch *G. scandens* occurs on most islands, but not on Hood, Tower or Culpepper. The large cactus ground-finch *G. conirostris* occurs only on these three latter islands.

(ii) The small ground-finch *G. fuliginosa* inhabits the arid zone of most islands, but is absent from Culpepper, Wenman and Tower. Only on these three latter islands does the sharp-beaked ground-finch *G. difficilis* occupy the arid zone.

Separation by Habitat: 2 Cases

(i) On the central islands *G. difficilis* breeds only in the humid forest, and *G. fuliginosa* only in the arid and transitional zones. (See also preceding section.)

(ii) On Albemarle and Narborough, the mangrove-finch *Camarhynchus helio-bates* breeds only in the coastal mangrove belt and the woodpecker-finch *C. pallidus* inland. (These two species are in the same subgenus.)

Separation by Feeding Habits: 2 Cases

(i) The cactus ground-finch *Geospiza scandens* breeds in the same habitat with the other ground-finches, *G. magnirostris*, *G. fortis* and *G. fuliginosa*, but its chief food is *Opuntia*, which they do not normally eat.

(ii) The vegetarian tree-finch *Camarhynchus crassirostris* feeds chiefly on leaves, fruits and buds, whereas the other species of *Camarhynchus* are mainly insectivorous.

Separation by Size of Beak, and Presumably by Food: 4 Cases

(i, ii) The large ground-finch *Geospiza magnirostris*, the medium *G. fortis* and the small *G. fuliginosa* occupy the same habitat but differ markedly in size of beak; their foods are partly different.

(iii, iv) A similar situation is presented by the insectivorous tree-finches, the large *Camarhynchus psittacula* and the small *C. parvulus*, with a third species of medium size, *C. pauper*, on Charles. The foods of these species have not been analysed.

. . .

ADAPTIVE DIFFERENCES BETWEEN SPECIES

To return to birds, it is clear from the foregoing discussion that, though adaptive differences between species may not be obvious, they must exist. There is no other way of accounting for the ecological isolation of each species. The nature of the adaptive differences presumably depends on the mortality factors which determine whether one or another species is to survive in a particular place. If food supply is limiting, then adaptations for feeding may be paramount, as in Darwin's finches. If predators are particularly important, then adaptive differences may concern means of escape, such as speed of flight or protective coloration. If many of the birds are killed by a climatic factor, then a migratory habit, or resistance to cold, or to drought, may assist in determining which species will persist in a particular region.

In a few cases, closely related bird species are known to differ adaptively. Thus when two species share the same habitat but differ in food, they often differ markedly in size of beak, as discussed in Chapter VI. An instance of an adaptive difference between species which occupy separate geographical regions is provided by the two Atlantic species of guillemot or murre. Brünnich's guillemot *Uria lomvia* is more northerly and also larger than the common guillemot *U. aalge*, a correlation which is in agreement with Bergmann's rule. An adaptive difference correlated with a difference in habitat is provided by the British pipits, the tree pipit *Anthus trivialis* having, like other tree-frequenting birds, a hind claw which is short as compared with that of the meadow pipit *A. pratensis*.

That adaptive differences are not known in a much larger number of cases is probably due simply to inadequate study. For example, it is not known what differences between the ground-finches *Geospiza fuliginosa* and *G. difficilis* make for the success of the former in the arid zone and the latter in the humid forest on the central Galapagos islands. But the distribution of these birds is explicable only on the view that such adaptive differences exist. Likewise adaptive differences must exist between the two species of chaffinch in the Canary Islands, though in this case also, the nature of the differences is not obvious. Where, as in the above instances, two bird species occupy separate habitats, there is no reason to think that competition between them is direct, meaning that one forcibly drives the other out. Even the most territorial birds attack members of other species only sporadically and ineffectively. A partial exception is provided by the severe competition for nesting sites often found between hole-nesting species, but this is a special case, and is not comparable with driving another species from an entire habitat. Probably the habitat differences between two species are normally brought about gradually by natural selection. Individuals of the one species survive better in those places where the other is at a comparative disadvantage, and vice versa, so that gradually each evolves a specific habitat preference.

In plants, as in birds, closely related species often occupy different habitats, and there is often a similar degree of difficulty in detecting adaptive differences between the species. But in plants the existence, though not the nature, of such differences is readily proved, for if one species is transplanted to the natural habitat of the other, it is usually eliminated. A parallel experiment under natural conditions is impracticable with birds, for on release they simply fly back to their natural homes.

While Robson and Richards (1936) seem mistaken in supposing that closely related species do not differ adaptively, they are correct to the extent that adaptive differences have not, in most cases, been described, and that the characters used by systematists to distinguish related species usually seem to be without adaptive significance. To the latter a prominent exception is provided by the specific recognition marks of many birds, also by the marked size differences often found between related species living in the same habitat. But many other specific differences seem very trivial, both in birds and other animals. That all such differences will eventually prove to be adaptive seems unlikely, but considerable further study is required before a conclusion can safely be reached on this point.

PRE-ADAPTATION

On Lord Howe Island, as already mentioned, occurred two species of white-eye, differing in size of beak and to some extent in food, and apparently derived from separate invasions of the island by the same Australian species *Zosterops lateralis* (see Fig. 24; also Mathews, 1928; Stresemann, 1981; Hindwood, 1940; Lack, 1944a). Presumably the earlier form had already become so large that, when the second arrived, the two did not compete sufficiently for one to eliminate the other. The size difference may well have been intensified later by natural selection, but some difference must have been present when they first met. In the same way the earlier form of chaffinch in the Canary

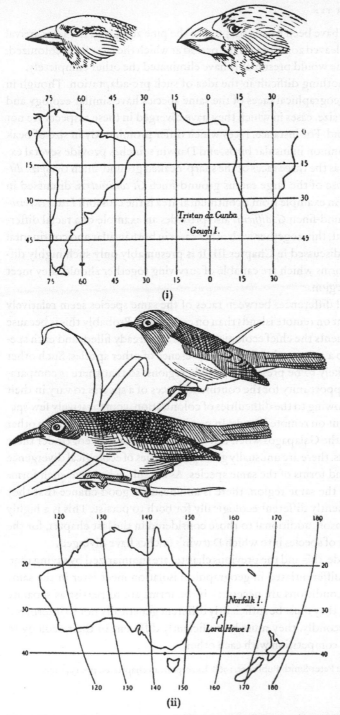

Fig. 24. Two species on the same island differing markedly in size. (i) *Nesospiza* on Tristan da Cunha (ii) *Zosterops* on Lord Howe I (after Grönvold)

Islands must have been better adapted to the pine zone, and the later arrival to the broad-leaved zone, before the period at which the later form colonized. Otherwise one would presumably have eliminated the other completely.

There is nothing difficult in the idea of such pre-adaptation. Though in many cases geographical races of the same species have similar ecology and are of similar size, cases in which they have diverged in these respects are not difficult to find. For instance, races which differ prominently in size of beak are not uncommon in insular birds, and Darwin's finches provide several examples, such as the three races of the sharp-beaked ground-finch *Geospiza difficilis* and those of the large cactus ground-finch *G. conirostris*, discussed in Chapter VI. An example from a continental area is the crossbill *Loxia curvirostra*. The ground-finch *G. difficilis* also provides an example of a racial difference in habitat, this, together with other cases in both insular and continental birds, being discussed in Chapter III. It is presumably only such highly differentiated forms which are capable of surviving together should they meet in the same region.

Ecological differences between races of the same species seem relatively more frequent on remote islands than on continents. Probably this is because on the continents the chief ecological niches are already filled, and each species is kept to a restricted niche by the presence of other species. Such other species are likely to be present over wide regions, so that there is comparatively little opportunity for the continental races of a species to vary in their ecology. But owing to the difficulties of colonization, comparatively few species are present on remote islands. In particular there were perhaps no other land birds in the Galapagos when Darwin's finches first arrived. Under these circumstances, there are unusually great possibilities of ecological divergence between island forms of the same species. As a result, when two such forms meet later in the same region, there is an unusually good chance that they will be sufficiently different ecologically for both to persist. This is a highly important reason, additional to those considered in the last chapter, for the large number of species into which Darwin's finches have diverged.

To conclude, this and the previous chapter are summarized by saying that, when forms differentiated in geographical isolation meet later in the same region, two conditions are necessary if the forms are to persist as separate species. First, they must be sufficiently different genetically not to interbreed freely, and secondly, they must be sufficiently different in their ecology to avoid serious competition with each other.

Gloucester, MA: Peter Smith Publisher, 1968. Excerpt from chapter 15, pp. 134–146.

Hybridization

Hybridization occurs when reproductive isolation between species is incomplete. Hybridization is first a measure of the degree of reproductive isolation. By quantifying hybridization, one quantifies the degree of genetic exchange between two species and consequently the degree to which they are evolutionarily independent. But hybridization is also a process that itself influences evolutionary dynamics of adaptation and diversification.

Hybridization was Darwin's big stumbling block. Natural selection may be the cause of differences among populations, but how does diversity persist if interbreeding among divergent groups keeps blending them back together? Without the benefit of Mendel's research on genetic transmission, Darwin was especially handicapped. In many ways, genetic inheritance resembles a process of blending. Offspring resemble both their parents, and for any particular trait they frequently have an intermediate phenotype. If such blending is the general rule, then variation will be diluted quickly. If one starts with red and white paint, then mixes them half-and-half, the result is pink paint. The red and white are gone for good. From differences, here red and white, homogeneity results—irrevocably. With this scenario of "blending inheritance," as populations continue to mix, diversity continues to be lost as intermediates dominate. Blending inheritance cannot explain how persistent differences among species evolve.

Darwin was much disturbed by this difficulty. Eventually, he proposed an alternative to blending inheritance, which he called "pangenesis." With pangenesis, "gemules" contained the inherited essences, each gemule representing some aspect of one parent. Gemules from each parent were transmitted to offspring, and rather than fusing together, they simply coexisted intact so that they could mobilize again in the next generation. That is, inheritance was "particulate" in the sense that the inherited material did not change to become intermediate in form, but retained its distinct quality. The red and the white particles, or gemules, remained red and white; they did not fuse and blend to pink. Darwin's theory of pangenesis was problematic in both its total lack of supporting evidence and its dependence on Lamarckian processes (the environment of the organism could change its gemules). Nevertheless, the insight of particulate inheritance was a major conceptual advance. In 1900 the rediscovery of Mendel's meticulous experimentation thoroughly eclipsed Darwin's speculative theory of pangenesis, but the idea of a particulate mode of inheritance was common to the work of both.

The most familiar outcome of hybridization or interbreeding is the formation of intermediate phenotypes. This occurs when neither parental allele is completely dominant to the other, such that both alleles contribute to the phenotype. This effect has been documented countless times, including in Darwin's finches. But hybridization has other important outcomes. First, hybridization is a form of gene flow across species; alleles from one species are transferred to another species and contribute to the genetic variation of that species. One study of Darwin's finches has estimated that hybridization increased genetic variation up to two orders of magnitude more than mutation occurring within a particular species has done. Moreover, hybridization altered genetic correlations in the finches; when correlations among traits were similar for both species, hybridization tended to increase the strength of correlations; but when correlations differed between species, hybridization resulted in weaker correlations. By increasing genetic variance and altering genetic correlations, hybridization in Darwin's finches has the potential to facilitate adaptation. The degree to which it has done so, enabling adaptive divergence as well as convergence, is still under investigation.

Second, hybridization can create novel genotypes and phenotypes not found in either parental population or species, sometimes exhibiting phenotypes more extreme than either parent. This is referred to as transgressive segregation. It occurs because different combinations of alleles are present in offspring than were present in either parent. The alleles have recombined. For example, suppose one parent has genotype *AAbb* and the other has genotype *aaBB*. Let the capital letters represent an allele that causes an increase in a trait, say larger bill size. The hybrids will have genotypes *AaBb*, and the next generation will have the full range of combinations, including the genotypes *aabb*, which will have a smaller bill than either parent, and *AABB*, which will have a larger bill than either parent. These combinations would never have been possible had hybridization not occurred and the alleles not recombined. Novel phenotypes can result not only from the dosage, or additive, effects just described, but can also occur when heterozygotes themselves have more extreme phenotypes (overdominance or underdominance) or when different combinations of alleles at different loci have nonadditive, or epistatic, effects on phenotypes. Either way, recombination can produce extreme phenotypes with values beyond those found in the parental lineages. This, of course, is a direct outcome of the particulate as opposed to blending inheritance of genes.

Transgressive segregation, by creating extreme phenotypes, may also result in recombinant offspring that are actually more highly adapted than either parent. In the above example, the extra-small bills may not be advantageous when only large seeds are abundant, but the extra-large bills may be. Recombining can therefore be a positive creative process. It has been told that the celebrated pioneering choreographer Merce Cunningham actually used a concept of recombination to generate dance sequences. He would compose several small dance phrases, transcribe them onto paper, and then put them all in a hat. The dancers would draw the phrases at random, piecing together the dance, the random combinations of the phrases coming from the luck of the draw. Admittedly, many such dance sequences may have been real stinkers. But a few were breathtakingly beautiful, and those are the ones that are remembered.

Hybridization is now considered, not always the ultimate disruptor of the processes of speciation, but a potential contributor to the process of diversification itself. With recombination between lineages, new extreme genotypes and phenotypes are possible and may even have a selective advantage under some ecological circumstances. In sunflowers, for instance, hybridization is reported to play a critical role in adaptive diversification to the point of speciation. Novel combinations of alleles have enabled new recombinant sunflowers to inhabit the harsh and xeric conditions of the deserts in the American Southwest. This has resulted in the adaptive divergence and ecological isolation of a new sunflower species, *Helianthus anomalus*. Hybridization therefore not only can keep populations and species from diverging but under some circumstances can also encourage adaptive divergence.

READINGS

The two readings in this chapter document hybridization in Darwin's finches and some of the consequences of hybridization for phenotypic divergence among species.

Hybrid pairings can be observed directly by following banded birds, recording interspecific pairings, and following the offspring produced from such pairings. This direct observation is likely to provide the most accurate information on the conditions under which hybridization occurs. Another method of quantifying hybridization is through the use of genetic markers. Genetic markers can reveal evidence for matings that were not observed directly. If offspring result from extrapair copulations between members of two different species, this sort of data can detect such infidelity and more accurately indicate the number of offspring obtained from particular matings and fertilizations. Genetic data can also be used to quantify genetic exchange between species over a longer time frame than can be observed directly. The first paper, by Peter Grant and colleagues, tests for hybridization between several species pairs, using genetic markers. The authors reasoned that if hybridization had occurred between two sympatric species, then those sympatric populations should share more alleles than would allopatric (necessarily nonhybridizing) populations. They found, in fact, evidence of hybridization between numerous sympatric species.

Hybridization having been documented, another question arises. What are the phenotypic and genetic consequences of genetic exchange between species? The second paper, by Peter and Rosemary Grant, provides a surprising case of incipient speciation that has apparently been facilitated by hybridization. Starting with a single peculiar-looking hybrid male immigrating into an established population, the events that led to increasing reproductive isolation included high mortality and inbreeding, cultural inheritance of song, and subsequent genetic divergence.

Hybridization in the Recent Past (2005)

❋ PETER R. GRANT, B. ROSEMARY GRANT, AND
KENNETH PETREN

Abstract—The question we address in this article is how hybridization in the recent past can be detected in recently evolved species. Such species may not have evolved genetic incompatibilities and may hybridize with little or no fitness loss. Hybridization can be recognized by relatively small genetic differences between sympatric populations because sympatric populations have the opportunity to interbreed whereas allopatric populations do not. Using microsatellite DNA data from Darwin's finches in the Galápagos archipelago, we compare sympatric and allopatric genetic distances in pairs of Geospiza and Camarhynchus species. In agreement with the hybridization hypothesis, we found a statistically strong tendency for a species to be more similar genetically to a sympatric relative than to allopatric populations of that relative. Hybridization has been studied directly on two islands, but it is evidently more widespread in the archipelago. We argue that introgressive hybridization may have been a persistent feature of the adaptive radiation through most of its history, facilitating evolutionary diversification and occasionally affecting both the speed and direction of evolution.

A resurgence of interest in the role of introgressive hybridization in evolution has occurred in the last 15–20 years (Arnold 1997, 2004; Dowling and Secor 1997; Ellstrand 2003). It is now known to be much more widespread taxonomically and geographically than formerly believed, and the genetic and fitness consequences have been determined in several plant and animal systems (Abbott 1992; Arnold 1992; Grant and Grant 1992; Dowling and DeMarais 1993; Rieseberg and Wendel 1993). A major contribution from breeding programs has been the demonstration of how novel and advantageous combinations of genes can be produced by transgressive segregation in hybridizing species (Rieseberg et al. 2003a, 2003b). A repeated finding from field studies is substantial introgression of some genes but not others (Cathy et al. 1998; Kim and Rieseberg 1999; Whitham et al. 1999; Sattler and Braun 2000; Martinsen et al. 2001; Sota 2002; Sullivan et al. 2002; Rieseberg et al. 2003b). Often the exchange of genes between species is unequal (Bacilieri et al. 1996; McDonald et al. 2001; Thulin and Tegelstrom 2002), with a tendency for genes to flow predominantly from common to rare species (Dowling et al. 1989; Taylor and Hebert 1993; Wayne 1993). Hybridization is therefore of interest to evolutionary biologists because partial and unequal gene exchange can have important effects on the dynamics of hybrid zones, speciation, and adaptive radiation (Chiba 1993; Grant 1998; Schluter 2000; Saetre et al. 2001; Bensch et al. 2002; Ortíz-Barrientos et al. 2002; Rieseberg et al. 2003b; Servedio and Saetre 2003; Gee 2004; Seehausen 2004).

Where phylogenies can be established with a system of neutral genetic markers, it is possible to determine the effects of introgression by reconstructing the transfer of genetic and phenotypic characters from one lineage to another (Dowling et al. 1989; DeMarais et al. 1992; McDade 1992; Andersson 1999; Matos and Schaal 2000; Sota et al. 2001; Allender et al. 2003; Feder et al. 2003). For groups of organisms that have recently diversified rapidly and are perhaps continuing to do so, establishing phylogenetic connections (in the absence of hybridization) is not easy because genetic differences are small and lineage sorting has not progressed far (but see Allender et al. 2003). Compounding the difficulties, introgressive hybridization occurring after speciation has begun can blur the small nascent genetic differences (Patton and Smith 1994; Clarke et al. 1998).

The question we address in this article is how hybridization in the recent past can be recognized with selectively neutral genetic data. The question is challenging. Two sympatric populations of a related group of organisms are genetically similar to a degree that may be a function of both common ancestry and genetic exchange after their initial divergence. The question then is how the consequences of genetic exchange can be distinguished from the effects of common ancestry. A solution we explore here lies in the comparison of sympatric and allopatric populations of the same species. Sympatric populations have the opportunity to interbreed; allopatric populations do not. In the absence of interbreeding, sympatric populations of two species should be no more similar to each other genetically than each one is to allopatric populations of the other. In contrast, the introgression hypothesis predicts that a species is more similar genetically to a sympatric relative than to allopatric populations of that relative, as a result of exchanging alleles.

Recent adaptive radiations are useful systems for adopting this approach to the question of detecting evidence of past introgressive hybridization. They typically comprise several independent populations of the same species, which is a useful feature for making multiple comparisons for statistical purposes (Schluter and Nagel 1995; Losos et al. 1998; Chiba 1999; Rüber et al. 1999; Allender et al. 2003; Gillespie 2004; McKinnon et al. 2004), and the species are often not completely reproductively isolated from each other (Grant and Grant 1997a; Price and Bouvier 2002; Mendelson 2003). Especially useful are the radiations in archipelagos where a prima facie case can be made for allopatric speciation on geographical grounds alone (Wagner and Funk 1996; Grant 1998; Schluter 2000). In this article, we use genetic data from Darwin's finch species on the Galápagos islands. We apply the method of comparing sympatric and allopatric populations to several pairs of species, find evidence of introgressive hybridization, and in the "Discussion" consider the evolutionary implications of hybridization and its relevance to sympatric speciation.

STUDY SYSTEM

Mitochondrial DNA and allozyme data suggest that Darwin's finches shared a common ancestor 2–3 million years ago (Grant 1999; Sato et al. 1999; Grant and Grant 2002a). Since then, a minimum of 14 species were formed. They have been classified on morphological grounds (Lack 1945, 1947) as ground finches (*Geospiza;* six species), tree finches (*Camarhynchus* sensu lato = *Camarhynchus* + *Cactospiza;* five species), warbler finch (*Certhidea;* one or two species; Petren et al. 1999; Grant and Grant 2002b), vegetarian finch (*Platyspiza;* one species), and Cocos finch (*Pinaroloxias;* one species). Studies of allozymes (Yang and Patton 1981), microsatellites (Petren et al. 1999), song (Bowman 1983), and breeding (Grant 1999) have largely supported the morphological classification, whereas mitochondrial DNA (*cyt b*) and nuclear internal transcribed spacer genes have provided ambivalent support (Freeland and Boag 1999; Sato et al. 1999). Populations occur on one to more than 20 islands in various combinations (Grant 1999). Interbreeding of some species has been suspected on grounds of morphological intermediacy (Lack 1945, 1947; Bowman 1961, 1983). Several populations of ground finches are unusually variable in beak dimensions (Lack 1945, 1947; Bowman 1961; Grant and Grant 2000), and some of these populations are known to hybridize (Grant 1999). Interbreeding of ground finch species has been studied on two islands for many years (Grant and Grant 1989, 2002c; Grant et al. 2004). The F_1 hybrid and backcross individuals have been found to experience high fitness under favorable environmental conditions, with no indication of sterility or inviability (Grant and Grant 1992, 1998). We concentrate on the ground finch species (except for the oldest species, *Geospiza difficilis,* because relationships among populations of this species are uncertain; Grant et al. 2000) and also on the unique population of *Geospiza conirostris* on Española (Petren et al., forthcoming). There are too few populations (<6) of individual tree finch species for statistical analyses by binomial tests, the principal test we use; therefore, we have combined the species in a single analysis.

METHODS

Details of collecting and genotyping procedures have been published elsewhere (Petren 1998; Petren et al. 1999; Grant et al. 2001). A total of 42 populations were studied, and sample sizes per population varied from four to 59 individuals (table 1), with a mean of 17.4. For comparing genetic differences between sympatric and allopatric populations of related species, we use Nei's genetic distance (D; Nei 1972) calculated from allele frequencies at 16 microsatellite loci. We use Nei's D rather than diversity-based distances such as Fst because, by taking allele identity into account, it is more appropriate when there has been recent admixture between species (Hedrick 1999). The genetic distance between species i and species j on island k, the ik-jk distance, is compared with allopatric distances ik-$jm \ldots p$ for species i and jk-$im \ldots p$ for species j. The subscript $m \ldots p$ refers to all other islands occupied by species i or j. The null hypothesis is that there is no difference between the sympatric distance ik-jk and the allopatric distances with which it is compared. Because we are concerned with the direction of differences between sympatric and allopatric distances and not with their magnitude, we use two-tailed binomial tests applied to pairs of species identified by morphology (Lack 1947; Schluter 1984), allozymes (Yang and Patton 1981), and microsatellites (Petren et al. 1999) as close relatives.

There are two ways to make the comparisons: a sympatric distance between two species may be compared with allopatric distances for each species separately or together. Each has the disadvantage of using some of the data twice, and therefore we present both sets of results. For the analysis of species considered one at a time, the genetic distance between a pair of sympatric populations (ik-jk) is used twice, once in the analysis for species i and once for species j. To adjust for this repetition we set α at 0.025, but we present all probabilities of <0.05. For the analysis of species i and j combined, each sympatric distance is used only once, but each allopatric distance (ik-jm) is sometimes used twice, once in comparison with the sympatric distance on island k (ik-jk) and once in comparison with the sympatric distance on island m (im-jm). This situation arises because congeneric species occur as often together on the same island as on separate islands. To adjust for this repetition, we serially deleted each allopatric distance once in each analysis and averaged the associated P values.

TABLE 1. Number of genotyped individuals (N), mean number of alleles (A), and mean observed heterozygosities (H_O)

Species and island	N	A	H_O
Geospiza fuliginosa:			
Daphne	28	9.6	.66
Santa Cruz	24	10.5	.72
Santiago	19	9.4	.74
Rábida	10	6.4	.68
Española	10	6.8	.68
Floreana	10	6.6	.72
San Cristóbal	21	8.7	.77
Pinta	10	5.8	.76
Isabela	13	7.6	.69
Geospiza fortis:			
Daphne	59	8.2	.63
Santa Cruz	39	10.4	.67
Santiago	9	6.2	.66
Marchena	17	6.1	.60
San Cristóbal	4	4.4	.76
Pinta	12	6.7	.74
Isabela	11	7.1	.71
Geospiza magnirostris:			
Daphne	54	6.2	.60
Santa Cruz	12	6.8	.61
Genovesa	32	6.6	.54
Rábida	5	4.2	.57
Marchena	10	5.1	.62
Pinta	7	4.1	.63
Fernandina	9	5.9	.60
Isabela	6	4.9	.60
Geospiza scandens:			
Daphne	52	6.1	.66
Santa Cruz	23	9.0	.61
Santiago	4	4.5	.77
Rábida	12	5.0	.61
Marchena	6	4.1	.55
San Cristóbal	6	4.9	.74
Genovesa	48	6.5	.59
Camarhynchus pallida:			
Santa Cruz	27	4.4	.38
Isabela	13	3.4	.41
Camarhynchus parvulus:			
Santa Cruz	18	5.9	.49
Floreana	22	6.1	.51
Isabela	7	4.4	.51

Species and island	N	A	H$_O$
Camarhynchus psittacula:			
Santa Cruz	5	3.1	.48
Pinta	8	2.8	.45
Fernandina	5	3.6	.51
Isabela	4	2.7	.45
Camarhynchus pauper:			
Floreana	19	5.1	.49
Camarhynchus heliobates:			
Isabela	12	2.6	.31

Note: The *G. conirostris* population on Genovesa has been included with *G. scandens* because of their genetic similarity (Petren et al., forthcoming). H_O at 16 microsatellite loci.

RESULTS

Each Species Considered Separately

Table 2 gives the details and results of statistical tests. Nei's *D* between sympatric populations (D_{sym}) can be compared with the mean and range of allopatric populations of the same species (D_{all}). In 21 of the 26 comparisons, including all involving *Geospiza fuliginosa*, D_{sym} is smaller than the average of allopatric *D* values. The next two columns in the table give the sample size of allopatric populations and the number of them for which $D_{sym} < D_{all}$ in parentheses. The last three columns give the results of the two-tailed binomial tests of the null hypothesis that the number of comparisons where $D_{sym} < D_{all}$ equals the number where $D_{sym} > D_{all}$.

Figure 1 illustrates a strong tendency for ground finch species to be genetically more similar to each other in sympatry than in allopatry. In table 2, 18 of 52 tests (32.7%) provide statistical evidence of greater similarity in sympatry, and three give the opposite result at $P < .05$. In six cases, both species *i* and *j* are significantly more similar in sympatry than either is to any allopatric population of the other. These are *Geospiza magnirostris* and *Geospiza scandens* (*conirostris*) on Genovesa, *G. fuliginosa* and *Geospiza fortis* on Santa Cruz and Isabela, *G. fuliginosa* and *G. scandens* on Santa Cruz and San Cristóbal, and *G. fortis* and *G. magnirostris* on Santa Cruz.

In the aggregate, the high frequency of significant results in one direction is not expected by chance; 18 results in one tail of a normal frequency distribution and only three in the other has an associated two-tailed binomial probability of 0.002. Adjusting α to 0.025 to allow for the fact that the same genetic distance *ik-jk* between sympatric species is used in two tests reduces the ratio of significant differences from 18 : 3 to 14 : 1. The associated two-tailed binomial probability is 0.001. Also not expected by chance is the high frequency of low probability values (0.008); five of the 11 tests (about 45%)

TABLE 2. Genetic distances of pairs of *Geospiza* species in sympatry (D_{sym}) compared with allopatry (D_{all})

Island of sympatry	Geospiza		Sympatric D_{sym}	Allopatric mean D_{all}	Allopatric range D_{all}	N allopatric populations ($D_{sym} < D_{all}$)		Probability[a]		
	Species 1	Species 2				Species 1	Species 2	Species 1	Species 2	Combined species
Daphne	magnirostris	scandens	.550	.632	.515–.736	7 (5)	7 (7)	.016	...	.012
Santa Cruz	magnirostris	scandens	.562	.655	.515–.829	7 (7)	7 (6)	...	.016	.002
Rábida	magnirostris	scandens	.726	.694	.585–.948	7 (1)	7 (5)	...	...	...
Marchena	magnirostris	scandens	.841	.718	.567–.937	7 (1)	7 (0)	–.016	...	–.002
Genovesa	magnirostris	scandens	.484	.701	.538–.824	7 (7)	7 (7)	.016	.016	<.001
Daphne	fuliginosa	fortis	.230	.294	.172–.448	8 (8)	6 (2)	...	.008	...
Santa Cruz	fuliginosa	fortis	.131	.218	.134–.373	8 (8)	6 (6)	.032	.008	<.001
Santiago	fuliginosa	fortis	.216	.250	.134–.480	8 (5)	6 (3)	...	...	...
San Cristóbal	fuliginosa	fortis	.248	.268	.150–.515	8 (6)	6 (2)	...	...	...
Pinta	fuliginosa	fortis	.269	.333	.201–.515	8 (3)	6 (6)	.032	...	...
Isabela	fuliginosa	fortis	.143	.223	.145–.378	8 (8)	6 (6)	.032	.008	<.001
Daphne	fuliginosa	scandens	.510	.533	.316–.681	8 (2)	7 (7)	.016	...	...
Santa Cruz	fuliginosa	scandens	.312	.459	.316–.666	8 (8)	7 (7)	.016	.008	<.001
Santiago	fuliginosa	scandens	.565	.571	.425–.710	8 (6)	7 (2)	...	...	...

Rábida	fuliginosa	.563	.555	.403–.743	8 (3)	7 (4)	...	...	...
San Cristóbal	fuliginosa	.459	.324	.330–.700	8 (8)	7 (7)	.016	.008	<.001
Daphne	fortis	.543	.529	.339–.791	6 (0)	7 (7)	.016	-.032	...
Santa Cruz	fortis	.445	.358	.333–.581	6 (5)	7 (6)	...	...	.022
Santiago	fortis	.570	.524	.344–.684	6 (4)	7 (4)	...	...	...
Marchena	fortis	.589	.685	.416–.711	6 (1)	7 (1)	...	...	...
San Cristóbal	fortis	.524	.434	.333–.676	6 (5)	7 (5)	...	...	...
Daphne	magnirostris	.350	.446	.208–.518	6 (0)	7 (2)	...	-.032	-.022
Santa Cruz	fortis	.251	.147	.171–.370	6 (6)	7 (7)	.016	.032	<.001
Marchena	fortis	.310	.238	.176–.450	6 (4)	7 (6)	...	...	...
Pinta	fortis	.418	.507	.289–.534	6 (2)	7 (1)	...	...	...
Isabela	fortis	.301	.250	.171–.440	6 (5)	7 (4)	...	...	...

Notes: Numbers of allopatric populations (N) for each species used in the comparisons with sympatric populations are given in columns 6 and 7, together with the number of comparisons (in parentheses) in which the genetic difference in sympatry (D_{sym}) is less than the allopatric difference (D_{all}). Probabilities associated with extreme frequencies of $D_{sym} < D_{all}$ are given in the last three columns, and a minus sign indicates extreme frequencies in the opposite direction $D_{sym} > D_{all}$. Ellipses indicate $P > .05$. The population of *Geospiza conirostris* on Genovesa has been included with *Geospiza scandens* owing to their close genetic similarity (Petren et al., forthcoming).

[a]Two-tailed binomial probabilities are shown for testing of the null hypothesis that the number of comparisons that yield $D_{sym} < D_{all}$ equals the number where $D_{sym} > D_{all}$.

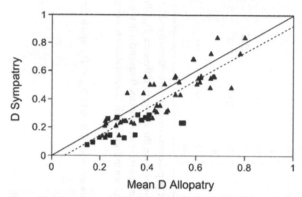

Fig. 1. Closely related species of Darwin's finches are often genetically more similar to a sympatric relative than to allopatric populations of that relative. For each population of each species, the genetic distances with all allopatric populations of the other species have been averaged. Equality of sympatric and mean allopatric distances is shown by the solid line, and the trend line for all data calculated by ordinary least squares regression is shown as a dashed line. *Geospiza* are ground finch species, and *Camarbyncbus* are tree finch species.

that could have produced such low values, because the sample size of eight was large enough, did so.

Contrary to the general trend, the genetic distance between *G. scandens* and *G. magnirostris* on Marchena is very large when compared with allopatric populations of *G. scandens* (table 2). We have not studied the species on Marchena in detail. Possibly one or both species colonized it recently.

Pairs of Sympatric Species Considered Together

Out of a total of 26 tests, nine pairs of populations (34.6%) are significantly more similar in sympatry than in allopatry, whereas two are significantly different in the opposite direction. In the aggregate, the associated two-tailed binomial probability of a 9 : 2 ratio approaches but does not reach statistical significance ($P = .066$). Noteworthy are six probabilities <0.001 associated with the trend. Against the trend, the genetic distance between *G. fortis* and *G. magnirostris* on Daphne Major is large compared with allopatric distances. *Geospiza magnirostris* colonized Daphne in 1982–1983 and has not hybridized with any of the species (Grant et al. 2001).

Tree Finches

Tree finches are not included in the above analyses for lack of adequate samples of populations of each pair of species for binomial tests; nevertheless, the few data are suggestive of similarity in sympatry in agreement with the hybridization hypothesis. For example, *Camarhynchus pauper* and *Camarhynchus parvulus* are believed to hybridize on Floreana on the basis of museum specimens of intermediate morphology (Lack 1945, 1947). *Camarhynchus pauper*, occurring nowhere else in the archipelago, is more similar to sympatric *C. parvulus* ($D = 0.076$) than it is to two allopatric populations of *C. parvulus* ($D = 0.113$–0.179). *Camarhynchus psittacula* and *C. parvulus* may also hybridize (Lack 1945, 1947). These species are more similar to each other on Santa Cruz ($D = 0.095$) and Isabela ($D = 0.147$) than either is to five out of six allopatric populations of the other species ($D = 0.083$–0.464). To circumvent sample size limitations, we have combined all data for tree finches in the genus *Camarhynchus* in figure 1 and have compared all sympatric populations with the average of the heterospecific allopatric populations. In 15 out of 16 comparisons, the genetic distance between sympatric populations is shorter than the average distance from the allopatric populations. The two-tailed binomial probability of this result is <0.001. The same comparison of ground finch populations gives a 37 : 15 ratio in the same direction (large sample binomial test with correction for continuity: $z = 2.91$, $P = .0036$).

DISCUSSION

We found a statistically strong tendency for a species to be more similar genetically to a sympatric relative than to allopatric populations of that relative. Genetic similarity of sympatric populations can be interpreted as evidence of a sympatric origin of species, evidence of introgressive hybridization, or both.

Sympatric Speciation

The logic of comparing sympatric and allopatric populations rests on the assumption of approximately constant rates of divergence in neutral alleles. Populations diverge in selectively neutral allele frequencies as a result of mutation and drift (assuming no selection on linked genes). Divergence is usually assumed to be stochastically time dependent, and the magnitude of independent change in the two lineages is assumed to be equal on average. However, allele frequency changes may occur faster early in the founding of a new population as a result of bottleneck effects on heterozygosity (Nei et al. 1975; Chakraborty and Nei 1977; Clegg et al. 2002; Grant 2002). If this happens, change will be greater in a lineage that has undergone several successive founder events in different environments (e.g., Clegg et al. 2002) than in a lineage that has remained in a single environment.

If species arise in a single location, and speciation occurs entirely sympatrically (e.g., Schliewen et al. 1994; Feder 1998; Via 2001), the sympatric populations are likely to be genetically more similar to each other at that location than are derived, bottleneck-affected populations whether they are sympatric or allopatric. However, introgression is an additional reason to expect genetic similarity because hybridization is most likely to occur early in the speciation history of two species (Grant and Grant 1997a).

One way to distinguish between these two causes of genetic similarity in sympatry is to use the fact that repeated patterns of genetic similarity in sympatry, in conjunction with differences between sympatric locations, are better explained by introgression than by any model of ancestral effects due to common ancestry or sympatric speciation (Matos and Schaal 2000; Shaw 2002). Thus, detecting genetic effects of introgression in a system where sympatric speciation occurs requires greater genetic similarity of species to be shown in at least two sympatric locations; if one is the location of origin, the other is a location of introgression since it is improbable that the same sympatric speciation would occur in two locations. In this study, we found two cases of members of a pair of species being most similar genetically in two sympatric locations. In two other instances, members of a pair of species were most similar genetically in one sympatric location. In these cases, sympatric speciation with bottleneck effects in derived populations could be a sufficient explanation. The populations in question are *Geospiza fortis* and *Geospiza magnirostris* on Santa Cruz and *G. magnirostris* and *Geospiza scandens* (*conirostris*) on Genovesa. Sympatric speciation has previously been invoked to explain an unusual distribution of beak sizes of *G. fortis* at the first location (Ford et al. 1973), although hybridization with *G. magnirostris* has been suspected (Grant 1999). Hybridization has been documented between the pair of species on Genovesa (Grant and Grant 1989).

Nevertheless, although some of the observations are consistent with a sympatric speciation explanation for the genetic similarities, there are two reasons to doubt that Darwin's finches have speciated in this way. The first is a general one. Sympatric speciation results from disruptive selection and genotype-dependent, perhaps frequency-dependent, mate choice (Arnegard and Kondrashov 2004; Seehausen and Schluter 2004; van Doorn et al. 2004). Given genetic variation in mate preference and genetic covariation with a preferred continuously varying trait, a species could diverge rapidly in the preferred and correlated traits (Kirkpatrick and Servedio 1999; Servedio 2000, 2004; Saetre et al. 2001; Servedio and Saetre 2003; Bolnick 2004). However, genetic variation in mate preferences is not known in birds (Bakker and Pomiankowski 1995), and if it exists, it is likely to be overwhelmed by uncorrelated environmental variation due to random effects and to learning. Like many other birds (Irwin and Price 1999; ten Cate and Vos 1999), Darwin's finches imprint on morphological and vocal stimuli from their parents or other adults (Bowman 1983; Grant and Grant 1997b, 1998). Their songs are learned, culturally transmitted traits. Until a cultural equivalent to the sympatric speciation model of genetic variation and covariation is developed (see Lachlan and Slater 1999; Lachlan and Feldman 2003), the plausibility of sympatric speciation in songbirds remains doubtful except in unusual breeding systems (e.g., Payne et al. 2002; Sorenson et al. 2003).

More specifically, mate preferences have not been detected within species of Darwin's finches; mating appears to be generally random with respect to phenotype (Grant and Grant 1989, 1996, 1997b; Keller et al. 2001). The most favorable case for sympatric speciation involved a population of cactus finches *G. scandens* (*conirostris*) on Genovesa. In 1978, we found an ecological and morphological subdivision in the dry nonbreeding season (Grant and Grant 1979), possibly indicative of disruptive selection and incipient speciation; however, this was transitory because in subsequent wet seasons mating was not assortative (Grant and Grant 1987b), except for a tendency for experienced birds to pair together (Grant and Grant 1987a). Similar findings for some African finches have been reported by Smith (1993).

Coyne and Price (2000) reviewed the literature on island birds and found little evidence for sympatric speciation. We conclude that genetic similarity in sympatry is best accounted for by introgression.

Introgressive Hybridization

Genetic similarities of populations on Daphne and on Genovesa are congruent with long-term observations, although the situation on Daphne is complex. *Geospiza fortis* has hybridized with *Geospiza fuliginosa* and *G. scandens* there for at least the last 30 years (Grant et al. 2004). Although rare, interbreeding with *G. fuliginosa* predominated in the first half of that period (Grant 1993). The pattern of introgression changed in the second half when *G. fuliginosa*, never common, became rarer, and the frequency of interbreeding with *G. fortis* declined. Interbreeding of *G. fortis* and *G. scandens* increased, with genes flowing initially more from *G. scandens* to *G. fortis* and then later in the reverse direction. The net result was a genetic and morphological convergence of these two populations (fig. 2). *Geospiza scandens* underwent the largest change (Grant et al. 2004).

As expected from these observations, *G. fortis* is more similar genetically to *G. fuliginosa* and *G. scandens* on Daphne than to their respective allopatric populations (table 2). On the other hand, *G. scandens* differs genetically more from *G. fortis* on Daphne than from all allopatric populations of *G. fortis*. We interpret this result to reflect a large effect of *G. fuliginosa* genes on *G. fortis* in the early part of the study. Data for table 2 were obtained before 1990; to be consistent with sampling on other islands, we included early samples preferentially. When the 2002–2003 samples are substituted for the earlier samples, the large genetic distance between *G. fortis* and *G. scandens* is greatly reduced (fig. 3). This demonstrates directly that the effect of introgressive hybridization is to diminish the genetic distance between sympatric species, both in absolute terms and relative to allopatric populations.

Despite providing strong support for the hypothesis, the method of comparing sympatric and allopatric populations is limited, as illustrated by a

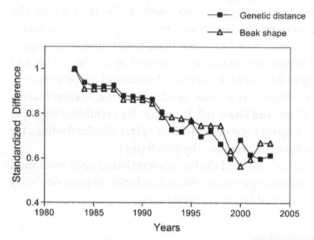

Fig. 2. Morphological and genetic convergence of *Geospiza fortis* and *Geospiza scandens* on Daphne Major Island. Standardization was achieved by giving a value of 1.0 to the difference between the species in 1982 in beak shape and Nei's *D* (from Grant et al. 2004).

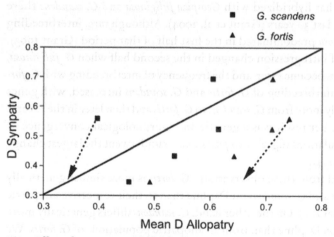

Fig. 3. Effect of contemporary hybridization on genetic distances between *Geospiza scandens* and *Geospiza fortis*. Data are taken from figure 1. The effect of substituting samples from Daphne Major Island in 2002–2003 for pre-1990 samples is shown by dashed lines with arrowheads. The sympatric distance is greatly reduced as a result of persistent but rare introgressive hybridization. Additionally, this alters the mean allopatric distances by a small amount.

fourth species on Daphne, *G. magnirostris*. It colonized Daphne as a breeding species in 1982–1983 and since then has not hybridized with any of the resident species (Grant et al. 2001); yet, it is genetically more similar to Daphne *G. scandens* than to any other population of *G. scandens* (table 2), while at the same time it is substantially different from the sympatric population of *G. fortis*. The two most plausible explanations for these observations involve, first, the multiple origins of the *G. magnirostris* colonists (Grant et al. 2001) and, second, the particular pattern of genetic exchange among the other resident species as described above. This is a good example of how three-way or four-way interactions can complicate both the expectations and interpretations of two-way interaction data.

Introgression and Adaptive Radiation

The evidence for introgression is clearest with the most recently formed *Geospiza* species at the twigs of the phylogenetic tree, and it involves several populations. There is no reason to think introgression is solely a modern phenomenon. Therefore, we suggest that when older members were themselves at the twigs, they may have hybridized as often as modern species do and with the same evolutionary consequences. Indeed, there is indirect evidence that the basal warbler finch still hybridizes, though very rarely, with the more recently evolved tree finches (Bowman 1983; Grant 1999). Thus, introgressive hybridization may have been prevalent throughout the evolutionary history of finches on Galápagos. It could have facilitated evolutionary diversification of Darwin's finches and initiated speciation through the occasionally high proportion of hybrids and backcrosses by chance among the colonists of new small islands that have been repeatedly produced by fluctuating sea levels in the last million years or more (Grant 1999).

In other systems, introgressive hybridization is known to influence the speed and direction of evolution in new environments by elevating levels of genetic variation (Lewontin and Birch 1966; Chiba 1993; Wang and Szmidt 1995; Abbott et al. 2003), relaxing genetic covariation among traits (Grant and Grant 1994), and producing novel genotypes (Anderson and Stebbins 1954; Svärdson 1970; Rieseberg et al. 2003b). It potentially facilitates evolutionary divergence and speciation by refueling populations and reorienting their evolution when their environments change.

A stronger role for hybridization in the Darwin's finch radiation was first suggested by Lowe (1936), based on the botanical work of Lotsey (1916). He proposed that Darwin's finches represent a hybrid swarm, which can be defined as a population comprising "an extremely variable mixture of species, hybrids, backcrosses, and later generation recombination types" (Grant 1971, p. 155). A modified version of this idea has been independently applied to the finches by Seehausen (2004) and was used as a model for adaptive radiations in general. In this version, the radiation is not a hybrid swarm but is the product of diversifying selection on the elements of a swarm. For the scheme to work, markedly different species have to be produced first, and then hybridization creates a swarm. Evidence for the early occurrence of hybridization is a mismatch between nuclear and mitochondrial reconstructions of phylogeny at the base of a branch of diversification (Arnold 1992), owing to unequal exchange of nuclear and mitochondrial alleles (Shaw 2002). Depending on whether nuclear or mitochondrial genes introgress the most, one set reflects phylogeny better than the other, and the other reflects superimposed effects of hybridization.

For example, the Cocos finch (*Pinaroloxias inornata*) is placed inconsistently on phylogenetic trees: at the base of the tree finch branch of the mtDNA tree (Sato et al. 1999; Burns et al. 2002) and at a much earlier point of origin on the microsatellite reconstruction (Petren et al. 1999). To account for this discrepancy, Seehausen (2004) proposed a hybrid origin of the tree and ground finches. However, this proposal necessitates a complex sequence of events involving dispersal of the Cocos finch from Galápagos to Cocos Island (600 km to the northeast) and divergence, followed by reversed dispersal to the Galápagos, interbreeding with a population related to the vegetarian finch (*Platyspiza crassirostris*), and eventual extinction on Galápagos. Alternatively, low levels of introgression could have affected microsatellite alleles but not mtDNA. This would have caused Galápagos species to converge genetically at nuclear loci, for which there is evidence, while the Cocos finch remained isolated and genetically distinct. Therefore, we doubt that hybridization was necessary for any part of the adaptive radiation of Darwin's finches as proposed under the hybrid swarm hypothesis, even though the effectiveness of directional selection would have been aided by introgressive hybridization.

The hybrid swarm hypothesis may have more applicability to the cichlid fish of the African Great Lakes, where many more species evolved in a much shorter time (Kornfield and Smith 2000; Seehausen 2000; Danley and Kocher 2001; Allender et al. 2003; Kocher 2004). Increasing attention has been given to hybridization as a contributing cause (Salzburger et al. 2002; Smith et al. 2003; Schliewen and Klee 2004; Seehausen 2004). There are several more adaptive radiations (Schluter 2000) with unambiguous phylogenies that would be worth examining for evidence of lasting effects of hybridization. Distinguishing between hybrid speciation and introgressive influences on speciation is a challenge for future research to confront.

ACKNOWLEDGMENTS

We thank the Galápagos National Parks Service and the Charles Darwin Research Station for permission to carry out the fieldwork and for logistical support. For field assistance, we thank P. T. Boag, K. T. Grant, O. Jennersten, L. F. Keller, G. C. Keys, I. J. Lovette, E. Monson, D. Moore, K. Petit, G. Retzlaff, G. Rosenquist, G. Seutin, K. Tarvin, and C. Valle. This research has been supported by grants from the National Science Foundation. Two reviewers helped clarify some issues and their presentation.

LITERATURE CITED

Abbott, R. J. 1992. Plant invasions, interspecific hybridization and the evolution of new plant taxa. Trends in Ecology & Evolution 7:401–405.

Abbott, R. J., J. K. James, R. I. Milne, and A. C. M. Gillies. 2003. Plant introductions, hybridization and gene flow. Philosophical Transactions of the Royal Society of London B 358:1123–1132.

Allender, C. J., O. Seehausen, M. E. Knight, G. F. Turner, and N. Maclean. 2003. Divergent selection during speciation of Lake Malawi cichlid fishes inferred from parallel radiations in nuptial coloration. Proceedings of the National Academy of Sciences of the USA 100:14074–14079.

Anderson, E., and G. L. Stebbins Jr. 1954. Hybridization as an evolutionary stimulus. Evolution 8:378–388.

Andersson, M. 1999. Hybridization and skua phylogeny. Proceedings of the Royal Society of London B 266:1579–1585.

Arnegard, M. E., and A. S. Kondrashov. 2004. Sympatric speciation by sexual selection alone is unlikely. Evolution 58:222–237.

Arnold, M. 1992. Natural hybridization as an evolutionary process. Annual Review of Ecology and Systematics 23:237–261.

———. 1997. Natural hybridization and evolution. Oxford University, New York.

———. 2004. Natural hybridization and the evolution of domesticated, pest and disease organisms. Molecular Ecology 13:997–1007.

Bacilieri, R., A. Ducousso, R. J. Petit, and A. Kremer. 1996. Mating system and asymmetric hybridization in a mixed stand of European oaks. Evolution 50:900–908.

Bakker, T. C. M., and A. Pomiankowski. 1995. The genetic basis of female mate preferences. Journal of Evolutionary Biology 8:129–171.

Bensch, S., A. J. Helbig, M. Salomon, and I. Seibold. 2002. Amplified fragment length polymorphism analysis identifies hybrids between two subspecies of warblers. Molecular Ecology 11:473–481.

Bolnick, D. I. 2004. Waiting for sympatric speciation. Evolution 58:895–899.

Bowman, R. I. 1961. Morphological differentiation and adaptation in the Galápagos finches. University of California Publications in Zoology 58:1–302.

———. 1983. The evolution of song in Darwin's finches. Pages 237–537 in R. I. Bowman, M. E. Berson, and A. E. Leviton, eds. Patterns of evolution in Galápagos organisms. American Association for the Advancement of Science, San Francisco.

Burns, K. J., S. J. Hackett, and N. K. Klein. 2002. Phylogenetic relationships and morphological diversity in Darwin's finches and their relatives. Evolution 56:1240–1252.

Cathy, J. C., J. W. Bickham, and J. C. Patton. 1998. Introgressive hybridization and nonconcordant evolutionary history of maternal and paternal lineages in North American deer. Evolution 52:1224–1229.

Chakraborty, R., and M. Nei. 1977. Bottleneck effects on average heterozygosity and genetic distance with the stepwise mutation model. Evolution 31:347–356.

Chiba, S. 1993. Modern and historical evidence for natural hybridization between sympatric species in *Mandarina* (Pulmonata: Camaenidae). Evolution 47:1539–1556.

———. 1999. Accelerated evolution of land snails *Mandarina* in the Oceanic Bonin Islands: evidence from mitochondrial DNA sequences. Evolution 53:460–471.

Clarke, B., M. S. Johnson, and J. Murray. 1998. How "molecular leakage" can mislead us about speciation. Pages 181–195 in P. R. Grant, ed. Evolution on islands. Oxford University Press, Oxford.

Clegg, S. M., S. M. Degnan, J. Kikkawa, C. Moritz, A. Estoup, and I. P. F. Owens. 2002. Genetic consequences of sequential founder events by an island colonizing bird. Proceedings of the National Academy of Sciences of the USA 99:8127–8132.

Coyne, J. A., and T. D. Price. 2000. Little evidence for sympatric speciation in island birds. Evolution 54:2166–2171.

Danley, P. D., and T. F. Kocher. 2001. Speciation in rapidly diverging systems: lessons from Lake Malawi. Molecular Ecology 10:1075–1086.

DeMarais, B. D., T. E. Dowling, M. E. Douglas, W. L. Minkley, and P. C. Marsh. 1992. Origins of *Gila seminuda* (Teleostei: Cyprinidae) through introgressive hybridization: implications for evolution and conservation. Proceedings of the National Academy of Sciences of the USA 89:2747–2751.

Dowling, T. E., and B. D. DeMarais. 1993. Evolutionary significance of introgressive hybridization in cyprinid fishes. Nature 362:444–446.

Dowling, T. E., and C. L. Secor. 1997. The role of hybridization and introgression in the diversification of animals. Annual Review of Ecology and Systematics 28:593–619.

Dowling, T. E., G. R. Smith, and W. M. Brown. 1989. Reproductive isolation and introgression between *Notropis cornutus* and *Notropis chrysocephalus* (family Cyprinidae): comparison of morphology, allozymes, and mitochondrial DNA. Evolution 43:620–634.

Ellstrand, N. C. 2003. Current knowledge of gene flow in plants: implications for transgene flow. Philosophical Transactions of the Royal Society of London B 358:1163–1170.

Feder, J. L. 1998. The apple maggot fly, *Rhagoletis pomonella:* flies in the face of conventional wisdom about speciation? Pages 130–144 in D. J. Howard and S. H. Berlocher, eds. Endless forms: species and speciation. Oxford University Press, New York.

Feder, J. L., S. H. Berlocher, J. B. Roethele, H. Dambroski, J. J. Smith, W. L. Perry, V. Gavrilovic, K. E. Filchak, J. Rull, and M. Aluja. 2003. Allopatric genetic origins for sympatric host-plant shifts and race formation in *Rhagoletis.* Proceedings of the National Academy of Sciences of the USA 100:10314–10319.

Ford, H. A., D. T. Parkin, and A. W. Ewing. 1973. Divergence and evolution in Darwin's finches. Biological Journal of the Linnean Society 5:289–295.

Freeland, J. R., and P. T. Boag. 1999. The mitochondrial and nuclear genetic homogeneity of the phenotypically diverse Darwin's ground finches. Evolution 53:1553–1563.

Gee, J. M. 2004. Gene flow across a climatic barrier between hybridizing avian species, California and Gambel's quail (*Callipepla californica* and *C. gambelii*). Evolution 58:1108–1121.

Gillespie, R. 2004. Community assembly through adaptive radiation in Hawaiian spiders. Science 303:356–359.

Grant, B. R., and P. R. Grant. 1979. Population variation and sympatric speciation. Proceedings of the National Academy of Sciences of the USA 76:2359–2363.

———. 1987a. Mate choice in Darwin's finches. Biological Journal of the Linnean Society 32:247–270.

———. 1989. Evolutionary dynamics of a natural population: the large cactus finch of the Galápagos. University of Chicago Press, Chicago.

———. 1996. Cultural inheritance of song and its role in the evolution of Darwin's finches. Evolution 50:2471–2487.

———. 1998. Hybridization and speciation in Darwin's finches: the role of sexual imprinting on a culturally transmitted trait. Pages 404–422 in D. J. Howard and S. H. Berlocher, eds. Endless forms: species and speciation. Oxford University Press, New York.

———. 2002a. Lack of premating isolation at the base of a phylogenetic tree. American Naturalist 160:1–19.

Grant, P. R. 1993. Hybridization of Darwin's finches on Isla Daphne Major, Galápagos. Philosophical Transactions of the Royal Society of London B 340:127–139.

———, ed. 1998. Evolution on islands. Oxford University Press, Oxford.

———. 1999. Ecology and evolution of Darwin's finches. Princeton University Press, Princeton, NJ.

———. 2002. Founder effects and silvereyes. Proceedings of the National Academy of Sciences of the USA 99:7818–7820.

Grant, P. R., and B. R. Grant. 1987b. Sympatric speciation and Darwin's finches. Pages 433–457 in D. Otte and J. Endler, eds. Speciation and its consequences. Sinauer, Sunderland, MA.

———. 1992. Hybridization of bird species. Science 256:193–197.

———. 1994. Phenotypic and genetic effects of hybridization in Darwin's finches. Evolution 48:297–316.

———. 1997a. Genetics and the origin of bird species. Proceedings of the National Academy of Sciences of the USA 94:7768–7775.

———. 1997b. Hybridization, sexual imprinting, and mate choice. American Naturalist 149:1–28.

———. 2000. Quantitative genetic variation in populations of Darwin's finches. Pages 3–40 in T. A. Mousseau, B. Sinervo, and J. Endler, eds. Adaptive genetic variation in the wild. Oxford University Press, New York.

———. 2002b. Adaptive radiation of Darwin's finches. American Scientist 90:130–139.

———. 2002c. Unpredictable evolution in a 30-year study of Darwin's finches. Science 290:707–711.

Grant, P. R., B. R. Grant, and K. Petren. 2000. The allopatric phase of speciation: the sharp-beaked ground finch (*Geospiza difficilis*) on the Galápagos islands. Biological Journal of the Linnean Society 69:287–317.

———. 2001. A population founded by a single pair of individuals: establishment, expansion, and evolution. Genetica 112/113:359–382.

Grant, P. R., B. R. Grant, J. A. Markert, L. F. Keller, and K. Petren. 2004. Convergent evolution of Darwin's finches caused by introgressive hybridization and selection. Evolution 58:1588–1599.

Grant, V. 1971. Plant speciation. Columbia University Press, New York.

Hedrick, P. 1999. Perspective: highly variable loci and their interpretation in evolution and conservation. Evolution 53:313–318.

Irwin, D. E., and T. Price. 1999. Sexual imprinting, learning and speciation. Heredity 82:998–1010.

Keller, L. F., P. R. Grant, B. R. Grant, and K. Petren. 2001. Heritability of morphological traits in Darwin's finches: misidentified paternity and maternal effects. Heredity 87:325–336.

Kim, S.-C., and L. H. Rieseberg. 1999. Genetic architecture of species differences in annual sunflowers: implications for adaptive trait introgression. Genetics 153:965–977.

Kirkpatrick, M., and M. R. Servedio. 1999. The reinforcement of mating preferences on an island. Genetics 151:865–884.

Kocher, T. D. 2004. Adaptive evolution and explosive speciation: the cichlid fish model. Nature Reviews of Genetics 5:288–298.

Kornfield, I., and P. F. Smith. 2000. African cichlid fishes: model systems for evolutionary biology. Annual Review of Ecology and Systematics 31:163–196.

Lachlan, R. F., and M. W. Feldman. 2003. Evolution of cultural communication systems: the coevolution of cultural signals and genes encoding learning preferences. Journal of Evolutionary Biology 16:1084–1095.

Lachlan, R. F., and P. J. B. Slater. 1999. The maintenance of vocal learning by gene-culture interaction: the cultural trap hypothesis. Proceedings of the Royal Society of London B 266:701–706.

Lack, D. 1945. The Galapagos finches (Geospizinae): a study in variation. Occasional Papers, California Academy of Sciences 21:1–159.

———. 1947. Darwin's finches. Cambridge University Press, Cambridge.

Lewontin, R. C., and L. C. Birch. 1966. Hybridization as a source of variation for adaptation to a new environment. Evolution 20:315–336.

Losos, J. B., T. R. Jackman, A. Larson, K. deQueiroz, and L. Rodriguez-Schettino. 1998. Contingency and determinism in replicated adaptive radiations of island lizards. Science 279:2115–2118.

Lotsey, J. P. 1916. Evolution by means of hybridization. M. Nijhoff, The Hague.

Lowe, P. R. 1936. The finches of the Galápagos in relation to Darwin's conception of species. Ibis 6:310–321.

Martinsen, G. D., T. G. Whitham, R. J. Turek, and P. Keim. 2001. Hybrid populations selectively filter gene introgression. Evolution 55:1325–1335.

Matos, J. A., and B. A. Schaal. 2000. Chloroplast evolution in the *Pinus montezumae* complex: a coalescent approach to hybridization. Evolution 54:1218–1233.

McDade, L. 1992. Hybrids and phylogenetic systematics. II. The impact of hybrids on cladistic analysis. Evolution 46:1329–1346.

McDonald, D. B., R. P. Clay, R. T. Brumfield, and M. J. Braun. 2001. Sexual selection on plumage and behavior in an avian hybrid zone: experimental tests of male-male interactions. Evolution 55:1443–1451.

McKinnon, J. S., S. Mori, B. J. Blackman, L. David, D. M. Kingsley, L. Jamieson, J. Chou, and D. Schluter. 2004. Evidence for ecology's role in speciation. Nature 429:294–298.

Mendelson, T. C. 2003. Sexual isolation evolves faster than hybrid inviability in a diverse and sexually dimorphic genus of fish (Percidae: *Etheostoma*). Evolution 57:317–327.

Nei, M. 1972. Genetic distance between populations. American Naturalist 196:283–292.

Nei, M., T. Maruyama, and R. Chakraborty. 1975. The bottleneck effect and genetic variability in populations. Evolution 29:1–10.

Ortíz-Barrientos, D., J. Reiland, J. Hey, and M. A. F. Noor. 2002. Recombination and the divergence of hybridizing species. Genetica 116:167–178.

Patton, J. L., and M. F. Smith. 1994. Paraphyly, polyphyly, and the nature of species boundaries in pocket gophers (genus *Thomomys*). Systematic Biology 43:11–26.

Payne, R. B., K. Hustler, R. Stjernstedt, K. M. Sefc, and M. D. Sorenson. 2002. Behavioral and genetic evidence of a recent population switch to a novel host species in brood-parasitic indigobirds *Vidua chalybeata*. Ibis 144:373–383.

Petren, K. 1998. Microsatellite primers from *Geospiza fortis* and cross-species amplification in Darwin's finches. Molecular Ecology 7:1782–1784.

Petren, K., B. R. Grant, and P. R. Grant. 1999. A phylogeny of Darwin's finches based on microsatellite DNA length variation. Proceedings of the Royal Society of London B 266:321–329.

Petren, K., B. R. Grant, P. R. Grant, and L. F. Keller. Forthcoming. Comparative landscape genetics and the adaptive radiation of Darwin's finches: the role of peripheral isolation. Molecular Ecology.

Price, T. D., and M. M. Bouvier. 2002. The evolution of F_1 postzygotic incompatibilities in birds. Evolution 56:2083–2089.

Rieseberg, L. H., and J. F. Wendel. 1993. Introgression and its consequences in plants. Pages 70–90 in R. G. Harrison, ed. Hybrid zones and the evolutionary process. Oxford University Press, Oxford.

Rieseberg, L. H., A. Widmer, A. M. Arntz, and J. M. Burke. 2003a. The genetic architecture necessary for transgressive segregation is common in both natural and domesticated populations. Philosophical Transactions of the Royal Society of London B 358:1141–1147.

Rieseberg, L. H., O. Raymond, D. M. Rosenthal, Z. Lai, K. Livingstone, T. Nakazato, J. L. Durphy, A. E. Schwarzbach, L. Donovan, and C. Lexer. 2003b. Major ecological transitions in wild sunflowers facilitated by hybridization. Science 301:1211–1216.

Rüber, L., E. Verheyen, and A. Meyer. 1999. Replicated evolution of trophic specializations in an endemic cichlid fish lineage from Lake Tanganyika. Proceedings of the National Academy of Sciences of the USA 96:10230–10235.

Saetre, G.-P., T. Borge, J. Lindell, T. Moum, C. R. Primmer, B. C. Sheldon, J. Haavie, A. Johnsen, and H. Ellegren. 2001. Speciation, introgressive hybridization and nonlinear rate of molecular evolution in flycatchers. Molecular Ecology 10:737–749.

Salzburger, W., S. Baric, and C. Sturmbauer. 2002. Speciation via introgressive hybridization in East African cichlids? Molecular Ecology 11:619–625.

Sato, A., C. O'hUigin, F. Figueroa, P. R. Grant, B. R. Grant, H. Tichy, and J. Klein. 1999. Phylogeny of Darwin's finches as revealed by mtDNA sequences. Proceedings of the National Academy of Sciences of the USA 96:5101–5106.

Sattler, G. D., and M. J. Braun. 2000. Morphometric variation as an indicator of genetic interactions between black-capped chickadees and Carolina chickadees at a contact zone in the Appalachian mountains. Auk 117:427–444.

Schliewen, U. K., and B. Klee. 2004. Reticulate sympatric speciation in Cameroonian crater lake cichlids. Frontiers in Zoology 1:5.

Schliewen, U. K., D. Tautz, and S. Paabo. 1994. Sympatric speciation suggested by monophyly of crater lake cichlids. Nature 368:629–632.

Schluter, D. 1984. Morphological and phylogenetic relations among the Darwin's finches. Evolution 38:921–930.

———. 2000. The ecological theory of adaptive radiation. Oxford University Press, Oxford.

Schluter, D., and L. M. Nagel. 1995. Parallel speciation by natural selection. American Naturalist 146:292–301.

Seehausen, O. 2000. Explosive speciation rates and unusual species richness in haplochromine fishes: effects of sexual selection. Advances in Ecological Research 30: 235–271.

———. 2004. Hybridization and adaptive radiation. Trends in Ecology & Evolution 19:198–207.

Seehausen, O., and D. Schluter. 2004. Male-male competition and nuptial-colour displacement as a diversifying force in Lake Victoria cichlid fishes. Proceedings of the Royal Society of London B 271:1345–1354.

Servedio, M. R. 2000. Reinforcement and the genetics of nonrandom mating. Evolution 54:21–29.

———. 2004. The evolution of premating isolation: local adaptation and natural and sexual selection against hybrids. Evolution 58:913–924.

Servedio, M. R., and G.-P. Saetre. 2003. Speciation as a positive feedback loop between postzygotic and prezygotic barriers to gene flow. Proceedings of the Royal Society of London B 270:1473–1479.

Shaw, K. L. 2002. Conflict between nuclear and mitochondrial DNA phylogenies of a recent species radiation: what mtDNA reveals and conceals about modes of speciation. Proceedings of the National Academy of Sciences of the USA 99:16122–16127.

Smith, P. F., A. D. Konings, and I. Kornfield. 2003. Hybrid origin of a cichlid population in Lake Malawi: implications for genetic variation and species diversity. Molecular Ecology 12:2497–2504.

Smith, T. B. 1993. Disruptive selection and the genetic basis of bill size polymorphism in the African finch *Pyrenestes*. Nature 363:618–620.

Sorenson, M. D., K. M. Sefc, and R. B. Payne. 2003. Speciation by host switch in brood parasitic indigobirds. Nature 424:928–931.

Sota, T. 2002. Radiation and reticulation: extensive introgressive hybridization in the carabid beetles *Ohomopterus* inferred from mitochondrial gene genealogy. Population Ecology 44:145–156.

Sota, T., R. Ishikawa, M. Ujiie, F. Kusumoto, and A. P. Vogler. 2001. Extensive transspecies mitochondrial polymorphisms in the carabid beetles *Carabus* subgenus *Ohomopterus* caused by repeated introgressive hybridization. Molecular Ecology 10:2833–2847.

Sullivan, J. P., S. Lavoué, and C. D. Hopkins. 2002. Discovery and phylogenetic analysis of a riverine species flock of African electric fishes (Mormyroidae: Teleostei). Evolution 56:597–616.

Svärdson, G. 1970. Significance of introgression in coregonid evolution. Pages 33–59 in C. C. Lindsey and C. S. Woods, eds. Biology of coregonid fishes. University of Manitoba Press, Winnipeg.

Taylor, D. J., and P. D. N. Hebert. 1993. Habitat-dependent hybrid parentage and differential introgression between neighboringly sympatric *Daphnia* species. Proceedings of the National Academy of Sciences of the USA 90:7079–7083.

Ten Cate, C., and D. R. Vos. 1999. Sexual imprinting and evolutionary processes in birds: a reassessment. Advances in the Study of Animal Behaviour 28:1–31.

Thulin, C.-G., and H. Tegelstrom. 2002. Biased geographical distribution of mitochondrial DNA that passed the species barrier from mountain hares to brown hares (genus *Lepus*): an effect of genetic incompatibility and mating behaviour? Journal of Zoology (London) 258:299–306.

Van Doorn, G. S., U. Dieckmann, and F. J. Weissing. 2004. Sympatric speciation by sexual selection: a critical reevaluation. American Naturalist 163:709–725.

Via, S. 2001. Sympatric speciation in animals: the ugly duckling grows up. Trends in Ecology & Evolution 16:381–390.

Wagner, W. L., and V. A. Funk, eds. 1996. Hawaiian biogeography: evolution on a hotspot archipelago. Smithsonian Institution, Washington, DC.

Wang, X.-R., and A. E. Szmidt. 1995. Hybridization and chloroplast DNA variation in a *Pinus* species complex from Asia. Evolution 48:1020–1031.

Wayne, R. K. 1993. Molecular evolution of the dog family. Trends in Genetics 9:218–224.

Whitham, T. G., G. D. Martinsen, K. D. Floate, H. S. Dungey, B. M. Potts, and P. Keim. 1999. Plant hybrid zones affect biodiversity: tools for a genetic-based understanding of community structure. Ecology 80:416–428.

Yang, S.-Y., and J. L. Patton. 1981. Genic variability and differentiation in Galápagos finches. Auk 98:230–242.

American Naturalist 166 (2005): 56–67. © 2005 by the University of Chicago Press.

The Secondary Contact Phase of Allopatric Speciation in Darwin's Finches (2009)

❋ PETER R. GRANT AND B. ROSEMARY GRANT

Abstract—Speciation, the process by which two species form from one, involves the development of reproductive isolation of two divergent lineages. Here, we report the establishment and persistence of a reproductively isolated population of Darwin's finches on the small Galápagos island of Daphne Major in the secondary contact phase of speciation. In 1981, an immigrant medium ground finch (Geospiza fortis) arrived on the island. It was unusually large, especially in beak width, sang an unusual song, and carried some Geospiza scandens alleles. We followed the fate of this individual and its descendants for seven generations over a period of 28 years. In the fourth generation, after a severe drought, the lineage was reduced to a single brother and sister, who bred with each other. From then on this lineage, inheriting unusual song, morphology, and a uniquely homozygous marker allele, was reproductively isolated, because their own descendants bred with each other and with no other member of the resident G. fortis population. These observations agree with some expectations of an ecological theory of speciation in that a barrier to interbreeding arises as a correlated effect of adaptive divergence in morphology. However, the important, culturally transmitted, song component of the barrier appears to have arisen by chance through an initial imperfect copying of local song by the immigrant. The study reveals additional stochastic elements of speciation, in which divergence is initiated in allopatry; immigration to a new area of a single male hybrid and initial breeding with a rare hybrid female.

One hundred and fifty years ago, Charles Darwin (1859) offered an explanation for the process of speciation by which an ancestral species gives rise to one or more derived species through adaptive evolutionary divergence (1). The explanation involved colonization of a new area, adaptive divergence in allopatry, and a barrier to interbreeding when differentiated populations encountered each other in sympatry. Darwin was much clearer on the early stages of speciation than on the later ones. He wrote to one of his many correspondents "... those cases in which a species splits into two or three or more new species ... I should think near perfect separation would greatly aid in the 'specification' to coin a new word" (2). Fortunately "specification" did not catch on, and we use the term "speciation" instead, but the fundamental importance of spatial (geographical) isolation for population divergence has persisted and is incorporated in most, although not all, current models of speciation (3–6).

When divergent populations subsequently meet, their respective members do not breed with each other, or if they interbreed, they do so rarely. Differences in signaling and in response systems that function when mates are chosen arise in allopatry and constitute a premating barrier to interbreeding in sympatry. The barrier may be fully formed in allopatry, in which case no interbreeding occurs in sympatry, or it may be strengthened by natural selection that causes further divergence in sympatry, in two ways. Offspring produced by interbreeding may be relatively unfit, either because the genomes of their parents are incompatible to some degree or because they are at an ecologically competitive disadvantage in relation to the parental populations. Discriminating among these three alternatives has been difficult, because it requires observations to be made in nature on patterns of mating at the time secondary contact is established and in subsequent generations.

We have been fortunate to witness such a secondary contact. Here, we report the origin and persistence for three generations of a premating barrier to interbreeding between two groups of Darwin's finches on one of the Galápagos islands. The barrier arose as a consequence of allopatric divergence in morphology, introgressive hybridization, and divergence of song in sympatry. The barrier has genetic and learned components. Morphology is genetically inherited, whereas song is culturally inherited. Especially noteworthy is the absence of evolutionary change in sympatry in one group in response to the other or to the ecological environment. Our example highlights a stochastic element in the process of speciation.

RESULTS

Immigration

A long-term study of Darwin's finch populations on the Galápagos island of Daphne Major was started in 1973, and by the beginning of 1981 >90% of the two species, *G. fortis* (medium ground finch) and *G. scandens* (cactus finch), had been measured and marked with a unique combination of colored and metal leg bands. In that year, after breeding had ceased, a medium ground finch male with exceptional measurements was captured. It weighed 29.7g, which is >5g heavier than any other *G. fortis* that had bred on the island, and is at the upper end of size variation of *G. fortis* on the neighboring large island of Santa Cruz (7). An analysis of alleles at 16 microsatellite loci with a no-admixture model in the program Structure (8–11) shows that the probability of this individual belonging to the resident Daphne population is 0.088, and of being a member of the conspecific population on Santa Cruz is 0.912. Therefore, we consider it to be an immigrant. Although it is most likely to have come from the large neighboring island of Santa Cruz, we cannot be certain of the exact source (see *Methods*). Morphologically, it is similar to *G. fortis*, but with a somewhat pointed beak profile like that of *G. scandens*, and therefore possibly of mixed genetic composition. In a second analysis, using an admixture model with samples of these two species from Santa Cruz, Structure assigned a greater fraction of its genome to *G. fortis* (0.659) than to *G. scandens* (0.341) (see *Methods*). It is therefore genetically heterogeneous, and we consider it to be a hybrid.

We have followed the survival and reproduction of this individual and all of its known descendants (Fig. 1), here termed the immigrant lineage, for seven generations (F_0 to F_6) spanning 28 years.

Interbreeding Followed by Inbreeding

The immigrant hybrid male (5110) (Fig. 2) carrying some *G. scandens* genes (see *Methods*) bred with a female *G. fortis* also carrying some *G. scandens* genes (Fig. 1). Their sons bred with members of the resident population of *G. fortis*; no breeding females were produced by 5110. One of the sons (15830) gave rise to the next five generations (F_2–F_6) along one line of descent. The male in generation F_3 along this line was not genotyped. We strongly suspect that it was a member of the lineage, because it was seen to be unusually large (11) and sang the characteristic song of the lineage (see *Reproductive Isolation*). Members of the subsequent two generations (F_4 and F_5) bred only with each other and were thus endogamous.

The mating pattern is indicated by direct observations of pairs. Pairs may not be biological parents, however, because extra-pair mating in *G. fortis* is known to occur on this island at a frequency of 15–20% (12). Genetic evidence of paternity is more reliable and confirms our observational assessment of parentage. Genetic analysis reveals that all 25 genotyped members of the lineage in generations F_4–F_6 are homozygous (*183/183*) at microsatellite locus *Gf.11*. The homozygote state at this locus is highly unusual. Of 249 genotyped *G. fortis* individuals on the island from 2002 onwards that were not in the lineage, but contemporary with generations F_3–F_6, 27 carried one copy of the *183* allele and one individual carried two copies. Given a frequency of the

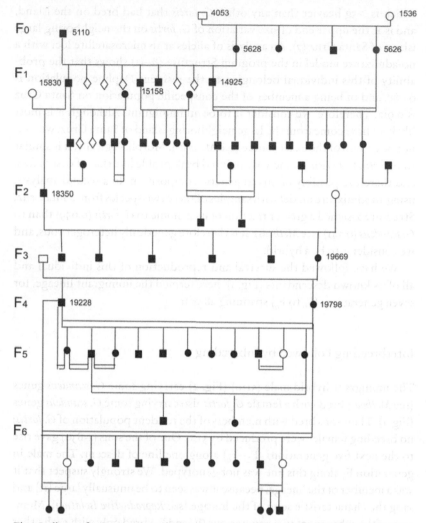

Fig. 1. Pedigree of an immigrant *G. fortis* male (5110) with a line of descent to an exclusively inbreeding (endogamous) group. For details of the construction of the pedigree, see *Methods.* Males are indicated by squares, females by circles, and birds of unknown sex by diamonds. Individuals of unknown genotype are indicated by open symbols, and filled symbols refer to genotyped birds. Salient individuals in the pedigree are indicated by their band numbers, e.g., the mate (5628) of the original immigrant (5110) is a backcross from *G. scandens.* Pairs of close relatives are connected by double lines. The frequency of inbreeding among close relatives in the immigrant lineage is exceptionally high. Keller et al. (12) analyzed 364 unique matings, where all four grandparents were known in the *G. fortis* population (including the immigrant lineage) up to 1992, and found that only three (0.8%) were the product of matings between first-degree relatives ($f = 0.25$). Two of the three are in the pedigree above.

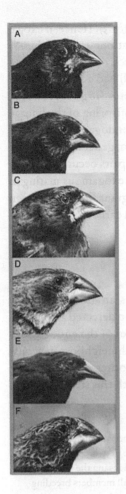

Fig. 2. The immigrant lineage contrasted with *G. scandens* and *G. magnirostris* on Daphne Major Island. (*A*) 5110, the original immigrant (generation F$_0$); (*B*) *G. fortis* 15830 (generation F$_1$), son of 5110; (*C*) *G. fortis* 19256 (generation F$_5$); (*D*) *G. fortis* 19566 (generation F$_6$); (*E*) *G. scandens* 15859; (*F*) *G. magnirostris* 17339.

183 allele of 0.056 (29/498), the expected frequency of the homozygotes with random mating is 0.0032, or one individual in ≈300. Homozygotes were equally rare before 2002.

The original immigrant male (5110) was a homozygote (*183/183*) and his son (15830), grandson (18350), and great grand-daughter (19669) in the line of descent (Fig. 1) were heterozygotes (*183/–*). The mate of 19669 must have carried at least one copy of the *183* allele, because their offspring were homozygous (*183/183*). This fact adds weight to the suggestion above that the mate of 19669 was also a member of the immigrant lineage (generation 3).

Although members of the lineage bred with each other (endogamy) in two or more generations, they might have also produced offspring by breeding with members of the resident population through extra-pair mating (exogamy). Cryptic exogamous mating can be tested by taking advantage of the fact that all endogamous parents are homozygous (*183/183*). Hence, if exogamous offspring are produced, they must carry at least one copy of the *183* allele at *Gf.11*. Twenty-eight individuals hatched in 2002 or later could have been produced by exogamous mating because they all had a *183* allele: One was a homozygote (*183/183*) and 27 were heterozygotes (*183/–*). However, all of these individuals were ruled out as exogamous offspring of the lineage because none of them matched any member of the endogamous group of breeders (generations F_4 and F_5) or the mother (19669; generation F_3) at all of the remaining 15 loci; mismatches of at least 4 base pairs occurred at 2–10 loci. Thus, we conclude there has been no detectable exogamous mating in the last two generations in eight years, and the immigrant lineage has been exclusively endogamous since 2002 and possibly much earlier.

Reproductive Isolation

A premating barrier to the exchange of genes thus exists; an additional intrinsic postmating barrier is unlikely because it has not been detected among any of the six *Geospiza* species (13). Furthermore, territories of the endogamous group formed spatially restricted clusters (Fig. 3) with neighbors in acoustic contact, which suggests that they recognize each other in the breeding season as members of the same group. Contrasting with this strong pattern,

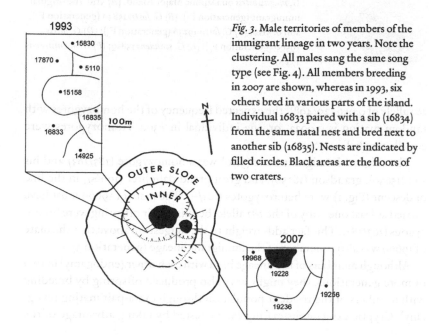

Fig. 3. Male territories of members of the immigrant lineage in two years. Note the clustering. All males sang the same song type (see Fig. 4). All members breeding in 2007 are shown, whereas in 1993, six others bred in various parts of the island. Individual 16833 paired with a sib (16834) from the same natal nest and bred next to another sib (16835). Nests are indicated by filled circles. Black areas are the floors of two craters.

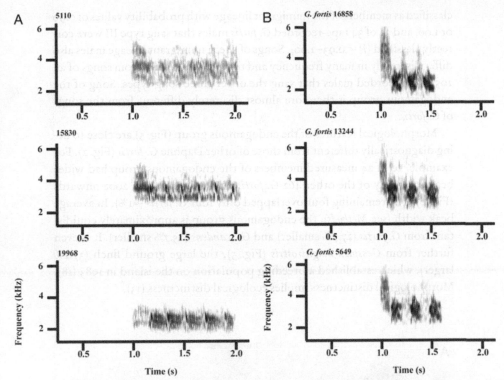

Fig. 4. Songs of the original immigrant (5110), a son (15830) and a fifth generation descendant (19668), compared with three Daphne *G. fortis* individuals that sang a standard form of type III. Immigrants differ from residents statistically in lower maximum frequency and higher note repetition rate (see *Discussion; Behavior and Speciation*). A wideband setting and a Hamming window with DFT 256 were used.

no more than two close relatives have been observed breeding in adjacent territories in the *G. fortis* population during 22 years (1976–98) of intensive study.

The barrier to interbreeding among *Geospiza* species has two elements, song and morphology (13). Specific features of both elements are learned during a short sensitive period early in life, while the young are dependent upon parents for food (14, 15). Male *G. fortis* sing only one song. There is individual variation on a *G. fortis* theme, which can be classified into four types on Daphne that are recognizable by sonograph and to the human ear (16, 17); females do not sing. Sixteen of 17 singing males in the lineage (94.3%), including the original immigrant (5110), sang a variant form of type III, also recognizable to the human ear: The seventeenth sang a type I song and did not breed. Eleven of them were tape-recorded and sonographed (Fig. 4). In a multiple discriminant function analysis (see *Methods*), all 11 were correctly classified as members of the immigrant lineage with probability values of 0.99 or 1.00, and 32 of 34 tape-recorded *G. fortis* males that sang type III were correctly classified ($P = 0.93–1.00$). Songs of the 11 immigrant lineage males also differ discretely in many frequency and temporal measures from songs of all 205 tape-recorded males that sang the other three song types. Song of the endogamous group is therefore almost discretely different from the songs of *G. fortis.*

Morphological features of the endogamous group (Fig. 5) are close to being diagnostically different from those of other Daphne *G. fortis* (Fig. 2). For example, 20 of 24 measured members of the endogamous group had wider beaks than any of the other 462 *G. fortis* on the island from 2002 onwards (Fig. 6). The remaining four overlapped only four *G. fortis* (≈1%). In average beak width (see *Methods*), the endogamous group is approximately equidistant from *G. fortis* (27.7% smaller) and *G. scandens* (25.7% smaller). It is even further from *Geospiza magnirostris* (Fig. 5), the large ground finch (37.2% larger), which established a breeding population on the island in 1983 (18). Morphological distinctness implies ecological distinctness (13).

Fig. 5. Immigrant and resident *G. fortis.* (*Upper*) 9807, member of the immigrant lineage (generation F_5). (*Lower*) 19181, contemporary member of the resident population of *G. fortis* on Daphne Major Island.

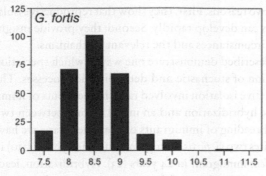

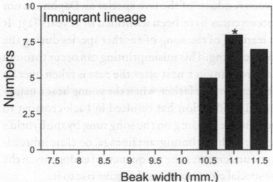

Fig. 6. Morphological contrast between the immigrant lineage ($n = 20$) with other *G. fortis* ($n = 280$) on Daphne Major Island in the years 2005–09. The position of the original immigrant (5110) is indicated by an asterisk.

DISCUSSION

The Tempo and Mode of Speciation

Charles Darwin believed that evolution took place too slowly to be observed, and therefore speciation, the evolution of a new species, would take immeasurably longer (1). Evolution by natural selection is now known to occur rapidly in a variety of taxa and environments (19), including Darwin's finches in the Galápagos (20, 21), but Darwin's opinion on the slowness of speciation remains the consensus view (3, 4, 6, 22). Although generally true, it may apply more often to postmating than to premating isolation, where behavior plays an important role. Behavior has the potential to change rapidly when learned as opposed to being genetically fixed, especially in vertebrates. The origin of a premating barrier between populations is a crucial component of speciation, because regardless of whether sympatric, closely-related species can or cannot produce viable and fertile offspring, the vast majority do not interbreed or do so very rarely. Our observations on the development of premating isolation of two divergent lineages after secondary contact are

therefore significant for two reasons. First, they show that reproductive isolation of small populations can develop rapidly. Second, they provide insight
into the environmental circumstances and the relevant mechanisms.

The events we have described demonstrate one way in which speciation
proceeds by a combination of stochastic and deterministic processes. The
development of reproductive isolation involved rare chance events of immigration and introgressive hybridization and an initial mating between two
hybrid individuals. The breeding of immigrants on Daphne is rare; we have
identified by microsatellites two *G. fortis* immigrants and one hybrid (5110) in
18 years and one *G. scandens* immigrant in 24 years (23). Hybridization, leading to gene exchange between residents of the two species on Daphne, is not
much more frequent; 13 occurrences have been identified in 21 years (23). It
generally results from the learning of the song of another species during the
early sensitive period of song learning. This misimprinting can occur through
cross-fostering, when an egg remains in a nest after the nest is taken over by
another species, or after the death of the father, when the young hear a neigh-
bor of another species (13). Hybridization has resulted in back-crossing to
one parental species or the other depending on the song sung by the hybrid's
father (15), but, unlike the case with the immigrant lineage, no close inbreeding has ensued. This outcome underscores the uniqueness of endogamy in the
immigrant lineage and the special circumstances that gave rise to it.

Ecology of Speciation

Stochastic elements are inevitably present in allopatric speciation, owing to
different mutations occurring at random in separate environments (24, 25).
In the words of Hermann Muller (24) "Thus a long period of non-mixing of
two groups is inevitably attended by the origination of actual immiscibility,
i.e., genetic isolation." There is also an inevitable element of determinism in
allopatric speciation arising from ecological differences between separate
environments, because no two environments can be exactly the same, and
therefore selection pressures must differ. Each class of factors, stochastic and
deterministic, could be vital or trivial in particular cases. There is no single
mechanism of speciation (3–6). The challenge for evolutionary biologists is to
identify and assign importance to each contributing factor when accounting
for the causes and circumstances of speciation in particular cases, and then to
seek generalizations. Theories serve as a guide.

One theory of speciation proposes a completely allopatric origin of a barrier to interbreeding with different emphases on random processes and selection (3, 5, 24, 25). A second theory proposes a major genetic change, shortly
after the founding of a new population by a few individuals, in which random
drift plays an essential, but not exclusive, role (26). Neither is applicable to

our study, because the barrier originated partly in sympatry with song (see *Behavior and Speciation*) but without genetic change in sympatry.

Instead, our observations are largely consistent with an ecological theory of speciation (27–30) in which a barrier to interbreeding arises as a behaviorally flexible correlated effect or by-product of adaptive divergence of an ecologically selected trait (22, 27, 28). Beak size, with known ecological function of food handling (31, 32), is also a key component of the barrier as it signals species identity in a reproductive context (33). Divergence in beak size of the immigrant lineage and the residents occurred in allopatry. The immigrant lineage did not diverge from the residents in beak size in sympatry, as would be expected if selection minimized ecological competition between them (character displacement theory; ref. 21) or minimized the probability of interbreeding (reinforcement theory; refs. 34, 35).

However, the resident population of *G. fortis* did diverge from the immigrant lineage, and divergence may have facilitated intra-group mating. Divergence was caused by natural selection during the drought of 2004, when small members of the *G. fortis* population survived best, in part because large-beaked members suffered in competition for food with *G. magnirostris* (21). The sole surviving brother and sister of the immigrant lineage (F_4 generation) bred with each other in 2005, when large members of the *G. fortis* population were scarce. Intra-group mate recognition and endogamy in generations F_4 and F_5 may have been facilitated by the morphological divergence of resident *G. fortis* from *G. magnirostris*. In the five years after the natural selection event in 2004, the endogamous group was almost completely separated in morphology from the residents (Fig. 4). Ecological differences associated with morphological separation probably contributed to their coexistence. The same has been argued for double-invasion species-pairs of birds on islands, in which a mainland species colonizes an island twice. Successful establishment of the second population depends upon prior divergence of the first in morphology and ecology and population-specific mating (36).

Behavior and Speciation

The other component of behavioral isolation is song. The song of the immigrant male 5110 was acquired initially by learning from early exposure to songs on Santa Cruz. Then it appears to have been modified by imperfect copying of the type III song of Daphne *G. fortis* during the crystallization phase of song production in his first breeding season in 1983. The alternative possibility of an allopatric origin is not supported by any of the >100 spectrograms of tape-recorded song in the published literature from Santa Cruz (14, 37) and other islands (14). We have not heard the type III song or 5110's at locations on both the north and the south coasts of the adjacent Santa Cruz Island or on any of the other major islands of the archipelago (Santa Fe, Floreana, and those listed in *Methods*).

The imperfect copying of a resident's song appears to be a stochastic element in the development of reproductive isolation, and a nonecological component of the barrier to interbreeding. An alternative possibility is that the particular characteristics of the immigrant song could be a correlated effect of allopatric divergence in beak morphology and therefore part of the syndrome of ecological speciation (38). Song characteristics can be affected by beak morphology for biomechanical reasons; the larger the beak the slower the production of repeated notes and the smaller the range of their frequencies (39). Thus, a large bird like 5110 and its descendants might sing a slower version of the type III song over a smaller range of frequencies. However, the former expectation is not upheld; the mean repetition rate of notes is not slower, but faster, in songs of the immigrant lineage ($n = 11$ birds) than in type III songs of *G. fortis* ($n = 34$ birds; $F_{1,43} = 19.45$, $P < 0.0001$). In contrast, and consistent with the biomechanics hypothesis, the maximum frequency is reduced in the first note of songs of the immigrant lineage (Fig. 3) compared with type III songs ($F_{1,43} = 62.59$, $P < 0.0001$) and so is the frequency range of this note ($F_{1,43} = 54.19$, $P < 0.0001$). However, reduction does not appear to be due to mechanical constraints, because the largest species of ground finch on Daphne, *G. magnirostris* (Fig. 5), sings a song with a large frequency range in its initial note (18) like the type III song of *G. fortis*.

We conclude that song features of the immigrant lineage are not a byproduct of beak divergence in allopatry. Reproductive isolation depended in part on ecological factors associated with beak size and in part on chance behavioral factors associated with song learning independent of ecology. The importance of beak size in mate choice has been emphasized in a study of assortative mating in a population of *G. fortis* on Santa Cruz Island (40). Chance factors have been invoked to explain the large differences in songs between populations of *Geospiza difficilis* on adjacent islands (41).

Future Prospects of Incipient Species

These observations provide important insight into the process of speciation at the beginning of the sympatric phase following divergence in allopatry. They also raise a question that is rarely if ever asked: How many generations of exclusively within-group mating are needed before the group is recognized as a separate species that deserves taxonomic status? There is no nonarbitrary answer. We treat the endogamous group as an incipient species because it has been reproductively isolated from sympatric *G. fortis* for three generations and possibly longer.

Many episodes of incipient speciation probably fail for every one that succeeds in reaching complete genetic isolation due to incompatibility factors. In the present case, it is too early to tell whether reproductive isolation is transitory or likely to be enduring. The odds would seem to be against long-term persistence of the immigrant lineage as a reproductively isolated population. First, numbers are small and stochastic fluctuations in population size may result in extinction. Second, the new populations might run the risk of competitive exclusion from *G. fortis* and/or *G. magnirostris* if the food environment changed. Third, it might disappear through interbreeding with *G. fortis* and/or *G. scandens*, an example of reproductive absorption of one species by another (30), initiated perhaps by extra-pair mating or misimprinted song. Fourth, it might suffer from inbreeding depression.

With regard to the last possibility, a small closed inbreeding population is expected to lose alleles by chance, leading to extreme homozygosity, which makes the population more vulnerable to extinction. However, the history of another episode of immigration shows that neither extreme homozygosity nor extinction is inevitable. A breeding population of *G. magnirostris* was established on Daphne in 1983 by two immigrant females and three males (18). Inbreeding depression was moderately severe two generations later (42, 43), but subsequent immigration alleviated the effects and the population has persisted (23, 42). In the present case genetic heterogeneity of both the immigrant male and his mate due to gene mixing with *G. scandens* (see Fig. 1 legend and *Methods*) makes it likely that the population is open to genetic input from resident or immigrant *G. fortis* and *G. scandens*. The outcome, fusion or persistence, will depend on rates of introgression and fates of introduced genes (23, 44, 45). Divergence in beak size has increased the chances of long-term ecological coexistence.

CONCLUSION

Our observations provide insight into speciation and hence, into the origin of a new species. They show how a barrier to interbreeding can arise behaviorally and without genetic change in sympatry. A necessary condition was prior ecological divergence, and introgressive hybridization was possibly another. Evidently it takes only a single diploid immigrant to start the process by breeding with a resident, and tolerance of the effects of inbreeding is needed to complete it.

METHODS

Assignment Tests

We used genotypic information from blood samples to identify individuals with version 2.2 (8) of the program Structure (9, 10). Individuals were assigned to specified groups with a probability estimated by a Bayesian analysis of frequencies of microsatellite alleles at 16 loci (11). We applied the majority rule ($P > 0.500$) to assign individuals to groups. Following the authors' recommendations we used a burn-in of 50,000 iterations, a run length of 100,000, and for each new analysis we repeated the procedure once to make sure results were consistent. We used a No Admixture model for questions about population membership of an individual and an Admixture model for questions about the fraction of an individual's genome attributable to each of two populations. The correlated alleles option was used throughout. An individual for assignment was given a value of zero in the Popflag column, and all individuals from defined islands were given a value of 1, which allowed repeated updating of allele frequencies of all groups except the targeted individual. We split the birds into an early (up to 1998) and a late group (1999–2008) because full pedigree information was available up to 1998, and only partial information was available afterward. There was almost no breeding between 1999 and 2001.

To identify F_1 hybrids and backcrosses we used an ancestry model with two prior generations. This procedure gives an estimated probability that an individual belongs to another species (generation 0), having a parent (generation –1) or having a grandparent (generation –2) from another species. The last two are almost equivalent to F_1 hybrid and a backcross generation (first or higher).

Source and Identity of the Original Immigrant

The individual 5110, captured in a mist net in 1981, was initially suspected of being an immigrant. It was much larger than any resident member of *G. fortis* on Daphne and more similar in size and proportions to *G. fortis* on other islands. Moreover, we could not identify potential parents at a time when 90% of *G. fortis* and *G. scandens* were banded. On geographical grounds the large, neighboring island of Santa Cruz is the most likely source (30). Furthermore, allele *183* at locus *Gf.11*, which is homozygous in 5110, has a frequency of 0.077 in the Santa Cruz population of *G. fortis,* but is not present in any of our admittedly small samples from other islands. Assuming Santa Cruz is correctly identified as the source, we performed an assignment test with alleles at 16 microsatellite loci using a no-admixture model in the program Structure (8–11). The probability of this individual belonging to the resident Daphne population ($n = 77$) was found to be 0.088, and the probability of being a member of the conspecific population on Santa Cruz ($n = 39$) was 0.912. Therefore, 5110 was a probable immigrant.

We used the following samples of *G. fortis* genotypes from defined populations in an attempt to identify the source island of 5110 with Structure:

Santa Cruz ($n = 39$), Santiago ($n = 9$), Rábida ($n = 3$), Marchena ($n = 17$), San Cristóbal ($n = 4$), Pinta ($n = 12$), Isabela ($n = 11$), and Daphne ($n = 77$). The defined Daphne population comprised only those individuals that hatched on the island. A sample of 12 birds captured on Daphne without bands and therefore potential immigrants, including 5110, comprised an undefined population. All but three were assigned to the Daphne population at $P > 0.950$. Assignment probabilities of 5110 were 0.427 to Isabela, 0.282 to Santiago, and 0.244 to Santa Cruz, but 0.000 to Daphne. Sequential deletions of the other populations with small samples (Rábida, San Cristóbal, Santiago) gave similar results with the probability of assignment to Isabela being the highest and to Daphne always being 0.000. These results support the immigration hypothesis. However, the analysis failed to identify the source island, probably because the *G. fortis* populations are too similar genetically (23).

Although clearly referable to *G. fortis*, 5110 has a somewhat pointed beak profile like that of *G. scandens* (Fig. 2) and is therefore possibly of mixed genetic composition. In an analysis using an admixture model with samples of these two species from Santa Cruz, Structure assigned a large fraction of its genome to *G. fortis* (0.659) and a smaller fraction to *G. scandens* (0.341). Therefore, 5110 is genetically heterogeneous. The homozygous condition of the *183* allele at locus *Gf.11* is further evidence of 5110 being a hybrid because the allele is at a much higher frequency in the Santa Cruz population of *G. scandens* (0.361) than in *G. fortis* (0.077). There is indirect genetic evidence of rare interbreeding on this island between *G. fortis* and *G. scandens* (46). These facts, together with the exceptional morphology, support the hypothesis that the immigrant was a *G. fortis* × *G. scandens* hybrid or backcross.

Identity of Birds Breeding with Members of the Immigrant Lineage

In generations $F_0–F_2$ (Fig. 1), two male and five female resident *G. fortis* that mated with members of the immigrant lineage were genotyped. Their identities were first established by their measurements (47), then assessed with a no-admixture model in Structure. In the analysis of 1423 *G. fortis* and 504 *G. scandens* present in 1978–98, a genotyped sib 5627 was entered in place of the missing 5628. Structure assigned six of the targeted individuals to *G. fortis* with probabilities of 0.998 or 1.00. The remaining two, 5626 and 5627, were assigned to *G. fortis* with probabilities of 0.858 and 0.794, and to backcrosses from *G. scandens* to *G. fortis* with probabilities of 0.142 and 0.205, respectively. The sib that was not genotyped (5628) was therefore probably genetically heterogeneous also. We have no reason to suspect extra-pair paternity, as all three sibs were similar morphologically. For example, the beak depths and widths of 5626, 5627, and 5628 were respectively 8.9 and 8.2, 8.0 and 7.9, and 8.6 and 8.5 mm. One of them (5628) bred with the immigrant male (5110) and another (5626) bred with a son (14925). Their father (4053) sang a *G. fortis* song (type I) but was not genotyped. He was considered to be an F_1 hybrid because his measurements were on the borderline between those of the two species (7), but he may have been a first generation backcross.

Parentage

Parentage was initially inferred by observing adults attending a nest. Adults were identified by their color bands or their observed large size (11). To construct the pedigree of the immigrant lineage, we used allele lengths at 16 microsatellite loci (11). We allowed 2*bp* differences (and no more) between offspring and presumptive parents as being within the range of scoring variation (11, 12, 43). Almost all offspring matched both parents at all loci.

Construction of Fig. 1

The line of descent to the endogamous group is as follows. The immigrant male (5110), F_0 generation, bred with a genetically heterogeneous *G. fortis* female (5628). Although her parents were not genotyped, she was most likely to have been a backcross from *G. scandens* (see *Identity of Birds*). A son (15830), F_1 generation, bred with a *G. fortis* female of unknown genotype. She could not have been a sib, because all of 15830's sibs that attempted to breed were males. A grandson (18350), F_2 generation, might have bred with a sib; the genotype of his mate is not known. In the next generation (F_3), a daughter (19669) bred with a male that was, like her, unusually large and therefore may have been a sib. He was not captured and genotyped but was observed (and heard) repeatedly. The female 19669 was originally thought to be an immigrant (11), but we have since discovered a complete genetic match at 16 loci between 19669 and 18350. Generations F_4–F_6 comprise the endogamous group.

Generation F_5 is shown at two levels corresponding to early (2006–07) and late (2008–09) production of offspring. The first level is known with certainty because only two members of the pedigree, 19228 and 19798, could have produced them. The parents of those at the second level could not be identified genetically, because more than one generation was present at that time. Because there are many offspring, it is likely the parents were members of both F_4 and F_5 generations. The unknown genotypes of two females in the fifth generation were inferred from the genotypes of their offspring and the offspring's known fathers. A third of unknown genotype was seen to be very large (11).

The immigrant 5110, after breeding with 5628, bred with two banded *G. fortis* females. In both cases their offspring did not breed. They have been omitted from the figure for simplicity.

Song and Morphology

Song was recorded with a Sony (TCM 5000) tape recorder and a Sennheiser AKG D900 microphone (17). Two to 15 songs per bird were recorded. Because songs remain unchanged throughout life (17), only the first one recorded for each bird was included in analyses of songs performed in Raven version 1.3, Beta version (48). The following were measured for each song: number of notes, number of notes/sec, central frequency, frequency at which maximum energy was produced, and for each of the first two notes the duration, minimum and maximum frequency, and frequency range. Five uncorrelated variables were entered simultaneously into a two-group discriminant function analysis performed in JMP 7.0 (49). These variables were number of notes, number of notes/sec, duration of first note, and maximum frequency of the first and second notes. The groups were 11 males of the immigrant lineage that sang the type III variant and 34 *G. fortis* that sang the type III song, which is the most similar type to the immigrant's song. All immigrant males were classified correctly ($P = 0.99$ or 1.00). All but two of the 34 type III songs were classified correctly ($P = 0.93$–1.00). The other two were misclassified as songs of the immigrant lineage at $P = 0.93$ and 0.97, respectively. For a test of the fit of the discriminant function, Wilks' lambda = 0.2010, Exact $F_{5,34} = 31.003$ ($P < 0.0001$). Songs of the immigrant group differed significantly from *G. fortis* type III songs in maximum frequency, frequency range, and note repetition rate (see *Results; Reproductive Isolation*), but did not differ in minimum frequency, central frequency, or frequency at which maximum energy was produced ($P > 0.1$).

Morphological measurements were made as described in ref. 47 and illustrated in ref. 13. Beak-width means in millimeters and standard deviations for the samples of birds on the island from 2002 onwards are 10.82 ± 0.432 for the immigrant lineage ($n = 21$), 8.47 ± 0.616 for all other *G. fortis* ($n = 462$), 8.61 ± 0.546 for *G. scandens* ($n = 291$), and 14.85 ± 0.870 for *G. magnirostris* ($n = 241$). Beak-depth means are slightly larger in each case.

ACKNOWLEDGMENTS

We thank the many assistants who have helped us on Daphne, and the Galápagos National Parks Service and Charles Darwin Foundation for logistical support. We are grateful to Paula Hulick for help with graphics, Dan Davison for help with statistical programming, and Trevor Price, Nathalie Seddon, and Margarita Womack for many helpful suggestions. The National Science Foundation and Princeton University's Class of 1877 endowment provided funding.

REFERENCES

1. Darwin C (1859) *On the Origin of Species by Means of Natural Selection* (J. Murray, London).

2. Darwin C (1878) Letter to K. Semper, Nov. 26 in *The Life and Letters of Charles Darwin. Including an Autobiographical Chapter. Vol. III,* ed Darwin, F. (D. Appleton and Co., New York, 1919), p 160.

3. Mayr E (1963) *Animal Species and Evolution* (Belknap, Cambridge, MA).

4. Coyne JA, Orr HA (2004) *Speciation* (Sinauer, Sunderland, MA).

5. Gavrilets, SS (2004) *Fitness Landscapes and the Origin of Species* (Princeton Univ Press, Princeton, NJ).

6. Price T (2008) *Speciation in Birds* (Ben Roberts, Greenwood, CO).

7. Grant PR (1993) Hybridization of Darwin's finches on Isla Daphne Major, Galápagos. *Philos Trans R Soc London Ser B* 340:127–139.

8. Pritchard JK, Wen X, Falush D (2007) *Documentation for Structure software: Version 2.2.* Available at http://pritch.bsd.uchicago.edu/software. Accessed May 23, 2008.

9. Pritchard JK, Stephens M, Donnelly P (2000) Inference of population structure using multilocus genotype data. *Genetics* 155:945–959.

10. Falush D, Stephens M, Pritchard JK (2003) Inference of population structure: Extensions to linked loci and correlated allele frequencies. *Genetics* 164:1567–1587.

11. Grant PR, Grant BR (2008) Pedigrees, assortative mating and speciation in Darwin's finches. *Proc R Soc London Ser B* 275:661–668.

12. Keller LF, Grant PR, Grant BR, Petren K (2001) Heritability of morphological traits in Darwin's finches: Misidentified paternity and maternal effects. *Heredity* 87:325–336.

13. Grant PR, Grant BR (2008) *How and Why Species Multiply* (Princeton Univ Press, Princeton, NJ).

14. Bowman RI (1983) in *Patterns of Evolution in Galápagos Organisms,* ed Bowman RI, Berson M, Leviton AE (AAAS Pacific Division, San Francisco), pp 237–537.

15. Grant BR, Grant PR (1998) in *Endless Forms: Species and Speciation,* eds Howard DJ, Berlocher SH (Oxford Univ Press, New York), pp 404–422.

16. Gibbs HL (1990) Cultural evolution of male song types in Darwin's medium ground finches, *Geospiza fortis. Anim Behav* 39:253–263.

17. Grant BR, Grant PR (1996) Cultural inheritance of song and its role in the evolution of Darwin's finches. *Evolution* 50:2471–2487.

18. Grant PR, Grant BR (1995) The founding of a new population of Darwin's finches. *Evolution* 49:229–240.

19. Hendry AP, McKinnon MT, eds (2001) *Microevolution. Rate, Pattern, Process* (Kluwer Academic, Boston).

20. Grant PR, Grant BR (2002) Unpredictable evolution in a 30-year study of Darwin's finches. *Science* 296:707–711.

21. Grant PR, Grant BR (2006) Evolution of character displacement in Darwin's finches. *Science* 313:224–226.

22. Dobzhansky T (1937) *Genetics and the Origin of Species* (Columbia Univ Press, New York).

23. Grant PR, Grant BR (2009) Conspecific versus heterospecific gene exchange between populations of Darwin's Finches. *Philos Trans R Soc London Ser B,* in press.

24. Muller H (1940) *The New Systematics,* ed Huxley JS (Clarendon, Oxford), pp 185–268.

25. Mani GS, Clarke B (1990) Mutational order—a major stochastic process in evolution. *Proc R Soc London Ser B* 240:29–37.

26. Mayr E (1954) in *Evolution as a Process,* ed Huxley J, Hardy AC, Ford EB (Allen and Unwin, London), pp 157–180.

27. Lack D (1947) *Darwin's Finches* (Cambridge Univ Press, Cambridge, U.K).

28. Grant PR (1986) *Ecology and Evolution of Darwin's Finches* (Princeton Univ Press, Princeton, NJ).

29. Schluter D (1996) Ecological causes of adaptive radiation. *Am Nat* 148(Suppl): S40–S64.

30. Schluter D (2009) Evidence for ecological speciation and its alternative. *Science* 323:737–741.

31. Schluter D, Grant PR (1984) Determinants of morphological patterns in Darwin's finch communities. *Am Nat* 123:175–196.

32. Herrel AJ, Podos J, Huber SK, Hendry AP (2005) Bite performance and morphology in a population of Darwin's finches: Implications for the evolution of beak shape. *Funct Ecol* 19:43–48.

33. Ratcliffe LM, Grant PR (1983) Species recognition in Darwin's finches (*Geospiza,* Gould) I: Discrimination by morphological cues. *Anim Behav* 31:1139–1153.

34. Johnson MS, Murray J, Clarke B (2000) Parallel evolution in Marquesan partulid land snails. *Biol J Linn Soc* 69:577–598.

35. Nosil P, Yukilevich R (2008) Mechanisms of reinforcement in natural and simulated polymorphic populations. *Biol J Linn Soc* 95:305–319.

36. Grant PR (2001) Reconstructing the evolution of birds on islands: 100 years of research. *Oikos* 92:385–403.

37. Huber S, Podos J (2006) Beak morphology and song features covary in a population of Darwin's finches. *Biol J Linn Soc* 88:89–498.

38. Seddon N (2005) Ecological adaptation and species recognition drives vocal evolution in neotropical suboscine birds. *Evolution* 59:200–215.

39. Podos J (2001) Correlated evolution of morphology and vocal signal structure in Darwin's finches. *Nature* 409:185–188.

40. Huber SK, De León LF, Hendry AP, Bermingham E, Podos J (2007) Reproductive isolation of sympatric morphs in a population of Darwin's finches. *Proc R Soc London Ser B* 274:1709–1714.

41. Grant BR, Grant PR, Petren K (2000) The allopatric phase of speciation: The sharp-beaked ground finch (*Geospiza difficilis*) on the Galápagos islands. *Biol J Linn Soc* 69:287–317.

42. Grant PR, Grant BR, Petren K (2001) A population founded by a single pair of individuals: establishment, expansion, and evolution. *Genetica* 112/113:359–382.

43. Keller LF, Grant PR, Grant BR, Petren K (2002) Environmental conditions affect the magnitude of inbreeding depression in survival of Darwin's finches. *Evolution* 56:1229–1239.

44. Bolnick DI, Caldera EJ, Matthews B (2008) Evidence for asymmetric migration load in a pair of ecologically divergent stickleback populations. *Biol J Linn Soc* 94:273–287.

45. Grant BR, Grant PR (2008) Fission and fusion of Darwin's finches populations. *Philos Trans R Soc London Ser B* 363:3821–3829.

46. Grant PR, Grant BR, Petren K (2005) Hybridization in the recent past. *Am Nat* 166:56–67.

47. Boag PT, Grant PR (1984) The classical case of character release: Darwin's finches (*Geospiza*) on Isla Daphne Major, Galápagos. *Biol J Linn Soc* 22:243–287.

48. Charif RA, Clark CW, Fristrup KM (2006) Raven 1.3 User's Manual (Cornell Laboratory of Ornithology, Ithaca, NY).

49. SAS Institute (2007) JMP 7.0 (SAS Institute Inc., Carey, NC).

Proceedings of the National Academy of Sciences USA 106 (2009): 20141–20148. Reprinted with permission from the National Academy of Sciences.

The Genetic Basis of Variation: Molecular Genetics, Development, and Evolution

Genetics are what make evolutionary biology a historical science. Heredity determines the continuity of phenotypes from one generation to the next, throughout the history of life. To understand evolutionary processes, it is essential to understand the substrate of evolutionary change. The field of genetics is in an exciting exploratory phase, not unlike the global explorations of natural historians in Darwin's time. Basic principles of transmission are being newly discovered, descriptions of fundamental genome structure newly generated, and processes whereby genotypes are translated into phenotypes are being characterized for the first time. Understanding genetic mechanisms of transmission and phenotypic expression is fundamental to understanding evolutionary dynamics. This final chapter explores the underlying genetic causes of heritable differences among individuals and species.

The causes of phenotypic variation and their contributions to evolutionary processes have been long debated, sometimes acrimoniously. With the rediscovery of Mendel's work, in fact, an outright feud erupted between the "biometricians," who concerned themselves with continuous or gradual variation, and the "mutationists," who argued that only large, discrete differences contributed to evolutionary processes. Hugo de Vries and William Bateson championed this latter view, which at the time was a major and persistent challenge to the hypothesis of continuous, gradual Darwinian evolution. Both schools had to contend with major obstacles. The biometricians faced the problem that many small differences were not inherited from one generation to the next, but rather seemed to be caused by unmeasurable differences in the environment alone. Indeed, early experiments that imposed artificial selection on lines that expressed these minor differences, like those conducted by Johanssen in 1903, showed no sign at all of any response to selection. It so happened that these lines were also homozygous—containing no genetic variation. Therefore, by confounding, or failing to distinguish between, small genetic versus environmentally induced differences, mutationists could argue that small differences were evolutionarily irrelevant. They had to face the problem, however, that if inheritable mutations are of large effect, they are likely to be "monstrous," disadvantageous, or so different from existing variants that they may prohibit reproduction. Later in the twentieth century, genetic studies were able to demonstrate that small differences, too, were inherited in a discrete, Mendelian manner and could therefore contribute to the evolutionary process. This was a major contribution of the "Evolutionary Synthesis."

The major goal of the fields of molecular genetics, molecular quantitative genetics, and genomics is to characterize the genetic basis of phenotypes. Researchers in these fields devote much effort to identifying the genes and genetic pathways necessary for the expression of particular traits. This endeavor has intrinsic interest in that it aims to discover in a most basic sense how things are put together and how they function. Each trait has its own story in its unique particulars.

A related line of inquiry aims to identify genes and pathways that *vary* within and between species. The distinction between genes necessary for phenotypic expression and those that vary is crucial, for it is the variable genes that contribute to the evolutionary process. By identifying the variable genes and the nature of their variation, we come closer to understanding the process of evolution.

How does the genetic basis of variation influence evolutionary processes? First, it determines the nature of differences among individuals and the conditions under which those differences are manifest. For example, if variation is primarily due to differences in gene products, then the altered gene product will be produced, and the two variants will differ, whenever the gene is expressed and the product formed. If variation is due primarily to differences in the timing, location, or quantity of gene expression, however, differences may be apparent only under more specific conditions. For example, if one variant expresses a gene under cool conditions and the other variant does not, then differences between the two variants will be apparent only at cool temperatures. Likewise, if one variant expresses a gene in the upper and lower mandible and the other expresses it only in the upper mandible, then the two variants will differ only in the lower mandible. Therefore, when gene expression (as opposed to gene products) differs, differences among variants can be more fine-scaled, and apparent only in specific structures or under specific environmental conditions. Thus a major question is the extent to which phenotypic variation among individuals is caused by variation in gene products—involving coding regions of DNA—or gene expression—involving regulatory regions of DNA. Also important is how gene products regulate the expression of other genes.

Second, the genetic basis of variation determines how independently traits can evolve. If one gene influences many traits, a phenomenon called pleiotropy, then genetic correlations among traits can be strong, and evolution in one trait will cause correlated evolution in other traits. Such correlations among traits can constrain adaptive evolution and even cause maladaptive evolution (see chapter 7). Conceivably, genetic correlations among traits may facilitate adaptive evolution if a more coordinated change in several traits simultaneously is advantageous. How frequently this may occur is not well known.

Third, the genetic basis of variation influences the magnitude of the differences between individuals. Does a single genetic difference cause a small or large difference in phenotype, and, similarly, are large phenotypic differences between individuals due to a few or many genetic differences? These questions are reminiscent of the debate between the biometricians and the mutationists. Theoretically, in populations that are already well adapted, large changes in phenotypes are much more likely to be maladaptive than small changes. If most genetic changes lead to large phenotypic effects, it will take longer for any favorable genetic variant to appear in those populations (although it would rise in frequency more quickly). Conversely, if genetic changes tend to be of small effect, then adaptive variants may appear more frequently.

In the field of "Evolution and Development," a.k.a. "Evo-Devo," these questions have acquired considerable traction. A major goal in this field is to analyze how genes determine developmental processes and resulting phenotypes. Ernst Haeckel's proposition that "ontogeny recapitulates phylogeny" gave impetus to the idea that developmental processes are important in evolutionary processes. His recapitulation theory argues that the developmental stages through which organisms pass are also ancestral evolutionary states—that is, an organism's embryonic phases resemble its remote adult ancestors. This theory has not withstood the test of data, but mounting evidence suggests that changes in developmental programs are associated with important morphological differences among taxa. In particular, the location, timing, and amount of the expression of particular genes involved in development determine major morphological outcomes, including the identity of organs. For example, spatial overlap in the expression of three classes of genes (A, B, and C) in flowers determines whether green sepals, colorful petals, male stamens, or female carpels will be formed. If the spatial overlap changes, the organs themselves change. Look for sepals on lilies and you won't find them, and all those chaste, double-petaled ornamental blooms—where are their anthers? Differences in the timing or duration of gene expression accounts for other major morphological differences too. Protracted expression and development of some floral genes produce nectar spurs—extra bumps and protuberances on petals that hold nectar. Conversely, early truncation of the expression of some genes and developmental pathways can lead to the retention of juvenile traits, a condition called neoteny. Such arrested development in the odd neotenic salamanders causes larval-looking adults: gilled baby-faces with fully developed testicles or ovaries. Big-headed, bare-skinned humans are thought to be another example of neoteny.

The field of developmental genetics characterizes the extant structure of developmental pathways: what genes are necessary to produce a particular phenotype, what regulates those genes and determines when and where they are expressed, what happens if one changes that expression experimentally? Does a monster result, or does developmental integration result in a functional yet novel form? The field of Evo-Devo characterizes how developmental pathways *differ* among taxa and thereby addresses the evolution of these pathways. Such comparative studies reveal what pathways, and sometimes what genes, have diverged between lineages.

Both studies of speciation and comparative studies of development face similar obstacles. Speciation studies are hampered because differences between species accumulate well after speciation has occurred, making it difficult to identify those genes associated with the speciation process itself. Similarly, evo-devo studies frequently compare quite distantly related taxa. Comparative studies of development using only distantly related taxa cannot identify which parts of the pathway diverged first and which most recently; nor can they determine which, if any, were associated with the process of speciation itself. For that, more closely related taxa must be compared. By comparing the same developmental pathways in distantly *and* closely related taxa, there is the possibility of piecing together the evolutionary sequence of developmental changes.

Yet comparisons of development across even closely related species still leaves an important gap: that gap between microevolution and macroevolution. Discovery of the nature of the link between adaptive microevolution and macroevolutionary differences thus awaits the characterization of variation in development and gene expression *within* species. A field of "micro-evo-devo" awaits its own development.

READINGS

The readings in this chapter identify genes and genetic pathways associated with differences in beak development among species of Darwin's finches. The first paper identifies a gene, *Bmp4*, involved in determining bill depth. Arhat Abzhanov and colleagues found that birds with deeper and broader bills have elevated levels of *Bmp4* expression, and that *Bmp4* is expressed earlier in development in these large-billed birds. Experimental manipulations of the expression of *Bmp4* verified the correlation between *Bmp4* expression and bill morphology. Therefore, the timing and overall level of expression of this gene differs among species and is associated with species differences in bill depth.

The second reading identifies a pathway involved in determining bill length: the calmodulin pathway. Unlike the study of *Bmp4*, in which the researchers already knew of genes that were likely to influence beak development and were able to quantify and manipulate their expression, this study uses genomics approaches to identify unknown genes that are likely to be involved in beak development. Comparing the levels of expression of a large number of genes in the different finch species during development, Abzhanov and colleagues found that the expression of calmodulin (CaM) differs among species with longer versus shorter bills. They then used methods similar to those described in the first paper to verify the correlation between CaM expression and bill length.

While these studies are intrinsically interesting in what they reveal about the molecular-genetic and physiological pathways of development, they also provide insight into evolutionary processes. First, they demonstrate that different genes regulate different aspects of bill shape—the trait that determines feeding efficiencies and diets and influences survival and species coexistence. Bill depth and bill length can evolve independently because different genes control these traits. Thus the molecular-genetic basis of traits influences genetic correlations among traits, one of the primary determinants of evolutionary trajectories.

Second, the studies show that variation in gene expression accounts for major differences among species. Difference in the timing, spatial patterns, and overall level of expression of particular genes can cause large changes in adaptively important phenotypes. When, where, and how much of those gene products are produced can account for a significant amount of variation observed among species.

Yet the patterns of expression of these genes may depend not only on changes in the regulatory regions of those genes themselves but also on changes in products of other genes that regulate gene expression. Biologists are still surprisingly far from understanding the full genetic basis of phenotypic variation in any trait. Such understanding requires elucidation of the entire genetic system of interacting genes—a complex ecology itself. Indeed, the conceptual and statistical methodology used to analyze ecological interactions among networks of species is indispensable in the construction of analytical methodology to assess how genes, environments, and timing interact to produce a phenotype.

~~~~~~

Bmp4 *and Morphological Variation of Beaks in Darwin's Finches* (2004)

~~~~~~

❋ ARHAT ABZHANOV, MEREDITH PROTAS, B. ROSEMARY GRANT, PETER R. GRANT, AND CLIFFORD J. TABIN

Editor's note: Figures S1, S3, S5, and S6 are not reprinted here. Figures 2, 3, S2, and S4 were originally printed in color but have been reproduced here in black and white. Arrows

in figure 3 (A–F) have been redrawn to show contrast. The contrast in figures 3L and 3O has been enhanced.

Abstract—Darwin's finches are a classic example of species diversification by natural selection. Their impressive variation in beak morphology is associated with the exploitation of a variety of ecological niches, but its developmental basis is unknown. We performed a comparative analysis of expression patterns of various growth factors in species comprising the genus Geospiza. We found that expression of Bmp4 in the mesenchyme of the upper beaks strongly correlated with deep and broad beak morphology. When misexpressed in chicken embryos, Bmp4 caused morphological transformations paralleling the beak morphology of the large ground finch G. magnirostris.

Darwin's finches are a group of 14 closely related songbird species on the Galápagos Islands and Cocos Island (*1–3*) collected by Charles Darwin and other members of the *Beagle* expedition in 1835 (*4*). Many biology textbooks use these birds to illustrate the history of evolutionary theory as well as adaptative radiation, natural selection, and niche partitioning (*5–7*). The diverse shapes and sizes of the finch beaks are believed to be maximally effective for exploiting particular types of food, including seeds, insects, and cactus flowers (*3, 7*). The external differences in beak morphology reflect differences in their respective craniofacial skeletons (*3, 8*). The specialized beak shapes are apparent at hatching (*3, 8*) and thus are genetically determined.

To study the craniofacial development of Darwin's finches, we first developed a staging system by which we could compare them to each other and to the chicken, the existing avian model system (fig. S1 [not reprinted]) (*9*). We used this system to compare beak development in six species of Darwin's finches belonging to the monophyletic ground finch genus *Geospiza*. The sharp-beaked finch *G. difficilis*, with a small symmetrical beak, is the most basal species (Fig. 1A) (*10*). The other species fall into two groups: three species with broad and deep beaks used for crushing seeds (small, medium, and large ground finches—*G. fuliginosa, G. fortis,* and *G. magnirostris*) and cactus finches with long pointed beaks used for reaching into cactus flowers and fruits (cactus and large cactus finches—*G. scandens* and *G. conirostris*) (Fig. 1A) (*7, 10*).

We compared beak development in embryos of all six species. Species-specific differences in the morphological shape of the beak prominence are first apparent by embryonic stage 26 (Fig. 1, B and C, and fig. S2). We therefore expected factors involved in directing the differential aspects of beak morphologies to be expressed at or before this time. We also expected such species-specific differences to reside in the mesenchyme on the basis of recent transplantation experiments between quail and duck embryos (*11*).

We analyzed expression patterns of a variety of growth factors, which are known to be expressed during avian craniofacial development (*12–14*),

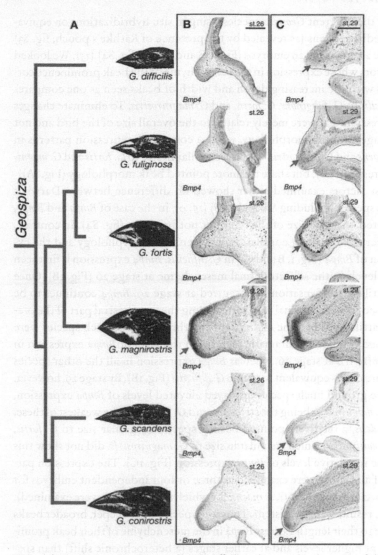

Fig. 1. (A) Previous studies suggest that *G. difficilis* is the most basal species of the genus *Geospiza*, and the rest of the species form two groups: ground and cactus finches, with distinct beak morphologies. (B) At stage (st.) 26, *Bmp4* is strongly expressed in a broad distal-dorsal domain in the mesenchyme of the upper beak prominence of *G. magnirostris* and at significantly lower levels in *G. fortis* and *G. conirostris*. No *Bmp4* was detected in the mesenchyme of *G. difficilis*, *G. fuliginosa*, and *G. scandens*. (C) At stage 29, *Bmp4* continues to be expressed at high levels in the distal beak mesenchyme of *G. magnirostris*. Broad domains of *Bmp4* expression are detectable around prenasal cartilages of *G. fuliginosa* and *G. fortis*. A small domain of strong *Bmp4* expression is also found in the most distal mesenchyme of *G. conirostris*, and weaker expression is seen in *G. scandens* and *G. fortis* (arrows). Scale bars: 1 mm in (B) and 2 mm in (C).

among the different *Geospiza* species, using in situ hybridizations on equivalent medial sections (as revealed by the presence of Rathke's pouch; fig. S3) of stage 26 and stage 29 embryos (Fig. 1, B and C, and fig. S4) (*15*). We looked for factors whose expression in the mesenchyme of the beak prominence correlated with the increasing depth and width of beaks seen as one compares *G. difficilis* to *G. fuliginosa*, *G. fortis*, and *G. magnirostris*. To eliminate changes in expression that were merely related to the overall size of the bird and not to changes in beak morphology, we also compared expression patterns in *G. scandens* and *G. conirostris*, which are similar in size to *G. fortis* and *G. magnirostris*, respectively, but share the more pointed beak morphology (Fig. 1A).

Most factors examined either showed no difference between Darwin's finches species (including *Shh* and *Fgf8*) (*15*) or, in the case of *Bmp2* and *Bmp7*, correlated with the size of the beak but not its shape (fig. S4). In contrast, we observed a striking correlation between beak morphology and the expression of *Bmp4* (Fig. 1, B and C). In *G. difficilis*, *Bmp4* expression is first seen at low levels in the subectodermal mesenchyme at stage 26 (Fig. 1B). Once the cartilage condensation has occurred at stage 29, *Bmp4* continues to be expressed in mesenchymal cells surrounding the most rostral part of the prenasal cartilage. When the embryos of the three ground finch species were examined, we noted a dramatic increase in the level of *Bmp4* expression in *G. magnirostris* at stage 26, whereas *Bmp4* expression in all the other species was more or less equivalent to that in *G. difficilis* (Fig. 1B). By stage 29, however, all three ground finch species displayed elevated levels of *Bmp4* expression, with *G. magnirostris* being the strongest and *G. fuliginosa* the weakest of these. *G. scandens*, a relatively pointed-beaked species of similar size to *G. fortis*, and *G. conirostris*, which is similar in size to *G. magnirostris*, did not show this increase in relative levels of *Bmp4* expression (Fig. 1C). The expression patterns of all factors were examined in three or four independent embryos for each species (except for *G. scandens*, for which two embryos were examined), and the results were consistent. Thus, the species with deeper, broader beaks relative to their length express *Bmp4* in the mesenchyme of their beak prominences at higher levels and at earlier stages (a heterochronic shift) than species with relatively narrow and shallow beak morphologies. Moreover, the differences in *Bmp4* expression are coincident with the appearance of species-specific differences in beak morphology. This observed correlation was specific to *Bmp4* expression in the upper beak, whereas expression of *Bmp4* in the lower beak remains constant in spite of the fact that lower beak morphology varies in concert with that of the upper beak (*15*).

We next tested whether the observed change in *Bmp4* expression could be partially responsible for the differences in beak morphology in ground finch species. *Bmp4* has been previously shown to be important for the production of skeletogenic cranial neural crest cells and capable of affecting patterning, growth and chondrogenesis in derivatives of the mandibular and maxillary prominences (*16–19*). However, the expression of *Bmp4* is quite dynamic during craniofacial development and might be expected to play different roles at various times.

During craniofacial development, *Bmp4* is first expressed in the epithelium of the maxillary and lateral frontonasal prominences in early embryos. The same factor is later expressed in the distal mesenchyme of the upper beak of embryos at stage 29 and later (Fig. 2, A and B). We took advantage of the ability to misexpress genes during chicken development with the RCAS replication-competent retroviral vector to test the effect of increasing BMP4 levels in both of these domains. Because the RCAS vector does not spread across basement membranes, we were able to confine misexpression to either the facial ectoderm or mesenchyme (Fig. 2, C and E). Infection of the facial ectoderm with the RCAS::*Bmp4* virus caused smaller and narrower upper beaks (fig. S5 [not reprinted]). Ectodermally infected beaks also showed a dramatic loss of chondrogenesis in the adjacent mesenchyme (Fig. 2, D and F, and fig. S5), indicating a role in epithelial-to-mesenchymal signaling early in head morphogenesis.

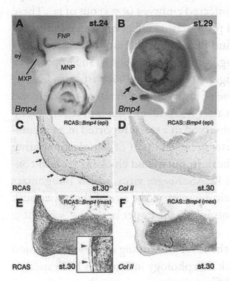

Fig. 2. (*A*) At stage 24, *Bmp4* is expressed in the ectoderm of the maxillary process (MXP) and lateral frontonasal process (FNP). (*B*) By stage 29, *Bmp4* expression expands into the distal mesenchyme of the upper and lower beaks (black arrows). (*C* and *E*) Misexpression of RCAS::*Bmp4* can be targeted to either the facial ectoderm [arrowheads in inset in (E)] or the mesenchyme (mes) of the frontonasal process of a chicken embryo, as revealed with in situ hybridization with a virus-specific probe. (*D*) No chondrogenesis is detected in the embryos whose epithelium (epi) was infected with RCAS::*Bmp4* virus. (*F*) In contrast, embryos whose FNP mesenchyme was infected with RCAS::*Bmp4* virus showed high levels of chondrogenesis as revealed with an anti–*Col II* riboprobe. MNP, mandibular process; ey, eye. Scale bars: 1 mm in (C) and (E).

In a second set of misexpression experiments designed to mimic the elevated levels of *Bmp4* seen in *G. magnirostris*, we injected RCAS::*Bmp4* virus into the mesenchyme of the frontonasal process of chicken embryos at stage 23 to 24. Because of the time required for viral infection and spread, this results in robust misexpression in the distal frontonasal process around stage 26 (*15*), which is the time when elevated *Bmp4* levels are first seen in *G. magnirostris*. The phenotypes we obtained were quite different from those resulting from epithelial misexpression, showing that *Bmp4* expression has distinct functions in the epithelium and mesenchyme. Rather than diminished beaks, beaks resulting from infection of the mesenchyme were reminiscent of those of the ground finches with deep and broad beaks. These morphological changes in beak morphology were observed before the onset of skeletogenesis, as revealed by *Col II* expression (*15*). By stage 36, the infected beaks ($n = 13$) were on average about 2.5 times as wide ($\pm21\%$) and 1.5 times ($\pm16\%$) as deep as uninfected control beaks ($n = 11; P < 0.003$) (Fig. 3, A, B, D, and E). The more massive *Bmp4*-infected beaks had a corresponding increase in the size of the skeletal core (Fig. 3, G and H, and fig. S6 [not reprinted]), again in parallel to a larger beak skeleton of *G. magnirostris*. This skeletal phenotype was observed in the majority of the infected embryos ($n = 11$ out of 13). These data suggest that BMP4 may have a proliferating effect on skeletal progenitors in the upper jaw. Indeed, we find that cell proliferation, as assessed by bromodeoxyuridine (BrdU) labeling, is highest in a zone of the upper beak process where *Bmp4* is expressed (Fig. 3, J to L; marked with arrow in J, asterisks in L and O). Moreover, this zone of high cell proliferation expands and shows a higher level of proliferation after RCAS::*Bmp4* misexpression (Fig. 3, M to O). A similar phenotype was observed in a study reported in an accompanying paper, where *Bmp4* was misexpressed as part of a study comparing its role in the development of the beak in ducks and chickens (*20*). In contrast, mesenchymal injection of the RCAS::*Noggin* virus, which antagonizes BMP2/4/7 signaling, led to a dramatic decrease in the size of the upper beak and to much smaller skeletal elements inside the upper beak ($n = 7$ out of 9; $P < 0.002$) (Fig. 3, C, F, and I).

We have identified variation in the level and timing of *Bmp4* expression that correlates with variation in beak morphology in Darwin's finch species. We are tempted to speculate that differences in the cis-regulatory elements of *Bmp4* may underlie the distinct expression patterns, although alternatively they could be explained by differences in the timing or amounts of upstream inductive factors or differences in the transduction of such signals. Two such potential upstream signals are *Sonic hedgehog* (*Shh*) and *Fibroblast growth factor 8* (*Fgf8*), which are expressed in the beak epithelium. Beak outgrowth occurs at the location where their expression domains meet, and SHH and FGF8 have been shown to synergistically drive outgrowth, and in the process to

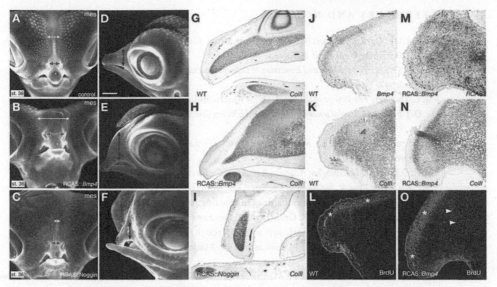

Fig. 3. (*A* and *D*) Ultraviolet pictures of wildtype stage 36 chicken embryonic heads. The width and depth of the beak are shown with white and black doubleheaded arrows, respectively. The width of the beak tip is indicated with a doubleheaded gray arrow. (*G*) Prenasal cartilage in wildtype chickens forms a narrow protruding skeletal rod. (*B* and *E*) RCAS:: *Bmp4* infection in the mesenchyme of the frontonasal process caused a significant increase in the width and depths of the beak. (*H*) These larger upper beaks contained more skeletogenic cells, as revealed with *Col II* in-situ hybridization. (*C* and *F*) In contrast, infection with RCAS::*Noggin* led to narrower and shallower upper beaks. (*I*) Ectopic *Noggin* produced smaller skeletal elements inside the upper beak. BrdU labeling reveals that the *Bmp4* expression domain [black arrow in (*J*)], which is rostral and dorsal to the developing prenasal cartilage (*K*), is closely associated with proliferating cells (*L*) of stage 30 chick embryos. The upper beaks of embryos infected with RCAS::*Bmp4* (*M*) by stage 30 develop larger cartilages (*N*), and there is an up-regulation of cell proliferation both around and within the developing cartilage (*O*). Scale bars: 2 mm in (D); 0.5 mm in (G); and 1 mm in (J).

induce expression of *Bmp4* in subjacent mesenchyme (*21, 22*). Also, we have not ruled out the possibility that genes expressed in other regions of the face are important for directing morphogenesis. In addition to the correlation between variation in *Bmp4* levels and the development of the beaks of Darwin's finches, we have also found that artificially increasing *Bmp4* levels in the beak mesenchyme is sufficient to alter beak morphology in the same direction as is seen in the larger ground finches. Thus, although polymorphism in other genes may also contribute to differences in beak morphology, we propose that variation in *Bmp4* regulation is one of the principal molecular variables that provided the quantitative morphological variation acted on by natural selection in the evolution of the beaks of the Darwin's finch species (*23*).

REFERENCES AND NOTES

1. D. Lack, *Darwin's Finches* (Cambridge Univ. Press, Cambridge, 1947).

2. R. I. Bowman, *Univ. Calif. Publ. Zool. 58*, 1 (1961).

3. P. R. Grant, *The Ecology and Evolution of Darwin's Finches* (Princeton Univ. Press, Princeton, NJ, 1999).

4. C. Darwin, *The Voyage of the Beagle* (New American Library, New York, 1988).

5. D. J. Futuyma, *Evolutionary Biology, Third Edition* (Sinauer Associates, Sunderland, MA, 1998).

6. S. Freeman, J. C. Herron, *Evolutionary Analysis, Third Edition* (Prentice-Hall Inc., Upper Saddle River, NJ, 2003).

7. D. Schluter, *The Ecology of Adaptive Radiation* (Oxford Univ. Press, Oxford, 2000).

8. P. R. Grant, *Proc. R. Soc. London Ser. B 212*, 403 (1981).

9. Materials and methods are available as supporting material on *Science* Online.

10. K. Petren, B. R. Grant, P. R. Grant, *Proc. R. Soc. London Ser. B 266*, 321 (1999).

11. R. A. Schneider, J. A. Helms, *Science 299*, 565 (2003).

12. R. A. Schneider, D. Hu, J. L. Rubenstein, M. Maden, J. A. Helms, *Development 128*, 2755 (2001).

13. P. A. Trainor, K. R. Melton, M. Manzanares, *Int. J. Dev. Biol. 47*, 541 (2003).

14. P. Kulesa, D. L. Ellies, P. A. Trainor, *Dev. Dyn. 229*, 14 (2004).

15. A. Abzhanov, M. Protas, B. R. Grant, P. R. Grant, C. J. Tabin, data not shown.

16. B. Kanzler, R. K. Foreman, P. A. Labosky, M. Mallo, *Development 127*, 1095 (2000).

17. S. Ohnemus *et al.*, *Mech. Dev. 119*, 127 (2002).

18. A. J. Barlow, P. H. Francis-West, *Development 124*, 391 (1997).

19. I. Semba *et al.*, *Dev. Dyn. 217*, 401 (2000).

20. P. Wu *et al.*, *Science 305*, 1465 (2004).

21. D. Hu, R. S. Marcucio, J. A. Helms, *Development 130*, 1749 (2003).

22. A. Abzhanov, C. J. Tabin, *Dev. Biol.*, in press.

23. T. D. Price, P. R. Grant, *Am. Nat. 125*, 169 (1985).

24. We thank field assistants J. Chavez, G. Castaneda, O. Perez, F. Brown, and A. Aitkhozhina; the Charles Darwin Research Station and the Galápagos National Park for permits and logistical support; M. Kirschner for discussions that led to the inception of this project; and P. Wu and C.-M. Chuong for sharing data before submission. A.A. was supported by the Cancer Research Fund of the Damon Runyon–Walter Winchell Foundation Fellowship, grant DRG1618. This project was funded by NIH grant PO1 DK56246 to C.J.T.

SUPPORTING ONLINE MATERIAL

Materials and Methods

Staging System for Analyzing Songbird Development

A straightforward staging system already exists for the domesticated chicken, described by Hamburger and Hamilton (*S1*), and we based our finch staging system on it. However, the chicken (a precocial bird, order Galliformes) is quite divergent from songbirds, such as the finches (altricial birds, order Passeriformes). Not only is their incubation period different, but also various aspects of embryogenesis proceed at distinct relative rates in the finches compared to the chicken. Therefore, we used DIG-labeled antisense RNA probes against various genes known to be involved in craniofacial development to establish stages when, in particular, the finch craniofacial primordia corresponded to various stages in chick development as reflected in expression patterns. To have access to large numbers of embryos, we made use of a readily available species of songbirds, the Zebra Finch (*Taeniopygia guttata*) that has an identical incubation period to Darwin's Finches. While we originally intended to isolate Finch probes for this purpose, we discovered that probes directed against chick genes readily cross-reacted with Zebra Finch embryos, so those were employed for all hybridizations. Based on these molecular data as well as morphological craniofacial features, we were able to develop a robust staging system for Finch craniofacial development (Fig. S1; S2). Subsequent examination of Darwin's Finch embryos verified that their craniofacial development was very similar to that of *T. guttata* by both morphological and molecular criteria (data not shown).

Collection and Treatment of Embryonic Material from Darwin's Finches

Under an agreement with the Galápagos National Park, we received quotas for collecting embryos of *Geospiza magnirostris*, *G. fortis*, *G. fuliginosa*, *G. scandens*, *G. conirostris*, *G. difficilis* and *Certhidea olivacea* on the islands of Santa Cruz and Genovesa. Singing males and their nests were identified at the beginning of the wet season. After breeding had begun nests were checked every day. Darwin Finches females lay clutches of 3–5 eggs, one per day. To avoid disrupting breeding, we collected only the third egg to be laid and incubated it at 100°F. The embryos were harvested at E5 (st.26) and E6.5 (st.29) according to our altricial avian development staging series. The staging series for songbird development will be described in detail elsewhere. Embryonic material was fixed in 4% paraformaldehyde in phosphate buffered saline (PBS) for 2–3 hours at ambient temperature and stored in RNAlater reagent (Ambion) at about 5°C for 2–5 weeks. The heads were rehydrated in PBS, frozen in OCT and sagittally cryosectioned medially (Fig. S3). Chick antisense riboprobes were prepared and used on Darwin's Finch embryos as previously described (S2) (Fig. S4). We analyzed 19 heads of Darwin's Finches: *G. magnirostris* ($N = 3$), *G. fortis* ($N = 4$), *G. fuliginosa* ($N = 4$), *G. conirostris* ($N = 3$), *G. scandens* ($N = 2$), and *G. difficilis* ($N = 3$). A chick *Bmp4* probe was used for in-situ hybridizations.

Chicken Embryo Manipulations and Statistical Analysis

Fertilized eggs were obtained from SPAFAS (Norwich, CT), incubated at 100°F, and the embryos were staged according to Hamburger and Hamilton (S3). The RCAS::*Bmp4* and RCAS::*Noggin* constructs have been previously described (S4, S5). To infect embryos for in vivo studies we either injected the distal part of the frontonasal process of st.24 chick embryos or pooled high titer concentrated virus into the semi-enclosed space surrounding the heads of stage 15 embryos. RCAS(B)::AP (alkaline phosphatase) virus of similar high titer displayed infection of the epithelium and underlying dermis of the head after 36 hours of infection (stage 20) and 48 hours (stage 22) (not shown) that ranged from patchy (10–20% of head surface) to thorough (60–70% of head surface). The infected embryos were collected at stages 30 and 36, fixed overnight in 4% paraformaldehyde in PBS and frozen in OCT for sagittal sectioning. The embryonic heads were photographed and measured in NIH Image 1.62. These arbitrary units were used for the Analysis of Variance function (ANOVA toolbox) in Excel X to calculate standard deviations and *P*-values for the data. For BrdU labeling eggs of stage 30 wild-type

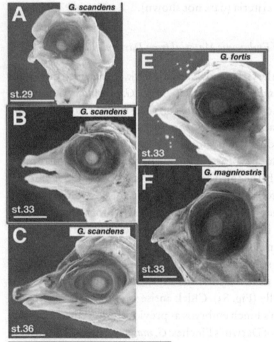

Fig. S2. (A–D) Species-specific differences appear relatively early during development and are maintained in the embryos of *G. scandens* as they develop long, shallow and pointed beaks, and these features are maintained through later developmental stages. *(B, E, F)* Species-specific characteristics are easily recognizable by stage 33 in embryos of *G. scandens (B), G. fortis (E),* and *G. magnirostris (F).* Scale bars are: B–G 5mm.

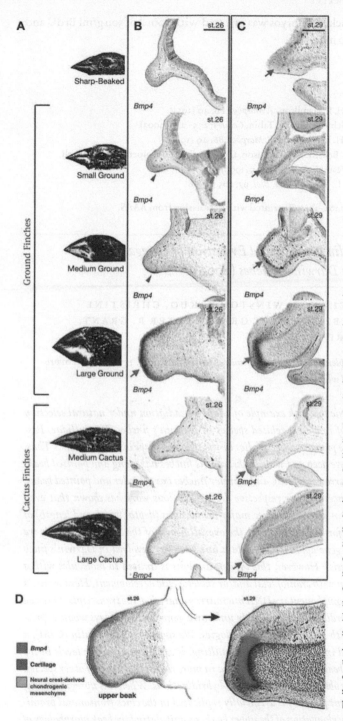

Fig. S4. Comparative analysis of *Bmp2* and *Bmp7* expression domains in species of *Geospiza*. (*A*) The six species of *Geospiza* display distinct beak shapes and sizes. (*B*) *Bmp2* was expressed in ventral epithelium and immediately adjacent areas of ventral mesenchyme at stage 26 embryos of *G. fortis* and *G. magnirostris*, and *G. scandens* and *G. conirostris*. The expression was strongest [in] *G. magnirostris* and *G. conirostris* embryos, the two largest species. (*C*) *Bmp7* was expressed in the ventral-most mesenchyme of the FNP of all the species sampled. (*D*) Domains of *Bmp4* expression in the upper beak prominence of *G. magnirostris* at stages 26 and 29. Scale bars: 1 mm in B, C.

and infected chicken embryos was injected with 300μl of 50μg/ml BrdU and incubated for 60 minutes.

References

S1. V. Hamburger, H. L. Hamilton, *J. Morphol. 88*, 49 (1951).

S2. A. E. Brent, R. Schweitzer, C. J. Tabin, *Cell 113*, 235–248 (2003).

S3. V. Hamburger, H. L. Hamilton, *J. Morphol. 88*, 49 (1951).

S4. D. Duprez, E. J. Bell, M. K. Richardson, C. W. Archer, L. Wolpert, P. M. Brickell, P. H. Francis-West, *Mech. Dev. 57*, 145 (1996).

S5. J. Capdevila, R. L. Johnson, *Dev. Biol. 97*, 205 (1998).

Science 305 (2004): 1462–1465. Reprinted with permission from AAAS.

The Calmodulin Pathway and Evolution of Elongated Beak Morphology in Darwin's Finches (2006)

✳ ARHAT ABZHANOV, WINSTON P. KUO, CHRISTINE HARTMANN, B. ROSEMARY GRANT, PETER R. GRANT, AND CLIFFORD J. TABIN

Editor's note: Supplementary materials not reprinted here. Figs. 1, 3, and 4 are reproduced in black and white.

Abstract—A classic textbook example of adaptive radiation under natural selection is the evolution of 14 closely related species of Darwin's finches (Fringillidae, Passeriformes), whose primary diversity lies in the size and shape of their beaks.[1-6] Thus, ground finches have deep and wide beaks, cactus finches have long and pointed beaks (low depth and narrower width), and warbler finches have slender and pointed beaks, reflecting differences in their respective diets.[6] Previous work has shown that even small differences in any of the three major dimensions (depth, width and length) of the beak have major consequences for the overall fitness of the birds.[3-7] Recently we used a candidate gene approach to explain one pathway involved in Darwin's finch beak morphogenesis.[8] However, this type of analysis is limited to molecules with a known association with craniofacial and/or skeletogenic development. Here we use a less constrained, complementary DNA microarray analysis of the transcripts expressed in the beak primordia to find previously unknown genes and pathways whose expression correlates with specific beak morphologies. We show that calmodulin (CaM), a molecule involved in mediating Ca²⁺ signalling, is expressed at higher levels in the long and pointed beaks of cactus finches than in more robust beak types of other species. We validated this observation with in situ hybridizations. When this upregulation of the CaM-dependent pathway is artificially replicated in the chick frontonasal prominence, it causes an elongation of the upper beak, recapitulating the beak morphology of the cactus finches. Our results indicate that local upregulation of the CaM-dependent pathway is likely to have been a component of the evolution of Darwin's finch species with elongated beak morphology and provide a mechanistic explanation for the independence of beak evolution along different axes. More generally, our results implicate the CaM-dependent pathway in the developmental regulation of craniofacial skeletal structures.

To understand the genetic basis of the species-specific beak morphologies, we previously performed a comparative candidate gene analysis with developmental genes known to be associated with craniofacial development. We found that a broader and earlier domain of bone morphogenetic protein 4 (BMP4) expression in the distal neural-crest-derived mesenchyme correlated with the very deep and wide beak morphology of the ground finches.[8] This expression difference was shown to be functionally significant by misexpression analysis in chick embryos.[8]

However, the candidate gene approach did not yield any candidates for pathways that could be involved in evolution of the longer beak morphology characteristic of the cactus finch species. To identify pathways involved in the evolution of long beaks, cDNA microarrays were used for a direct comparison of the gene expression profiles of several thousand transcripts in stage 26 frontonasal processes (which give rise to the upper beak) of five species of genus *Geospiza*: the sharp-beaked finch (*Geospiza difficilis*), the medium and large ground finches (*G. fortis* and *G. magnirostris*), and the cactus and large cactus finches (*G. scandens* and *G. conirostris*) (Fig. 1a; Methods). We first used hierarchical clustering to inspect whether the overall expression profiles clustered according to species (Methods). The resultant tree illustrates that most of the individual expression profiles clustered by species. For example, most individuals of either *G. scandens* (4 of 5) or *G. conirostris* (4 of 4) clustered together with individuals of the same species (Fig. 1b). The expression profile of one particular individual of *G. scandens* (G.s.5) clustered more closely with those of *G. conirostris*, probably reflecting a certain degree of morphological (and thus developmental) overlap between these two species (Fig. 1d). *G. scandens* and *G. conirostris* collected for this study live on different islands (Santa Cruz and Genovesa, respectively), thus excluding the possibility of species misidentification. We obtained very similar results when we compared *G. fortis* and *G. magnirostris* individuals (largely distinct expression profiles but some individuals clustering with the other species; not shown). It therefore seems that the analysis is sensitive to phylogenetic differences between different species of Darwin's finches.

We then clustered the measurements of signal ratios and intensities for different transcripts to identify genes that were upregulated or downregulated in all individuals of a particular morphology compared with the basal *G. difficilis* reference (Fig. 1, and Supplementary Fig. S1 [not reprinted]; Methods). All of 100 candidates from the resulting final cluster of cactus finch

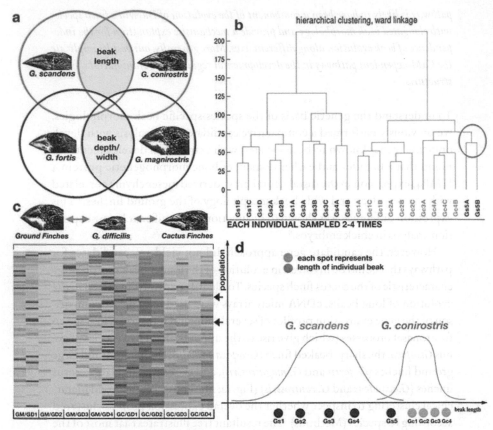

Fig. 1. Microarray analysis in different finch species. *a*, Clustering strategy to isolate transcripts whose expression correlated with beak morphology. *b*, The Ward linkage tree showed that most of the individual samples clustered by species. Each individual was sampled two to four times. The *y*-axis is euclidian distance between branches. *c*, The final clusters of transcripts, which were upregulated in the comparison between cactus finches and the sharp-beaked finch and were downregulated or remained unchanged in the ground finches compared with the sharp-beaked finch. *d*, Individual expression profiles clustered by species except for the occasional individual profile, probably reflecting a certain overlap in morphology or development between the species. Each spot represents the length of an individual beak.

morphology-specific genes were sequenced to reveal their identity. These genes were screened to produce candidates that were expressed at moderate or high levels of signal intensity on the microarray and were expressed at least fivefold higher in the cactus finches. We found two microarray spots carrying probes with identical sequences for CaM (Supplementary Fig. S2 [not reprinted]) that were both at a much higher level of average signal in the cactus finch beaks than in the reference sharp-beaked finch (*G. difficilis*) (Supplementary

Information). Because these clones represented the most differentially expressed gene that was not an enzyme, a housekeeping gene or a ribosomal gene and the only representative of a signalling transduction pathway, we focused on CaM as our primary candidate associated with the elongated cactus finch beak morphology. CaM is a Ca^{2+}-binding protein that can bind to and regulate many different protein targets and is a key component of a Ca^{2+}-dependent signal transduction pathway.[9] It has not been previously characterized in either craniofacial or skeletal development.

Our microarray data indicate a possible correlation between elevated levels of CaM and the more elongated beak morphology of the cactus finches. To validate this suggestion we performed a comparative *in situ* hybridization on embryos of Darwin's finches (Fig. 2). We found that CaM was indeed expressed at detectably higher levels in the distal-ventral mesenchyme of the frontonasal processes in the cactus finches ($n = 3$) and large cactus finches ($n = 3$) than in similar-sized processes of the ground finches (Fig. 2). Thus, higher expression of CaM is indeed associated with the elongated beak morphology of the cactus finches.

To address whether differential levels of CaM-dependent signalling might be important in the development of distinct beak morphologies in different species of Darwin's finches, we wished to elevate the level of CaM-dependent signalling in the developing chick beak prominence. To accomplish this we used a constitutively activated form of a downstream effector of CaM, CaM kinase kinase (CaMKII; M. J. Taschner, S. Schnaiter and C.H., unpublished observations). In one of the major cellular responses to increased Ca^{2+} levels, CaM activates CaMKII. CaMKII, in turn, activates downstream kinases, enabling them to phosphorylate and hence activate various targets, such as the transcription factors cAMP-response-element-binding protein (CREB), serum response factor and CREB-binding protein.

Both CaM and CaMKII are expressed at low levels in the mesenchyme of the chicken upper beak process (data not shown). To stimulate higher CaM-dependent CaMKII signalling, we used an avian retroviral vector carrying the constitutively active form of CaMKII (RCAS::CA-CaMKII) to infect the distal mesenchyme of early stage-24 chick frontonasal processes. Although virus injection was targeted to the distal upper beak prominence with relative ease, the resultant infection was somewhat variable. In more than half of the cases, the infection was limited to the distal and ventral mesenchyme surrounding the developing cartilage and did not spread to the skeletal element (Fig. 3d; $n = 9$ of 16; not shown). Because these infected regions closely approximate the domain of elevated CaM expression in the cactus finches (Fig. 2), we restricted our analysis to these embryos. In chicks in which activated CaMKII was misexpressed specifically in the distal/ventral mesenchyme, we observed a significant increase in the length of the beaks (length in arbitrary

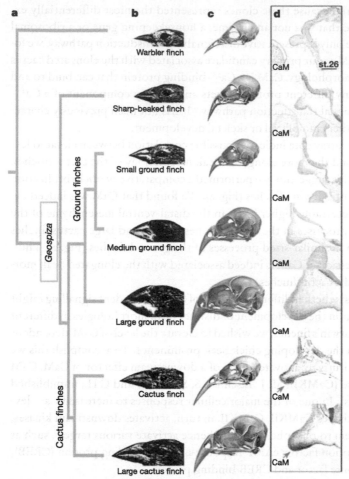

Fig. 2. Comparative analysis of CaM expression in finches. *a, b, Geospiza* group species displaying distinct beak morphologies form a monophyletic group. *c,* The differences in beak morphology are skeletal. *d,* CaM is expressed in a strong distal-ventral domain in the mesenchyme of the upper beak prominence of the large cactus finch, *G. conirostris,* somewhat lower levels in cactus finch, *G. scandens,* and at significantly lower levels in the large ground finch and medium ground finch, *G. magnirostris* and *G. fortis,* respectively. Very low levels of CaM were detected in the mesenchyme of *G. difficilis,* G. *fuliginosa* and the basal warbler finch *Certhidea olivacea.* CaM expression domains are indicated with short arrows in *d.* Scale bar, 1 mm in *b.* The molecular tree is from ref. 23; images of skulls are from ref. 6, with permission from the author.

units: 0.78 ± 0.05 (±s.d.); $P < 0.0056$, $n = 9$; more than 10% beak elongation relative to control embryos (Fig. 3a–d)), whereas the beak width and depth were not affected (Fig. 3, and Supplementary Fig. S3 [not reprinted]).

For a better understanding of the phenotype induced by activated CaM-KII in beaks, we studied the expression of markers for skeletal cell types, proliferation and cell death (Fig. 3, and Supplementary Fig. S4 [not reprinted]). *In situ* hybridization analysis with probes directed against skeletogenic genes verified that the cartilage element was indeed longer in the distal/ ventrally infected embryos (Fig. 3). We found that cell death, as revealed by the TUNEL (TdT-mediated dUTP nick end labelling) assay, was essentially unchanged relative to the wild-type condition (not shown). Staining for anti-PCNA (proliferating cell nuclear antigen) antibody and labelling with bromodeoxyuridine indicate that proliferation might have been upregulated in the distal half of the pre-nasal cartilage element of infected beaks (not shown). The expression of cell-cycle regulators cyclin D1 and the activator protein 1 (AP1) family members c-Jun and c-Fos was upregulated in the distal

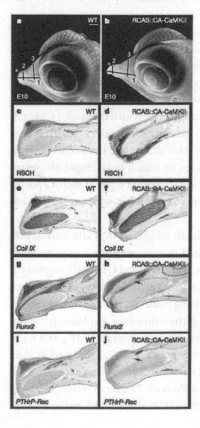

Fig. 3. Functional analysis of CaM-dependent pathway in beak development. *a*, *b*, Whole head views of embryonic day 10 (E10; HH stage 36) wild-type (*a*) and RCAS::CA-CaMKII-infected (*b*) chicken embryos. The length of the beak is shown with a [horizontal] line; the depth of the beak at the base and the depth of the beak at the tip are shown with [vertical lines]. *c–j*, We used RSCH (*c*, *d*), *Coll IX* (*e*, *f*), *Runx2* (*g*, *h*) and *PTHrP-Rec* (*i*, *j*) probes to reveal RCAS infection (RSCH), chondrocytes (*Coll IX*) and early osteoblasts (*Runx2*). *c*, *e*, *g*, *i*, Wild type; *d*, *f*, *h*, *j*, RCAS::CA-CaMKII infected. The star indicates an egg-tooth. Scale bar, 2 mm in *a*.

parts of pre-nasal cartilage and its perichondrium (arrow in Supplementary Fig. S4J). As these skeletal tissues were not infected with the retrovirus, this effect was indirect. The same genes were downregulated in the mesenchyme, which forms the premaxillary dermal bone (arrowheads in Supplementary Fig. S4I). Consistent with our observations, parallel studies in the laboratory of one of us (C.H.) have also indicated that CaMKII signalling leads to the elongation of limb skeletal elements in a non-cell-autonomous manner (M. J. Taschner, S. Schnaiter and C.H, unpublished observations), in a similar manner to our observation in the beak.

There are few examples of the identification and characterization of developmental pathways responsible for evolutionary morphological change.[10–15] Previous work on Darwin's finches has shown that selection on adult beak variation results in changes in growth in the next generation, as a result of genetic correlations between adult and juvenile expression of the same morphological characters.[16,17] Here, using a microarray approach, we found evidence that a higher level of CaM-dependent signalling is both biologically relevant and functionally important for the morphogenesis of the longer beaks of cactus finches used in a specialized manner to probe cactus flowers and fruit for nectar.

The avian beak varies between species, and indeed between individuals of the same species, along at least three different axes: length, depth and width. In Darwin's finches it can be seen that there is both linkage and independence in the variation along these different axes. For example, in the two species of cactus finches there is a much weaker genetic correlation between adult beak length and depth or width than between depth and width themselves. In contrast, in *G. fortis* all three correlations are strong.[18,19] The data reported here provide at least a partial explanation for this difference between the finch species. Embryos of the cactus finch species, in which beak length shows independence from width and length, strongly express CaM (a factor specifically affecting beak length) during beak morphogenesis, whereas *G. fortis* has minimal CaM expression (Fig. 4).

In considering the independence of beak length from width and depth, it is particularly intriguing that, here and in our previous study, we have shown that two different factors (BMP4 and activated CaMKII), expressed in a similar domain in the beak prominence, result in changes in growth along different dimensions of the developing beak (width/depth and length, respectively) without negatively affecting the other axes, explaining the observed independence of these traits. Moreover, analysis of BMP4 and CaM expression in chick embryos infected with RCAS::CA-CaMKII and RCAS::BMP4, respectively, showed that these two molecules do not regulate each other during beak development (data not shown). These two molecules vary independently of each other in some species as demonstrated by their recip-

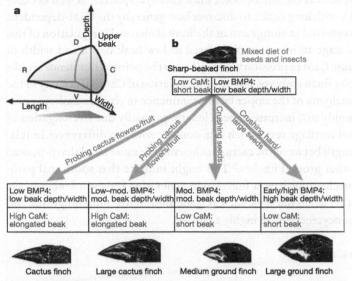

Fig. 4. BMP- and CaM-dependent signalling regulates growth along different axes, facilitating the evolution of distinct beak morphologies in Darwin's finches. *a*, Developing avian beak is a three-dimensional structure that can change along any of the growth axes. *b*, A beak of the sharp-beaked finch reflects a basal morphology for *Geospiza*. The model for BMP4 and CaM involvement explains development of both elongated and deep/wide beaks of the more derived species. Abbreviations: C, caudal; D, dorsal; R, rostral; V, ventral.

rocal expression levels in the two medium-sized species with thick and with long beak morphologies, whereas they are expressed together in the developing beak of the large cactus finch in good correlation with its robust yet elongated beak morphology (Figs 1, 3 and 4).[6,8] Theoretically, such modular developmental regulation need not have been necessary to construct a beak. A common set of growth-promoting pathways could, in principle, have led to outgrowth of the frontonasal process along all three dimensions. Indeed, a single factor, BMP4, seems to stimulate growth of the beak along two dimensions, promoting deeper and wider beaks, explaining the linkage in their variation. Using a common signal for outgrowth in different axes still permits evolution of form, for example through the action of localized antagonists. However, the developmental programme using a distinct pathway for the long axis of the beak enhances the "evolvability" along this dimension, by allowing independent variation along the different axes. Thus, the dissociable regulation of growth in different beak axes is an example of the organization of the biology of the organism facilitating the generation of variation on which natural selection can act.[20]

It will be of interest to sample other species of Darwin's finches with elongated beaks such as the woodpecker finch *Cactospiza pallida*, as well as other avian species with long beaks, to discover how generally the CaM-dependent pathway is involved in elongation of the beak skeleton. Upregulation of this pathway at stage 26 is not simply related to low beak depth and width of beaks, because CaM expression was very low in the pointed but slender beaks of the warbler finch embryos (Fig. 3). Upregulation of CaMKII activity in the distal mesenchyme of the upper beak prominence in chicken embryos produced a roughly 10% increase in beak length, primarily due to elongation of the pre-nasal cartilage rod, which is in concert with the differences in relative beak length between the cactus finches and the more basal sharp-beaked finch and other ground finches.[6] This might indicate that additional pathways and mechanisms have a function in generating the even longer beaks found in other avian lineages, such as some shorebirds, hummingbirds and Hawaiian honeycreepers, all highly adaptive.

METHODS

Collection and Treatment of Embryonic Material from Darwin's Finches

Under an agreement with the Galápagos National Park, we received quotas for collecting embryos of *G. magnirostris*, *G. fortis*, *G. fuliginosa* and *G. scandens* from Santa Cruz Island, and *G. conirostris*, *G. difficilis* and *C. olivacea* from Genovesa Island. To avoid causing nest defection, only the third egg was collected shortly after it was laid; it was then incubated at 100°F (37.8°C). Embryonic material was fixed in 4% paraformaldehyde in PBS for 2 h at ambient temperature and stored in RNAlater reagent (Ambion) at about 5°C for two to five weeks. Chick antisense riboprobes (*CaM*, *Coll IX*, *Runx2* and *PTHrP-Rec*) were prepared and used on Darwin's finch embryos as described previously.[8] We analysed 26 heads of Darwin's finches: large ground finch ($n = 3$), medium ground finch ($n = 5$), small ground finch ($n = 4$), large cactus finch ($n = 4$), cactus finch ($n = 3$), sharp-beaked finch ($n = 3$) and warbler finch ($n = 4$).

Chicken Embryo Manipulations and Statistical Analysis

Fertilized eggs (Spafas) were incubated at 100°F. Embryos were staged as described in ref. 21. The RCAS::CA-CaMKII construct will be described elsewhere (M. J. Taschner, S. Schnaiter and C.H., unpublished observations). To infect embryos for studies *in vivo* we injected the distal part of the frontonasal process of HH stage-24 chick embryos. The infected embryos were collected at embryonic day 10, fixed overnight in 4% paraformaldehyde in PBS, and frozen in OCT (optimal cutting temperature) medium for sectioning. Extent of viral infection was assayed by hybridization *in situ* with a virus-specific RSCH probe. The embryonic heads were photographed and measured in NIH Image 1.62. These arbitrary measurements were used for the analysis of variance function (ANOVA toolbox; *t*-test) in Microsoft Excel X to calculate standard deviations and *P* values for the data.

Darwin Finch Microarray Production and Usage

A DNA microarray (21,168 spots) was printed from a non-normalized poly(A)-primed cDNA library made from RNA isolated from multiple (12 individuals) frontonasal processes of stage-26 and stage-29 embryos of the medium ground finch, *G. fortis*. We used Cy5-labelled probes made from individual frontonasal processes of the four derived species of *Geospiza* for direct comparisons against a common Cy3-labelled reference sample made from pooled RNA of several (nine individuals) embryos of more basal *G. difficilis* (Fig. 1a, and Supplementary Fig. S1). In most cases we compared four unrelated individuals from each of the derived cactus finch and ground finch species (*G. scandens*, *G. conirostris*, *G. magnirostris* and *G. fortis*) against the pooled common reference (Figs 1a and 4, and Supplementary Fig. S1). RNA from each individual finch beak prominence was independently amplified and labelled in triplicate with a control dye swap. We used the two highest-quality sets of microarray data from each triplicate for clustering. Raw.gpr files were generated with GenePix 3.0 (Molecular Devices). Normalization and statistical analysis of the GPR data files were performed in MatLab (The Math Works). Data were normalized with the Lowess algorithm.[22] Only spots with a signal intensity exceeding the median background +2 s.d. were considered, which left 7,369 spots. The data was $\log_2$ transformed.

Microarray Cluster Analysis

Clustering analysis and visualization were computed in MatLab. Agglomerative hierarchical clustering was performed by using the euclidean distance measure: the average linkage and Ward heuristics were used to connect the gene clusters. For *k*-means clustering, the *k*-means algorithm partitioned the genes into *k* discrete clusters on the basis of their expression. The number *k* (50) was preselected. The resultant tree illustrates that duplicates of the amplification/labelling experiments from the same individuals clustered together (Fig. 1b).

Measurements of signal ratios and intensities for different transcripts were clustered to identify genes that were upregulated or downregulated in all individuals of a particular species compared with the basal *G. difficilis* reference.

Species-specific clusters were further cross-compared to reveal transcripts that were consistently upregulated in frontonasal processes of all individuals of the cactus-finch beak morphology and that remained unchanged or were downregulated in beak primordia of the ground finches. Conveniently, the cactus finch has an upper beak depth similar in size to that of the medium ground finch, whereas the large cactus finch's beak is more similar in size to that of the large ground finch while differing strongly in shape. This allowed us to separate transcripts exhibiting the size-specific regulation from those with shape-specific regulation.

Data Analysis: Hierarchical Clustering

Gene expression patterns of nine experimental samples representing five *G. scandens* and four *G. conirostris* samples were analysed by hierarchical clustering with Ward linkage. The samples were divided into two groups on the basis of differences in gene expression. Figure 2 shows a cluster dendrogram of the two groups, *G. scandens* (blue) [black] and *G. conirostris* (red) [gray]. Within each branch the siblings were found to cluster tightly together on a sub-branch.

ACKNOWLEDGEMENTS

We thank all field assistants and participants of the field collecting trips— M. Protas, J. Chavez, G. Castaneda, O. Perez, F. Brown, A. Aitkhozhina, M. Gavilanes, M. Paez, K. Petren, J. Podos and S. Kleindorfer—for their advice with species identification and other advice; Charles Darwin Research Station on Santa Cruz Island and The Galápagos National Park for permits and logistical support; and M. Kirschner for discussions that led to the inception of this project. A.A. was supported by the Cancer Research Fund of a Damon Runyon–Walter Winchell Foundation Fellowship. This project was funded by a program project grant from the NIH to C.J.T.

AUTHOR CONTRIBUTIONS

A.A. performed embryonic material collection, microarray probe preparation, microarray hybridizations and scanning, sectioning of material, *in situ* hybridizations and CaMKII functional analysis in chicken embryos. W.P.K. conducted all relevant bioinformatics analyses. C.H. constructed the RCAS virus carrying the constitutively active version of CaMKII. B.R.G. and P.R.G. provided logistics and secured permits for the fieldwork on the Galápagos Islands. C.J.T. conceived and supervised the project. A.A., B.R.G., P.R.G. and

C.J.T. co-wrote the manuscript. All authors discussed the results and commented on the manuscript.

AUTHOR INFORMATION

The sequence of Darwin's finch CaM has been deposited at GenBank under accession number DQ386479, and the microarray data have been filed with ArrayExpress under the accession number E-MEXP-702.

ENDNOTES

1. Darwin, C. R. *Journal of Researches into the geology and Natural History of the various countries visited during the voyage of H.M.S. Beagle, under the command of Captain FitzRoy, R.N.* 2nd edn (John Murray, London, 1845).

2. Lack, D. *Darwin's Finches* (Cambridge Univ. Press, Cambridge, 1947).

3. Grant, P. R. *The Ecology and Evolution of Darwin's Finches* (Princeton Univ. Press, Princeton, New Jersey, 1999).

4. Futuyma, D. J. *Evolutionary Biology* 3rd edn (Sinauer Associates, Sunderland, Massachusetts, 1998).

5. Freeman, S. & Herron, J. C. *Evolutionary Analysis* 3rd edn (Prentice Hall, Englewood Cliffs, New Jersey, 2003).

6. Bowman, R. I. Morphological differentiation and adaptation in the Galapagos finches. *Univ. Calif. Publ. Zool. 58*, 1–302 (1961).

7. Grant, P. R. & Grant, B. R. Unpredictable evolution in a 30-year study of Darwin's finches. *Science 296*, 707–711 (2002).

8. Abzhanov, A., Protas, M., Grant, R. B., Grant, P. R. & Tabin, C. J. Bmp4 and morphological variation of beaks in Darwin's finches. *Science 305*, 1462–1465 (2004).

9. O'Day, D. H. CaMBOT: profiling and characterizing calmodulin-binding proteins. *Cell. Signal. 15*, 347–354 (2003).

10. Burke, A. C., Nelson, C. E., Morgan, B. A. & Tabin, C. Hox genes and the evolution of vertebrate axial morphology. *Development 121*, 333–346 (1996).

11. Averof, M. & Patel, N. H. Crustacean appendage evolution associated with changes in Hox gene expression. *Nature 388*, 682–686 (1997).

12. Abzhanov, A. & Kaufman, T. C. Novel regulation of the homeotic gene Scr associated with a crustacean leg-to-maxilliped appendage transformation. *Development 126*, 1121–1128 (1999).

13. Abzhanov, A. & Kaufman, T. C. Crustacean (malacostracan) Hox genes and the evolution of the arthropod trunk. *Development 127*, 2239–2249 (2000).

14. Shapiro, M. D. *et al.* Genetic and developmental basis of evolutionary pelvic reduction in threespine sticklebacks. *Nature 428*, 717–723 (2004).

15. Kimmel, C. B. *et al.* Evolution and development of facial bone morphology in threespine sticklebacks. *Proc. Natl. Acad. Sci. USA 102*, 5791–5796 (2005).

16. Grant, P. R. Inheritance of size and shape in a population of Darwin's finches. *Proc. R. Soc. Lond. B 212*, 403–432 (1983).

17. Price, T. D. & Grant, P. R. The evolution of ontogeny in Darwin's Finches: a quantitative genetics approach. *Am. Nat. 125*, 169–188 (1985).

18. Grant, B. R. & Grant, P. R. *Evolutionary Dynamics of a Natural Population. The Large Cactus Finch of the Galápagos* (Univ. Chicago Press, Chicago, Illinois, 1989).

19. Grant, P. R. & Grant, B. R. Phenotypic and genetic effects of hybridization in Darwin's finches. *Evolution Int. J. Org. Evolution 48*, 297–316 (1994).

20. Kirschner, M. W. & Gerhart, J. C. *The Plausibility of Life: Resolving Darwin's Dilemma* (Yale Univ. Press, New Haven, Connecticut, 2005).

21. Hamburger, V. & Hamilton, H. L. A series of normal stages in the development of the chick embryo. *J. Morphol. 88*, 49–92 (1951).

22. Dudoit, S., Yang, H., Callow, M. J. & Speed, T. P. Statistical methods for identifying genes with differential expression in replicated cDNA experiments. *Statist. Sin. 12*, 111–139 (2002).

23. Petren, K., Grant, B. R. & Grant, P. R. A phylogeny of Darwin's finches based on microsatellite DNA length variation. *Proc. R. Soc. Lond. B 266*, 321–329 (1999).

Bibliography

GENERAL INTRODUCTION

Barrett, P. H., ed. 1960. A transcription of Darwin's first notebook [B] on "Transmutation of species." *Bulletin of the Museum of Comparative Zoology, Harvard* 122 (April): 245–296.

Darwin, C. R. 1837. Notebook [B] on "Transmutation of species." 1837–1838. Transcribed by D. Kohn. In *Charles Darwin's Notebooks, 1836–1844*, edited by P. H. Barrett, P. J. Gautrey, S. Herbert, D. Kohn, and S. Smith. Cambridge: Cambridge University Press, 2008.

———. 1886. *The Formation of Vegetable Mould through the Action of Worms with Observations on Their Habits*. New York: D. Appleton and Company.

Grant, P. R., and B. R. Grant. 2008. *How and Why Species Multiply: The Radiation of Darwin's Finches*. Princeton, NJ: Princeton University Press.

Lack, D. 1947. *Darwin's Finches*. Gloucester, MA: Peter Smith, 1968.

Lawrence, K. T., L. Zhonghui, and T. D. Herbert. 2006. Evolution of the eastern Tropical Pacific through Plio-Pleistocene glaciation. *Science* 312:79–83.

Lewinton, R. C., and J. L. Hubby. 1966. A molecular approach to the study of gene heterozygosity in natural populations: II, Amount of variation and degree of heterozygosity in natural populations of *Drosophila pseudoobscura*. *Genetics* 54:595–609.

Snodgrass, R. E., and E. Heller. 1904. Papers from the Hopkins-Stanford Galapagos Expedition, 1898–1899. *Proceedings of the Washington Academy of Sciences* 5:231–372.

Wara, M. W., A. C. Ravelo, and M. L. Delaney. 2005. Permanent El Niño–like conditions during the Pliocene warm period. *Science* 309:758–761.

CHAPTER 1: GROUNDING A LEGEND

Barlow, N. 1935. Charles Darwin and the Galapagos Islands. *Nature* 136:891.

Darwin, C. R. 1859. Letter 2608 to Charles Lyell, 27 December. In C. R. Darwin, *The Correspondence of Charles Darwin*, volume 7, *1858–1859*, edited by Frederick Burkhardt and Sydney Smith. Cambridge: Cambridge University Press, 1991.

Darwin, F., ed. 1909. *The Foundations of The Origin of Species: Two Essays Written in 1842 and 1844*. Cambridge: Cambridge University Press.

Hooker, J. D. 1844. Letter 723 to C. R. Darwin, 12 December 1843–11 January 1844. Letter 793: Charles R. Darwin to Leonard Jenyns, 25 [November 1844]. In C. R. Darwin, *The Correspondence of Charles Darwin*, volume 2, *1837–1843*, edited by Frederick Burkhardt and Sydney Smith. Cambridge: Cambridge University Press, 1986.

Larson, E. J. 2001. *Evolution's Workshop: God and Science in the Galapagos Islands*. New York: Basic Books.

Owen, R. 1844–1846. *History of British Fossil Mammals and Birds*. London: John van Voorst Paternoster Row.

CHAPTER 2: A HISTORY OF PLACE

Darwin, C. R. 1839. *Journal of Researches into the Geology and Natural History of the Various Countries Visited by H. M. S. Beagle*. London: Henry Colburn.
————. 1845. Galapagos Archipelago. In *The Voyage of the Beagle*, 2nd edition, chapter 17. New York: Modern Library.
————. 1909. Essay of 1844. In *The Foundations of The Origin of Species: Two Essays Written in 1842 and 1844*, edited by F. Darwin. Cambridge: Cambridge University Press.
Larson, E. J. 2001. *Evolution's Workshop: God and Science in the Galapagos Islands*. New York: Basic Books.
Slevin, J. R. 1959. The Galapagos Islands: A history of their exploration. *Occasional Papers of the California Academy of Science* 25:1–54, 75–88, 105–139.

CHAPTER 3: LAND: A THOUSAND ACCIDENTS

Brown, J. 1983. *The Secular Ark*. New Haven, CT: Yale University Press.
Darwin, C. R. 1837. Notebook [B] on "Transmutation of species." Transcribed by D. Kohn. In *Charles Darwin's Notebooks, 1836–1844*, edited by P. H. Barrett, P. J. Gautrey, S. Herbert, D. Kohn, and S. Smith. Cambridge: Cambridge University Press, 2008.
————. 1839. Galapagos Archipelago. In *Journal of Researches into the Geology and Natural History of the Various Countries Visited by H. M. S. Beagle*, facsimile reprint of the 1st edition, chapter 19. New York: Hafner Publishing Company, 1952.
————. 1958. *The Autobiography of Charles Darwin, 1809–1888*. Edited by N. Barlow. New York: W. W. Norton and Company.
Fitzroy, R. 1839. *Narrative of the Surveying Voyages of His Majesty's Ships* Adventure *and* Beagle, *between the Years 1826 and 1836, Describing Their Examination of the Southern Shores of South America, and the* Beagle's *Circumnavigation of the Globe*. Vol. 2. London: Henry Colburn.
Larson, E. J. 2001. *Evolution's Workshop: God and Science in the Galapagos Islands*. New York: Basic Books.
Lyell, C. 1830–1833. *Principles of Geology*. Vol. 2. London: John Murray.
Rudwick, M. J. S. 1997. *Georges Cuvier, Fossil Bones, and Geological Catastrophes: New Translations and Interpretations of the Primary Texts*. Chicago: University of Chicago Press.

CHAPTER 4: A CONFUSION OF FINCHES

Bowman, R. I. 1961. Morphological differentiation and adaptation in the Galápagos finches. *University of California Publications in Zoology* 58:1–302.
Gould, J. 1841. Part 3, *Birds*. In C. R. Darwin, *The Zoology of the H.M.S. Beagle*. New York: New York University Press.
Grant, P. R., and B. R. Grant. 2008. *How and Why Species Multiply: The Radiation of Darwin's Finches*. Princeton, NJ: Princeton University Press.
Lack, D. 1945. The Galapagos finches (Geospizinae): A study in variation. *Occasional Papers of the California Academy of Sciences* 21:1–152.
Mayr, E. 1942. *Systematics and the Origin of Species*. New York: Columbia University Press.
Paynter, R. A. Jr. 1970. Subfamily Emberizinae. In *Checklist of Birds of the World*, edited by R. A. Paynter Jr., 13:3–214. Cambridge, MA: Museum of Comparative Zoology.

Petren, K., B. R. Grant, and P. R. Grant. 1999. A phylogeny of Darwin's finches based on microsatellite DNA length variation. *Proceedings of the Royal Society of London B* 266:321–330.

Ridgway, R. 1896. Birds of the Galapagos Archipelago. *Proceedings of the U.S. National Museum* 19:459–670.

Rothschild, W., and E. Hartert. 1899. A review of the ornithology of the Galapagos Islands, with notes on the Webster-Harris Expedition. *Novitates Zoologicale* 6.

Sclater, P. L., and O. Salvin. 1870. Characters of new species of birds collected by Dr. Habel in the Galapagos. *Proceedings of the Zoological Society of London* 38:322–323.

Snodgrass, R. E., and E. Heller. 1904. Papers from the Hopkins-Stanford Galapagos Expedition, 1898–1899. *Proceedings of the Washington Academy of Sciences* 5:231–372.

Swarth, H. S. 1931. The avifauna of the Galapagos Islands. *Occasional Papers of the California Academy of Science* 18.

Vincek, V., C. O'hUigin, Y. Satta, N. Takahata, P. T. Boag, P. R. Grant, B. R. Grant, and J. Klein. 1996. How large was the founding population of Darwin's finches? *Proceedings of the Royal Society B* 264:111–118.

Zink, R. M. 2002. A new perspective on the evolutionary history of Darwin's finches. *Auk* 119:864–871.

INTRODUCTION TO SECTION 2: ADAPTATION AND THE EVOLUTION OF DIVERSITY

Hutchinson, G. E. 1959. A homage to Santa Rosalia, or Why are there so many kinds of animals? *American Naturalist* 93:145–159.

CHAPTER 5: WHAT MATTERS? VARIATION AND ADAPTATION

Baur, G. 1890–1891. The Galápagos Islands. *American Antiquarian Society, Proceedings*, n.s., 7:418.

Larson, E. J. 2001. *Evolution's Workshop: God and Science in the Galapagos Islands*. New York: Basic Books.

Lowe, P. R. 1936. The finches of the Galapagos in relation to Darwin's conception of species. *Ibis* 6 (2): 310–321.

Lurie, E. 1988. *Louis Agassiz: A Life in Science*. Baltimore: Johns Hopkins University Press.

Wright, S. 1932. The roles of mutation, inbreeding, crossbreeding and selection in evolution. *Proceedings of the 6th International Congress on Genetics* 1:356–366.

CHAPTER 6: DIVERSITY AS ADAPTATION

Connell, J. H. 1980. Diversity and the coevolution of competitors, or the ghost of competition past. *Oikos* 35:131–138.

Gause, G. F. 1934. *The Struggle for Existence*. Baltimore: Williams and Wilkins.

Gifford, E. W. 1919. Field notes on the land birds of the Galapagos Islands and of Cocos Island, Costa Rica. *Proceedings of the California Academy of Sciences*, series 4, 2:189–258.

Mittelbach, G. G., D. W. Schemske, H. V. Cornell, A. P. Allen, J. M. Brown, M. B. Bush, S. P. Harrison, A. H. Hurlbert, N. Knowlton, H. A. Lessios, C. M. McCain, A. R. McCune, L A. McDade, M. A. McPeek, T. J. Near, T. D. Price, R. E. Ricklefs, K. Roy, D. F. Sax, D. Schluter, J. M. Sobel, and M. Turelli. 2007. Evolution and the latitudinal diversity gradient: Speciation, extinction and biogeography. *Ecology Letters* 10:315–331.

Van Valen, L. 1973. A new evolutionary law. *Evolutionary Theory* 1:1–30.

CHAPTER 7: DARWIN'S FINCHES AS A CASE STUDY OF
NATURAL SELECTION

Bumpus, H. C. 1898. The variations and mutations of the introduced sparrow, *Passer domesticus*. Pp. 1–15 in *Biological Lectures Delivered at the Marine Biological Laboratory of Wood's Hole, 1896–7*. Boston: Ginn and Company.

Galton, F. 1889. *Natural Inheritance*. London: Macmillan.

Lande, R. 1979. Quantitative genetic analysis of multivariate evolution, applied to brain: Body size allometry. *Evolution* 33:402–416.

Lande, R., and S. J. Arnold. 1983. The measurement of selection on correlated characters. *Evolution* 37:1210–1226.

Pearson, K. 1893. Contributions to the mathematical theory of evolution. *Proceedings of the Royal Society* 54:329–333.

Provine, W. B. 1971. *The Origins of Theoretical Population Genetics*. Chicago: University of Chicago Press.

Weldon, W. R. F. 1895. An attempt to measure the death-rate due to the selective destruction of *Carcinus moensa* with respect to a particular dimension. *Proceedings of the Royal Society* 57:360–379.

CHAPTER 8: SEXUAL SELECTION IN DARWIN'S FINCHES

Bowman, R. I. 1979. Adaptive morphology of song dialects in Darwin's finches. *Journal of Ornithology* 120:353–389.

Darwin, C. R. 1871. *The Descent of Man and Selection in Relation to Sex*. London: John Murray.

Fisher, R. A. 1930. *The Genetical Theory of Natural Selection*. Oxford University Press.

Grant, B. R., and P. R. Grant. 1987. Mate choice in Darwin's finches. *Biological Journal of the Linnean Society* 32:247–270.

———. 1996. Cultural inheritance of song and its role in the evolution of Darwin's finches. *Evolution* 50:2471–2487.

Lande, R. 1981. Models of speciation by sexual selection on polygenic traits. *Proceedings of the National Academy of Science USA* 78:3721–3725.

Price, T. D. 1984. Sexual selection on body size, territory and plumage variables in a population of Darwin's finches. *Evolution* 38:327–341.

Ratcliff, L. M., and P. R. Grant. 1983. Species recognition in Darwin's finches (*Geospiza*, Gould): I, Discrimination by morphological cues. *Animal Behavior* 31:1139–1153.

CHAPTER 9: MICROEVOLUTION AND MACROEVOLUTION:
DOES ONE EXPLAIN THE OTHER?

Arnold, S. J., M. E. Pfrender, and A. G. Jones. 2001. The adaptive landscape as a conceptual bridge between micro- and macroevolution. *Genetica* 112/113:9–32.

Carson, H. L. 1968. The population flush and its genetic consequences. In *Population Biology and Evolution*, edited by R. C. Lewontin, 123–137. Syracuse, NY: Syracuse University Press.

Carson, H. L., and A. R. Templeton. 1984. Genetic revolutions in relation to speciation phenomena: The founding of new populations. *Annual Review of Ecology and Systematics* 15:97–131.

Dobzhansky, T. 1937. *Genetics and the Origin of Species*. New York: Columbia University Press.

Eldredge, N., and S. J. Gould. 1972. Punctuated equilibria: An alternative to phyletic gradualism. In *Models in Paleobiology*, edited by T. J. M. Schopf, 82–115. San Francisco: Freeman Cooper.

Gould, S. J., and N. Eldredge. 1992. Punctuated equilibrium comes of age. *Nature* 366:223–227.

Jones, A. G., S. J. Arnold, and R. Bürger. 2003. Stability of the G-matrix in a population experiencing mutation, stabilizing selection, and genetic drift. *Evolution* 57:1747–1760.

Rice, W. R., and E. E. Hostert. 1993. Laboratory experiments on speciation: What have we learned in 40 years? *Evolution* 4:1637–1653.

Schluter, D. 1996. Adaptive radiation along genetic lines of least resistance. *Evolution* 50:1766–1774.

Steppan, S. J., P. C. Phillips, and D. Houle. 2002. Comparative quantitative genetics: Evolution of the G matrix. *Trends in Ecology and Evolution* 17:320–327.

CHAPTER 10: THE EVOLUTION OF REPRODUCTIVE ISOLATION

Coyne, J. A. 1992. Genetics and speciation. *Nature* 355:511–515.

Coyne, J. A., and H. A. Orr. 1998. The evolutionary genetics of speciation. *Philosophical Transactions of the Royal Society of London B* 353:287–305.

Grant, B. R., and P. R. Grant. 1979. Darwin's finches: Population variation and sympatric speciation. *Proceeding of the American Academy of Sciences USA* 76:2359–2363.

———. 1983. Fission and fusion in a population of Darwin's finches: An example of the value of studying individuals in ecology. *Oikos* 41:530–547.

Mayr, E. 1942. *Systematics and the Origin of Species*. New York: Columbia University Press.

CHAPTER 11: HYBRIDIZATION

Darwin, C. R. 1868. *The Variation of Animals and Plants under Domestication*. London: John Murray.

Grant, P. R., and B. R. Grant. 1994. Phenotypic and genetic effects of hybridization in Darwin's finches. *Evolution* 48:297–316.

Grant, P. R., B. R. Grant, J. A. Markert, L. F. Keller, and K. Petren. 2004. Convergent evolution of Darwin's finches caused by introgressive hybridization and selection. *Evolution* 58:1588–1599.

Mendel, G. 1966. Experiments on plant hybrids. In *The Discovery of Genetics: A Mendel Source Book*, edited by C. Stern and E. R. Sherwood. Translated by E. R. Sherwood. San Francisco: W. H. Freeman. First published as "Versuche über Pflanzen-Hybriden" in *Verhandlungden des naturforschenden Vereines in Brunn* 4 (1865), Abhandlungen pp. 3–47.

Rieseberg, L. H., M. A. Arntz, and J. M. Burke. 2003. The genetic architecture necessary for transgressive segregation is common in both natural and domesticated populations. *Philosophical Transactions of the Royal Society of London* 358:1141–1147.

Rieseberg, L. H., O. Raymond, D. M. Rosenthal, Z. Lai, K. Livingstone, T. Nakazato, J. L. Durphy, A. E. Schwarzbach, L. A. Donovan, and C. Lexer. 2003. Major ecological transitions in wild sunflowers facilitated by hybridization. *Science* 301:1211–1216.

CHAPTER 12: THE GENETIC BASIS OF VARIATION: MOLECULAR
GENETICS, DEVELOPMENT, AND EVOLUTION

Carroll, S. B. 2005 *Endless Forms Most Beautiful: The New Science of Evo Devo and the Making of the Animal Kingdom.* New York: W. W. Norton.

Coen, H. S., and E. M. Meyerowitz. 1991. The war of the whorls: Genetic interactions controlling flower development. *Nature* 353:31–33.

Haeckel, E. H. 1892. *The History of Creation, or the Development of the Earth and Its Inhabitants by the Action of Natural Causes.* New York: Appleton and Company.

Johannsen, W. 1903. Über Erlichkeit in Populationen und in Reinen Linien. In *Selected Readings in Biology for Natural Sciences,* edited by the Staff of Natural Sciences 3. Partially translated by Harold Gall and Elga Putschar. Chicago: University of Chicago Press, 1955.

Orr, H. A. 2005. The genetic theory of adaptation: A brief history. *Nature Reviews Genetics* 6:119–127.

Provine, W. B. 1971. *The Origins of Theoretical Population Genetics.* Chicago: University of Chicago Press.

Yule, G. U. 1903. Professor Johannsen's experiments in heredity. *New Phytologist* 2:235–242.

Printed and bound by CPI Group (UK) Ltd, Croydon, CR0 4YY

27/10/2024

14580401-0005